Molecular Immunology
SECOND EDITION

EDITED BY

B. David Hames

*Department of Biochemistry
and Molecular Biology
University of Leeds*

and

David M. Glover

*Cancer Research Laboratories
Department of Anatomy and Physiology
University of Dundee*

 IRL PRESS

—at—

OXFORD UNIVERSITY PRESS

Oxford New York Tokyo

Oxford University Press, Walton Street, Oxford OX2 6DP

Oxford   New York

Athens   Auckland   Bangkok   Bombay
Calcutta   Cape Town   Dar es Salaam   Delhi
Florence   Hong Kong   Istanbul   Karachi
Kuala Lumpur   Madras   Madrid   Melbourne
Mexico City   Nairobi   Paris   Singapore
Taipei   Tokyo   Toronto

and associated companies in
Berlin   Ibadan

Oxford is a trade mark of Oxford University Press

Published in the United States
by Oxford University Press Inc., New York

A catalogue record for this book is available from the British Library

Library of Congress Cataloging in Publication Data
Molecular immunology/edited by B. David Hames and David M. Glover. – 2nd ed.
(Frontiers in molecular biology ; 11)
Includes bibliographical references and index.
1. Molecular immunology. 2. Immunochemistry. I. Hames, B. D.
II. Glover, David M. III. Series.
QR185.6.M655   1995   574.2'9—dc20   95–21147

ISBN 0 19 963379 7 (Hbk)
ISBN 0 19 963378 9 (Pbk)

Typeset by
Footnote Graphics, Warminster, Wilts.
Printed in Great Britain by
The Bath Press, Avon

Preface

Our understanding of the molecular functioning of the immune system has undergone radical expansion and revision in recent years. This completely updated and very substantially extended edition of *Molecular Immunology* follows from the success of the original book published in 1988. It is unique not only for its breadth and depth of coverage of the immune system but also for the quality of its authorship.

The first chapter, by Fred Alt and his colleagues, focuses on B cell differentiation, reviewing our current knowledge of the organization of immunoglobulin genes and control of the gene rearrangement events involved in their assembly and expression. Many new advances have occurred in this field since the subject was covered in the first edition. In Chapter 2, Mark Davis and Yueh-Hsiu Chien describe in detail the structure, rearrangement, expression and selection of T cell receptor (TCR) genes as well as recognition of peptide/MHC complexes. This is followed, in Chapter 3, by an authoritative contribution by Cox Terhorst and his colleagues that concentrates on the molecular details of pathways of T lymphocyte signal transduction. The central role of the MHC in cell-mediated immune responses is covered in Chapters 4 and 5; Chapter 4, by David Jewell and Ian Wilson, describes the structure of MHC class I and II molecules and their peptide binding specificities while John Monaco, in Chapter 5, reviews the molecular mechanisms of antigen processing, the generation of MHC-binding peptides and their delivery to the appropriate MHC molecules. B cell activation is the topic of Chapter 6 by Gerry Klaus. It describes the major cell-surface molecules involved in regulating the responses of mature B cells to antigens, the mechanisms of signal transduction and our current knowledge of the events that occur during T-cell–B-cell interaction. In the intervening years since the first edition of this book, the explosion of knowledge and interest in molecules of the immune system, coupled with the powerful techniques of recombinant DNA technology, has spawned an exciting new field of research, antibody engineering, leading to the creation of antibodies and antibody-derived molecules designed for specific applications in medicine and industry. Progress has been truly astonishing in this area in the last six years or so. In Chapter 7, Martine Verhoeyen and John Windust review the brief history of antibody engineering and present the latest advances in techniques, approaches and applications. Finally, Chapter 8 by Ken Reid provides a comprehensive description of the complement system, covering the molecular mechanisms and control of activation by both the classical and alternative pathways.

The immune system is a truly complex field of study and hence we are indebted to all of the authors for such comprehensive yet lucid accounts of its many facets. We are certain that the quality of the contributions will be widely recognised by their many fellow researchers in immunology. To bring such a book to fruition with very busy authors was no mean task and so we thank them sincerely for their hard work and enthusiasm. Their reward will be for this book to be as widely read and quoted as its predecessor.

Leeds   David Hames
Dundee   David Glover
September 1995

Contents

List of contributors   xiv

Abbreviations   xvi

1 Mechanism and control of immunoglobulin gene rearrangement   1

RUSTY LANSFORD, AMI OKADA, JIANZHU CHEN, EUGENE M. OLTZ,
T. KEITH BLACKWELL, FREDERICK W. ALT, and GARY RATHBUN

1. **Introduction**   1
2. **Overview of B cell differentiation**   1
3. **The antibody molecule**   3
   3.1 Structure of the Ig molecule   3
   3.2 Structure of Ig genes and proteins   4
4. **Mechanism of Ig gene assembly**   7
   4.1 Overview of the V(D)J recombination mechanism   8
   4.2 Recombination recognition sequences   10
   4.3 Deletion versus inversion   12
   4.4 Unusual aspects of the V(D)J recombination mechanism   15
   4.5 Lymphoid-specific activities involved in V(D)J recombination   17
   4.6 Generally expressed components of recombinase   19
5. **Organization of H chain variable region gene segments**   21
   5.1 The murine *IgH* locus   21
   5.2 The human *IgH* locus   24
   5.3 The murine κ LC locus   26
   5.4 The human κ LC locus   26
   5.5 The murine λ LC locus   27
   5.6 The human λ LC locus   27
   5.7 Surrogate LC genes   28
6. **H chain gene expression**   30
   6.1 Ig gene transcription and translation   30
7. **Generation of the antibody repertoire**   31
   7.1 Primary diversification mechanisms   31

    7.2 Other diversification mechanisms   32

8. Regulation of Ig gene expression   32

    8.1 HC gene expression   32

    8.2 LC gene expression   33

    8.3 Expression of incompletely assembled H chain genes   33

9. Regulation of Ig gene assembly   35

    9.1 Ordered assembly and expression of Ig genes   35

    9.2 H chain gene assembly   37

    9.3 L chain gene assembly   37

    9.4 κ versus λ LC gene assembly (LC isotype exclusion)   38

    9.5 Molecular characterization of early B cell stages based on surface marker expression   39

    9.6 Potential role of SLC receptors in regulating precursor B cell development   40

    9.7 Allelic exclusion   40

    9.8 Ig receptor editing   43

    9.9 Potential restrictions in the primary murine V_H repertoire associated with the V(D)J rearrangement process   43

10. Molecular mechanisms that regulate antigen receptor gene assembly   45

    10.1 Accessibility   45

    10.2 Model systems to study control of V(D)J recombination   46

    10.3 Molecular mechanisms of accessibility   47

    10.4 Abnormalities in recombinational accessibility   52

11. IgH class switching   52

    11.1 Overview   52

    11.2 Organization of Ig C_H genes   53

    11.3 Mechanism of S region mediated CSR   55

    11.4 Non-deletional CSR   58

    11.5 Control of CSR   59

Acknowledgements   64

References   64

2 T cell antigen receptor genes   101

MARK M. DAVIS and YUEH-HSIU CHIEN

1. Introduction   101

2. TCR polypeptides   102

3. *TCR* genes   103

4. **Rearrangement and expression of T cell receptor genes in the thymus**   106

    4.1 Programmed γδ *TCR* rearrangements   108

    4.2 Allelic exclusion   108

    4.3 Choice of γδ versus αβ lineage   109

5. **TCR repertoire selection in the thymus**   109

    5.1 Failure to rearrange an expressible TCR heterodimer   110

    5.2 Negative selection   110

    5.3 Positive selection   111

6. **αβ TCR recognition of peptide/MHC complexes**   113

7. **Affinity of αβ TCRs for peptide/MHC ligands**   116

8. **Superantigens**   118

9. **γδ T cell recognition**   119

    9.1 Specificity of diverse γδ T cells   119

    9.2 Function of diverse γδ T cells   122

    9.3 Monomorphic γδ T cells   122

References   123

3 **T lymphocyte signal transduction**   132

COX TERHORST, HERGEN SPITS, FRANK STAAL, and MARK EXLEY

1. **Introduction**   132

2. **Antigen recognition by T cell receptors**   132

3. **Structure of the T cell receptor for antigen**   137

    3.1 The components of the TCR/CD3 complex   138

    3.2 Assembly in the endoplasmic reticulum   139

4. **Signal transduction through the TCR/CD3 complex**   145

    4.1 Signal transduction molecules associated with CD3-ζ and CD3-ε   147

    4.2 On the role of CD3-ζ and -ε in transgenic and mutant mice   154

5. **The primary co-stimulatory events**   160

    5.1 The role of CD4 and CD8   161

    5.2 Signal transduction events induced by triggering of CD28 and CTLA-4   166

    5.3 Alternative T cell co-stimulatory structures   170

    5.4 The phosphotyrosine phosphatase CD45   172

6. Conclusions 174
Acknowledgements 175
References 175

4 Structure and function of MHC class I and class II antigens 189

DAVID A. JEWELL and IAN A. WILSON

1. Introduction 189
2. Gene arrangement of MHC 191
3. Antigen presentation 191
4. Structure of MHC class I antigens 194
5. Peptide–MHC class I interactions 198
6. Class I vs class II structures 204
Acknowledgements 208
References 208

5 Molecular mechanisms of antigen processing 212

JOHN J. MONACO

1. Introduction 212
2. The class I (endogenous) antigen processing pathway 214
   2.1 MHC-linked transport protein (*Tap*) genes 214
   2.2 MHC-linked proteasome subunit genes 221
   2.3 Molecular chaperones in the class I pathway 223
   2.4 Specificity and polymorphism in the endogenous antigen processing pathway 224
3. The class II (exogenous) pathway 226
   3.1 Proteases involved in the class II pathway 226
   3.2 The role of invariant chain (I_i) 227
   3.3 Class II antigen processing mutants 232
   3.4 Presentation of endogenous antigen by MHC class II molecules 233
4. Conclusions 234
Acknowledgements 235
References 235

6 B cell activation   248

GERRY G. B. KLAUS

   1. Introduction   248
       1.1 B cell response to specific antigens   248
       1.2 Responses to polyclonal activators   248
       1.3 Cell surface glycoproteins on B cells   249
   2. Surface immunoglobulin (sIg) receptors   250
       2.1 The effects of anti-Ig on mature B cells   250
       2.2 Signal transduction via sIg receptors   251
   3. The CD19/CD21 complex   256
   4. FcγRII (CD32)   257
   5. CD23   259
   6. Class II MHC   260
   7. CD45 (leucocyte common antigen, murine Ly5)   261
   8. CD40   262
       8.1 Functional role of CD40   262
       8.2 The CD40 ligand   263
   9. Other signalling molecules on B cells   263
       9.1 CD72 (murine Lyb2)   263
       9.2 CD20   263
       9.3 CD22 (formerly Lyb8 in the mouse)   264
       9.4 CD38   264
   10. T cell-dependent B cell activation   265
       10.1 MHC-restricted and unrestricted B cell activation   265
       10.2 Events occurring during T cell–B cell interaction   266
       10.3 Cytokine-driven B cell proliferation and differentiation   269
   11. Conclusions   271
   References   272

7 Advances in antibody engineering   283

MARTINE E. VERHOEYEN and JOHN H. C. WINDUST

   1. Introduction   283
   2. Antibody structure   283
   3. Antibody gene cloning   285

4. Recombinant antibody gene expression   285

4.1 Expression in animal cells   285

4.2 Expression in *Escherichia coli*   287

4.3 Expression in other microorganisms   291

4.4 Expression in plants   293

5. Combinatorial libraries and phage display   294

5.1 Combinatorial libraries   294

5.2 Phage display of antibody fragments   296

6. Applications of engineered antibodies   301

6.1 Structure–function studies   301

6.2 Antibodies for *in vivo* use   302

6.3 Bispecific and bivalent antibodies   306

6.4 Bifunctional antibodies   307

6.5 Antibody domains as scaffolds for insertion of functional sites   308

6.6 Molecular recognition units (MRUs)   309

6.7 Catalytic antibodies   309

7. Conclusions   310

References   311

8 The complement system   326

KENNETH B. M. REID

1. Introduction   326

2. Activation of the classical pathway   330

2.1 Structure, activation, and control of C1   330

2.2 Activation of the classical pathway via mannan binding protein (MBP) 333

2.3 Molecular cloning of C1q   334

2.4 Molecular cloning of C1r and C1s   335

2.5 Molecular cloning of mannan binding protein (MBP)   336

2.6 Molecular cloning of mannan binding protein-associated serine
protease (MASP)   336

2.7 Molecular cloning of C1 inhibitor (C1-Inh)   337

3. Activation of the alternative pathway   338

3.1 Factor D   338

3.2 Properdin—a positive regulator of the alternative pathway   339

4. Complement class III products of the MHC   339

4.1 Component C4: structure and activation   340

4.2 Component C2 and factor B: activation and role in formation of
   the C3 convertases   340

4.3 Molecular genetics of C4   343

4.4 Molecular genetics of component C2 and factor B   346

4.5 Molecular map of the MHC class III region   349

5. **Components C3 and C5**   349

   5.1 Component C3   349

   5.2 Component C5   354

6. **Regulation of complement activation by control of the C3 and
   C5 convertases. Factor 1 and the members of the RCA
   (regulator of complement activation) cluster**   354

   6.1 Factor 1   356

   6.2 Plasma control proteins: C4b-binding protein and factor H   358

   6.3 Membrane-associated regulatory proteins   360

7. **The C3a, C4a, and C5a anaphylatoxins**   364

8. **Complement receptor type 3 and p150,95: members of the
   LFA-1 family of cell adhesion molecules designated the
   leucocyte-specific β_2 integrins**   364

9. **The terminal attack complex C5b–9 and the relationship of C8
   and C9 to perforin from killer lymphocytes**   366

   References   370

Index   383

Contributors

FREDERICK W. ALT
Howard Hughes Medical Institute, Children's Hospital, Department of Genetics, Harvard University Medical School and Center for Blood Research, 300 Longwood Avenue, Boston, MA 02115, USA.

T. KEITH BLACKWELL
Howard Hughes Medical Institute, Children's Hospital, Department of Genetics, Harvard University Medical School and Center for Blood Research, 300 Longwood Avenue, Boston, MA 02115, USA.

JIANZHU CHEN
Howard Hughes Medical Institute, Children's Hospital, Department of Genetics, Harvard University Medical School and Center for Blood Research, 300 Longwood Avenue, Boston, MA 02115, USA.

YUEH-HSIU CHIEN
The Department of Microbiology and Immunology, Stanford University Medical School, Stanford, CA 94305–5402, USA.

MARK M. DAVIS
Howard Hughes Medical Institute and The Department of Microbiology and Immunology, Stanford University Medical School, Stanford, CA 94305–5402, USA.

MARK EXLEY
Immulogic Pharmaceutical Corporation, 610 Lincoln Street, Waltham, MA 02154, USA.

DAVID A. JEWELL
Department of Molecular Biology, MB13, The Scripps Research Institute, 10666 North Torrey Pines Road, La Jolla, CA 92037, USA.

GERRY G. B. KLAUS
Laboratory of Cellular Immunology, National Institute for Medical Research, Mill Hill, London NW7 1AA, UK.

RUSTY LANSFORD
Howard Hughes Medical Institute, Children's Hospital, Department of Genetics, Harvard University Medical School and Center for Blood Research, 300 Longwood Avenue, Boston, MA 02115, USA.

JOHN J. MONACO
Howard Hughes Medical Institute and Department of Molecular Genetics, Biochemistry, and Microbiology, University of Cincinnati, 231 Bethesda Avenue, Cincinnati OH 45267–0524, USA.

AMI OKADA
Howard Hughes Medical Institute, Children's Hospital, Department of Genetics, Harvard University Medical School and Center for Blood Research, 300 Longwood Avenue, Boston, MA 02115, USA.

EUGENE M. OLTZ
Howard Hughes Medical Institute, Children's Hospital, Department of Genetics, Harvard University Medical School and Center for Blood Research, 300 Longwood Avenue, Boston, MA 02115, USA.

GARY RATHBUN
Howard Hughes Medical Institute, Children's Hospital, Department of Genetics, Harvard University Medical School and Center for Blood Research, 300 Longwood Avenue, Boston, MA 02115, USA.

KENNETH B. M. REID
MRC Immunochemistry Unit, Department of Biochemistry, University of Oxford, South Parks Road, Oxford OX1 3QU, UK.

HERGEN SPITS
Laboratory of Molecular Immunology, Dana–Farber Cancer Institute, Harvard Medical School, Boston, MA 02115, USA.

FRANK STAAL
Laboratory of Molecular Immunology, Dana–Farber Cancer Institute, Harvard Medical School, Boston, MA 02115, USA.

COX TERHORST
Laboratory of Molecular Immunology, Dana–Farber Cancer Institute, Harvard Medical School, Boston, MA 02115, USA.

MARTINE E. VERHOEYEN
Department of Immunology, Unilever Research, Colworth House, Sharnbrook, Bedfordshire MK44 1LQ, UK.

IAN A. WILSON
Department of Molecular Biology, MB13, The Scripps Research Institute, 10666 North Torrey Pines Road, La Jolla, CA 92037, USA.

JOHN H. C. WINDUST
Department of Immunology, Unilever Research, Colworth House, Sharnbrook, Bedfordshire MK44 1LQ, UK.

Abbreviations

ABC	ATP-binding cassette
AFC	antibody-forming cell
ALL	acute lymphoblastic leukaemia
A-MuLV	Abelson murine leukaemia virus
APC	antigen-presenting cell
ARH	antigen receptor homology
BCR	B-cell antigen receptor
BrdU	bromodeoxyuridine
C	constant
CBHI	cellobiohydrolase I
CCP	complement control protein
CD	cluster of differentiation
CDR	complementarity-determining region
CEA	carcinoembryonic antigen
CFTR	cystic fibrosis gene
CHO	Chinese hamster ovary
CLIP	class II-associated invariant chain peptide
CLL	chronic lymphatic leukaemia
CMV	cytomegalovirus
CRD	carbohydrate recognition domain
CS	circumsporozoite surface
CSR	class switch recombination
CT	cytotoxic T lymphocyte
D	diversity
dAbs	domain antibodies
DAF	decay acceleration factor
dEC	dendritic epidermal cells
DHFR	dihydrofolate reductase
DN	double negative
DP	double progeny
DSBR	double-strand break repair
DTH	delayed-type hypersensitivity
EBV	Epstein–Barr virus
EGF	epidermal growth factor
EGFR	epidermal growth factor receptor
FDC	follicular dendritic cells
FR	framework region
GS	glutamine synthetase

HAMA	human–antimouse antibody
HANE	hereditary angioneuritic (o)edema
HBcAg	hepatitis B virus core antigen
HEL	hen-egg lysozyme
HMFG	human milk fat globule
HRF	homologous restriction factor
Hsp	heat-shock protein
IEL	intraepithelial lymphocyte
IFN	interferon
Ig	immunoglobulin
I_i	invariant chain
IL	interleukin
IMAC	immobilized metal-ion chromatography
IP_3	inositol 1,4,5-triphosphate
ITAM	immune receptor tyrosine-based activation motif
J	joining
LAD	leucocyte adhesion deficiency
LCA	leucocyte common antigen
LDL	low-density lipoprotein
LDLr	low-density lipoprotein repeat
LHR	long homologous repeat
LPS	lipopolysaccharide
mAb	monoclonal antibody
MAC	membrane attack complex
MASP	MBP associated serine protease
MBP	mannan-binding protein
MCP	membrane co-factor protein
MMTV	mouse mammary tumour virus
MRU	molecular recognition unit
MSX	methionine sulfoximine
MTOC	microtubular organizing centre
MTX	methotrexate
MuLV	murine leukaemia virus
NF-AT	nuclear factor of activated T cells
NK	natural killer
NMR	nuclear magnetic resonance
NP	4-hydroxy-3-nitrophenylacetyl
PBL	peripheral blood lymphocyte
PCR	polymerase chain reaction
PFGE	pulsed-field gel electrophoresis
PI-3K	phosphatidylinositol 3-kinase
PIP_2	phosphatidylinositol 4,5-bisphosphate
PKC	protein kinase C
PLAP	placental alkaline phosphatase

PLC	phospholipase C
PMA	phorbol myristic acetate
PNH	paroxysmal nocturnal haemoglobinuria
PTK	protein tyrosine kinase
PTPase	phosphotyrosine phosphatase
RAG	recombination activating gene
RCA	regulator of complement activation
RFLP	restriction fragment length polymorphism
rmsd	root mean square difference
RS	recombination signal
RSS	recombination signal sequence
RSV	respiratory syncytial virus
scFv	single chain Fv
SCID	severe combined immune deficiency
SCR	single chain repeat
SDM	site-directed mutagenesis
serpin	serine protease inhibitor
sIg	surface immunoglobulin
SLE	systemic lupus erythematosus
Slp	sex-limited protein
SP	single progeny
SRBC	sheep red blood cells
SRP	signal recognition particle
TAP	*t*ransporter *a*ssociated with *a*ntigen *p*rocessing or *t*ransporter of *a*ntigenic *p*eptides
TCR	T-cell antigen receptor
TD	T (cell) dependent
TdT	terminal deoxynucleotidyl transferase
TGF	transforming growth factor
T_H	T helper
TI	T (cell) independent
TK	thymidine kinase
TNF	tumour necrosis factor
V	variable
VSV	vesicular stomatitis virus
YAC	yeast artificial chromosome

1 | Mechanism and control of immunoglobulin gene rearrangement

RUSTY LANSFORD, AMI OKADA, JIANZHU CHEN,
EUGENE M. OLTZ, T. KEITH BLACKWELL,
FREDERICK W. ALT, and GARY RATHBUN

1. Introduction

The seminal studies of Landsteiner led to the notion that immunoglobulins can interact with an unlimited array of antigens (1). B lymphocytes produce immuno-globulins (Ig) which are composed of heavy (H) and light (L) polypeptide chains. At the amino terminus of H and L chains are regions of highly variable amino acid sequences (variable region) which pair with each other to form the antigen-binding site. The vast diversity of the antibody repertoire derives largely from the fact that H and L chain variable regions are encoded by a portion of the H or L chain gene that is assembled somatically from component germline gene segments (2). The T cell antigen receptor (TCR) is composed of immunoglobulin like polypeptides, which are also encoded by genes that are assembled somatically by the same enzymatic mechanisms (3). Ig heavy chain genes can also undergo a separate type of recombination event, termed 'class switch recombination', which permits clonal lineages of B cells to express the same variable region with a different constant region effector function (4). This chapter will focus on our current knowledge of the organization of Ig genes in mice and humans and on the mechanistic aspects and the control of the gene rearrangement events involved in their assembly and expression.

2. Overview of B cell differentiation

In mammals, B lymphocytes are generated in the liver during fetal life, but shortly after birth the bone marrow becomes the site of lymphopoiesis and remains so throughout adult life (5, 6). In the bone marrow, B lymphocytes arise from a small self-renewing population of stem cells, which lie within a matrix of stromal cells that produce factors which immature B lymphocytes require for growth. Ig heavy chain (HC) variable region genes are assembled from component variable (*V*), diversity (*D*) and joining (*J*) gene segments and Ig light chain (LC) variable region genes from component *V* and *J* gene segments during this differentiation process, which generates 'virgin' B lymphocytes that produce surface Ig and, presumably, have not yet been stimulated by antigens. Virgin B lymphocytes migrate out of the bone marrow to peripheral lymphoid tissues such as the lymph nodes and spleen, where antigen contact and further differentiation, including Ig heavy chain class switching, may occur (Fig. 1).

The clonal selection mechanism of antibody production is a fundamental aspect of the generation of a specific immune response. The clonal selection theory (Fig. 1) states that the immune response to an antigen is initiated when that antigen is recognized by a B lymphocyte surface receptor of a pre-existing specificity (9). In this context, each B lymphocyte displays antigen receptors on its surface that consist of a unique species of membrane-bound Ig. When a particular B lympho-

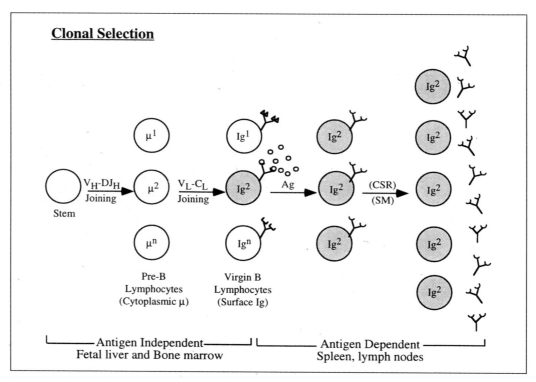

Fig. 1 Clonal selection and B lymphocyte differentiation. The antigen-independent stages of differentiation generate a vast number of independent B lymphocyte clones, each of which bears a surface receptor with a unique antigen-recognition site. Binding of antigen (Ag) by the receptor of a particular clone triggers clonal expansion and differentiation into plasma cells which secrete antibody molecules that display the same antigen binding site. Upon stimulation cells may also undergo class switching (CSR) and/or somatic mutation (SM). μ denotes the unique cytoplasmic μ molecules that are expressed in individual pre-B cells and Ig denotes the unique immunoglobulin molecules that are expressed in B cells (adapted from refs 7, 8).

cyte binds antigen, it is induced to proliferate. B lymphocytes within this selected clone can differentiate into plasma cells, which secrete large amounts of the selected Ig molecule.

The clonal selection mechanism implies that each Ig molecule produced by a given B lymphocyte displays an identical set of antigen-binding specificities. Otherwise, antigenic triggering of a B lymphocyte clone might lead to the production of multiple different antibodies, thereby diluting out the selected antibody, along with producing potentially harmful antibodies (for example autoreactive antibodies). Each B lymphocyte produces Ig molecules with identical antigen-binding specificities because it expresses only the products of a single H and a single L chain gene, a principle which is referred to as allelic exclusion (10, 11).

3. The antibody molecule

3.1 Structure of the Ig molecule

A monomeric Ig molecule (Fig. 2) is composed of two identical H and two identical L chains that are linked together by disulfide bonds (12). Both H and L chains are organized into domains that are defined by homology and that are approximately 110 amino acids in length (13). Each domain forms a conserved structure known as the 'antibody fold', which is stabilized by an internal disulfide linkage that forms

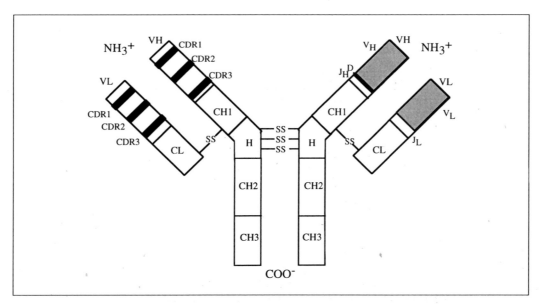

Fig. 2 Structure of a mouse IgG molecule. The typical antibody molecule consists of two identical H chains and two identical light chains linked to each other by disulfide bonds (ss). The V_H, C_H, V_L, and C_L homology domains are shown as boxes and the hinge region is denoted H. The approximate boundaries of CDR regions and of sequences encoded by V_H, D, J_H, V_L, and J_L segments are indicated by different shadings (adapted from refs 7, 8).

a loop of about 65 amino acids (14). The amino terminal domain of H and L chains comprises the variable region, which is 107 to 118 amino acids in length (Fig. 2) (2). Within the variable region are three areas of greatest sequence variability (hypervariable regions) that are separated by regions of relatively constant amino acid sequence (framework regions) (15). This overall structure has been conserved within H and L chain variable regions among species as diverse as humans and sharks. H and L chain hypervariable regions together form the potential antigen-binding site, and are therefore referred to as complementarity-determining regions (CDRs) (16).

Amino acid sequences within the remainder of H and L chains (Fig. 2) are relatively invariable, but deviate among different constant (C) region types (isotypes) that were originally classified according to reactivity to specific antisera (15). H chain C region isotypes define the classes and subclasses of mammalian Igs (12). Thus, in the mouse, the eight Ig classes and subclasses of IgM, IgD, IgG3, IgG1, IgG2b, IgG2a, IgE, and IgA are defined by H chain isotypes that are denoted by their respective Greek letters (for example μ, δ, γ, ϵ, and α). Mammalian H chain C regions mediate immunologic effector functions, such as complement fixation, placental transfer, and binding to cell-surface Fc receptors, that are specific to particular isotypes (17). Functional differences have not been identified for the two isotypes of mammalian L chains, kappa (κ) and lambda (λ). H chain C regions contain between two and four domains that are distantly homologous to each other and to each L chain C region domain. Certain H chain isotypes also contain a hinge region (Fig. 2) that may facilitate antigen binding by increasing H chain flexibility (18).

H chains can be produced in either a membrane-bound or a secreted form, which are distinguished by specific sequences at their carboxy termini (19–21). The Ig molecule is expressed on the cell surface as a monomer, but Ig molecules can be secreted as monomers (IgG and IgE) or, in conjunction with the J chain protein, as dimers (IgA), or as pentamers (IgM) (22).

3.2 Structure of Ig genes and proteins

The H chain variable region is encoded by an exon that is assembled upstream of C region coding sequences by the joining of component V_H, D, and J_H segments (Fig. 3A) (2). A V_H segment encodes a 5′ hydrophobic leader sequence and approximately 98 amino acids of the variable region that extend from the amino terminus to CDR3, and include CDR1 and CDR2 (Fig. 2). CDR3 is encoded by the D segment (3–7 amino acids) and by sequences at the V_H–D and D–J_H junctions (Fig. 4). The remainder of a J_H segment encodes the fourth framework region, which consists of 12–17 amino acids. The variable region of each L chain isotype is encoded by an upstream exon that is assembled from analogous V_L and J_L segments which are joined directly to each other; there are no L chain D segments (Figs 2 and 3) (2). The basic $V–(D)–J$ structure of these segments has been conserved among species, but their number and organization varies widely. In mice and humans, the V, D, and J segments are present as clusters with multiple members (Fig. 3).

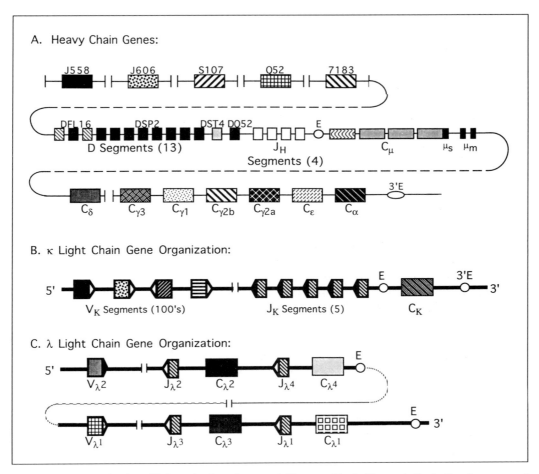

Fig. 3 *Ig* gene organization. (A) Organization of mouse *IgH* genes. Shaded boxes indicate the chromosomal locations of five of the V_H segment families. The *D* segment families and the J_H segments are shaded differently and the intronic enhancer (E) is shown just 5' of C_μ. Separate H chain C region exons are shown for the C_μ sequences only. The C_H genes are shown in their proper order along with the 3' IgH enhancer (3' E_H). (B) Organization of the mouse κ light chain locus. V_κ segments are shaded differently to indicate that there are separate V_κ families. The C_κ region is flanked by the intronic κ enhancer and the 3' κ enhancer elements (3' E) (adapted from ref. 8). (C) Organization of the mouse λ light chain. The four λ gene families are shown in their proper order along with their corresponding enhancer elements (E).

Each C region domain is encoded by a separate set of exons. H chain C regions are encoded by exon clusters that also encode the hinge region and alternate carboxy-terminal sequences (2, 20). In mice and humans, the C_μ region exon cluster is located closest to the J_H segments, and is followed approximately 2.5 kb downstream by the C_δ exons, then, much further downstream, by the exon clusters encoding the other H chain isotypes (Fig. 3A) (23–25). B lymphocytes express μ H chains first during differentiation (26).

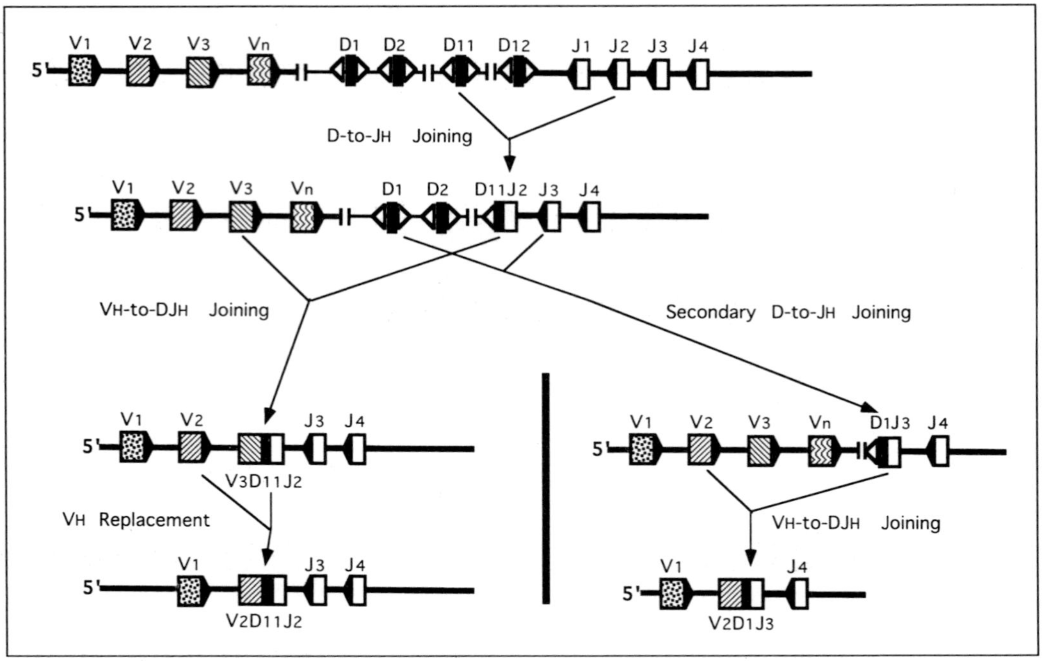

Fig. 4 Assembly of the V region of a H chain gene. Open triangles denote recognition elements separated by a 23 bp spacer and filled triangles denote recognition elements separated by a 12 bp spacer. The solid triangle shown within the V1 in the V_H replacement portion represents the heptamer that is conserved near the 3' end of V_H segments (see text).

Differential splicing of primary RNA transcripts joins V and C region exons, thereby permitting the simultaneous expression of μ and δ H chains that contain the same variable region (Fig. 5) (2, 27, 28). Most virgin B lymphocytes express both IgM and IgD (28, 29) on their surface. The function of IgD remains unclear; it was proposed that it might be a particularly effective antigen receptor because it contains an unusually large hinge region (28). However, mice engineered to lack a functional C_δ gene show no major impairments in B-cell development or immune responsiveness (30, 31); although they may show a delay in affinity maturation during the early primary T cell dependent immune response (32).

During an immune response, a B lymphocyte can express a different H chain isotype (i.e. $C_{\gamma 1}$ or C_ϵ), by replacing the C_μ coding sequences with those of a different C region by a DNA recombination mechanism (class switching) that is distinct from the mechanism involved in assembly of the variable region exon (33–35). This process allows the progeny of a given B lymphocyte to produce antibodies that retain their original specificity but that can perform different effector functions (33, 34, 36). In the mouse, the six different C_H isotypes are ordered 5'–γ3–γ1–γ2b–γ2a–ϵ–α–3' and lie from 50 to 150 kb downstream of the C_δ gene (Fig. 3; also see Fig. 12) (25).

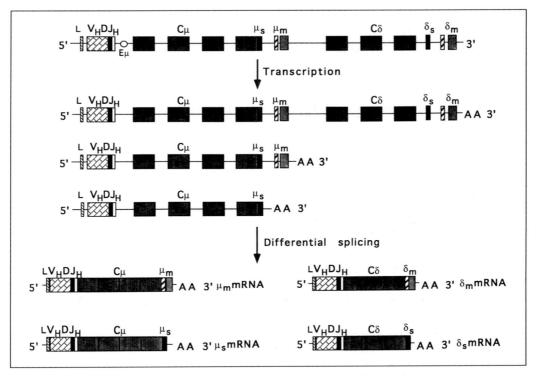

Fig. 5 Expression of the μ and δ heavy chain genes. Depicted is transcription from the V_H promoter through the leader (L), V_H, D, J_H, C_μ, and C_δ gene segments along with differential splicing which gives rise to both the membrane and secreted form of IgM and IgD (27). B lymphocytes which express both μ and δ H chains produce transcripts which contain both C_μ and C_δ sequences, but in plasma cells transcription is apparently terminated upstream of the C_δ exons (27).

4. Mechanism of Ig gene assembly

Early information concerning the mechanism of the V(D)J recombination process came primarily from comparison of the structure of germline variable region gene segments to that of their rearranged counterparts in transformed cell lines and tumours such as myelomas (2). Based on this information, it has been evident for some time that the assembly process probably involves:

- the recognition of recombination signal sequences (RSSs) that flank each variable region gene segment
- the creation of double-stranded breaks between the involved segments and their RSS
- the precise ligation of the RSSs, and the imprecise joining of coding ends including potential removal or *de novo* addition of nucleotides at these junctions (2, 37).

In recent years, additional important insights into the V(D)J recombination mechanism have been derived. Until recently, more dynamic studies of the

mechanism relied on transformed pre-B cell lines, typified by A-MuLV transformed pre-B lines, that continued to express V(D)J recombination activity during
propagation in culture. Because these lines actually express the activities required
for V(D)J recombination, they have also been extremely useful for gene transfer
experiments as hosts for recombinant plasmid or viral vectors (recombination
substrates) containing cloned germline variable region gene segments (38, 39).
Under appropriate conditions, these introduced gene segments undergo recombination in the transfected cell, making it possible to design recombination substrate experiments that test various hypotheses concerning the mechanism and
regulation of Ig gene assembly (38–43). In particular, extrachromosomal V(D)J
recombination substrates (43) that facilitate the study of different combinations
of variable region gene segments in various orientations with either normal or
mutated elements have been extremely useful in dissecting novel aspects of the
mechanism. A general description of recombination substrates will be given below
in the context of deletional and inversional modes of V(D)J recombination.

Recent work has indicated that the V(D)J recombination process involves both
specific activities that are expressed only in the appropriate precursor lymphocytes
as well as more generally expressed activities that are recruited to carry out their
functions in the context of V(D)J recombination. The isolation of the recombination
activating genes (*Rag*) which encode the tissue specific activities necessary and
sufficient to confer V(D)J recombination activity to any cell type (44, 45) was a
major breakthrough in this field and has allowed the generation of many new
approaches to studying various aspects of V(D)J recombination.

4.1   Overview of the V(D)J recombination mechanism

The process of V(D)J recombination is apparently initiated by the generation of a
double-stranded break that occurs on both recombining partners at the junction of
the heptamer and coding sequences (Fig. 6A). At least in some cases, the + and
− strands of the coding region are covalently linked to form 'hairpin' structures
(Fig. 6C); while the majority of liberated RSSs are directly ligated heptamer to
heptamer (Fig. 6B). The covalently closed coding ends probably are resolved by
cleavage of the loop near the hairpin (Fig. 6C, top), with the site of cleavage

Fig. 6 V(D)J recombination mechanism. (A) The V(D)J recombination is initiated by the generation of double-
stranded breaks. Coding sequences are enclosed in boxes; recombination signal sequences (RSSs) are
represented by triangles; arrows depict sites of the double-strand break. The most RSS-proximal 5 bp from
each coding segment are shown. ☐ denotes mutations that have been found to affect various steps.
(B) Liberated RSSs are directly ligated heptamer to heptamer. (C) Resolution of the coding join. The cleavage
of the coding -RSS junction probably results in the formation of a hairpin by the covalent joining of the +
and − strands (top). Cleavage of this hairpin results in the addition of the nucleotides from the opposite
strand as P-nucleotides (in bold; bottom). (D) Processing of the coding join in the presence of TdT activity.
Nucleotides can be deleted from this now open end. ↻ represents exonucleases (top). N regions are
denoted above with a square dot centred over the nucleotide (·). The coding ends are filled in and ligated (denoted with open boxes). (E) Processing of the coding join in the absence of TdT. Homologous nucleotides near
the coding segment of the two joining ends are aligned (top). ↻ represents exonucleases (top). The coding
ends are filled in and ligated (denoted with boxes; bottom). 'Overlapping' sequences are denoted with ⌐‗⌐.

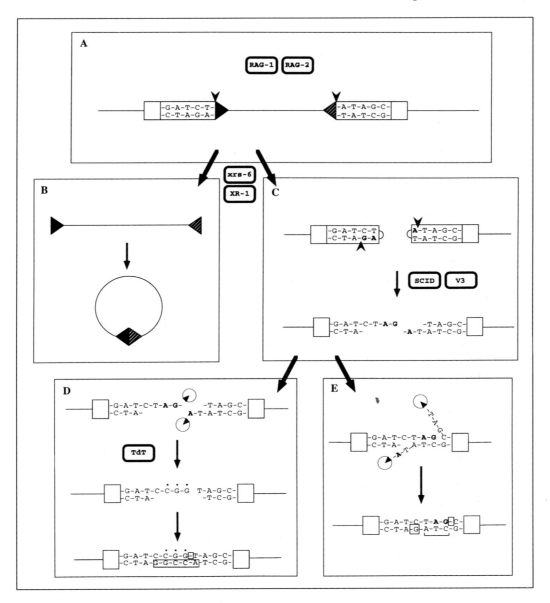

determining the number of bases that can be added to the coding junction as P (palindromic) nucleotides (Fig. 6C, bottom). If the cleavage occurs at the apex of the hairpin, no P-nucleotides are added; whereas cleavage at sites increasingly distant from the site of hairpin formation results in the addition of more P-nucleotides. Nucleotides may also be deleted from this now open end (Fig. 6D and 6E, top). Extra nucleotides (referred to as 'N regions') also may be added to the ends of the coding regions (Fig. 6D, centre) to generate non-germline encoded sequence diversity at these junctions. 'Overlapping' nucleotides on the overhanging 5' ends of the involved coding sequences may be aligned to direct the joining process (Fig. 6E, bottom); P-nucleotides and N-regions that have been added on to these ends probably can participate in such an alignment process (Fig. 6E, top). Gaps in the joining region are filled in and the processed coding ends are ligated (Fig. 6D and E, bottom). This subject has been reviewed by Gellert (44).

4.2 Recombination recognition sequences

The site-specific joining of Ig (and TCR) variable region gene segments by V(D)J recombinase is mediated by recognition of the conserved recombination recognition or signal sequences (RSSs) that immediately flank each gene segment (2, 47, 48). RSSs consist of a palindromic heptamer with the consensus sequence CACAGTG and an AT-rich nonamer with the consensus sequence of ACAAAAACC (2, 49). The heptamer and nonamer are separated by a non-conserved spacer of either 12 ± 1 or 22 ± 1 base pairs (bp) (47). While the actual sequence composition does not appear to be relevant, the number of nucleotides between the heptamer and nonamer is crucial for the adjacent DNA segment to be an efficient substrate for recombination (47, 49–51). The 12 or 23 spacer lengths permit approximately one or two complete turns of the DNA helix, respectively, so that the helical grooves within recognition elements are aligned on the DNA molecule in the same rotational orientation (3, 47, 49). It is conceivable that *trans*-acting heptamer and nonamer binding factors need to interact for effective recombination to occur. Putative RSS recognition proteins that bind to the heptamer or nonamer sequences have been isolated, and the cDNAs which encode them have been cloned (52–56). However, to date, no role of these proteins in V(D)J rearrangement has been established.

V(D)J recombination is normally mediated only between variable region gene segments which are flanked by RSSs that contain spacers of different length. This phenomenon is referred to as the '12–23' rule (Fig. 7A) (47). With respect to Ig HC variable region gene segments, the J_H and V_H segments have an RSS with a 23 bp spacer while the D segments are flanked on both their 5' and 3' sides by RSSs with 12 bp spacers. Therefore, assembly of a complete V(D)J heavy chain variable region gene necessarily involves D to J_H and V_H to DJ_H joining, as V_H segments cannot join to the J_H segments because of the 12–23 rule. For light chains, where the variable region is assembled from only V and J segments, either the V and J families, respectively, have 12 or 23 bp spacers to permit direct V to J joining (Fig. 3 and Fig. 7B) (47, 57, 58).

The V(D)J recombinase system and its cognate RSS have been conserved throughout evolution, so that transgenic chicken or rabbit Ig variable region gene segments can be accurately joined in lymphoid cells of mice (59, 60). Systematic analysis of the heptamer and nonamer requirements using recombination substrates have revealed three coding segment proximal bases of the heptamer to be the most important for the recognition process (49). Recombination substrates with mutations in these nucleotides rearrange less efficiently compared to substrates

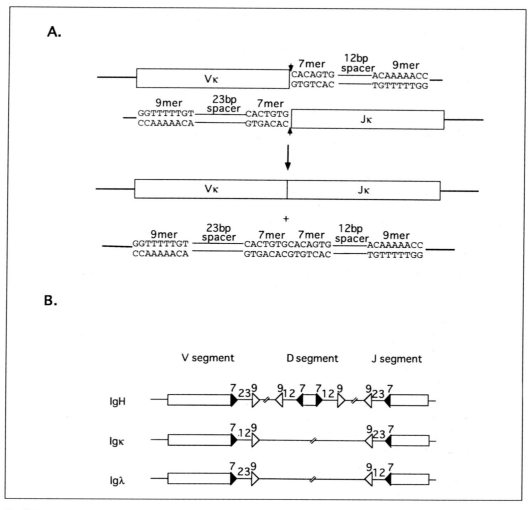

Fig. 7 The '12–23' rule. (A) Diagram of recombination signal sequences (RSSs) with 23 and 12 bp spacer sequences in between the consensus heptamer (7mer) and nonamer (9mer) that respectively flank the V_κ and J_κ coding segments (denoted as open boxes). V(D)J rearrangement results in the joining of sequences flanking the RSS with the 23 bp spacer with sequences adjacent to RSS with the 12 bp spacer. (B) Spacing between the heptamer and the nonamer of RSSs flanking different Ig HC and LC V-region gene segments. The heptamer is denoted as a closed triangle, while the nonamer is denoted as an open triangle.

with consensus sequences (49, 50, 61). Although the nonamer can vary significantly from its consensus sequence, a G (but not T) replacement in position 6 or 7 of the consensus nonamer dramatically affects the recombination efficiency of the substrate (49).

Sequences that are flanked by a complete RSS can also be joined to sequences flanked by a heptamer alone (62–68). The deletion of C_κ that occurs in many λ LC producing murine and human B lymphocytes is often mediated by the rearrangement of a complete RSS 3′ of C_κ (named *recombination signal* or *kappa deleting element*), to sequences adjacent to an isolated heptamer in the J_κ–C_κ intron (64, 65, 68). V gene replacement events result from the recombination of a germline V_H to a rearranged $V_H(D)J_H$; the partner to the RSS of the germline V_H segment is a heptamer conserved in the third framework region of most murine and human V_H segments (66, 67, 69, 70). Heptamer-mediated rearrangements have been reported to occur at low frequencies in recombination substrates (49), and many lymphomas have aberrant translocations that apparently are products of a V(D)J recombinase-mediated rearrangement of a *TCR* or *Ig* gene segment to spurious heptamers (71).

The previous examples demonstrate that the nonamer is not an absolute requirement for V(D)J recombination. Thus, recognition of heptamer-like sequences (which occur frequently in the genome) must not be the only restraint that confines the activities of the recombinase to appropriate target variable region gene segments. There may be lineage or stage-specific mechanisms that promote the occurrence of such joins in V_H replacements or κ deletion event, by relaxing the specificity of the V(D)J recombination reaction or by utilizing, as yet, unrecognized motifs in place of the nonamer. Finally, assays with rearrangement substrates that contain combinations of defective signal sequences have demonstrated that one RSS is not sufficient to stimulate the recombination machinery, whereas an RSS in *cis* with even a defective heptamer can be targeted for sequence-specific double-stranded breaks (72, 73). Thus, interaction of two RSSs appears to be necessary for the initiation of the cleavage reaction.

4.3 Deletion versus inversion

In general, both the coding segments and the RSS of the recombining segments are joined during V(D)J recombination (37). As a result, the V(D)J recombination mechanism allows sequences between the recombining segments to be either deleted or inverted depending upon their relative orientation (Fig. 8A; ref. 37):

1. When the recombining segments are aligned on the chromosome with their respective recognition sequences facing each other, recombination between

Fig. 8 (A) Inversional, deletional, or pseudonormal V(D)J joining. Striped or grey squares represent V_κ or J_κ coding segments, closed triangles represents heptamer sequences, open triangles represent nonamer sequences. (B) Stable V(D)J recombination substrates. The incorporation of genes that confer drug resistance (*gpt* or *TK*, denoted as rectangles) into recombination substrates stably integrated in the genome of cell lines can be used to select for cells that have recombined the substrate using mycophenolic acid and HAT, BrdU, respectively, as shown. (C) Extrachromosomal recombination substrates. Extrachromosomal recombination substrates designed by Hesse *et al.* (43) can be transiently introduced into Rag-1,2$^+$ cell lines to assay for rearrangement. These substrates harbour a prokaryotic promoter (p*lac*) and chloramphenicol resistance (*cat*) gene, separated by the prokaryotic transcription terminator from bacteriophage λ (oop). V(D)J rearrangement of RSSs that flank the oop sequence results in the formation of a plasmid that can now actively transcribe the *cat* gene in bacteria. The extrachromosomal substrates are isolated from the cells and introduced into bacteria; the ratio of chloramphenicol and ampicillin doubly-resistant bacteria colonies to ampicillin resistant colonies reflects the efficiency of rearrangement of the cell line. The relative orientation of the RSSs in the unrearranged substrate determines whether the 'signal' or 'coding' joins are assayed.

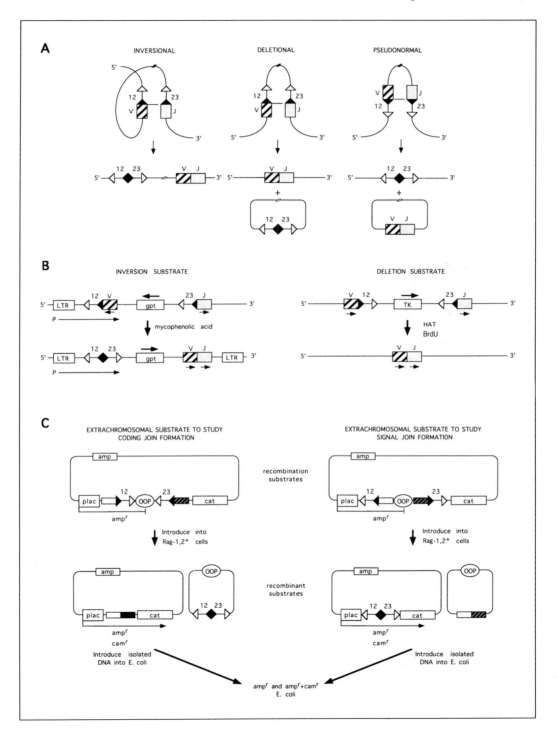

them excises the intervening DNA as a circle containing the fused heptamers (deletional join; Fig. 8A, centre).

2. When the recognition sequences of the recombining elements are facing in the same direction, both the joined segments and the fused heptamers are retained in the chromosome, and the intervening DNA is inverted (inverted join; Fig. 8A, left) (37, 63).

3. Finally, if the recombining segments are aligned on the chromosome with their respective recognition sequences pointing away from each other, the fused recognition sequences are retained and the fused coding sequences and intervening DNA are deleted as a circle (pseudonormal join; Fig. 8A, right) (37).

Examples of each of these types of rearrangements have been observed to occur in the recombination of endogenous variable region gene segments (37, 63, 74–80).

The general significance of deletional versus inversional joining pathways in normal physiology is still not clear. For example, in the Ig κ *LC* locus the orientation of some *V* gene segments relative to the *J* segments results only in inversional joining (Fig. 3B and see below). D_H segments should be able to undergo both inversional and deletional joins to J_H segments, since they are flanked by a 12 bp recombination signal sequence on both 5' and 3' sides. Yet the vast majority of endogenous DJ_H joins utilize the 3' D_H RSS and result in deletion. From observations made using extrachromosomal substrates with D_H segments in normal or inverted orientation with respect to J_H segments, this preference has been suggested to be at least partly due to influences from the nucleotide sequence of the coding sequences on the recombination process, rather than an inherent predisposition for the occurrence of deletional over inversional joins (81–83).

The three modes of V(D)J recombinational joining have been exploited for the construction of recombination substrate vectors that have allowed examination of various aspects of the V(D)J recombination mechanism. Inversional recombination substrates rely on the inversional mode of V(D)J joining between two target gene segments in the substrate to activate a selectable marker gene such as *gpt* or *neo*[r] that lies between the two target substrates (Fig. 8B, left) (39, 44, 84). These substrates have been particularly useful in stable transfection (or infection) experiments. Deletional or pseudonormal modes of V(D)J recombination have been particularly useful in the context of transient V(D)J recombination substrates (43). In these vectors, V(D)J recombination between two target RSSs leads to the deletion of a transcriptional stop sequence and, as a result, transcriptional activation of a chloramphenicol transacetylase gene (*cat*). The basic strategy for their use involves the initial transient introduction into mammalian cells where V(D)J recombination may occur and the subsequent rescue and assay for V(D)J recombination events based on the generation of chloramphenicol resistance following introduction into bacteria (Fig. 8C). Deletional and inversional substrates have also been used in stable (38, 42) and transient (43, 85) substrates, respectively.

4.4 Unusual aspects of the V(D)J recombination mechanism

4.4.1 Double-stranded breaks and hairpins

For V(D)J rearrangement to take place, all four DNA strands involved in the re-arrangement process must be cleaved during the reaction. This site-specific recombination reaction was proposed to be initiated by a double-strand break (DSB) at the junction of the coding and signal sequences (37). The finding of DNA segments with double-strand breaks adjacent to TCR_δ signal sequences provided direct evidence for a DSB intermediate in V(D)J rearrangement (86). These molecules, specific to the thymus where rearrangement of the $TCR\delta$ locus occurs, were identified as double-strand cleavage products by their restriction map and sensitivity to exonuclease. More recently, double-strand breaks adjacent to the D_H, J_H, and J_κ RSS have been identified (87); analysis of these products revealed that the double-strand breaks resulting from cleavage by the V(D)J recombinase are blunt and phosphorylated at the 5' end.

The murine scid defect involves an impairment in the ability to join liberated coding ends (see below). Analyses of the cleaved coding sequence intermediates of TCR_δ rearrangements in the thymus of mice with the severe combined immune deficiency (scid) defect demonstrated that such ends could be isolated in the form of covalently closed hairpin structures (88). These hairpins apparently result from the covalent ligation of the terminal nucleotides of the + and − strands of the coding segments after double-strand cleavage, and have been suggested to be a normal intermediate in the recombination process (88). In the scid thymus, these hairpin intermediates may accumulate as a result of the defect, rather than as a direct product of the scid defect (see Section 4.6.1) (88). The mechanism of the formation of these hairpin structures generated during V(D)J recombination may be similar to the mode of generation of hairpin DNA structures in site-specific recombination processes in bacteria and yeast (reviewed in refs 46, 89, 90). Obviously, such hairpins must be opened to continue the joining process; this step will be further considered in the context of P-elements and the scid defect (88, 90).

4.4.2 N-regions

It was proposed that CDR3 diversity is enhanced by the addition of non-templated nucleotides (N-regions) to the 5' ends of the liberated coding sequences before they are joined (37). The joining region of the Ig and TCR *V*-region segments encodes CDR3, which is one of the contact points of the receptor with the antigen. Thus, the addition of random N-region sequences can increase the range of antigens recognized by receptors encoded by each *V*-region gene segment. It was further proposed that the enzyme terminal deoxynucleotidyl transferase (TdT), a lymphoid specific enzyme that can add nucleotides on to primers in a template independent manner (91), was responsible for N-region addition activity (37). This proposal was recently confirmed by gene-targeted mutational analyses (see Section 4.5.2 for more detail) (92, 93). Other mechanisms proposed to generate N-regions such as 'capture' of oligonucleotides (94, 95) are apparently not of major importance in this process. N-regions may also be added to RS joins. However, N-regions in RS joins do not appear to occur as frequently or to contain as many nucleotides (on average) as those of coding joins (74, 76–80, 95–98). Finally, the addition of N-regions is a developmentally regulated process that appears to be controlled by the presence or absence of TdT expression (see below) (99–116).

4.4.3 Homology-mediated joining

The absence of N-regions in fetal-derived lymphocytes has enabled the detection of homology-mediated joining between the coding segments (112, 114, 117–119). If two or three base-pair homologies exist between the ends of the rearranging segments, they are often preferentially used in the recombination process. The resulting join has characteristic 'ambiguous' nucleotides that could be ascribed to either involved V-region gene segment (37, 112, 114, 119–121). Thus, the diversity of variable regions formed from the rearrangement of V, D, J gene segments in the absence of N-region addition may be significantly restricted by this process (112, 114, 119).

Homology between the joining segments is not an essential feature for V(D)J recombination to occur. Sequences that do not have short homologies rearrange efficiently, and sequences that have short homologies do not always utilize them in the rearrangement process (92, 122). The sequence composition around potential homologies between joining segments probably significantly influence the contribution of the homology to the joining reaction (92). Importantly, N-region addition may obviate the use of germline encoded homologies, offering a means to regulate the use of this pathway (see below). On the other hand, N-regions may also participate in such homology-mediated joining pathways, although it is not possible to identify such participation due to the 'random' nature of N-regions.

4.4.4 Hairpins and palindromic sequences (P-nucleotides)

Nucleotides distinct from N-regions, termed P-nucleotides, can also be added at the junctions of coding segments (99, 110). P-nucleotides are short (usually one or two) nucleotide additions found adjacent to intact coding sequences that form palindromes with the terminal sequences of the coding region. For example, a gene segment ending in -5'-GATG-3' could have adjacent P-nucleotides of -CA or -C, so that the sequence at a join using this segment becomes -5'-GATGCA-3' or -5'-GATGC-3'. P-nucleotides were first identified, and are particularly evident, as recurring nucleotide insertions in TCRγ and δ joins from fetal and newborn thymus where N-region addition is infrequent (99). P-nucleotide addition can be responsible for the addition of specific nucleotides to the junction of coding sequences. This process is most likely responsible for generating an invariant serine in the D_H–J_H junctions of arsonate binding antibodies by adding a TC next to a terminal GA of a D-segment (99). P-nucleotide addition also probably generates the C or A residue often found in V_λ to J_λ Ig LC rearrangements in chicken (110).

Most P-nucleotides observed in joins of variable gene rearrangements from normal animals are no more than 2 or 3 bp long, whereas extraordinarily long

P-nucleotides of up to 15 bp have been found in the junctions of resolved *TCRγ* rearrangements from scid mice (123, 124) and from hamster cells that harbour a mutation analogous to the murine *scid* mutation (125). In this regard, the murine scid V(D)J recombinational defect may be based on the inability to resolve the hairpin structure generated at coding sequence ends (88). Thus, the accumulation of hairpin coding ends from the *TCRδ* locus in the thymus of scid mice can be attributed to a defect in the ability to properly resolve the hairpin structure. The addition of abnormally long P-nucleotides found in scid rearrangements (125, 124, 126) might result from the resolution of the hairpin by random nicking which could occur further from the site of the initial double-strand break than the nicks generated by the normal V(D)J recombination machinery (88, 90).

4.5 Lymphoid-specific activities involved in V(D)J recombination

4.5.1 Recombination activating genes

The recombination activating genes (*Rag*) 1 and 2 (45, 127) were isolated on the basis of genomic DNA transfection experiments which led to the ability to activate an inversional V(D)J recombination substrate in fibroblasts and thereby provide a selectable drug resistance (44, 45, 127). The *Rag*s encode nuclear phosphoproteins that do not share substantial homology with other known gene products, but do contain motifs found in certain proteins implicated in transcription and/or recombination (128–130). Various lines of evidence suggest that *Rag-1* and -2 have a direct role in V(D)J recombination (131). However, the actual function of these genes in V(D)J rearrangement has not yet been unequivocally elucidated. Thus, while *Rag-1* and/or -2 may encode specific components of the V(D)J recombinase, it remains possible that they could encode transcription factors that turn on the V(D)J recombinase genes, or proteins that in some other way activate the V(D)J recombinase machinery (reviewed in refs 129 and 132). Likewise, it is also possible that these gene products could serve more than one function; recent transgenic studies have suggested that Rag products may function to negatively influence aspects of peripheral T cell development (133).

Co-expression of *Rag-1* and -2 occurs at significant levels only in precursor lymphocytes or their transformed derivatives. The independent expression of *Rag-1* and -2 was detected in non-lymphoid tissues, leading to speculation of potential roles in functions other than V(D)J recombination, such as development of the central nervous system (134). However, the introduction of homozygous gene-targeted disruptions of either *Rag-1* or *Rag-2* into mice results in an identical phenotype: the lymphoid compartments are depleted of mature T and B cells, while the animals, to date, have been found to be otherwise normal (135, 136). Therefore, the *Rag* appear to have functions required only for the development of lymphocytes.

B and T lymphocyte development in *Rag-1* or *Rag-2* deficient mice is blocked at the early stage when endogenous TCR or Ig gene assembly normally initiates (135, 136). Thus, the *Rag-1* and *-2* deficient mice generate populations of precursor B or T cells that harbour unrearranged endogenous antigen receptor loci due to an inability to initiate V(D)J rearrangement. Transfection of a functional *Rag-1* or *Rag-2* gene into the appropriate *Rag*-deficient A-MuLV transformed pre-B cell lines led to the rapid onset of rearrangement of the endogenous J_H locus (G.R. and F.A., unpublished data). As A-MuLV transformants from wild-type mice normally have rearrangements of both J_H alleles at the time of isolation, the complementation results imply that the IgH chromosomes in the *Rag*-deficient pre-B cells are poised to undergo rearrangement and that the *Rag* deficiency is the only block in this process. Additional experiments also confirmed that the only block to lymphocyte development in *Rag*-deficient mice is the inability to initiate V(D)J recombination. Thus, it is possible to restore relatively normal numbers of mature lymphocytes in the *Rag*-deficient mice by introducing functionally assembled TCR or Ig transgenes into the *Rag*-deficient background (133–136, 137–140). Together, these findings indicate that the primary function of *Rag-1* and *Rag-2* is to encode (or activate) tissue-specific components of the V(D)J recombinase that are required for initiation of V(D)J recombination.

The conservation of the structure and organization of the *Rag* in evolution is rather striking. In all vertebrate species examined, the two genes lie adjacent to each other in opposite transcription orientation within 8 kb. Both have two exons, with the protein-encoding sequences in one large exon (45, 127, 141). In this context, it is notable that sequences in the RSS, whose recognition and/or cutting appears to be dependent on *Rag* expression, are also highly conserved among different species (45, 127, 129, 142). The putative function and genomic structure of the *Rag-1* and *-2* sequences have been suggested to be highly reminiscent of virally-encoded genes involved in transposition (45). Together with the conserved target RSSs, these observations have generated speculation that a recombination system of an infecting virus or fungus may have been opportunistically adopted during evolution by a primordial vertebrate as a way of bringing together gene segments in multiple different combinations (45).

4.5.2 Terminal deoxynucleotidyl transferase (TdT)

TdT was initially identified, in partially purified DNA polymerase preparations from calf thymus, as an activity that can polymerize deoxynucleotides on to a 3'-hydroxyl terminus without a template (91). TdT activity was found in tissues rich in early lymphocyte stages from bone marrow and thymus but not in circulating mature lymphocytes, consistent with a role in lymphocyte differentiation (143). On the basis of these characteristics of the enzyme and its expression pattern, it was proposed that N-regions were added by TdT activity (37).

Support for this proposal was generated by correlations between *TdT* expression in cell lines or normal lymphocyte populations and N-region addition to V(D)J joins generated by such cells. Thus, TdT is found in postnatally-derived immature lymphocytes where N-region addition is observed, but not in fetal lymphoid tissues, where N-region addition occurs less frequently (99–116). Likewise, N-region addition to endogenous or transfected substrates in A-MuLV transformed pre-B lines correlated with the presence or absence of endogenous TdT expression (41, 95, 144). Strong support for this proposal came from the finding that introduction of *TdT* expression vectors into pre-lymphocytes and *Rag-1* and *-2* expressing fibroblasts resulted in the increased addition of N-regions to the V(D)J joins of co-introduced rearrangement substrates (145, 146).

While the studies outlined above strongly indicated that TdT can, in fact, generate N-regions, none addressed the proposal that N-regions can be added by TdT-independent mechanisms. The physiological role of TdT in N-region addition was definitively proven by the gene-targeted ablation of the functional *TdT* gene (92, 93). TdT-deficient lymphocytes were found to undergo normal rearrangement of the TCR and Ig variable region gene segments, but their V(D)J joins showed a nearly total lack of N-regions. Consistent with the interpretation that homology-mediated joins are favoured by the absence of N-region addition (112, 114, 119, 121), homology-mediated V(D)J joins were detected more frequently in the adult TdT-deficient lymphocytes than in normal adult lymphocytes (92, 93).

4.6 Generally expressed components of recombinase

4.6.1 *scid* Mice

The first evidence for the involvement on non-lymphocyte specific components of the V(D)J recombination reaction came from the analysis of mice homozygous for the autosomal recessive *scid* mutation. Such mice generally do not develop mature T and B cells due to a defect in V(D)J recombination (147). At this level, the defect appears similar to that of the *Rag*-deficient animals except that *scid* mice do develop some mature B and T cells with time (i.e. their mutation is 'leaky') whereas *Rag*-deficient animals do not (135, 136). In this context, the leakiness of the homozygous *scid* mutation can be attributed in large part to the aspect of the V(D)J recombination process that it impairs. Thus, unlike the *Rag* mutations, the homozygous *scid* mutation does not affect initiation of the V(D)J recombination reaction. Rather, the *scid* mutation results in a selective impairment in ability to join liberated coding sequences while leaving the joining of the RSSs relatively unimpaired (147–153). These manifestations of the *scid* mutation on V(D)J joining provide clear support for the proposal that the coding and RSS joins are handled differently by the V(D)J recombination system (37).

The fact that the initiation of the V(D)J recombination reaction proceeds normally in the context of the homozygous *scid* mutation can largely explain the leakiness of this mutation. Thus, coding segments are specifically liberated by the recombination system; at low frequency these gene segments can be joined (apparently by illegitimate recombination pathways) to generate normal junctions which ultimately allow a very low level generation of functional lymphocytes (151, 153, 154). Because the homozygous *Rag* mutations do not allow initiation of the reaction, those defects cannot be leaky in this fashion.

The *scid* defect also affects functions other than V(D)J recombination. Specifically, fibroblast and lymphoid cell lines generated from *scid* mice have increased sensitivity to X-irradiation and other agents that induce double-stranded breaks (70, 155, 156). In yeast and prokaryotes, mutations that affect the ability to repair DSB often also influence site-specific and general recombination (reviewed in refs 157, 158). The correlation between double-strand break repair (DSBR) and recombination suggests that factors involved in DSBR may be generally recruited to participate in rearrangement reactions that have DSB as intermediates (158). Thus, the *scid* gene may encode a generally expressed product that participates in the repair of double-stranded DNA breaks.

4.6.2 Other non-lymphoid factors involved in V(D)J rearrangement

To search for additional genes that encode activities that impact on the V(D)J recombination process, Chinese hamster ovary (CHO) cells selected for mutations of genes involved in DNA repair were analysed for their ability to undergo V(D)J recombination. When *Rag-1* and *-2* plus an extrachromosomal recombination substrate were introduced together into the mutant cell lines, UV-sensitive CHO lines (defective in excision repair) were normal in ability to undergo V(D)J recombination (159, 160). However, three of six ionizing radiation sensitive, double-strand break repair mutants (*Xrs-6*, *XR-1*, and *V-3*) were found to be defective in their ability to undergo V(D)J recombination (125, 159, 160). All three of these mutants appeared capable of initiating the reaction. Two (*Xrs-6* and *XR-1*) were defective in their ability to join both coding and RSSs, while another (*V-3*) was defective primarily in joining coding (but not RS) sequences. Somatic cell genetic analyses indicated that the hamster *V-3* mutation affects the same genetic complementation group as the murine *scid* mutation, and thus may involve the same gene (*V3*, see Fig. 6) (125). As the three CHO mutations belong to different complementation groups, two distinct from scid, these findings indicate that at least three genes exist that encode generally expressed factors whose activities (or lack of activity) can impact on the V(D)J recombination mechanism. The shared role of these gene products in DSB repair and V(D)J recombination suggests a scenario in which the specific components of V(D)J recombinase evolved the ability to recruit these general activities to 'repair' the double-strand breaks that are specifically introduced during the V(D)J recombination reaction (159).

Recent experiments have strongly indicated that the *Xrs-6* mutation involves a gene which encodes a DNA end-binding protein referred to as Ku (161). The Ku protein is a complex of 70 kDa and 80 kDa subunits that was first defined as an autoantigen. Several groups showed that *Xrs-6* cells were deficient in an end-binding activity (162, 163). In addition, the gene encoding the 80 kDa subunit of Ku was mapped to a region of human chromosome 2 (164) to which the *Xrs-6* gene had also been mapped (165). Finally, complementation of Xrs-6 cells with expression constructs encoding the human 80 kDa Ku subunit substantially reverted the X-ray sensitivity of these cells and also conferred upon them the ability to undergo normal V(D)J recombination.

While the function of Ku in double-strand break repair and V(D)J recombination is unknown, it seems possible that it may be involved in binding to the double-strand breaks where it could either serve a protective function or activate additional steps. In the latter context, Ku has been shown to be the DNA binding subunit of the DNA dependent protein kinase (166, 167). Strikingly, recent experiments have provided strong evidence that the DNA-PK catalytic subunit gene is the target of the murine *scid* mutation (647, 648).

5. Organization of H chain variable region gene segments

Immunoglobulin HC and LC loci have hundreds to thousands of *V* gene segments spanning several megabases. The large numbers of germline *V* gene segments contributes to the diversity of the antibody repertoire.

5.1 The murine *IgH* locus

The murine *IgH* locus lies on chromosome 12. There are four J_H segments which are located approximately 8 kb upstream of the C_μ constant region gene (Fig. 3A). There are 13 known *D* segments and these are located from approximately 1 to 80 kb 5' of the J_H cluster (2). The *D* segments can be divided into four families based on flanking sequence homology: the single Q52 type *D* segment (*DQ52*) is located about 750 bp 5' of J_H1, the nine SP2-type *D* segments lie between 10 and 80 kb upstream of *DQ52*, and the two known FL16-type *D* segments are located on either side of the most 5' *DSP2* segment (2). Recently, an additional *D* segment, *DST4*, that resides about 16.5 kb upstream of *DQ52* has been identified (168). Based on sequences of $V_H(D)J_H$ junctions, there do not appear to be many other unidentified *D* segments.

In mouse, the V_H segments are located an unknown distance 5' of the *D* cluster, but probably within 100–200 kb of the J_H locus. The number of individual V_H segments vary with the strain of mouse and range from 100 to as many as several thousand (169). All characterized murine V_H segments are members of one of 14 families that are defined by nucleotide sequence homology within the coding region (169–172, and reviewed in ref. 173). Family members were initially defined as being at least 80 per cent homologous while genes with less than 70 per cent homology were considered to be members of different families. Some of the more recently defined V_H families are considered to be borderline by this definition (i.e. they differ from other families by less than 80 per cent, but exhibit 70 per cent or greater similarity to previously defined V_H families; ref. 174). Murine V_H families have been placed into three general groups on the basis of nucleotide and amino acid sequence similarities. Group I contains the *Q52*, *36–60*, *3609*, and V_H12 families; Group II contains *J558*, *GAM3–8*, and *SM7*; Group III contains *7183*, *X-24*, *J606*, *S107*, V_H10, V_H11, and V_H13 (Fig. 9A) (15, 173, 175).

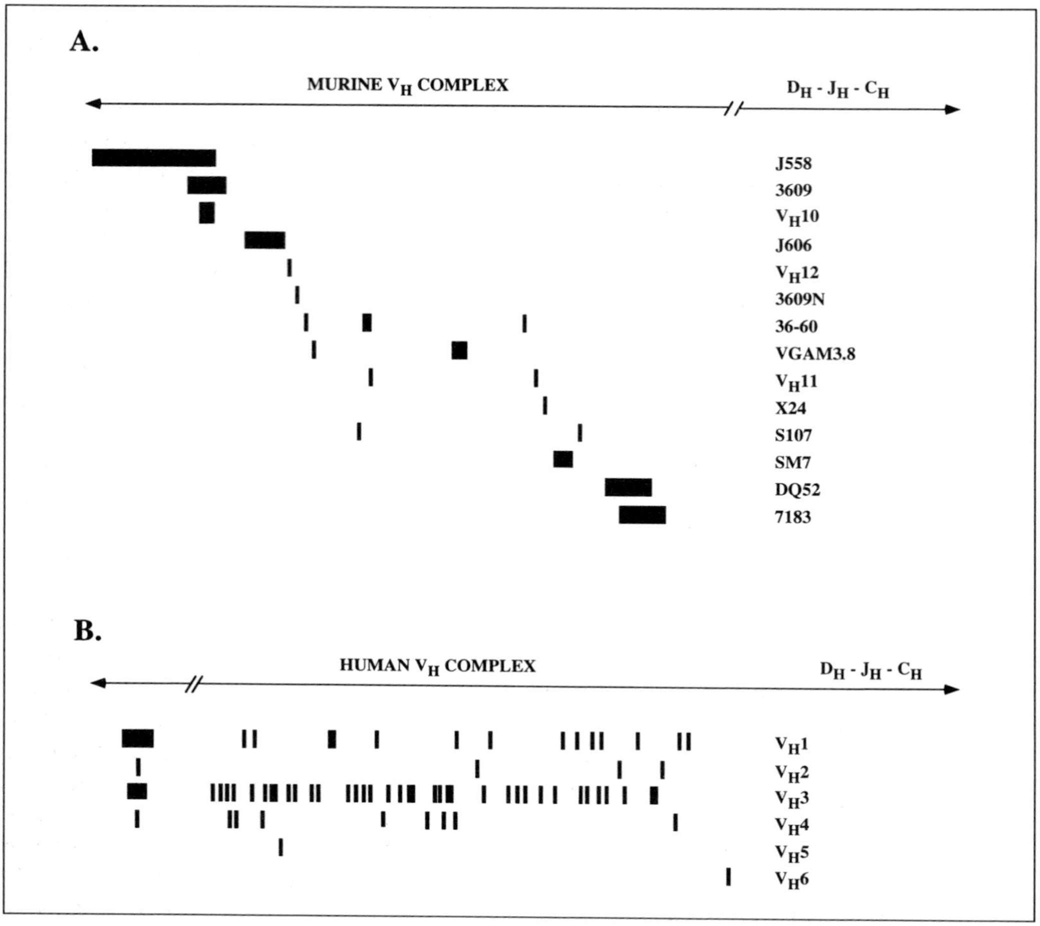

Fig. 9 (A) Organization of the murine V_H complex. The solid black boxes represent various V_H families which are defined on the right. The *J558* V_H family may contain 500–1000 V_H segments and has been shown to be interspersed with at least two different V_H families. The murine V_H complex has not yet been physically linked to the D_H region (indicated by break //). This map is a consolidation from several studies (105, 174, 176). The design of the map was adapted from refs 170, 177. (B) Organization of the human V_H complex. Each thin vertical solid box in the contiguous portion of the V_H locus represents one V_H gene from the represented human V_H family, denoted on the right. The map of this 0.8 Mb region has been determined by cloning, and is adapted from ref. 209. The J_H distal region of the human V_H locus has not been linked with the body of the locus (as denoted by break //). This region has been determined by deletion mapping and the map is adapted from ref. 177. Boxes in this region represent clusters of V-segments of each denoted family member.

Most murine V_H families are extremely polymorphic in different inbred strains of mice both in terms of restriction fragment lengths (RFLP) as well as in the numbers of cross-hybridizing sequences. Since the majority of murine V_H gene segments appear to be in the same transcriptional orientation as the J_H segments, V_H to DJ_H joining most often results in deletion of intervening DNA. The loss of this intervening DNA in $V_H(D)J_H$ rearrangements in myelomas, B lymphomas, B cell hybridomas, and A-MuLV transformed pre-B lines has been used to 'deletion map' the order of the 14 V_H families (105, 170, 174, 176, 178, 179).

These types of studies demonstrated that the murine V_H families are generally organized as a continuum of intermingled or, in certain cases, dispersed clusters (Fig. 9A). In general, V_H groups I and III exhibit considerable intermingling of member V_H families and are J_H proximal, while the majority of Group II V_H families are more J_H distal (170). The most J_H distal V_H families (*J558* and *3609*) and J_H proximal families (*Q52* and *7183*) exhibit substantial interspersion (105, 170, 174, 176, 178–180 and reviewed in ref. 181). Discrete subfamilies of Group II *J558* V_H genes (defined as V_H gene segments of at least 90 per cent nucleotide sequence similarity) have been partially mapped within the *J558* family in A/J mice (182); these subfamilies are clustered, intermingled, and dispersed. In this regard, the *J558* family may represent a more recently evolved V_H family and represent a potential future source for the evolution of additional V_H families. The general pattern of V_H organization outlined above seems to be conserved in most mouse strains (183). Strain-specific differences appear to result from the presence or absence of specific gene segments or blocks of V_H genes and the occasional dispersion of a given V_H gene segment (176, 179, 180).

The actual sequence complexity of murine V_H segments may be significantly less than their actual number would suggest. V_H coding regions sometimes differ by only one or a few base pairs (169, 182, 184). This phenomenon is particularly striking in the V_H*J558* family in BALB/c mice, which may contain as many or more members (500–1000) as all the other V_H families combined, and which appears to consist of at least 35–45 subsets of 10–20 closely related members each (169). The size of this family varies greatly among different inbred strains and isolates of wild mice, suggesting that the murine V_H locus has expanded and contracted relatively recently (185). V_H segments that are identical (particularly within their CDRs) do not contribute significantly to germline encoded antibody diversity. Thus estimates of germline V_H repertoire must reflect only the complexity of the locus (the number of unique gene segments) and so results in values that are as low as a few hundred (or even fewer) uniquely contributing gene segments (181). In addition, estimates of the proportion of murine V_H segments that are non-functional pseudogenes range up to 40 per cent, although this percentage appears to be lower in some families (180, 186–189). Therefore, the actual complexity of the murine V_H locus may be fewer than 100 sequences.

Many V_H pseudogenes contain only a small number of mutations and are as highly conserved within framework regions as are functional V_H segments, which indicates that they might have a physiological role. Regions of homology consistent with gene conversion or unequal crossover events have been observed among germline V_H and V_L segments in several species, suggesting that V_H and V_L pseudogenes might serve as a 'reservoir' for short regions of homology to be transferred by these mechanisms during evolution (190–195).

5.2 The human *IgH* locus

The *IgH* locus in humans is located on chromosome 14. A battery of techniques, including deletion mapping, and analyses of bacterial phage, cosmid, and YAC clones coupled with one- and two-dimensional pulse-field gel electrophoresis, have led to the generation of a rather detailed map of the human V_H, D_H, and J_H regions (Fig. 9B). Thus, our understanding of these regions in the human, particularly with regard to the J_H proximal portion of the V_H complex now surpasses that of the mouse (196, 197).

As in the mouse, the J_H and D_H regions are clustered. In humans, a DQ52-related *D* segment is located about 100 bp 5' of six functional J_H segments (190). Approximately 30 D_H gene segments that fall into seven unrelated families (including DQ52) occupy about 50 kb of chromosomal sequence just 5' of the J_H (120, 198–202). D_H gene segments associated with V_H segments that appear to be functional have also been found dispersed into the human V_H complex, apparently in J_H distal regions (200–202). However, no evidence has been found for their usage in $V_H(D)J'_H$ rearrangements (177) and no firm evidence exists for their presence in the conventional V_H complex.

The J_H proximal D_H gene families are organized into four 9 kb repeated units each of which generally contains 5–6 D_H gene segments from different families. These repetitive units have been ordered as *D4–D1–D2–D3–DQ52–J$_H$'* with the *D3* unit beginning approximately 25 kb upstream of *DQ52* (120, 198, 200, 202). Twelve separate *D* segments have been linked within a 15 kb region of DNA (120). Two of these (belonging to the same *D* family) were designated DIR sequences (D_H gene containing irregular spacer sequences) indicating that they are flanked by different spacer lengths of 12 and 23 bp in the RSS. This spacer organization gives these D_H gene segments the ability to undergo D–D joining; such rearrangement events have been identified (120, 177) and predicted to occur at a high frequency (177). By contrast, D–D joining events are very rare in the mouse and are not mediated by a normal RSS (203). A D–D join was characterized in which the two *D* segments were apparently joined in opposite polarity to each other (120). In contrast to the situation in the mouse in which a single D_H reading frame is predominately utilized (see below), human D_H segments are apparently used equally in all three reading frames (120). Thus, it appears that the human Ig HC CDR3 region may have more diversity contributed from the germline *D* locus than that of the mouse as a result of an increased number and diversity of germline D_H gene segments, greater plasticity with respect to joining events, and similar usage of all three potential reading frames.

Between 50 and 150 human V_H segments, up to 50 per cent of which are pseudogenes, can be divided into six distinct homology based families (V_H1–V_H6) (193, 204–211). A seventh 'family' closely related (about 80 per cent) to V_H1 has recently been identified (212). The human V_H complex is thought to occupy 1.0–3.0 megabases (208, 209, 211). Although many human V_H segments are closely related to murine V_H segment families, others are not (reviewed in ref. 181).

Human V_H gene family members are extensively intermingled (Fig. 9B) (reviewed in ref. 213). The organization of the J_H proximal 150 kb of the human V_H complex has been physically linked and shown to contain five or six genes depending on the 'haplotype' (see below and ref. 214). The single member V_H6 family is the most J_H proximal (208) and has been linked approximately 80 kb 5' of J_H and about 25 kb upstream of the D4 cluster (199, 214). The V_H genes in the D proximal region have been ordered using overlapping cosmid clones: $V_H2–V_H4–V_H1–V_H1–V_H6–D4$ (214). Of substantial interest was the identification of an allelic polymorphism in this region that contained V_H gene segments whose sequences were nearly identical to two previously defined autoantibodies (214). No particular specificity has been yet assigned to antibodies whose Ig HC variable region employs a V_H6 gene segment.

The majority of human V_H gene segments also appear to reside in the same transcriptional orientation as the D_H region (177, 208, 209, 214), although it has been suggested that some quite distal V_H segments may be found in opposite polarity (177). The organization of the human V_H complex was initially generated by one- and two-dimensional pulse-field gel electrophoresis (PFGE) mapping (201, 208, 211, 214) and supported initial reports of the substantial interspersion in the V_H complex suggested by analysis of cloned human V_H segments (205, 206). Higher resolution maps of the human V_H complex was accomplished by deletion mapping analyses of cloned Epstein–Barr virus CEBV immortalized B cell lines from adults (177, 215–218).

While the distal 'end' of the human V_H complex has not yet been identified, utilization of YAC and cosmid cloning has generated a V_H map that extends to approximately 800 kb upstream of J_H (209, 214). Within this region, 64 V segments have been identified, 31 of which are pseudogenes. As is the situation in the mouse $J558$ V_H family, most of these pseudogenes are structurally sound and highly conserved with functional counterparts (209). Two-dimensional PFGE studies have mapped V_H gene segments at about 1.1 Mb from J_H (177). Sequence analyses, deletion mapping, PFGE, and cloning studies all suggest that the human V_H complex occupies approximately 1.0 Mb and contains about 50 functional V_H gene segments (209, 211, 217, 219).

The overall organization of the human V_H complex appears relatively constant in the limited studies involving different individuals (177). Several polymorphic regions of insertions or deletions of regions of DNA that contain several V segments exist within the V_H complex (209, 211, 215). These polymorphisms are used to define human V_H 'haplotypes' (209, 211, 214). Additionally, small regions of V segment duplications have been identified (209, 211, 217). Together with the presence of well-conserved pseudogenes, these polymorphic characteristics suggest that the human V_H complex is a dynamic region that is subject to a multiplicity of evolutionary diversification mechanisms that alter its content (210).

Several V_H and D_H gene segments have been dispersed to other chromosomes (200). Some of these segments, termed orphons, appear to be functional, implying that their dispersion was a relatively recent event in V_H evolution. Unlike dispersed

λ LC related genes (see below), the participation of these orphons in any aspect of the immune response has yet to be demonstrated.

5.3 The murine κ LC locus

The murine κ locus is located on chromosome 6 (Fig. 3B). There are five murine J_κ segments ($J_\kappa 1$–5) of which one ($J_\kappa 3$) is non-functional because it lacks a donor splice sequence (2). The J_κ gene segments are located approximately 2.5 kb upstream of the single C_κ gene (57, 58). At least 200 V_κ segments, which can be grouped into 19 homology-based families, lie at an unknown distance 5' of the J_κ segments (220 and reviewed in ref. 173). V_κ families exhibit a higher degree of interfamilial relatedness than V_H family counterparts and have been placed provisionally into three broad groups (173). Because V_κ gene segments exist in both transcriptional orientations relative to the C_κ gene (221), deletion mapping cannot be readily used to order V_κ families due to frequent retention of intervening DNA in inversional V_κ–J_κ rearrangements (221). RFLP analyses have, therefore, been utilized in inbred recombinant mice to determine V_κ organization (173). Some V_κ family members clearly are clustered in a continuum of several $V_\kappa 21$ members (222). However, other studies also suggest the presence of interspersed and dispersed V_κ gene segments (173).

5.4 The human κ LC locus

The human κ LC locus has been characterized in great detail. About 85 human V_κ segments, less than 50 of which are functional, occupy a 1.5–2.0 Mb stretch upstream of five functional J_κ segments and the single C_κ gene on chromosome 2 (221, 223–225). Human V_κ gene segments have been divided into four subgroups, the definition of which roughly corresponds to that of V_H families (221). Like human V_H segments, related human V_κ segments are extensively interspersed with unrelated V_κ gene segments, suggesting that this locus has been expanded after intermingling of family members (224, 225).

The most J_κ proximal region of the V_κ locus (termed region B) contains two linked V_κ gene segments (a $V_\kappa III$ and $V_\kappa IV$ segment) in opposite transcriptional polarity to the J_κ cluster; the most J_κ proximal V_κ, a $V_\kappa IV$ gene segment, is approximately 23 kb 5' of J_κ (221). Polarity of the V_κ complex switches in the V_κ gene segments 5' of $V_\kappa III$ and continues for the next megabase in the same transcriptional orientation as J_κ. This region contains the linked subregions L, A, and O and constitutes the major duplication unit in the V_κ locus. There are two copies of these subregions per haploid genome; the two copies bear striking sequence and restriction site similarity (221). Thirty V_κ genes have been characterized in the O region and 20 V_κ genes have been characterized in the A region (226, 227). The L region (low repetitive region) flanks the clusters of V_κ gene segments in the O and A regions. The J_κ distal duplication unit is in opposite transcriptional polarity to J_κ; rearrangements in the J_κ proximal unit results in deletion of intervening sequences while J_κ

rearrangements that involve distal region V_κ segments generate megabase-sized chromosomal inversions (228). The V_κ complex is, therefore, organized as follows: centromere–(L–A–O)–/–(O–A–L)–B–J_κ (221). The two duplication units (designated with parentheses) have not been physically linked. A third V_κ region of approximately 250 kb, designated W, contains at least 11 V_κ gene segments (221); nine of these have been sequenced and determined to be pseudogenes. The W region contains a 40 kb repeat structure and the V_κ genes within this repeat are opposite in their polarity to the remaining W-associated V_κ gene segments. The W region, while located on chromosome 2, has not been linked to either large duplication unit and it has been speculated that W may lie outside the defined V_κ complex (221). Several V_κ orphons have been identified and all are pseudogenes (221).

5.5 The murine λ LC locus

Lambda LCs account for only about 5 per cent of the serum LC in the mouse. Accordingly, the murine λ locus is less complex (in terms of numbers of unique V segments than the murine κ locus (Fig. 3C) (229). In BALB/c mice, three V_λ segments and four C_λ region exons (each C region having a J_λ gene segment in close proximity to it) have been identified (175, 230 and reviewed in ref. 231). The $J_\lambda 4$ segment is non-functional, apparently because of an abnormal splice donor sequence (2). The λ C region exons have been amplified to different copy numbers in separate populations of feral mice, suggesting that expansion and contraction events in this locus can also occur relatively rapidly (232).

The BALB/c Ig λ light chain locus is apparently organized as two duplication units that span approximately 200 kb (233–235). The gene order was determined as $V_\lambda 2$–V_x–$C_\lambda 2$–$C_\lambda 4$––$V_\lambda 1$–$C_\lambda 3$–$C_\lambda 1$. $V_\lambda 2$ (the most distal V gene segment) is separated from Vx by 19 kb and is approximately 190 kb upstream of the $C_\lambda 3$–$C_\lambda 1$ cluster. Vx is about 55 kb 5' of the $C_\lambda 2$–$C_\lambda 4$ cluster, the latter of which has been mapped about 92 kb 5' of $V_\lambda 1$. $V_\lambda 1$ (the most proximal V gene segment) has been localized 14 kb upstream of the $C_\lambda 3$–$C_\lambda 1$ cluster. There is some evidence to suggest preferential $V_\lambda 1$ usage in V_λ rearrangements observed in an A-MuLV transformed cell line (236); this may be related to putative enhancer-like elements associated with the $V_\lambda 1$–$C_\lambda 3$–$C_\lambda 1$ region that have been partially characterized (237, 238). In the mouse, greater than 90 per cent of murine serum L chains are of the κ isotype. However, the proportion of κ and λ varies among animals and can be reversed (for instance in whales and cows, ref. 17). The relative contribution of the λ LC locus to the LC repertoire probably is related significantly to the relative complexity of the locus in a given species (i.e. the number of unique germline V_λ segments) as outlined below.

5.6 The human λ LC locus

The human λ LC locus is on chromosome 22 and has been mapped more precisely to the 22q11.2 region (239). The C_λ complex spans about 34 kb and contains seven

C_λ genes, each associated with a 5' J_λ gene segment (240–242). Three of these genes, $C_\lambda4$, 5, and 6 are pseudogenes (241, 242). Spacing between C_λ genes varies from about 2–5 kb; J_λ–C_λ intron lengths (with the exception of $J_\lambda4$ in which a deletion has removed most of the intron) are approximately 1.3–1.5 kb (241).

Approximately 40 per cent of human serum L chains are of the λ type and, correspondingly, the human V_λ locus is more complex than that of the mouse (240). Amino acid and nucleotide sequences of human λ polypeptides and cDNA clones indicate that human V_λ segments can be divided into at least four (and probably more) homology-based subgroups (15, 242). These V_λ gene segments are organized in an intermingled fashion in the V_λ locus, and apparently in the same transcriptional polarity as the C_λ complex (242). Six V_λ gene segments, only one of which appears functional, occupy a stretch of approximately 60 kb in length directly upstream of $C_\lambda1$; these have been linked to each other and the C_λ locus (242).

The $G_\lambda1$ (or X6) sequence has been mapped 4.5 kb 5' of $C_\lambda1$ (243, 244). This 'gene', except in two very short regions, is not homologous to any V gene segment and has no RSS. The $G_\lambda1$/X6 nucleotide sequence is, however, extremely similar to exon I in the λ-like 14.1 (mouse λ5 homologue) gene (243) described below. Although $G_\lambda1$/X6 contains an open reading frame and a consensus RNA splice donor sequence, no function or product has been associated with this sequence. The only apparently functional gene segment within about 60 kb of $C_\lambda1$, a $V_\lambda III$ gene, is about 10 kb upstream of $G_\lambda1$/X6 (239, 242). Four irregularly spaced pseudo-genes occupy an approximately 30 kb stretch 5' of the functional $V_\lambda III$ gene segment (242).

5.7 Surrogate LC genes

Two λ-related genes play a crucial role in early B cell maturation (239, 245–253). In the mouse, these genes have been termed λ5 and $V_{preB}1$ (249, 254); the human counterparts are 14.1 and VpreB, respectively (239, 243, 245–248, 253). The λ5 and $V_{preB}1$ genes were initially identified on the basis of specific expression of their transcripts in pro- and pre-B cells (249, 254); human homologues 14.1 and VpreB have been shown to exhibit a similarly restricted pattern of expression (246, 247, 253, 255).

In human and mouse, these non-rearranging λ-related genes are located on the same chromosome but quite distal to the respective Ig λ LC loci (239). Murine λ5 is organized as a three exon structure separated by introns that are about 1.3 kb in length (256). The N-terminal 63 amino acids encoded by λ5 exon I display homology with V region sequences while the remainder has no homology to any V gene (256). However, the 39 amino acids encoded by exon II exhibit substantial homology over half its length to J_λ gene segments and surrounding regions. Exon III, which contains the C-terminal 104 amino acids, is highly related to C_λ genes. The organization of exons II and III, separated by a 1.3 kb intron, thus shows a remarkable resemblance to J_λ–C_λ structure and suggests that the λ5 locus may have arisen from a duplication–transposition event that involved the conventional Ig λ locus (252).

The relationship of the human λ5 homologue, *14.1*, to human Ig λ is more obvious. As in the mouse, *14.1* is organized in a three exon structure (239, 245, 247). Exon I is separated from exon II by an intron of 5 kb and there is an intron of approximately 1.3 kb between exons II and III. Strikingly, exon I is about 80 per cent similar to the $G_\lambda 1/X6$ sequence located about 4.5 kb 5' of $J_\lambda 1$ (239, 243, 244). Exons II and III are homologous to $J_\lambda 1$ and $C_\lambda 1$ and the intron between the latter is about 1.4 kb in length (239, 243, 244). Thus, *14.1* appears to be a duplication of approximately 6 kb of DNA that encompasses $G_\lambda 1/X6–J_\lambda 1–C_\lambda 1$ (239, 243). Two other λ-like genes, the pseudogene *18.1/F_λ1* and *16.1* (for which no function has been assigned), that contain only exons II and III, have also been characterized (239, 245). These genes are about 95 per cent similar to *14.1*. The *18.1/F_λ1* and *16.1* genes are also located on chromosome 22 but have not yet been linked to each other or to *14.1* (239).

Two V_{preB} genes have been identified in mouse. These genes, termed $V_{preB}1$ and $V_{preB}2$, do not rearrange and exhibit greater than 95 per cent similarity to each other (249). $V_{preB}1$, whose protein product is utilized in the surrogate light chain of the pre-B receptor, is located about 4.5 kb 5' of λ5. The location of the second V_{preB} gene is unknown. $V_{preB}1$ (and $V_{preB}2$) resembles a conventional V gene segment; it contains a 19 amino acid leader sequence that is separated from the coding exon (142 amino acids) by an intron of 87 bp (249). The V_{preB} genes are equally homologous to λ and κ V segments over most of the coding region exon. However, the most C-terminal region of about 25 amino acids shows no identity to Ig sequences (249).

A human homologue to murine V_{preB} has been identified that exhibits about 75 per cent overall homology to the mouse counterpart (246). The mouse and human similarity is particularly striking in the second and third framework regions. The genomic configuration of human V_{preB} is similar to that of the mouse. As in the mouse, human V_{preB} also contains a C-terminal region with no obvious identity to any Ig gene (246). Human V_{preB} is apparently on chromosome 22 but has not been linked to *14.1* (246).

The λ5 (*14.1*) and V_{preB} gene products, respectively, form covalent and non-covalent bonds with μ_m Ig heavy chain to provide the surrogate light chain that allows assembly of the so-called pre-B receptor. It has been reported that this receptor is expressed on the surface of late stage (small) pre-B cells (239, 248, 250, 255, 257–260). However, whether or not the HC/surrogate LC complex must be expressed on the cell-surface to effect any physiological functions is currently unresolved. In addition, other studies imply that the surrogate LC proteins may be expressed on earlier stage pro-B cells, possibly in association with another 'surrogate' HC (252). Gene-targeted mutation experiments have clearly demonstrated that λ5 expression is required for efficient differentiation of pro-B cells to later stages in the pathway (251). In the context, it has been suggested that expression of this receptor either in the cytoplasm and/or on the surface of pre-B cells may be involved in signalling Ig heavy chain rearrangement, allelic exclusion, or onset of LC rearrangement (see below).

6. H chain gene expression

6.1 Ig gene transcription and translation

Each V_H segment is proceeded upstream by a transcriptional promoter. Following assembly of a $V_H(D)J_H$ variable region, transcripts of the assembled μ H chain gene are initiated from the promoter of the rearranged V_H segment (Fig. 5) (2, 261, 262). Cleavage and polyadenylation of the H chain transcript between the μ_m and μ_s coding sequences leads to formation of a μ_s mRNA (19, 20, 263, 264). A μ_m mRNA is generated by termination (or perhaps also by cleavage) downstream of the μ_m exons, which are spliced to sequences encoding the C_H4 domain (19, 20, 263, 264). In IgM-secreting plasma cells, transcription appears to be terminated before the polymerase reaches C_δ sequences so that μ_s mRNA can be produced efficiently (29, 265, 266). Following V_LJ_L assembly, L chain RNA transcripts are initiated from analogous conserved V_L promoters, and terminated downstream of the C_L exon (267).

Translation of H and L chain mRNA begins at a translational initiation codon at the 5' end of the V gene sequences. This initiation codon defines the translational reading frame of the message (2). Because of the imprecision of the recombination mechanism, this variable region reading frame may be joined either in- or out-of-frame with the single translatable reading frame of the attached J segment and (through defined RNA splicing mechanisms) C region exons (37). If these reading frames have been joined in-frame, the rearrangement is referred to as 'productive'; because mRNA expressed from it can encode a full-length H or L chain protein. Of note, D segments can theoretically be employed in all three reading frames, as long as the junctional diversification mechanisms place the V and J in-frame (but see below).

Out-of-frame ('non-productive') rearrangements are often transcribed and processed into full-length mRNA, but translation of these mRNA species results in premature termination and production of protein fragments (268–270). In mature B lineage cell lines that have both a productive and a non-productive $V_H(D)J_H$ rearrangement, mRNA expressed from the non-productive allele is often present at a lower steady-state level (270). However, both productive and non-productive alleles are often transcribed at similar levels in pre-B cells. mRNA from non-productive alleles might be less abundant in more mature cells because it is less stable due to a lack of ribosomal protection of the untranslatable portion (271, 272).

If the recombination mechanism assorts reading frames randomly, it would be predicted that non-productive rearrangements would occur at a frequency of two in three. On the other hand, this frequency might underestimate the rate at which non-productive $V_H(D)J_H$ rearrangements occur *in vivo*, because of the existence of a large proportion of pseudo V_H genes (see above) and the fact that many murine

D segments have one or more potentially untranslatable reading frames due to the presence of translational stop codons (273). Early studies also estimated the *in vivo* frequency of non-productive $V_H(D)J_H$ and V_LJ_L rearrangements in splenic B cells to be about two-in-three (274, 275), and other studies suggested a similar frequency of non-productive rearrangements was also observed for $V_H(D)J_H$ rearrangements that occurred during culture of certain A-MuLV transformants (276). However, more recent studies (outlined above) have shown that homologies between DJ, VD, or VJ junctional sequences may promote the occurrence of joins in particular reading frames. Clearly, such a biased joining process could influence the relative levels of productive versus non-productive rearrangements for particular combinations of germline genes. Although the overall significance of this phenomenon remains to be determined, such biased rearrangement patterns may be a feature of particular repertoires (for example those formed in the absence of TdT activity).

7. Generation of the antibody repertoire

Having discussed the organization and structure of Ig variable region gene loci, the mechanisms involved in V(D)J assembly, the role of different gene segments with respect to encoding antigen binding portions of the antibody, and the basic mechanisms involved in Ig gene expression, we can now consider in general terms, the contribution of the various germline encoded and somatic factors to the generation of diversity.

7.1 Primary diversification mechanisms

Newly-generated virgin B lymphocytes express a 'primary' repertoire of antigen-binding specificities which presumably has not been influenced by contact with antigens or other selective forces. The wide range in specificities of this primary Ig repertoire in mouse and human derives from aspects of the somatic V(D)J assembly process (2). This assembly mechanism provides a vast diversity of potential specificities in the antibody repertoire expressed by newly generated B cells (the primary antibody repertoire). The reservoir of hundreds of germline V_H and V_L gene segments provides a primary source of diversity in CDR1 and CDR2. The assortment of different V_H, D, and J_H sequences or V_L and J_L sequences provides an even greater source of germline encoded diversity in CDR3. However, as described above, the process of V(D)J rearrangement has evolved to even further maximize the diversity in the CDR3 region. Thus, mechanisms that promote heterogeneity in the junctions of coding segments (junctional diversity) lead to great variations in the amino acid sequence of CDR3, even among variable region genes generated from the same V, (D), and J segments (277–280). The pairing of different combinations of H and L chains further diversifies the primary antibody repertoire which, based on the mechanisms outlined above, has been estimated to be as large as 10^{11} different antibodies or greater (281). The V_H to $V_H(DJ)_H$ joining mechanism could theoretically further contribute to H chain diversity in CDR3, although the frequency at which this reaction occurs in normal physiology is still unknown (Fig. 4). Finally, it is important to note that the degree to which many of these mechanisms contribute to diversification of the HC variable region gene repertoire appears to differ significantly with respect to fetal and adult or primary versus peripheral repertoires (discussed below).

7.2 Other diversification mechanisms

A well-documented source of diversification of variable region genes in mammals involves the somatic mutation of assembled variable region genes that results in changes in individual nucleotides (2). The somatic mutation process can occur over the entire assembled variable region gene. Thus, this process can generate diversity in all CDRs. Somatic mutation is most active late in an immune response, during antigen-driven clonal expansion, and is associated with generation of variant antibody molecules that have an altered affinity for antigen (282). The mechanism of the somatic mutation process still remains elusive, although new systems using Ig transgenes in mice may yield some insights (283–289).

An additional somatic mechanism, gene conversion, has been implicated as the primary generator of Ig H and L chain diversity in chickens (290–293). Gene conversion refers to a process by which a portion of sequences within one gene is replaced by sequences from another gene, while both structures remain intact along the chromosome (294–295). Although it has been considered (296, 297), somatic gene conversion has yet to be clearly implicated as a general mechanism repertoire diversification in murine B lymphocytes (298).

An in-depth review of somatic hypermutation and gene conversion is beyond the scope of this chapter. However, several comprehensive reviews of these subjects have recently appeared (299–302).

8. Regulation of Ig gene expression
8.1 HC gene expression

Control of H chain gene transcription involves both the V_H promoter and a transcriptional enhancer element which is located within the J_H-C_μ intron (Fig. 5) (303–306). The Ig H chain intronic enhancer (E_μ), like other transcriptional enhancers, functions in *cis*, and can do so in either orientation and at a distance of several kb (305–308). This enhancer element is active specifically in lymphoid cells, and apparently functions synergistically with the V_H promoter, which is also tissue-specific (303–307, 309, 310).

The E_μ element and the transcription factors that bind to it have been studied in great detail (reviewed in refs 311, 312). Of note, a number of studies have suggested that Ig HC genes can be transcribed following deletion of the intronic E_μ element (313–316). Such findings have been interpreted in the context of other potential regulatory elements in the *IgH* locus, suggesting a more important role for the E_μ element in promoting rearrangement than transcription, or a role for E_μ in establishing, but not maintaining, transcriptional competence of the *IgH* locus (317). Although activity of the promoters of rearranged V_H segments seems to depend, at least in part, upon the presence of the nearby E_μ element, at least some germline V_H segments are transcribed during the early stages of B cell differentiation where they are presumably under the control of other, yet to be defined, regulatory elements (318). Finally, an enhancer-like element has been found in the very 3' region of the Ig *HC* locus, approximately 15 kb downstream of the last C_H gene exons (319). This enhancer apparently is not required for variable region gene assembly or for expression of Ig μ heavy chains (320).

8.2 LC gene expression

L chain gene expression appears to be regulated by mechanisms that are analogous to those that control H chain gene expression. V_κ and V_λ promoters are structurally similar to V_H promoters, and are most active in lymphoid cells (321). Activity of the promoter of a rearranged V_κ segment depends, at least in part, on a tissue-specific enhancer located within the J_κ-C_κ intron (Fig. 3B), and which seems to act synergistically with the V_κ promoter (310, 322, 323). Replacement mutations of this enhancer element in mice appear to inhibit V_κ to J_κ rearrangement (see below) (126). There is an additional, quite strong, enhancer element located approximately 5 kb downstream of the C_κ gene; the role of this 3' κ enhancer' is not totally clear (324). It may have a role in κ variable region gene assembly, κ gene expression, or κ gene deletion (see below). Of particular interest, recent studies have implicated the activity of both the intronic and the 3' κ enhancer as being important for the somatic hypermutation of associated κ transgenes (285, 289).

In addition, two B-cell specific enhancers have been identified 3' of each λ constant region cluster (Fig. 3C) (325). The $E_{\lambda 2-4}$ enhancer is located 15.5 kb 3' of $C_\lambda 2$-$C_\lambda 4$ and the $E_{\lambda 3-1}$ enhancer is located 35 kb 3' of $C_\lambda 3$-$C_\lambda 1$. To date, no enhancer element has been found in the introns of the λ genes. DNA:protein interactions studies have identified several factors that interact with λ enhancer elements (326, 327). The identification of λ specific enhancers supports the model of independent regulation of λ and κ gene expression (325).

8.3 Expression of incompletely assembled H chain genes

8.3.1 Transcription of the J_H locus

The murine J_H–C_μ region is transcribed prior to assembly of a $V_H(D)J_H$ rearrangement. Germline transcripts that include the J_H cluster and C_μ region are also present in certain T cell lines (270, 328). Transcription of the J_H locus occurs before it is rearranged. These germline transcripts appear to be initiated from just 5' to the *DQ52* segments (329, 330). Such transcripts are found in *Rag*-deficient pre-B cells that cannot initiate the V(D)J recombination process (315). Another species of germline μ-specific transcript, designated 'I_μ' initiates in the J_H–C_μ intron just downstream of the enhancer (331, 332). I_μ transcripts appear to be present early during B lymphocyte ontogeny. They are present in many pre-B and B cell lines and also in certain T cell lines (270, 328, 331, 333). A processed I_μ transcript consists of a 5' exon of approximately 700 bp spliced to the μ coding exons and either the μ_m or μ_s sequences (331). Because the 5' exon contains multiple termination codons in all three reading frames, I_μ transcripts could encode a polypeptide containing C_μ sequences only if translation were initiated at an alternative codon downstream. However, such polypeptides have not been detected *in vivo* or *in vitro* (331). Thus, the I_μ transcripts appear to be be 'sterile' (i.e. not capable of encoding proteins). The function of sterile μ-specific transcription has not been determined. However, transcripts analogous to the I_μ transcripts have been shown to derive from all downstream C_H genes where they appear to have a role in the regulation of Ig heavy chain class switching. By analogy, the I_μ transcript may also have a role in this process (see below).

8.3.2 DJ$_H$ transcripts

DJ$_H$ rearrangements are also transcribed (334). These 'D$_\mu$' transcripts initiate from 5' D promoters and end at the μ_m or μ_s sites (270, 334). D$_\mu$ transcripts contain an ATG encoded by sequences upstream of the germline D segment (334). If this ATG is in phase with the μ C region coding sequences, a D$_\mu$ mRNA can be translated into a μ polypeptide that has a complete C region but lacks a standard variable region. D$_\mu$ polypeptides are generally expressed in the cytoplasm of differentiating B lymphocytes at much lower levels than are full-length μ H chains. Their amino termini are homologous to consensus leader polypeptides, suggesting that they might be present in the endoplasmic reticulum (334). Expression of D$_\mu$ proteins appears to abort B cell development and be responsible, in part, for the bias in D segment reading frame (335, 336). However, the function of this process remains speculative and it apparently does not occur in differentiating human pre-B cells (120). Based on cell-line studies, it was proposed that the D$_\mu$ protein may interact with surrogate LC proteins and a surrogate HC variable region protein to form an even earlier form of a pre-Ig receptor (337). However, the existence of such a receptor on (or in) normal cells has not been shown and its speculated roles, therefore, remain controversial.

8.3.3 Germline V$_H$ transcription

Germline V_H segments are transcribed in immature B lymphocytes during the stage of V_H-to-DJ$_H$ rearrangement (318). These transcripts are generally initiated at the normal V_H promoter and terminated downstream of the V_H coding sequences at various polyadenylation sites. In mice, transcripts from a variety of different germline $V_H J558$ genes are detectable in early B-lineage cells, but transcripts from germline V_H segments of other families have been difficult to detect. Expression of analogous germline V_H transcripts of the three $V_H 5$ family members has been found in human ALL and CLL B lineage tumour cell lines. In early B lineage cells, appropriately-spliced germline V_H transcripts are present in the cytoplasm, but it still remains to be determined as to whether they are translated into proteins (338). Translation of a germline V_H mRNA would yield a leader polypeptide followed by a single Ig domain, a structure similar to that of VpreB protein.

Germline V_H transcripts are present at high levels in fetal liver during the time at which large numbers of immature B lymphocytes are present, but are barely detectable in newborn liver and are absent in adult spleen, both of which contain B lymphocytes that express full-length μ mRNA (318). Germline V_H transcripts generally have not been detected in mature B lineage cell lines. Overall, it appears that germline V_H transcription is restricted to the early stages of B lymphocyte differentiation. It has been speculated that such transcription might be associated with control of V_H to DJ_H rearrangement (318) although direct evidence for this point is still lacking. Likewise, putative V_H proteins have been speculated, along with D_μ, λ5, and VpreB proteins, to be part of an early pre-Ig receptor (252, 339), but there is currently no direct evidence to support this notion.

8.3.4 Germline Ig κ LC transcription

The germline J_κ–C_κ transcript that is induced along with κ enhancer activation has been detected as an 8 kb polyadenylated nuclear RNA species that is initiated 5' of the J_κ segments and is terminated 3' of the C_κ coding region (340–342). During B lymphocyte differentiation, germline J_κ–C_κ transcription appears to be specifically induced prior to V_κ-to-J_κ rearrangement, for example, by LPS treating pre-B cell lines (340, 341, 343). More recently, it has been shown that such germline V_κ transcripts are induced by the expression of transgenic Ig HC proteins in pre-B cells of *RAG*-deficient mice, supporting the notion that Ig *HC* gene expression somehow activates the germline Ig *LC* locus (139, 140). Germline V_κ segments generally appear to be transcriptionally silent in mature κ-producing B lymphocytes. There have been some isolated examples of germline V_κ expression in cell lines, but it remains unclear to what extent they are transcribed in normal differentiating precursor lymphocytes (344, 345). It has also not been determined whether germline V_λ and J_λ segments are transcribed in cells that are undergoing V_λ-to-J_λ joining, but low-level transcription of a germline $V_\lambda 1$ segment has been observed in κ-producing murine myelomas (346), and in A-MuLV cells that undergo V_λ to J_λ rearrangement (347).

9. Regulation of Ig gene assembly

9.1 Ordered assembly and expression of Ig genes

Findings from several different experimental systems indicated that B cell differentiation generally proceeds in an ordered programme in which Ig H chains are rearranged and expressed before those of Ig L chains (Fig. 10). This ordered rearrangement programme may have evolved to facilitate regulatory mechanisms such as allelic and isotype exclusion and also to permit differential HC and LC V(D)J junctional diversification mechanisms to operate.

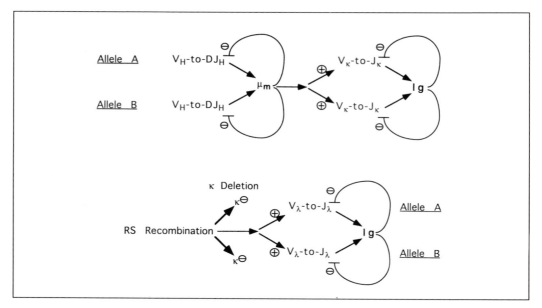

Fig. 10 Schematic diagram of the regulated models for control of *Ig* gene assembly. Formation of a productive $V_H(D)J_H$ rearrangement on either H chain allele leads to production of μ_m polypeptide which inhibits further H chain gene assembly and initiates V_κ-to-J_κ joining. V_λ-to-J_λ joining is observed in cells that have undergone non-productive V_κ-to-J_κ rearrangement and/or RS recombination, which deletes C_κ sequences. Formation of a functional L chain gene and production of an Ig molecule inhibits further L chain gene assembly (adapted from ref. 348).

Both A-MuLV transformants and murine fetal liver hybridomas often express intracellular μ H chain in the absence of a L chain, a finding which predicted the existence of a μ-only precursor-B (pre-B) cell (349, 350). This cell was later identified in normal fetal liver, in which it was demonstrated that μ H chain expression precedes L chain expression during ontogeny (351, 352). Subsequent analyses of Ig gene rearrangements in pre-B cell lines revealed that, during early differentiation, H chain genes generally are assembled and L chain genes are not, indicating that H chain genes are usually assembled before L chain genes (2, 353–355). Similarly, cell-sorting studies have determined that during differentiation of normal B cells, H chain genes are assembled and expressed in cells that have not begun to assemble L chain genes, and that L chain gene assembly follows and appears to be the rate-limiting step in expression of surface Ig (356–358). Therefore, it is clear that the majority of developing B lymphocytes first assemble *HC* genes and subsequently assemble *LC* genes. Nevertheless, recent gene-targeted mutational experiments which have led to the generation of mice that lack the ability to assemble a functional Ig *HC* gene have shown that, at least in a minor population of developing B cells, *LC* gene rearrangement and expression can occur in the absence of Ig HC gene expression (359–361).

9.2   H chain gene assembly

Most A-MuLV transformants derived from normal sources have DJ_H rearrangements at both H chain alleles at the time they are established (144, 180, 275, 276, 353, 362, 363). Many of these cell lines undergo V_H-to-DJ_H joining, and can also replace a DJ_H rearrangement by undergoing a 'secondary' D-to-J_H joining event, in which an upstream D segment is joined to a downstream J_H segment (Fig. 4). Initial D-to-J_H joining events preferentially utilize the most 3' D segment ($DQ52$) (180, 364–366). In contrast, the most 5' (V_H-proximal) D segments are used preferentially in secondary D-to-J_H rearrangements in A-MuLV transformants (180, 367). Thus, although usage of D segments may be biased toward more 3' segments initially, during formation of secondary D-to-J_H rearrangements, the recombinase appears to be directed toward more upstream D segments, perhaps by molecular mechanisms that evolved to direct rearrangement activity to the V_H cluster. In this regard, in A-MuLV transformants, V_H to DJ_H rearrangements appear to occur more frequently than secondary D to J_H rearrangement (180). This might be advantageous; otherwise secondary D to J_H rearrangements might lead to a predominance of $V_H(D)J_H$ rearrangements that utilize the most 3' J_H segment.

Direct joining of a V_H segment to a germline D segment is permitted by the 12/23 rule, but apparently does not occur at a significant level relative to D-to-J_H and V_H-to-DJ_H rearrangement events (275). In mutant pro-B cells that lack J_H gene segments, V_H to D rearrangements do not occur at detectable frequency (360). V_H-to-D joins have been observed occasionally in human B lymphocytes. However, most cases appear to involve joining of a V_H segment to a D segment located immediately 5' of a DJ_H rearrangement (365, 368). As described above, joining of D segments to each other in the mouse has been postulated, but not demonstrated to occur at significant frequency (203, 273). Thus, H chain gene assembly generally involves two stages: DJ_H rearrangements are assembled on both chromosomes prior to V_H-to-DJ_H joining (Fig. 4).

9.3   L chain gene assembly

A-MuLV transformants that undergo or have undergone L chain gene assembly usually produce a complete μ H chain, or at least appear to have done so at one time and have since lost μ production. These observations led to the suggestion that the μ polypeptide has a dual regulatory role: it can mediate H chain allelic exclusion and can facilitate the initiation of Ig L chain gene assembly (Fig. 10) (353). The hypothesis that the μ protein facilitates the initiation of L chain gene rearrangement initially derived support from analyses of A-MuLV transformants which progressed from having two DJ_H rearrangements to producing surface Ig (276). In subclones of these lines, V_κ to J_κ joining initiated following completion of a productive $V_H(D)J_H$ rearrangement and generally occurred only in subclones that produce full-length μ H chain.

Subsequently, transgenic expression of a construct encoding a μ_m, but not μ_s

polypeptide was found to lead to the generation of cells that assemble endogenous L chain genes (369, 370). Likewise, in A-MuLV transformants that contain two non-productive $V_H(D)J_H$ rearrangements, V_κ to J_κ joining could be initiated by the expression of a productive μ_m protein. The latter phenomenon can occur by a V_H replacement event or, in the absence of endogenous H chain production, by introduction of an expression vector that encodes a μ_m polypeptide, but not by an analogous construct that only encodes a μ_s polypeptide (369, 371–373). These findings suggested that the membrane-bound form of μ polypeptide is necessary to mediate its putative regulatory role in this process.

A number of studies have identified murine or human pre-B cell lines that expressed LC proteins in the absence of Ig HC proteins (149, 343, 374, 375), leading to the suggestion that at least some B cells may be generated through a developmental pathway in which LC gene rearrangement precedes that of HC gene rearrangement (374). Although the full significance of this proposal remains unclear, additional support for the notion came from findings that 2–5 per cent of pro-B cells in mice harbouring homozygous deletions of the μ_m exon or J_H segments (and therefore unable to express an Ig HC) had κ LC gene rearrangements (360, 361, 376). These findings indicate that LC gene rearrangement does not absolutely require the rearrangement and expression of a μ H chain. However, it remains possible that rearrangement of LC genes observed in such mutant cells may have been promoted by the developmental block imposed by the mutation. In any case, it is clear that the majority of normal developing B cells rearrange LC genes subsequent to the rearrangement and expression of HC genes.

9.4 κ versus λ LC gene assembly (LC isotype exclusion)

B cells either express κ LC genes or λ LC genes but not both, a phenomenon referred to as LC isotype exclusion (Fig. 10). In this context, LC gene assembly also appears to be ordered, with the assembly of κ genes preceding assembly of λ genes: in mature murine and human B-lineage cell lines that produce κ-containing Ig molecules, V_λ and J_λ segments usually are in germline configuration, but in lines that produce λ chains the J_κ–C_κ locus is generally either deleted or rearranged (269, 377, 378). Similarly, in A-MuLV transformants that have assembled κ genes, λ genes are in germline configuration, and the assembly of λ genes, which is rare in A-MuLV transformants, is often associated either with V_κ-to-J_κ rearrangement or with deletion at the κ locus (150, 347, 379).

Two general models have been proposed to explain the apparent assembly of κ genes prior to the assembly of λ genes:

1. a regulated mechanism that sequentially acts at κ and then at λ genes;
2. a stochastic mechanism in which κ and λ rearrangements are initiated simultaneously, but there is a much greater probability of κ gene rearrangement than λ gene rearrangement (19).

In the mouse, the number of κ^+ B cells exceeds that of λ^+ B cells by about 20-fold (i.e. 95 per cent κ^+ B cells and 5 per cent λ^+ B cells (ref. 380)). Mice heterozygous for a $J_\kappa C_\kappa$, C_κ, or intronic κ enhancer deletion have a pre-B pool size similar to that of wild-type but have an apparent 2-fold increase in production of λ^+ B cells (126, 381, 382). Mice homozygous for such mutations have an only slightly expanded pre-B pool size but a 10-fold increase in λ^+ B cell production (126, 381, 382). These findings have been interpreted as supporting an ordered model since it appears that the frequency of λ gene rearrangements has increased in the pool of pre-B cells harbouring homozygous κ deletions (381). However, other interpretations are possible (382) and additional experiments will be necessary to resolve the various models for regulation of κ versus λ gene rearrangement.

In λ-producing B cells, the C_κ locus is often deleted by the rearrangement of a κ RS (recombining sequence) that lies 25 kb 3' of C_κ, to either the J_κ–C_κ intronic heptamer or the V_κ RSS (63–65, 68). The common DNA region deleted in such rearrangements extends from the intronic heptamer to the 3'κ RSS, and includes intronic E_κ, C_κ, and 3' E_κ (347). The κ RS recombination has been proposed to activate or delete sequences required for λ rearrangement (Fig. 10). (68, 347). However, λ-producing B cells in mice harbouring J_κ–C_κ or intronic κ enhancer deletions do not show detectable κ RS rearrangements (126, 245, 381) and the level of RS rearrangements is reduced in λ-producing B cells harboring C_κ deletions (382). These findings demonstrate that 3' κ RS rearrangement, *per se*, is not required for λ gene assembly and expression, although such germline deletions might in some way function similarly to normal 3' κ RS rearrangements in facilitating λ gene assembly.

9.5 Molecular characterization of early B cell stages based on surface marker expression

Ordered H and L chain gene rearrangement and expression events have been correlated with phenotypic changes in the expression of certain surface markers on developing B cells (357–359, 361). These markers have been extremely useful in analysing the stages of normal differentiation at which particular rearrangement events occur. Based on expression of the B220 and CD43 surface markers, B lineage cells in normal bone marrow can be resolved into large, $B220^+CD43^+$ pro-B cells that mostly have only DJ_H rearrangement at H chain loci and small $B220^+CD43^-$ pre-B cells that have complete $V_H(D)J_H$ rearrangement and which are in the process of L chain gene rearrangement (356, 358). Consistent with a functional role for HC gene rearrangement and expression in promoting this transition, B cell differentiation in mice harbouring a homozygous *Rag-2* or *Rag-1* gene mutation (135, 136) or homozygous deletion of J_H gene segments (360, 361, 383) is blocked at the large, $B220^+CD43^+$ pro-B cell stage. Correspondingly, introduction of a functionally assembled μ H chain gene into these homozygous mutant backgrounds promotes the differentiation of pro-B cells into small, $B220^+CD43^-$ pre-B cells which express significant levels of germline κ gene transcripts (139, 140, 384). An issue that still remains is whether the μ chain expression actually induces κ gene rearrangement or promotes differentiation (survival) of pro-B cells to a stage in which rearrangements are tolerated (385, 386).

9.6 Potential role of SLC receptors in regulating precursor B cell development

The findings that the μ_m chains expressed at early stages of B cell differentiation can associate with the $\lambda5$ and VpreB surrogate light chains and probably accessory signal transducing molecules MB-1 and B29 have led to speculation that this complex may function to mediate regulatory signals during B cell differentiation (6, 249, 252, 254, 257–259, 339, 387–389). In this context, MB-1 and B29 were first identified in association with IgM and IgD on mature B cells (6, 252, 339, 388–390). MB-1 and B29 contain signal transduction motifs similar to those found in components of CD3 co-receptor complex associated with TCR on mature T cells, strongly suggesting that MB-1 and B-29 molecules are involved in signalling events related to antigen-binding to the Ig molecules (391). Support for the notion that the μ_m/ SLC complex may function in signalling critical events in early B cell development was provided by gene targeted mutational studies. B cell development was impeded at the large, $B220^+CD43^+$ pro-B stage in mice harbouring homozygous deletions of the μ_m exon or the $\lambda5$ gene (359, 376). Assembly and expression of Ig LC has been proposed to result in the displacement of $\lambda5$ and VpreB from the putative μ/SLC-complex and the expression of the complete Ig.

Recently, it was found that SLC is expressed on pro-B cells of *Rag*-deficient mice and its surface expression is down-regulated on pre-B cells of *Rag*-deficient mice complemented with functionally assembled μ heavy chain genes (139, 140, 392). SLC expressed on pro-B cells appears to be associated with a 130 kDa (p130) protein (393, 394), suggesting that the SLC complex may have other potential functions during this stages of B cell differentiation, such as signalling the initiation of IgH rearrangement. However, the precise function of this complex remains an enigma.

9.7 Allelic exclusion

9.7.1 Models of allelic exclusion

The recombination mechanism that assembles Ig genes operates in an inherently random fashion (subject to the possible homology-mediated restraints discussed above), and is therefore capable of rearranging either allele of a H or L chain gene productively. Theoretically, a B lymphocyte could assemble productive $V_H(D)J_H$ rearrangements at both H chain alleles, and by doing so might produce H chains of two different variable region specificities. Similarly, a B cell could theoretically assemble multiple productive L chain genes of κ or λ isotype. Studies that utilized allotypic markers to determine the allelic origin of H and L chains have demonstrated that greater than 99 per cent of B lymphocytes express the products of a single HC gene and a single LC gene as surface Ig, indicating that allelic (and isotypic) exclusion is a general characteristic of B lymphocytes (10, 11).

Various models have been proposed to explain how allelic exclusion is imposed (reviewed in ref. 7). The possibility that each B lymphocyte might assemble only a single H and a single L chain gene has been eliminated by the finding that rearrangement occurs at both H chain alleles and generally at more than one L chain allele. Imposition of allelic exclusion by the control of expression has also been ruled out by the observation that these multiple rearranged genes are transcribed, regardless of whether they are productive or not. According to 'stochastic' models, the Ig gene assembly process is not regulated, but productive H and L chain genes are formed very infrequently, and so the probability that a single B lymphocyte will assemble more than one productive H or L chain rearrangement is extremely low. Such stochastic models appear unlikely given that the rate of formation of a productive H or L chain rearrangement is apparently close to the theoretical one in three (see above), and by the finding that in the majority of B lymphocytes, 'silent' H or L chain alleles have been 'frozen' in DJ_H or germline configuration (274–276, 379). These latter observations strongly suggested that allelic exclusion is imposed by controlling Ig gene assembly. According to 'regulated' models (discussed in more detail below) Ig genes are assembled in an ordered process which is controlled to prevent assembly of multiple H and L chain alleles that could be expressed together in an Ig molecule (100, 269, 395, 396). The strongest support for these models has been derived from gene-targeted mutational studies (ref. 376 and see below).

9.7.2 HC allelic exclusion

Each differentiating B cell has two initial chances of assembling a productive $V_H(D)J_H$ rearrangement. As DJ_H rearrangements usually occur on both chromosomes in differentiating precursor B cells, it was proposed that the V_H to DJ_H rearrangement is the regulated step in the context of Ig HC allelic exclusion (ref. 275 and see below). Regulated models of H chain allelic exclusion propose that if the first V_H-to-DJ_H joining event is productive, then V_H-to-DJ_H joining on the other H chain allele is prevented by a feedback mechanism (Fig. 10) (270, 275, 353, 395). Because both productive and non-productive rearrangements are transcribed, the only apparent signal that could identify assembly of a productive rearrangement would be the μ polypeptide itself. It has, therefore, been proposed that the presence of μ H chain directs the cell to terminate H chain gene assembly. If the initial $V_H(D)J_H$ rearrangement is non-productive, then the cell has a chance to assemble a productive $V_H(D)J_H$ rearrangement on the second H chain allele.

The regulatory effect of μ H chain expression was tested by transgenic mouse experiments (reviewed in ref. 397). Expression of μ polypeptides from a transgenic H chain gene, which included both $μ_m$ and $μ_s$ sequences, appeared to inhibit the rearrangement of endogenous HC genes (398). Likewise, expression in transgenic mice of the membrane-bound form of a human or mouse μ H chain polypeptide appeared to inhibit expression of endogenous μ protein in normal B lymphocytes, most of which expressed the transgene as surface IgM (and IgD) (369, 398–401). In contrast, transgenic expression of a μ_s polypeptide did not appear to affect the assembly and expression of endogenous H chain genes (369, 400). Therefore, these findings were consistent with the proposal that expression of μ_m chains regulates the allelic exclusion process by feeding back to prohibit further V_H to DJ_H joining, although various potential effects mediated at the level of cellular selection of the transgenic population somewhat complicate the overall interpretation of such experiments.

More recently, the regulated model of Ig HC allelic exclusion was directly tested by the generation of heterozygous mutant mice that could express μ_s but not μ_m chains from one of their *IgH* loci (376). Significantly, rearrangement and expression of the mutant allele did not exclude the occurrence and expression of $V_H(D)J_H$ rearrangements on the wild-type allele, strongly supporting the regulatory role of the μ_m chain in the allelic exclusion process. A continuing question in the context of this regulated model (that also applies to LC allelic and isotype exclusion) is how a differentiating B cell can rearrange only one allele at a time in order to test for productive HC expression. Several possible mechanisms to permit such testing have been proposed (132).

9.7.3 LC allelic exclusion

Imposition of L chain allelic exclusion has both similarities and differences with that of Ig HC allelic exclusion. In the latter context, some B-lineage cell lines produce more than one species of full-length L chain polypeptide. However, in these lines only a single L chain species is associated with the H chain in an Ig molecule (268, 269, 402, 403). Thus, in addition to formation of a productive rearrangement, L chain gene assembly apparently must meet the additional requirement of creating a 'functional' gene which encodes an L chain that can associate with the pre-existing cellular HC to form a complete Ig molecule. According to the regulated model, L chain allelic exclusion is imposed by feedback from the Ig molecule itself, which signals termination of V_L to J_L joining (Fig. 10) (269). Thus, after L chain gene assembly is initiated in a differentiating B cell, rearrangement can continue until a functional L chain gene is generated and the expressed LC protein has associated with the μ protein.

This regulated model for L chain allelic exclusion has been strongly supported by transgenic mouse studies (397, 404, 405). For example, high-level expression of a transgenic κ polypeptide apparently prevents endogenous κ gene assembly; while endogenous κ gene assembly was observed only in cells in which the transgenic κ polypeptide was either not expressed or did not associate with an endogenous H chain (370, 406). The expression of complete Ig molecules apparently signals the cessation of V(D)J recombination. Consistent with this conclusion, cross-linking of surface IgM on V(D)J-recombinase-expressing B cell tumours (407) or cross-linking of TCR on $CD4^+CD8^+$ thymocytes leads to down-regulation of *Rag* expression (408, 409). However, as discussed in the next section, transgenic

IgM$^+$ immature B cells upon contact with antigen may be able reactivate *Rag* expression and, thus, V(D)J recombinase and rearrange their endogenous LC genes, leading to a change of Ig specificity (410–412).

9.8 Ig receptor editing

Non-productive rearrangements, or rearrangements that encode antigen receptors with undesirable specificities may be 'salvaged' (or 'edited') by productive V_H to $V_H(D)J_H$ replacement events (66, 67) or by V_κ to downstream J_κ replacement events (413), as has been observed in cell lines. Recently, V_H to $V_H(D)J_H$ replacement events have been observed at a low frequency *in vivo* in a mouse in which a *T15* coding $V_H(D)J_H$ gene segment replaces the endogenous J_H elements. Additional studies have suggested the possibility that receptor editing may be a regulated process which is associated with re-expression of *Rag* in sIgM-positive B lymphocytes (410–412). These studies, carried out in mice that carry self-reactive transgenic Ig genes, suggest that autoreactive B cells may escape becoming tolerized by assembling endogenous light chain genes whose products outcompete the transgenic light chains for H chain association, thereby altering the specificity of the B cell antigen receptor. Again, however, the overall impact of such mechanisms on normal physiology is not yet totally clear.

9.9 Potential restrictions in the primary murine V_H repertoire associated with the V(D)J rearrangement process

Analyses of B-lineage cell lines and of normal lymphoid tissues have shown that, in BALB/c and other murine strains, the most J_H-proximal V_H segments are utilized preferentially in V_H to DJ_H rearrangements (8, 180, 355, 415–420), although there may not be a strict preference among the most 3' V_H family segments (416, 421). This preferential rearrangement process leads to significant biases with respect to representation of different V_H genes in the primary repertoire. However, even given preferential utilization of 3' V_H gene segments, mechanisms exist to permit utilization of segments from the more 5' regions of the several megabase V_H locus.

At a gross level, the family distribution of V_H segment usage is apparently random (for example representation of family members corresponds roughly to family complexity) among B lymphocytes in peripheral lymphoid organs such as adult spleen, as determined by analyses of hybridomas and normal spleen tissue (296, 415, 422–424). The difference between the biased primary repertoire and the random ('normalized') adult repertoire could arise by various mechanisms. The pattern of V_H segment utilization by the recombination mechanism might shift during ontogeny, so that J_H-distal V_H segments would be utilized by differentiating B lymphocytes more frequently in adult bone marrow than in fetal liver (355). However, current evidence indicates that differentiating B lymphocytes in bone marrow also utilize J_H-proximal V_H segments preferentially (180, 416, 419, 422, 425), indicating that the recombination mechanism may also be 'biased' during adult life. It has been suggested that V_H replacement might contribute to randomization of the repertoire (66, 67) but V_H replacement appears to occur primarily among nearest neighbours, suggesting that hundreds of successive V_H replacements would be required to generate a $V_H(D)J_H$ rearrangement that used a relatively 5' V_H segment (66, 67). Under these conditions, it is difficult to envisage that V_H replacement events could significantly affect the distribution of V_H segment utilization. Although programmed shifts in the rearrangement frequency of V_H gene segments may contribute to repertoire normalization, it seems clear that normalization occurs, at least in part, via cellular selection mechanisms, possibly mediated by antigenic selection and/or interactions with immunoregulatory cells (415, 419, 423, 425, 426).

In addition to the preferential utilization of certain V_H segments, the neonatal heavy chain variable region gene repertoire is also restricted with respect to junctional diversification due to limited expression of TdT in fetal as opposed to adult B cell precursors discussed earlier (112, 113, 144, 427). Increased TdT activity diversifies the adult repertoire both through increased N-region addition and decreased frequency of homology-mediated joins. Thus, the murine fetal HC variable region repertoire could be substantially limited compared to the peripheral repertoire in the adult due to restricted V_H segment utilization, decreased junctional diversification, and increased frequency of certain V(D)J junctions as a result of homology-mediated joins.

The biological purpose of this two-tiered system for repertoire generation is not clearly understood. Potential roles of preferential V_H gene utilization still remain speculative (196, 197, 428), although several recent studies have begun to address this issue (429, 430). Likewise, the effect of N-regions on the TCR and Ig repertoires is not completely understood. N-region addition in the adult repertoire may have evolved to meet the need of the adult organism for a greater repertoire of antigen receptors than could be provided by the germline encoded sequences (112). However, TdT-deficient mice apparently are not grossly immunocompromized and appear to respond normally to several different antigens (93). The less diverse antigen receptors generated in the absence of TdT may have a specific role in the fetus not required or even detrimental in the adult (431). If such receptors recognized self-antigens, the prevention of their expression in adults may be of the utmost importance. Longer term observation of TdT-deficient mice and the generation of TdT over-expressing mice may yield insight into such potential roles.

Homology-mediated joining has been suggested to play an important role in establishing a particular and limited antigen receptor repertoire during early development (114). Thus, 'homology-mediated' rearrangements have been suggested to have evolved to form specific receptors frequently used by the organism in order to circumvent the wasteful process of selection (114, 117, 118). The use of homology-mediated joining has also been found to be a prominent feature of certain D_H to J_H junctions in normal fetal lymphocytes (114) and Ly1$^+$ (B1) B cells (112). However, the potential physiological significance of these junctions, if any, is not known. 'Homology-mediated' joins have been studied more extensively in the context of T cell development than B cell development (reviewed in refs 119 and 121), although the evolutionary significance of this phenomenon in T cells also has not yet been determined.

9.9.1 Utilization of human V_H segment in V(D)J rearrangements

V_H segments used in rearrangements in human lymphoblastoid cell lines (217) appeared to be predominantly from within the most J_H proximal 800–1000 kb of the V_H complex. The significance of J_H proximal utilization of only $\frac{1}{2}$–$\frac{1}{3}$ of the V_H complex is unclear and it is not yet known whether this pattern of rearrangement is also found in normal repertoires. Some evidence suggested preferential utilization of V_H6 in fetal as opposed to adult repertoires (365, 432). However, in studies of V_H gene usage in fetal development, different sets of V_H gene segments, localized to different regions within about 1000 kb, appear to be utilized as the fetus matures (177, 214, 216, 366, 432–434). However, none of these genes appeared to map to a distance greater than about 1.0 Mb 5' of J_H. The collective evidence suggests that this distance may, in fact, be the approximate size of the entire human V_H complex. Thus, it is not clear that the phenomenon of preferential rearrangement of 3' V_H gene segments will be as prominent a feature of the V(D)J recombination mechanism in human pre-B cells as in their murine counterparts.

9.9.2 Utilization of V_κ LC gene segments

The utilization of V_κ genes by the V(D)J recombination process has not been studied in as much detail as V_H utilization, in part because the organization of the locus has been less clearly defined (435, 436). However, current results suggest that although some V_κ segments may be preferentially rearranged, there is no obvious correlation with proximity to the J_κ region (437). In addition, N-region addition is not normally found in $V_\kappa J_\kappa$ or $V_\lambda J_\lambda$ joins. This probably results from the differential expression of TdT at the $V_H(D)J_H$ assembly as opposed to the $V_L J_L$ assembly stage of precursor B cell differentiation (357).

10. Molecular mechanisms that regulate antigen receptor gene assembly

10.1 Accessibility

As described above, the assembly of antigen receptor variable region genes is controlled at several levels (2, 3, 438). These include the regulation of tissue specificity (Ig variable region genes are assembled in pre-B, but not pre-T cells), the stage of lymphocyte differentiation (Ig HC chain genes usually are assembled before Ig LC chain genes), and assembly of only one productive allele per cell (allelic exclusion, see above). Although distinct enzymatic activities (recombinases) that would each target a specific locus were presumed to exist in early models of the regulation of V(D)J recombination (for example ref. 439), studies with artificial recombination substrates demonstrated that constructs composed of TCR sequences could rearrange in pre-B cell lines at comparable rates (41). These, and subsequent observations (for instance identification of *Rag*) strongly implicated the existence of a common V(D)J recombinase activity responsible for the assembly of all antigen receptor gene families (41, 45, 127, 318, 440).

Rag expression and V(D)J recombinase activity are present throughout pre-B and pre-T cell development. Thus, the tissue, stage, and allelic specificities observed for rearrangements must be controlled by modulating the availability of various *Ig* and *TCR* loci as substrates for the common recombinase; a concept commonly referred to as the accessibility model (reviewed in refs 7, 396). The cell-type specific accessibility of *Ig* and *TCR* loci to the V(D)J recombinase activity is particularly evident in non-lymphoid cells manipulated to express *Rag-1* and *-2*. Fibroblasts transfected with *Rag* expression vectors possess recombinase activity and rearranged both chromosomally-integrated and extrachromosomal recombination substrates (45, 127, 441). However, rearrangement of endogenous *V* gene segments was not observed (129, 441). Thus, the artificial substrates are distinguished from the endogenous antigen receptor loci by being accessible to the recombination machinery.

10.2 Model systems to study control of V(D)J recombination

The study of mechanisms responsible for controlling locus accessibility to V(D)J recombination has been hampered by the lack of model systems that are readily amenable to manipulation. A significant amount of information has been derived from studies of aspects of the endogenous Ig loci, such as transcriptional and methylation patterns in normal or transformed cell lines in which a particular locus was undergoing rearrangement (318, 341, 342, 345, 346, 442–445). Other more detailed studies have relied on the use of cloned V(D)J recombination substrates that consist of unrearranged *V, D*, or *J* segments in the presence or absence of various sequences to be tested for a controlling influence on rearrangement. Such substrates have been tested either by introduction into cell lines that express V(D)J recombination activity (131, 446) or into normal developing lymphocytes in the form of transgenes (441, 447, 448). More recent studies have used gene targeted mutational approaches to test the role of particular elements in controlling recombination in the endogenous Ig or TCR loci (126, 315, 312, 449).

In normal pre-lymphocytes, the assembly of antigen receptor loci appears to be completed within a few cell divisions (6). In these cells, substrate accessibility, not recombinase activity, appears to be the rate-determining factor in the rearrangement process of a particular gene segment. Cell systems designed to study the regulation of V(D)J recombination should ideally mimic these features. The transgenic and gene targeted mutational studies clearly provide a more physiologically relevant approach. However, the use of transfected substrates in cell lines offers a more rapid method to test the role of specific elements in controlling accessibility, for example by a mutational analysis of an implicated element.

Analyses of recombination substrates in A-MuLV transformed pre-B cells has provided numerous insights into the general principles of V(D)J recombination control (38, 41, 42, 446). However, several characteristics of A-MuLV transformed lines limit their usefulness as recipients of test substrates that are designed to define elements that control accessibility (131, 450, 451) including low and variable levels of V(D)J recombination activity which thus usually requires drug selection or PCR methods for detection of rearrangement events (38, 41, 451, 452). Such problems may lead to factors other than substrate accessibility influencing the rate of rearrangement of introduced substrates. The recent development of cell lines that can be induced to express very high levels of V(D)J recombination activity from very low baseline levels may circumvent some of the difficulties attributed to the use of A-MuLV transformants (131, 132).

With respect to cell line approaches, the manner in which recombination substrates are presented to the cell must also be considered. Although extra-chromosomal substrates have been effective for studying the mechanistic features of V(D)J recombination, such substrates probably do not accurately reflect constraints imposed on chromosomally integrated genes and, therefore, may not reflect accessibility mechanisms imposed on endogenous antigen receptor loci (453). Likewise, similar problems may affect the interpretation of stably integrated recombination substrate studies. In cells that are constitutively active for V(D)J recombination, test substrates may rearrange before they integrate into the chromosome. Thus, the rearrangement state of these chromosomally integrated substrates may not reflect normal rearrangement control mechanisms (131).

10.3 Molecular mechanisms of accessibility

10.3.1 Gene segment transcription correlates with active rearrangement

The original observation that pre-B, but not mature B cells, express transcripts originating from unrearranged V_H segments was a prelude for the generation of a large body of data that linked unrearranged V gene segment expression with recombinational accessibility (318). Transcription of both endogenous and recombination substrate V-region gene segments has been correlated with their accessibility to V(D)J recombinase (42, 318, 330, 341–343, 346, 445, 448, 454–456). Although, the exact relationship between germline V_H transcription and recombination is still not clear, the transcription of germline or incompletely rearranged V region gene segments has been proposed to reflect (or perhaps mediate) accessibility of these segments to V(D)J recombinase. Notably, the IgH region is also transcribed in pre-B that have not undergone D–J_H rearrangements (135, 315, 329, 451). In accordance with the accessibility model, such cells readily undergo D_{Q52} to J_H rearrangements in the presence of endogenous or introduced recombinase activity (G. R. and F. A., unpublished data) (330, 451).

Germline V_H transcripts originating from the $J558$ gene family are generally expressed by differentiating B cells that are at the stage of V_H-to-DJ_H joining, but are not expressed by differentiating T cells (318). In the context of an accessibility model, the lack of germline V_H segment transcription in differentiating T-cells is consistent with the fact that they do not undergo V_H to DJ_H joining events. However, transcription of unrearranged V_H segments from families other than V_H558 has not yet been detected. Thus, the role of germline V_H transcription with respect to accessibility still remains speculative. In this context, it remains of interest to determine whether the chromatin structure and germline transcription of all V_H segments are equal during early B-cell development.

The ordered nature of Ig gene rearrangement, coupled with the occurrence of Ig HC allelic exclusion at the V_H to DJ_H rearrangement stage, suggests that productive H chain rearrangement, either directly or indirectly, results in the decreased accessibility of the V_H locus and enhanced accessibility of the κ LC locus to the V(D)J recombinase. Transcription of the unrearranged κ locus strongly correlates with the accessibility of that locus for V(D)J recombination. Germline transcripts originating from the $J_κC_κ$ locus have been detected in pre-B cells (330, 341, 342). Furthermore, treatment of a pre-B cell line with bacterial lipopolysaccharide (LPS) results in activation of germline $C_κ$ transcription and, correspondingly, $V_κ$ to $J_κ$ rearrangement in the cell population (343). More recently, the transcriptional onset of the germline κ locus has been coupled with production of μ protein (139, 140). The introduction of an expressed Ig HC transgene (μ) into *Rag-2* deficient mice ($Rag\text{-}2^{-/-}$) results in the detection of germline $C_κ$ expression in the pre-B cells of the $μ/Rag\text{-}2^{-/-}$ animals; pre-B cells of $Rag^{-/-}$ animals do not transcribe the κ locus.

In A-MuLV transformed cells that express a productive endogenous or transgenic H chain locus, the unrearranged V_H segments often continue to be transcribed (451). Therefore, assuming normal levels of V(D)J recombinase, these cells could continue to undergo V_H to DJ_H rearrangement if transcription of an unrearranged *V*-region gene segment solely directly reflected accessibility of the segment to the V(D)J recombinase. Persistence of V_H segment accessibility may allow for the mechanisms that lead to V_H gene replacement (66). However, although transcription of *V* gene segments may be required for V(D)J recombinational accessibility, it may not be sufficient as additional factors may restrict the V to DJ rearrangement step (457) (see below).

10.3.2 Chromatin structure and methylation status of *V*-region gene segments correlates with rearrangement

In general, transcriptionally active loci are hypomethylated when compared with silent loci (458, 459). The methylation status of several transgenic recombination substrates were found to significantly affect the efficiency of their recombination (441, 447). Rearrangement of these transgenic substrates were restricted primarily to hypomethylated transgene copies. The methylation status of transiently introduced recombination substrates is also reduced by methylation (460).

Methylation could potentially affect the accessibility of *V*-region segments in several ways. The nucleosome structure of hypomethylated CpG islands, which are presumably transcriptionally active, apparently is different from that of bulk chromatin (461). A protein activity that binds to methylated DNA has been identified (462). This protein activity can inhibit transcription from weak promoters, yet can be neutralized by an enhancer element positioned in *cis*. Notably, the Igκ intronic transcriptional enhancer was recently shown to target tissue- and stage-specific demethylation to *Igκ* miniloci in B-cell lines (463), suggesting a possible mechanism by which it may influence accessibility for rearrangement. Thus, the cumulative data suggest that methylated loci do not undergo efficient V(D)J rearrangement. However, targeted replacement mutation of the E_μ element indicated that hypomethylation also may not be sufficient to render the endogenous J_H locus fully accessible for V(D)J recombination (315). Therefore, hypomethylation, like transcription, may be necessary but not always sufficient to promote full V(D)J recombinational accessibility.

The chromatin structure in the vicinity of *V*-region gene segments also appears to reflect the accessibility of the segment to V(D)J recombinase. Both endogenous and introduced *V*-region gene segments that are undergoing rearrangement have been found to be more sensitive to nuclease digestion in comparison to segments that do not rearrange in the cells (41, 42, 464, 465). Similarly, the DNase sensitivity of chromatin has been found to correlate with transcriptional activity (71, 341, 460, 466). It is not yet clear whether the changes in the chromatin structure, as measured by the accessibility of the DNA to nucleases, directly reflects the accessibility of the *V*-region gene segment to the recombinase, or is a reflection of the transcriptional activity of these regions.

10.3.3 Transcriptional control elements regulate recombinational accessibility

The correlation between transcription of unrearranged *V*-region gene segments and the onset of their rearrangement suggested that transcriptional control elements may play a role in regulating the accessibility of variable region gene segments. Transgenic recombination substrates were used to demonstrate that *cis*-acting sequences associated with known transcriptional control elements (for example enhancer elements) target V(D)J recombination in developing pre-lymphocytes (Fig. 11). Rearrangement and expression of transgenic substrates composed of *TCRβ* V, D, and J gene segments were shown to be completely dependent on the presence of a DNA fragment containing the Eμ element (448). Subsequently, the TCRβ and TCRα enhancers (E_β and E_α) were also found to confer accessibility of this substrate in transgenic animals (457, 467). In addition, the intronic Igκ enhancer (E_κ) and the viral SV40 enhancer also conferred transcriptional and recombinational activity to this TCRβ recombination substrate in a B cell line (131). In all these cases, substrate D_β-to-J_β rearrangement and transcription of the unrearranged construct gene segments were activated in normal lymphocytes or appropriate lymphoid cell lines when positioned adjacent to an active enhancer.

Transcriptional enhancer elements also appear capable of providing signals that regulate tissue- and temporal-specific rearrangement of *V*-region gene segments. For example, the rearrangement and expression of constructs harbouring TCR E_α elements displayed tissue specificity in mice, rearranging only in T lineage cells.

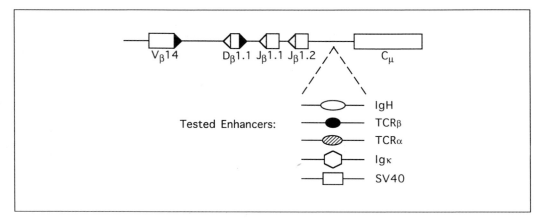

Fig. 11 Schematic of a $V_\beta D_\beta J_\beta C_\mu$ recombination substrate. The approximately 22 kb minilocus $V_\beta D_\beta J_\beta C_\mu$ and its derivatives were assembled from genomic sequences harbouring the *TCRβ*, $V_\beta 14$, $D_\beta 1.1$ and $J_\beta 1.1$ and *1.2* segments, and the IgH μ constant (C_μ) region. Regions spanning the coding regions of this segment are denoted as open rectangles. Open triangles represent RSS with 12 bp spacer sequences, while closed triangles represent RSS with 23 bp spacer sequences (adapted from refs 448, 457).

Furthermore, the transgenic E_α or E_β substrates were rearranged and expressed during thymic ontogeny in a pattern similar to the endogenous *TCRα* and *TCRβ* loci (467). The TCRδ enhancer has similarly been shown to direct the V–D to J rearrangement of a *TCRδ* locus-based transgenic recombination substrate (468). These observations strongly support the notion that transcription elements associated with the endogenous antigen receptor loci provide signals that regulate stage and/or tissue specific recombination.

Gene-targeted mutation of the endogenous E_μ and E_κ regions further confirmed that the integrity of enhancer elements is important for mediating rearrangement of their respective receptor loci (126, 315, 449). Removal of a 1 kb DNA fragment that contained E_μ (449), or its replacement with an expressed *neo^r* gene (315), resulted in a *cis*-acting inhibition of J_H locus rearrangement and germline transcription (315). However, the replacement or deletion of E_μ element did not completely abolish V(D)J recombination at the *IgH* locus. Therefore, it appears that the IgH intronic enhancer element is important for conferring accessibility, but that additional elements may exist and act in concert to generate full recombinational accessibility control.

Two additional enhancer elements within the *IgH* locus have been identified. Theoretically, these could be possible candidates to help confer full recombinational accessibility for V(D)J recombination. A B cell-specific enhancer element has been identified at the 3′ end of the *IgH* locus (469, 470). The 3′ enhancer appears to control some aspects of downstream germline C_H transcription (and class switching), but its deletion does not appear to result in a significant impairment in V(D)J rearrangement (320) (see Section 11.5.7). Another B cell-specific enhancer element has recently been identified in proximity to the *DQ52* gene segment (471, 472). The potential role of this element in promoting recombination accessibility is not yet known.

10.3.4 Additional elements beyond transcription and transcription control elements may regulate V(D)J rearrangement

A strong correlation has been found between the recombination and transcription potential of *V*-region segments. However, whether transcription of the segments affects recombination accessibility or is a reflection of the accessible state of the region to V(D)J recombinase remains an open question. Studies using transgenic recombination substrates provided evidence to suggest that TCRβ D and J recombination events are promoted by Ig or TCR enhancer functions identical to or closely associated with the transcriptional enhancing activities (457). However, such studies also indicated that the V to DJ joining event (the allelically excluded event) may be more stringently regulated by additional positive or negative control elements in order to ensure specificity (457).

As outlined above, various studies have shown that tissue specific rearrangement and allelic exclusion are regulated at the V to (D)J rearrangement step for both *IgH* and *TCRβ* loci (275, 473–475). The tissue specificity of the Vβ to DJβ rearrangement event has been reproduced in a *TCRβ* transgenic minilocus (Fig. 11) (448, 457). The minilocus undergoes IgH or TCRβ enhancer dependent D to J$_β$ rearrangements in most developing B and T cells (448, 457), whereas the minilocus $V_β$ to $DJ_β$ rearrangements were limited to T lineage cells (445, 457, 467). Significantly, the $V_β$ segment in these constructs is actively transcribed in B lineage cells throughout developmental stages when endogenous *IgH* genes rearrange (457). This observation suggests that transcriptional activation of a gene segment itself may not be sufficient to confer accessibility to a substrate V gene segment and that factors other than, or in addition to, transcription elements are required for tissue-specific regulation of V(D)J recombination. Similar observations have been made with respect to the transcription and recombinational activity of class switch regions (476). A candidate for such controlling elements has been identified; sequences that act to silence V(D)J recombination in the intron between the $J_κ$–$C_κ$ segments in chicken (477).

10.3.5 The potential role of various factors in the control of V(D)J recombination

How transcription *per se* or transcriptional control elements direct recombinase activity to specific loci remains unclear. Transcription may confer recombinational accessibility to substrate gene segments by promoting necessary interactions between recombination and transcription complexes or by altering the structural state of the DNA. Active transcription has been found to correlate with the increased rate of homologous recombination (478–480). Transcription of topologically-closed DNA can induce B to Z conformational changes (481). Such altered conformations may be recognized by factors involved in recombination. Transcription also induces positive supercoils ahead of, and negative supercoils behind, the RNA polymerase complex (482–487). Topoisomerases can resolve these topological changes by inducing breaks in the DNA (488). Negative supercoiling of a template DNA segment generated by transcription can induce site-specific recombination by the bacterial transposase $\gamma\delta$/Tn3 (489). By analogy, transcription of the germline V-region gene segments may similarly stimulate V(D)J recombination.

Cis-acting elements that control V(D)J rearrangement may overlap with elements that control transcription of V-region gene segments. Protein factors that regulate transcription and/or recombination could interact with accessible target sequences. Enhancers may also have independent roles in controlling transcription and recombination. To this effect, the finding that the intronic Ig core enhancer can confer accessibility of nearby DNA segments to nuclear factors independent of transcription is provocative (466). Finally, transcriptional enhancers may also act as nuclear matrix attachment regions, potentially providing binding or entry sites for proteins that are involved in recombination (490–494).

10.4 Abnormalities in recombinational accessibility

Many human B and T cell tumours harbour translocations that appear to have been mediated by inappropriate targeting of V(D)J recombinase activity. In certain human T cell leukaemias, bands 11 and 32 of chromosome 14 are translocated, or the region between them is inverted (495, 496). Many of these rearrangements have been generated by joining a TCR $J\alpha$ segment to an Ig V_H segment. Sequences immediately adjacent to the rearranged V_H segment are transcribed in T cells, consistent with the idea that these rearrangements result from aberrant accessibility control (71, 496).

Other chromosomal translocations appear to use 'pseudo' RSSs located in the vicinity of proto-oncogenes. Ig J_H and TCR $J\alpha$ segments have been found recombined with heptamer- and nonamer-like elements located in regions flanking the oncogenes (497, 498). Many of these joins contain apparent N-regions, suggesting that they occurred early in pre-lymphocyte development when TdT is expressed. In many cases of the endemic form of Burkitt's lymphoma, sequences flanked by apparent RSSs adjacent to the c-*myc* gene have been joined to a J_H segment, placing c-*myc* under the influence of H chain regulatory sequences (71). Similarly, a number of human B cell leukaemias and lymphomas are characterized by analogous, recombinase-mediated translocations between J_H segments and the *bcl-2* oncogene (499, 500). Inappropriate overexpression of c-*myc* and *Bcl-2* together in lymphocytes of transgenic mice leads to accelerated formation of lymphocytic tumours (501, 502) suggesting that the observed translocations are directly involved in development of the disease (503 and reviewed in ref. 504.)

11. IgH class switching
11.1 Overview

Class switch recombination (CSR) allows a clonal B cell to produce antibodies that retain their antigen specificity while acquiring a secondary effector function. Class switching involves juxtaposition of the expressed IgH V(D)J variable region gene to a downstream C_H gene which encodes the portion of the Ig that determines effector functions. CSR is mechanistically distinct from V(D)J recombination and usually occurs at a later stage in B cell differentiation. In general, the mechanism and regulation of class switching seems to be conserved between mouse and human. We will focus primarily on what is known with respect to the murine system since more information is available for this system.

11.2 Organization of the Ig C_H genes

The murine IgH constant region gene segments are localized across an approximately 200 kb region 3' of the J_H segments and are ordered 5'–C_μ–C_δ– –$C_\gamma 3$–$C_\gamma 1$–$C_\gamma 2b$–$C_\gamma 2a$–C_ϵ–C_α–3' (Fig. 12) (25, 505). Each C_H gene (except δ which does not undergo classical CSR) is preceded on the chromosome by a set of repetitive sequence elements that are the site of class switch recombination (termed S regions) (506–508). All of these C_H genes are also organized into transcription units in which a specific promoter and a short, non-coding exon (termed I region) lie upstream of the S region.

11.2.1 S regions

Terminally differentiated plasma cells (or hybridomas) that have switched HC isotype expression are universally characterized by CSR events involving the S region just upstream of the expressed C_H gene. As such recombination events, unlike V(D)J recombination events, occur within introns and not coding regions, they do not affect any translational reading frames. Individual S regions are located approximately 2 kb 5' of the corresponding C_H genes and range from 2.5 (S_ϵ) to 10 kb ($S_\gamma 1$) in length (506, 508–510). S regions are composed mainly of the pentameric repeats GAGCT and GGGGT (510, 511). The S_μ region can be divided into two patterns of sequence repeats: the 5' end is made up of the repeat sequence YAGGTTG (Y = pyrimidine) and the 3' end is mainly composed of the pentameric repeats $[(GAGCT)_n GGGGT]_m$, where n ranges from 1 to 7 and m is approximately 150 (512, 513). The IgG switch sequences are made up of 49 bp repeats that are between 45–82 per cent homologous within a particular S region (508, 510, 513–516). The S_ϵ and the S_α sequence repeats tend to be 40 bp and 80 bp long, respectively (506, 510, 517).

By considering the prevalence of the $(GAGCT)_n GGGGT$ repeats within the S regions, S_ϵ and S_α show the highest degree of homology with S_μ followed by $S_\gamma 3 > S_\gamma 1 > S_\gamma 2b > S_\gamma 2a$. It is notable that even though the S_ϵ region exhibits the highest degree of homology to the S_μ region, C_ϵ is the least used isotype (510, 518). Thus, the frequency of CSR to particular S regions is probably not determined simply by sequence homology. On the other hand, this observation does not eliminate homology-mediated recombination as a candidate for the CSR mechanism since other factors, such as accessibility, may influence the recombination rate (see below). Finally, although there have been some reports of consensus sequences adjacent to CSR sites within particular S regions (519), the potential role of such

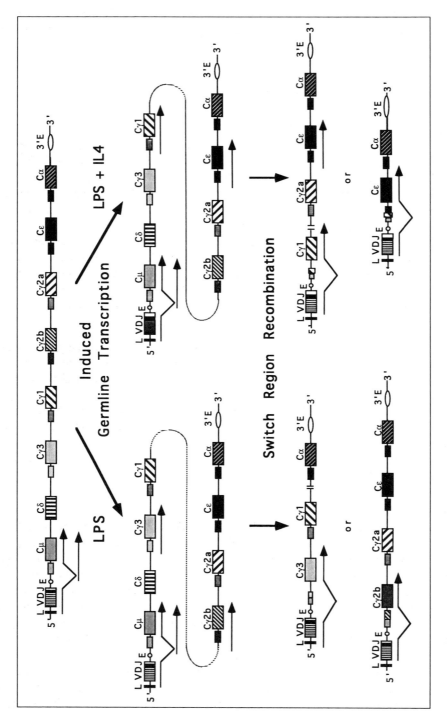

Fig. 12 Directed heavy chain class switch recombination. Stimulation of B cells with with mitogens/lymphokines directs transcription of and class switching to particular isotypes. Each C_H gene (rectangles) (except δ which does not undergo classical CSR) is preceded on the chromosome by an S region (small shaded rectangles) which is the site of class switch recombination (note that the figure is not drawn to scale). All of these C_H genes are also organized into transcription units in which a specific promoter and a short, non-coding exon (termed I region) lie upstream of the S region. The I region promoters that are induced by LPS or LPS/IL4 stimulation are shown by arrows (see text). E is the intronic enhancer; 3' E is the 3' enhancer.

sequences in mediating CSR is unknown. In any case, CSR does not appear to involve the high degree of sequence specificity employed in V(D)J recombination (520).

11.3 Mechanism of S region mediated CSR

11.3.1 Topology of CSR

B cells that have undergone a CSR event usually have deleted all of the C_H gene segments between the S_μ region and the S region of the C_H gene that is expressed (505, 521–524). Several groups have found extra chromosomal circles that were generated by S–S mediated deletions providing clear evidence that CSR frequently involves looping out and deletion of the intervening DNA (525–529). There is also some evidence that intervening DNA may be deleted as a linear molecule (530) but it is not clear how frequently this occurs.

Despite clear mechanistic differences between CSR and V(D)J recombination, there are some analogies between these processes in that both involve recognition (which must have some type of specificity with respect to CSR), endonucleolytic excision, juxtaposition, and re-ligation of two physically separate recombination sites. Furthermore, as in V(D)J recombination, cleaved S regions apparently can be resolved in several ways including:

1. There can be intra-S region ligation in a process somewhat analogous to open/shut joins in V(D)J recombination; such a process probably results in intra-S region deletions commonly observed in the S_μ region and other S regions (512, 531, 532).
2. The intervening DNA can be inverted; inversion events within and between S regions have been described both in A-MuLV transformants (525, 531, 533) and in hybridomas from activated splenic B cells (534, 535).
3. As described above, the ends of the deleted DNA can be ligated to form a circle (looping out and deletion).

11.3.2 Sequential switching

A variety of evidence has indicated that CSR can occur in a sequential mode. In other words, once a B cell has switched to a particular C_H gene, it can, upon appropriate stimulation, undergo an additional switch to a downstream C_H gene. Cellular analyses have supported the notion that this process can occur in normal spleen cells (535–537). At a molecular level, the first clear evidence for sequential CSR came from findings that some CSR sites contained fusions of three separate S regions (Fig. 12, bottom right) (for example the S_μ region linked to S_ϵ by a segment of S_γ (506, 538, 539). More recently, the generality of this phenomenon has been extended by the finding of sequentially switched S regions in recombination products in extrachromosomal circles isolated from *in vitro* stimulated splenic B cells (537) and *in vivo* from the splenic B cells of parasite-infected mice (528). Relevant to this finding, it has been demonstrated that, following CSR, the germline I_μ

promoter continues to be actively transcribed in association with the downstream C_H gene, generating a 'hybrid' germline transcript consisting of the I_μ exon spliced to the C_H exons of the switched C_H gene (540). In the context of the potential involvement of germline transcripts in CSR, such a hybrid transcript could be important for sequential CSR.

Sequential switching appears to be a physiologically relevant process, but its precise role has yet to be determined. It is clear that sequential CSR is not a required pathway. For example, mice harbouring a deletion of the $I_\gamma 1$ exon and promoter still undergo switching to the C_ϵ locus even though they are unable to switch to the $C_\gamma 1$ locus first (541).

11.3.3 Role of S regions in CSR

Precisely how S regions promote or enhance recombination remains controversial. Switch recombination does not seem to be sequence specific since recombination sites have been found throughout the various S regions and occasionally outside of them (520). However, CSR events frequently involve the short tandemly repeated sequences of the S regions. In this context, human hypervariable minisatellite DNA sequences (SAT) also comprise short tandemly repeated DNA sequences which seem to enhance recombination of both homologous and non-homologous sequences located nearby (542). The sequence of the human consensus 'minisatellite core' is $(\underline{GGAGGTGGGCAGGA}XG)_n$ (542), which is quite homologous to the pentameric repeats ($\underline{GAGCTGGGGT}$), commonly found in S regions. Minisatellite sequences have been proposed to enhance recombination much the same way Chi sequences do in *Escherichia coli*, by promoting the nicking and unwinding of nearby DNA (542, 543). The sequence similarities between Chi and switch region DNA has been previously observed (544). Additionally, both satellite and switch sequences (like Chi sequences) are highly GC-rich and exhibit a striking purine–pyrimidine strand bias. DNA molecules that contain purines on one strand can form triple helices by interacting simultaneously with two pyrimidine strands by Hoogstein bonding (545). The S region triple helices could be formed by three DNA strands or two DNA strands and an RNA strand (546, 547). These triple helical structures could enhance recombination by acting as nucleating sites for the switch recombinase(s) or by stabilizing the invasion of the S region DNA duplex by single-stranded DNA.

It is a matter of debate whether CSR is carried out by site-specific, homologous, illegitimate, or homeologous recombination. Site-specific recombination, typified by V(D)J recombination occurs between specific DNA sites and is catalysed by recombination machinery that has the ability to specifically recognize these sites. Homeologous recombination requires lengthy stretches of nucleotide homology to properly align DNA molecules and is catalysed by the general recombination machinery of the cell (543). Illegitimate recombination involves most of the other types of DNA exchange events that seem not to require sequence specificity (548, 549). Homeologous recombination occurs between similar, but not identical sequences (550). In yeast, homeologous recombination has been shown to occur between sequences that share only 52 per cent identity (550). The rate of recombination is proportional to the length of homology between sequences. Yet, sequencing data has shown that recombination junctions can occur between DNA strands with only 2–21 nucleotides in register. Since S regions share common pentameric repeats, yet are dissimilar on the grand scale, CSR could be mediated by homeologous recombination in that every S region has significant stretches of DNA that are over 52 per cent identical and also has short stretches of absolute nucleotide homology. Since the S region pentameric repeats can span several kilobases, one could imagine that numerous opportunities exist for the nucleation of a recombination event in or around the S regions. However, data obtained from sequences of numerous recombined S regions do not support a role for short homology in switch recombination (520), although if homologous regions are used in priming recombination, they may not have been detected (see Section 11.3.5).

11.3.4 Factors that interact with S regions

One approach to elucidating the mechanism of CSR is to identify and characterize *trans*-acting factors that interact with S regions. Several groups have reported DNA binding factors that interact with S region sequences. To date, DNA binding proteins have been identified *in vitro* that interact with S_μ (551, 552), $S_\gamma3$ (519, 553, 544), $S_\gamma1$ (554, 555), $S_\gamma2b$ (R.L. and F.A., unpublished data), S_ϵ (P. Rothman, unpublished data), and S_α (554, 556, 557). The well-known transcription factors Oct (555), NF–κB (553), and BSAP (Pax-5) (427, 551) have been identified as binding to S regions in these studies. As with similar efforts to study RS binding factors, there is still no demonstration of functional significance in switch recombination of the detected protein:DNA interactions within S regions.

11.3.5 Role of DNA replication in switch recombination

Several studies have suggested that CSR requires DNA synthesis. Stimulation of B cell proliferation with polyclonal activators, such as LPS and anti-IgM, induces switching from IgM to IgG (558). Conversely, treatment of activated B cells with DNA replication inhibitors, such as bromodeoxyuridine, thymidine, and hydroxy-urea, prevent B cell proliferation and IgM to IgG switching (559). Additionally, LPS/IL-4 stimulated B cells fail to proliferate or switch to IgG1 when co-cultured with aphidicolin, a DNA synthesis inhibitor, yet are able to proliferate and switch to IgG1 once the aphidicolin has been removed (560). Furthermore, studies of the cell-cycle kinetics of LPS-stimulated splenocytes has correlated switch region rearrangements with the S phase of the cell cycle (561). Finally, models for the linkage of CSR to DNA replication have been proposed on the basis of findings that recombined switch regions often contain deletions, duplications, and point mutations that are absent in germline switch regions (520, 562, 563). Nevertheless, while compelling, the data supporting a role for DNA replication in CSR is indirect and awaits more conclusive experimental verification.

11.3.6 CSR substrate studies

Class switch recombination substrate studies, based on the use of substrates containing S regions (analogous to those used to define V(D)J recombination) have provided some evidence that CSR may involve B cell specific factors (564, 565). In these studies, a much higher frequency of substrate CSR was detected in A-MuLV transformed pre-B lines (which undergo endogenous CSR events) than in T cell lines or in fibroblast cell lines. However, because of the nature of the substrates used (deletional), these studies were limited to only a few cell lines and thus the generality of the observations could not be fully assessed. In addition, it remains unclear why A-MuLV transformants undergo CSR since this event is not associated with normal pre-B lymphocytes. Also, it is important to note that the frequency of CSR in the A-MuLV transformants was still orders of magnitude lower than V(D)J recombination frequencies in such lines (38, 564). To date, it has been difficult to obtain CSR events with inversional CSR substrates. Therefore, studies of CSR and the activities involved have lagged behind those of V(D)J recombination. While some studies exist (566–573), the generation of new cell lines that more accurately represent normal B cells with respect to the stage and rate of switching would be an important advance with respect to our understanding of this process.

Other CSR studies (574, 575) have employed transient CSR substrates analogous to those used to study V(D)J recombination (43). As for V(D)J recombination, such substrates may prove to be extremely useful for dissecting the CSR mechanism in the long run. However, many recombination events in such substrates occur outside of the S regions and similar substrates have been found to undergo recombination at the same rates in lymphoid and non-lymphoid cells (R.L. and F.A., unpublished data). Therefore, at this point, it is not clear whether the recombination events detected by such experiments are actually relevant to CSR.

The most promising data with respect to CSR substrates have been obtained from transgenic mice that contain a human Ig minilocus (576, 577). The human Ig minilocus contains unrearranged human V_H, D_H, and J_H sequences, along with the μ intronic enhancer, I_μ, S_μ, C_μ, $I_\gamma 1$, $S_\gamma 1$, and $C_\gamma 1$ sequences, and finally rat 3' E_H sequences. This construct was shown to undergo both V(D)J and class switch recombination. Future modifications of this or other transgenic CSR substrates may allow a more detailed examination of the mechanisms that control the CSR process.

11.4 Non-deletional CSR

The dual expression of IgM and IgD on the surface of B cells is a well-characterized phenomenon. However, various studies have indicated that B cells can express other downstream isotypes on their surface along with IgM, although the mechanism and general physiological significance of this process remains to be elucidated (578). Several mechanisms have been proposed to account for double isotype positive cells. It is known that B lymphocytes are transiently double-positive while switching from IgM to IgG (579); the result of residual surface IgM still present even after its template C_μ gene has been deleted in order to express the down-stream C_γ gene and generate IgG. Others have argued that double-positive B lymphocytes are stably maintained, perhaps in the context of memory (578–582). Sister chromatid exchange between the active and the non-active allele has been proposed to explain the observation of double isotype positive cells (538). However, this seems to be more the exception than the rule (583). Some evidence has supported the notion that downstream C_H genes could be expressed without recombination via differential processing of a long transcript, analogous to the situation for dual IgM and IgD expression (570, 571, 584, 585). Another proposed mechanism for double isotype expression is that *trans*-splicing can fuse the V(D)J transcript to the germline transcripts of the downstream C_H genes (33, 586, 587). *Trans*-splicing has been shown to occur in trypanosomes, nematodes, plants, and, potentially, mammals (588–590). Suggestive, but not unequivocal support for *trans*-splicing in B cells has been found in the context of a transgenic mouse model system (591–593) and in cell lines (585).

11.5 Control of CSR

11.5.1 B cell specificity

CSR appears to be a strictly B cell specific event. Such specificity could be provided by a B cell specific component of the CSR recombination system that is S region specific or by specifically modulating the accessibility of S regions in B cells to a more generalized recombination system. If the apparent B cell specificity observed with the chromosomal recombination substrate studies can be generalized, it would suggest that accessibility, at least as defined by transcription, may not be the simple answer because these substrates are actively transcribed in non-B cell lines. As discussed later, gene-targeted mutation experiments suggest the existence of additional accessibility mechanisms beyond transcription (476). In this context, if the recombination events observed in extrachromosomal CSR substrates were, in fact, shown to represent the true CSR mechanism, then the apparent lack of cell type specificity observed with these would imply the utilization of a more general recombination mechanism with accessibility determining the specificity. These issues will be discussed further below in the context of gene-targeted mutation studies.

11.5.2 Induction of CSR

A variety of evidence has strongly indicated that CSR is a directed process that occurs during antigen stimulated B cell proliferation. Evidence that CSR is directed comes from findings that normal or transformed murine B cells that have undergone CSR events on both chromosomes most frequently switch to the same S region on both (594, 595). Likewise, pre-B or B cell lines that undergo switching events in culture tend to repeatedly switch to one or a few S regions (567, 573).

Activation of class switching in B cells can occur through several pathways. There are two T cell independent pathways, the TI-1 pathway mimicked *in vitro* by bacterial lipopolysaccharide (LPS) treatment and the TI-2 pathway mimicked by treatment of B cells *in vitro* with anti-IgD-dextran (559, 596). The particular C_H genes to which CSR occurs can be directed by the presence of particular lympho-kines during the activation process. For example, treatment of splenic B cells *in vitro* with LPS in the absence of lymphokines leads to the generation of IgG3 and IgG2b expressing cells (Fig. 12) (597), while treatment of splenic B cells with LPS and interleukin-4 (IL-4) leads to the appearance of IgG1 and IgE producing cells (598, 599). Likewise, stimulation with LPS and other lymphokines (for example γ-IFN or TGF-β) leads to the appearance of cells that have switched to other isotypes (for instance IgG2a or IgA) (600, 601). Various lines of evidence indicated that, in most cases, the regulation of isotype production observed in these contexts was due to the induction of CSR in these cells rather than selection of cells that had already switched (602). Importantly, *in vivo* studies have confirmed that directed class switching is a normal means of regulating the humoral immune response (602).

A major *in vivo* class switch activation pathway is the T cell dependent (TD) pathway. The key molecules involved in T-dependent B cell activation are now known to be CD40L (gp39) expressed on the cell-surface of activated T cells and the CD40 receptor expressed on the B cell-surface membrane (reviewed in ref. 603). Many of these studies have been carried out with human cells. *In vitro*, CD40L (or α-CD40 mAbs) is able to synergize with IL-4 to induce B cell prolifera-tion and class switching to IgE (604–606) and IgG1 (607), with IL-4 and IL-5 to drive switching to IgG1, IgE, and IgG3, with IL-10 and TGF-β to actuate switching to IgA (608, 609), and with IL-10 to increase expression of IgM and IgG (608). The essential role that the CD40L:CD40 plays in regulating class switching has been underscored by the finding that patients with X-linked hyper-IgM syndrome (a malady in which normal numbers of B cells express only the IgM and/or IgD isotypes) have defective CD40L:CD40 interactions due to a mutation in the gene for CD40L (610–612). Recently, the *CD40* gene has been inactivated by gene targeting giving rise to mice that have IgM⁺IgD⁺ B cells, but unable to switch to any of the other isotypes (613, 614).

11.5.3 Accessibility model for control of CSR

Correlations between the germline transcription of and switching to particular C_H genes were first made in the context of A-MuLV transformed pre-B lines and a B cell lymphoma line that undergo spontaneous switching in culture (531, 615). Subsequent studies confirmed the physiological importance of these findings by demonstrating that mitogen/lymphokine treatments of certain transformed B lineage cell lines and normal splenic lymphocytes differentially regulate germline transcription of particular C_H genes precisely in parallel to their effects on regulation of CSR to those genes (616–620). For example, LPS treatment of splenic B cells induces expression of germline γ2b and γ3 transcripts, while the addition of IL-4 along with the LPS suppresses transcription of germline *γ2b* and *γ3* genes, but induces $\gamma 1$ and ϵ germline transcription (616). Thus, the induction of CSR to a particular C_H gene is preceded by the prior transcriptional activation of that gene in its germline configuration.

By analogy to the earlier findings on the control of V(D)J recombination, the correlation between germline transcription of C_H genes followed by subsequent CSR to those genes was interpreted in the context of an 'accessibility' model for the control of CSR (531, 615). According to this model, a general CSR activity would be directed to particular switch regions by factors that correlated with their germline transcription. Unlike V(D)J recombination, it has been difficult to generate CSR recombination substrates (probably for technical reasons; see above) that allow unequivocal demonstration of the existence of a single CSR 'recombinase'. However, several recent gene-targeted mutational studies described below have provided strong support for this general model of CSR control.

11.5.4 Germline transcription of C_H genes

All germline C_H genes, except C_δ which does not undergo CSR, are organized into characteristic germline transcription units of conserved structure. In these units, transcription heterologously initiates from a promoter that lies 5' of the I region, runs through the S region and undergoes termination and polyadenylation at the normal sites downstream of the C_H exons. The murine I_μ (331, 332), $I_\gamma 3$ (621, 622), $I_\gamma 1$ (618, 623, 624), $I_\gamma 2b$ (625), $I_\gamma 2a$ (626), I_ϵ (627–629), and I_α (619, 628) regions have been cloned and their resulting transcripts analysed. Subsequent to generation of the primary transcript, RNA splicing mechanisms join the I exon to the C_H exons, deleting out the intervening S region-derived RNA sequences as well as other introns. A number of the promoters of the germline C_H genes have been defined in some detail. In some cases, sequences capable of conferring LPS/lymphokine mediated transcriptional regulation to test substrates have been identified (556, 630). However, the overall sequences required for the expression of the germline transcription units during normal development have not been defined.

The precise role of germline transcription and/or transcripts in CSR is unknown. Most I exons contain multiple stop codons in the three reading frames and therefore cannot encode a peptide of significant length. Thus, despite poly(A)$^+$ germline transcripts being detected in the cells, they probably do not function by encoding proteins (623, 627). In addition, there is no significant conservation of the sequences within the different germline C_H transcripts and transcription units. However, the overall structural organization of both the C_H transcripts and transcription units is highly conserved for the different C_H genes and in diverse mammalian species. Together, these observations strongly imply that germline transcription and/or germline transcripts function in the regulation of CSR.

The relationship between germline transcription units and regulation of CSR has been clearly confirmed by gene targeted mutation experiments. Initial studies have shown that mutation (deletion or replacement by a *neo*^{r} gene) of the $I_\gamma 1$ or $I_\gamma 2b$ exons and their promoters results in the inhibition of class switching to the corresponding genes (532, 631). Both the $I_\gamma 1$ and the $I_\gamma 2b$ gene replacements affected switching in a *cis*-acting fashion, preventing isotype-specific switching on the replaced chromosome while having no effect on switching to the other allele. Likewise, these specific I region mutations did not affect switching to any of the other C_H genes located upstream or downstream on the same chromosome. Thus, these experiments showed that the integrity of the I region/promoter was essential for efficient CSR.

The gene-targeted mutational experiments outlined above did not address potential roles of germline transcription versus germline transcripts in regulating the class switch process. However, similar experiments provided some insight into this issue by replacing the LPS plus IL-4 inducible I_ϵ promoter and exon in normal lymphocytes (476) or a pre-B line (632) with an LPS-inducible E_μ enhancer/V_H promoter expression cassette. In contrast to normal B cells, heterozygous and homozygous mutant B cells for the I_ϵ replacement had substantial transcription through the ϵ switch recombination region (S_ϵ) following treatment with LPS alone. Furthermore, following LPS stimulation, both heterozygous and homozygous mutant cells underwent low level switching to IgE production (10–100 fold less than LPS/IL-4 stimulated wild-type cells). Although heterozygous mutant cells underwent switching to IgE at essentially wild-type levels when stimulated with LPS and IL-4 (on their normal chromosome as the mutation was again *cis*-acting), homozygous mutant cells still showed extremely low levels of switching to IgE upon LPS and IL-4 stimulation (despite continued high level transcription through the S_ϵ region from the $E_\mu V_H$ promoter). These findings suggest that transcription *per se* can generate a low level of class switch recombination in the absence of I region sequences. However, the results also imply that the optimal efficiency of the process requires the presence of the intact I region and/or I region promoter in *cis*. Thus, although transcription through the S region may be a requirement for CSR, additional factors, associated with the presence of an intact I region and/or its promoter, appear to regulate the efficiency of the process.

11.5.5 Potential functions of transcription *per se*

By analogy to the transcriptional enhancement of homologous recombination (478, 480), it has been speculated that transcription through the S region promotes CSR, perhaps by inducing recombinagenic nicks and/or changes in the DNA configuration (reviewed in refs 128 and 489). As described above, transcription of the DNA molecule causes local and temporal changes in the supercoiling of the DNA template (see Section 10.3.5 and ref. 477). Considering that topoisomerases act differently on positive and negative supercoiled DNA (623), this may offer a mechanism by which recombinase could specifically nick an activated S region. Such mechanisms may explain the 'baseline' constitutive CSR levels observed in constitutively-transcribed S regions that lack a functional I region (476, 631).

11.5.6 Potential functions of I region sequences in CSR

The mounting evidence that implicates the integrity of the I region for efficient CSR further suggests that the germline transcripts themselves may function in this process. The germline transcripts may act as a signal for the cell to turn on class switch machinery, or they may take a direct part in the recombination mechanism. In support of the latter notion, antisense oligonucleotides for the $I_\gamma 2b$ and I_α germline transcripts have been shown to affect CSR *in vitro* (634, 635). Furthermore, S_α transcripts generated in a plasmid-based system have been proposed to stabilize a less supercoiled conformer of the corresponding S_α DNA by forming RNA–DNA triple helices; the supposition being that the less supercoiled state may be important for accessibility of switch recombinase(s) to the relaxed S_α region (546, 547).

The *cis*-acting decrease in CSR generated by the I_ϵ replacement, which was still transcribed through the S region, suggests that normal germline transcripts may promote CSR through specific interactions that involve the I region of the targeted genomic DNA. Elements in the genomic I regions may also be used to promote interactions between the targeted downstream C_H gene with the C_μ gene. Since the C_μ gene is organized into a standard germline transcription unit (I region, S region, C region) (331), the I_μ region/transcript may play a complementary role to the corresponding downstream sequences in the CSR mechanism. In this context, a large targeted deletion that extends from the J_H segments through a portion of I_μ has been shown to significantly impair CSR (636). However, the affected elements that potentially cause this impairment (transcriptional promoters, the E_μ element, or the I_μ exon) remain to be elucidated. Finally, it also is possible that nuclear factors might bind to the I region DNA/promoter to target or mediate the rearrangement process.

11.5.7 Additional elements that regulate CSR

Various lines of evidence suggested the potential existence of additional downstream regulatory elements involved in the control of CSR (313, 314, 317, 324, 637–641). A sequence with transcriptional enhancer activity was identified 3' of the C_α gene based on transfection type assays (319, 469, 470). Its potential role in regulation of CSR was tested by generating normal lymphocytes either heterozygous or homozygous for a mutation in which a 5 kb DNA segment encompassing the 3' E_H element was replaced with a *neo*^{r} gene (320). The homozygous 3' E_H mutation did not inhibit V(D)J recombination at the *IgH* locus, nor did it effect B cell proliferation or IgM secretion following *in vitro* stimulation. However, *in vitro* LPS or LPS plus IL-4 stimulated splenic B cells derived from homozygous mutant ES cells showed marked inhibition in switching to IgG3, IgG2b, IgG2a, and IgE, but not to IgG1. *In vivo* the 3' E_H mutation appeared to be associated with a deficiency in IgG3 and IgG2a. The class switch defects correlated with corresponding defects in germline transcription of the affected C_H genes.

The replacement mutation of the 3' E_H and flanking sequences affected switch recombination and transcription of four different C_H genes spread over a 120 kb locus. It was proposed that the 3' E_H region may function in the context of a locus control region (642) involved in regulating class switching in the context of certain (namely LPS) activation pathways (320). It is important to note that the 3' E_H

enhancer replacement mutation may have exerted its effects by deletion of a critical control element and/or via a promoter competition mechanism for the activity of a separate control element. The latter phenomenon has been observed by targeted mutation of the β-globin LCR (643). Finally, the ability of B cells harbouring the 3' E_H mutation to switch to some C_H genes clearly indicates the existence of pathways (potentially the CD40/CD40L pathway) for activation of CSR that function independently of the control elements affected by the 3' E_H mutation. Candidate control elements for such pathways include the enhancer associated with the $C_\gamma 1$ gene (551), a weak enhancer found just 3' of the C_α gene (644, 645) or the DNA elements corresponding to DNaseI hypersensitive sites located 3' of the 3' E_H (646).

Acknowledgements

This work was supported by the Howard Hughes Medical Institute and by the National Institute of Health grants AI20047, CA42335, and UO1 AI31541 to FWA. We thank Dr Wes Dunnick, Dr Dan Holmberg, and members of the Alt laboratory for critically reading this manuscript.

References

References marked * are recommended for further reading.

1. Landsteiner, K. (1962). *The specificity of serological reactions*. Dover Publications, New York.
*2. Tonegawa, S. (1983). Somatic generation of antibody diversity. *Nature*, **302**, 575.
3. Kronenberg, M., Siu, G., Hood, L., and Shastri, N. (1986). The molecular genetics of the T-cell antigen receptor and T-cell antigen recognition. *Annu. Rev. Immunol.*, **4**, 529.
*4. Coffman, R. L., Lebman, D. A., and Rothman, P. R. (1993). The mechanism and regulation of immunoglobulin isotype switching. In *Advances in immunology* (ed. F. J. Dison), Vol. 54, p. 229. Academic Press, New York.
5. Kincade, P. W. (1987). Experimental models for understanding B lymphocyte formation. *Adv. Immunol.*, **41**, 181.
*6. Rolink, A. and Melchers, F. (1991). Molecular and cellular origins of B lymphocyte diversity. *Cell*, **66**, 1081.
7. Blackwell, T. K. and Alt, F. W. (1989). Mechanism and developmental program of immunoglobulin gene rearrangement in mammals. *Annu. Rev. Gen.*, **23**, 605.
8. Alt, F. W., Blackwell, T. K., and Yancopoulous, G. D. (1987). Development of the primary antibody repertoire. *Science*, **238**, 1079.
*9. Burnet, F. M. (1957). A modification of Jerne's theory of antibody production using the concept of clonal selection. *Aust. J. Sci.*, **20**, 67.
10. Pernis, B. G., Chiappino, G., Kelus, A. S., and Gell, P. G. H. (1965). Cellular localization of immunoglobulins with different allotype specificities in rabbit lymphoid tissue. *J. Exp. Med.*, **122**, 853
11. Cebra, J., Chiappino, G., Kelus, A. S., and Gell, P. G. H. (1966). Rabbit lymphoid cells differentiated with respect to a, g, and m heavy chain polypeptide chains and to allotypic markers AA1 and AA2. *J. Exp. Med.*, **123**, 547.

12. Gally, J. A. (1973). Structure of immunoglobulins. In *The antigens* (ed. M. Sela), Vol. 1, p. 162. Academic Press, New York.

13. Edelman, G. M. (1970). The covalent structure of a human gG-immunoglobulin XI. Functional implications. *Biochemistry*, **9**, 3197.

14. Amzel, L. M. and Poljak, R. J. (1979). Three-dimensional structure of immuno-globulins. *Annu. Rev. Biochem.*, **48**, 961.

15. Kabat, E. A., Wu, T. T., Reid-Miller, M., Perry, H. M., and Gottesman, K. S. (1987). *Sequences of proteins of immunological interest*. US Dept. Health & Human Services, Washington, D.C.

16. Kabat, E. A. (1982). Antibody diversity versus antibody complimentary. *Pharm. Rev.*, **34**, 23.

17. Nissinoff, A., Hopper, J. E., and Spring, S. B. (1975). *The antibody molecule*. Academic Press, New York.

18. Poljak, R. J. (1978). Correlations between three-dimensional structure and function of immunoglobulins. *CRC Crit. Rev. Biochem.*, **5**, 45.

19. Alt, F. W., Bothwell, A. L., Knapp, M., Siden, E., Mather, E., Koshland, M., and Baltimore, D. (1980). Synthesis of secreted and membrane-bound immunoglobulin mu heavy chains is directed by mRNAs that differ at their 3' ends. *Cell*, **20**, 293.

20. Rogers, J., Early, P., Carter, C., Calame, K., Bond, M., Hood, L., and Wall, R. (1980). Two mRNAs with different 3' ends encode membrane-bound and secreted forms of immunoglobulin μ chain. *Cell*, **20**, 303.

21. Singer, P. A., Singer, H. H., and Williamson, A. R. (1980). Different species of messenger RNA encode receptor and secretory IgM μ chains differing at their carboxy termini. *Nature*, **285**, 294.

22. Koshland, M. E. (1989). The immunoglobulin helper: The J chain. In *Immunoglobulin genes* (ed. T. Honjo, F. W. Alt, and T. H. Rabbits), p. 345. Academic Press, New York.

23. Liu, C.-P., Tucker, P. W., Mushinski, F., and Blattner, F. R. (1980). Mapping of the heavy chain genes of mouse immunoglobulins M and D. *Science*, **209**, 1348.

24. Tucker, P. W., Lui, C.-P., Mushinski, F. and Blattner, F. R. (1980). Mouse immuno-globulin D: Messenger RNA and genomic DNA sequences. *Science*, **209**, 1353.

25. Shimizu, A., Takahashi, N., Yaoita, Y., and Honjo, T. (1982). Organization of the constant-region gene family of the mouse immunoglobulin heavy chain. *Cell*, **28**, 499.

26. Alt, F. W., Blackwell, T. K., DePinho, R. A., Reth, M. G., and Yancopoulos, G. D. (1986). Regulation of genome rearrangement events during lymphocyte differentiation. *Immunol. Rev.*, **89**, 5.

27. Maki, R., Roeder, W., Trauhecker, A., Sidman, C., Wabl, M., Rascke, L., and Tonegawa, S. (1981). The role of DNA rearrangement and alternate RNA processing in the expression of immunoglobulin δ genes. *Cell*, **24**, 353.

28. Blattner, F. R. and Tucker, P. W. (1984). The molecular biology of immunoglobulin D. *Nature*, **307**, 417.

29. Mather, E. L., Nelson, K. J., Haimovitch, J., and Perry, R. P. (1984). Mode of regula-tion of immunoglobulin μ- and δ-chain expression varies during B-lymphocyte maturation. *Cell*, **36**, 329.

30. Roes, J. and Rajewsky, K. (1991). Cell autonomous expression of IgD is not essential for the maturation of conventional B cells. *Int. Immunol.*, **3**, 1367.

31. Nitschke, L., Kosco, M. H., Kohler, G., and Lamers, M. C. (1993). Immunoglobulin D-deficient mice can mount normal immune responses to thymus-independent and -dependent antigens. *Proc. Natl Acad. Sci. USA*, **90**, 1887.

32. Roes, J. and Rajewsky, K. (1993). Immunoglobulin D (IgD)-deficient mice reveal an auxiliary receptor function for IgD in antigen-mediated recruitment of B cells. *J. Exp. Med.*, **177**, 45.

33. Shimizu, A. and Honjo, T. (1984). Immunoglobulin class switching. *Cell*, **36**, 801.

34. Radbruch, A., Burger, C., Klein, S., and Muller, W. (1986). Control of immunoglobulin class switch recombination. *Immunol. Rev.*, **89**, 69.

35. Harriman, W., Volk, H., Defranoux, N., and Wabl, M. (1993). Immunoglobulin class switch recombination. *Annu. Rev. Immunol.*, **11**, 361.

36. Snapper, C. M. and Finkelman, F. D. (1993). Immunoglobulin class switching. In *Fundamental immunology* (ed. W. E. Paul), p. 837. Raven Press, New York.

*37. Alt, F. W. and Baltimore, D. (1982). Joining of immunoglobulin heavy chain gene segments: implications from a chromosome with evidence of three D–J_H fusions. *Proc. Natl Acad. Sci. USA*, **79**, 4118.

38. Blackwell, T. K. and Alt, F. W. (1984). Site-specific recombination between immunoglobulin D and J_H segments that were introduced into the genome of a murine pre-B cell line. *Cell*, **37**, 105.

39. Lewis, S., Gifford, A., and Baltimore, D. (1984). Joining of V_κ to J_κ segments in a retroviral vector introduced into lymphoid cells. *Nature*, **308**, 425.

40. Lewis, S., Gifford, A., and Baltimore, D. (1985). DNA elements are asymmetrically joined during the site-specific recombination of kappa immunoglobulin genes. *Science*, **228**, 677.

*41. Yancopoulos, G. D., Blackwell, T. K., Suh, H., Hood, L. E., and Alt, F. W. (1986). Joining between introduced but not endogenous T-cell receptor segments in pre-B cells: evidence that B and T cells use a common recombinase. *Cell*, **44**, 251.

42. Blackwell, T. K., Moore, M. W., Yancopoulos, G. D., Suh, H., Lutzker, S., Selsing, E., and Alt, F. W. (1986). Recombination between immunoglobulin variable region gene segments is enhanced by transcription. *Nature*, **324**, 585.

*43. Hesse, J. E., Lieber, M. R., Gellert, M., and Mizuuchi, K. (1987). Extrachromosomal DNA substrates in pre-B cells undergo inversion or deletion at immunoglobulin V–(D)–J joining signals. *Cell*, **49**, 775.

*44. Schatz, D. G. and Baltimore, D. (1988). Stable expression of immunoglobulin gene V(D)J recombinase activity by gene transfer into 3T3 fibroblasts. *Cell*, **53**, 107.

*45. Oettinger, M. A., Schatz, D. G., Gorka, C., and Baltimore, D. (1990). *Rag-1* and *Rag-2*, adjacent genes that synergistically activate V(D)J recombination. *Science*, **248**, 1517.

*46. Gellert, M. (1992). V(D)J recombination gets a break. *Trends Genet.*, **8**, 408.

47. Early, P., Huang, H., Davis, M., Calame, K., and Hood, L. (1980). An immunoglobulin heavy chain variable region gene is generated from three segments of DNA, V_H, D and J_H. *Cell*, **19**, 1981.

48. Sakano, H., Kurosawa, Y., Weigert, M., and Tonegawa, S. (1981). Identification and nucleotide sequence of a diversity DNA segment (D) of immunoglobulin heavy-chain genes. *Nature*, **290**, 562.

49. Hesse, J. E., Lieber, M. R., Mizuuchi, K., and Gellert, M. (1989). V(D)J recombination: a functional definition of the joining signals. *Genes Dev.*, **3**, 1053.

50. Akira, S., Okazaki, K., and Sakano, H. (1987). Two pairs of recombination signals are sufficient to cause immunoglobulin V–(D)–J joining. *Science*, **238**, 1134.

51. Wei, Z. and Lieber, M. R. (1993). Lymphoid V(D)J recombination. Functional analysis of the spacer sequence within the recombination signal. *J. Biol. Chem.*, **268**, 3180.

52. Aguilera, R. J., Akira, S., and Sakano, H. (1987). A pre-B cell nuclear protein that specifically interacts with immunoglobulin V–J recombination sequences. *Cell*, **51**, 909.

53. Shirakata, M. (1991). HMG-1 related DNA-binding protein isolated with V–(D)–J recombination signal probes. *Mol. Cell. Biol.*, **11**, 4528.

54. Li, M., Morzycka-Wrobleska, E., and Desiderio, S. V. (1989). NBP, a protein that specifically binds an enhancer of immunoglobulin gene rearrangement: purification and characterization. *Genes Dev.*, **3**, 1801.

55. Matsunami, N., Hamaguchi, Y., Yamamoto, Y., Kuze, K., Kanagawa, K., Matsuo, M., Kawaichi, M., and Honjo, T. (1989). A protein binding to the J kappa recombination sequence of immunoglobulin genes contains a sequence related to the integrase motif. *Nature*, **342**, 934.

56. Wu, L. C., Mak, C. H., Dear, N., Boehm, T., Foroni, L., and Rabbits, T. H. (1993). Molecular cloning of a zinc finger protein which binds to the heptamer of the signal sequence for V(D)J recombination. *Nucleic Acids Res.*, **21**, 5067.

57. Sakano, H., Juppi, K., Heinrich, G., and Tonegawa, S. (1979). Sequences at the somatic recombination sites of immunoglobulin light-chain genes. *Nature*, **280**, 288.

58. Max, E. E., Seidman, J. G., and Leder, P. (1979). Sequences of five potential recombination sites encoded close to an immunoglobulin kappa constant region gene. *Proc. Natl Acad. Sci. USA*, **76**, 3450.

59. Bucchini, D., Reynaud, C.-A., Ripoche, M.-A., Grimal, H., Jami, J., and Weill, J. C. (1987). Rearrangement of a chicken immunoglobulin gene occurs in the lymphoid lineage of transgenic mice. *Nature*, **326**, 409.

60. Goodhardt, M., Cavelier, P., Akimenko, M. A., Lutfulla, G., Babinet, C., and Rougeon, F. (1987). Rearrangement and expression of rabbit immunoglobulin κ light chain gene in transgenic mice. *Proc. Natl Acad. Sci. USA*, **84**, 4229.

61. Hendrickson, E. A., Liu, V. F., and Weaver, D. T. (1991). Strand breaks without DNA rearrangement in V(D)J recombination. *Mol. Cell. Biol.*, **11**, 3155.

62. Seidman, J. G. and Leder, P. (1980). A mutant immunoglobulin light chain is formed by aberrant DNA- and RNA-splicing events. *Nature*, **286**, 286.

63. Hochtl, J. and Zachau, H. G. (1983). A novel type of aberrant recombination in immunoglobulin genes and its implication for V–J joining mechanism. *Nature*, **302**, 260.

64. Durdick, J., Moore, N. W., and Selsing, E. (1984). Novel kappa light-chain gene rearrangements in mouse lambda light chain-producing B lymphocytes. *Nature*, **307**, 749.

65. Siminovich, K. A., Bakhshi, A., Goldman, P., and Korsmeyer, S. J. (1985). A uniform deleting element mediates the loss of kappa genes in human B cells. *Nature*, **316**, 260.

66. Reth, M., Gehrmann, P., Petrac, E., and Wiese, P. (1986). A novel V_H to V_HDJ_H joining mechanism in heavy-chain-negative (null) pre-B cells results in heavy-chain production. *Nature*, **322**, 840.

67. Kleinfeld, R., Hardy, R., Tarlinton, D., Dangl, J., Herzenberg, L., and Weigert, M. (1986). Recombination between an expressed immunoglobulin heavy-chain gene and a germline variable gene in a Ly 1+ B-cell lymphoma. *Nature*, **322**, 843.

68. Moore, M. W., Durdik, J., Persiani, D. M., and Selsing, E. (1981). Deletions of κ chain constant region genes in mouse λ-producing B cells involve intrachromosomal DNA recombinations similar to V–J joining. *Proc. Natl Acad. Sci. USA*, **82**, 6211.

69. Covey, L. R., Ferrier, P., and Alt, F. W. (1990). V_H to V_HDJ_H rearrangement is mediated by the internal V_H heptamer. *Int. Immunol.*, **2**, 579.

70. Usuda, S., Takemori, T., Matsuoka, M., Shirasawa, T., Yoshida, K., Mori, A., Ishizaka,

K., and Sakano, H. (1992). Immunoglobulin V gene replacement is caused by the intramolecular DNA deletion mechanism. *EMBO J.*, **11**, 611.

71. Haluska, F., Tsujimoto, Y., and Croce, C. (1987). Mechanisms of chromosome translocations in B- and T-cell neoplasia. *Trends Genet.*, **3**, 11.

72. Hendrickson, E. A., Qin, X.-Q., Bump, E. A., Schatz, D. G., Oettinger, M., and Weaver, D. T. (1991). A link between double-strand break-related repair and V(D)J recombination: The scid mutation. *Proc. Natl Acad. Sci. USA*, **88**, 4061.

73. Lewis, S. M. and Hesse, J. E. (1991). Cutting and closing without recombination in V(D)J joining. *EMBO J.*, **10**, 3631.

74. Korman, A. J., Maruyama, J., and Raulet, D. H. (1989). Rearrangement by inversion of a T-cell receptor delta variable region gene located 3' of the delta constant region gene. *Proc. Natl Acad. Sci. USA*, **86**, 267.

75. Iwashima, M., Green, A., Davis, M. M., and Chien, Y.-S. (1988). Variable region (Vd) gene segment most frequently utilized in adult thymocytes is 3' of the constant (Cδ) region. *Proc. Natl Acad. Sci. USA*, **85**, 8161.

76. Malissen, M., McCoy, C., Blanc, D., Trucy, J., Devaux, C., Schmidt-Verhulst, A.-M., Fitch, F., Hood, L., and Malissen, B. (1986). Direct evidence for chromosomal inversion during T cell receptor B-gene rearrangements. *Nature*, **319**, 28.

77. Fujimoto, S. and Yamagishi, H. (1987). Isolation of an excision product of T-cell-receptor α-chain gene rearrangement. *Nature*, **327**, 242.

78. Okazaki, K., Davis, D. D., and Sakano, H. (1987). T cell receptor β gene sequences in the circular DNA of thymocyte nuclei: Direct evidence for intramolecular DNA deletion in V–D–J joining. *Cell*, **49**, 477.

79. Iwasato, T. and Yamagishi, H. (1992). Novel excision products of T cell receptor γ gene rearrangements and developmental stage specificity implied by the frequency of nucleotide insertions at signal joins. *Eur. J. Immunol.*, **22**, 101.

80. Steinmetz, M., Altenberger, W., and Zachau, H. G. (1980). A rearranged DNA sequence possibly related to the translocation of immunoglobulin gene segments. *Nucleic Acids Res.*, **8**, 1383.

81. Gauss, G. H. and Lieber, M. R. (1992). The basis for the mechanistic bias for deletional over inversional V(D)J recombination. *Genes Dev.*, **6**, 1553.

82. Gerstein, R. M. and Lieber, M. R. (1993). Coding end sequence can markedly affect the initiation of V(D)J recombination. *Genes Dev.*, **7**, 1459.

83. Boubnov, N., Wills, Z. P., and Weaver, D. (1993). V(D)J recombination coding junction formation without DNA homology: processing of coding termini. *Mol. Cell. Biol.*, **13**, 6957.

84. Yancopoulos, G. D., Nolan, G. P., Pollock, R., Prockop, S., Li, S. C., Herzenberg, L. A., and Alt, F. W. (1990). A novel fluorescence-based system for assaying and separating live cells according to VDJ recombinase activity. *Mol. Cell. Biol.*, **10**, 1697.

85. Lieber, M. R., Hesse, J. E., Mizuuchi, K., and Gellert, M. (1988). Studies of V(D)J recombination with extrachromosomal substrates. *Curr. Top. Microbiol. Immunol.*, **137**, 94.

*86. Roth, D. B., Nakajima, P. B., Menetski, J. P., Bosma, M. J., and Gellert, M. (1992). V(D)J recombination in mouse thymocytes: double-strand breaks near T cell receptor delta rearrangement signals. *Cell*, **69**, 41.

87. Schlissel, M., Constantinescu, A., Morrow, T., Baxter, M. and Peng, A. (1993). Double-strand signal sequence breaks in V(D)J recombination are blunt, 5'-phosphorylated, *Rag*-dependent and cell cycle regulated. *Genes Dev.*, **7**, 2520.

88. Roth, D. B., Menetski, J. P., Nakajima, P. B., Bosma, M. J., and Gellert, M. (1992). V(D)J recombination: broken DNA molecules with covalently sealed (hairpin) coding ends in scid mouse thymocytes. *Cell*, **70**, 983.

89. Ferguson, S. E. and Thompson, C. B. (1993). A new break in V(D)J recombination. *Curr. Biol.*, **3**, 51.

90. Lieber, M. R. (1991). Site-specific recombination in the immune system. *FASEB J.*, **5**, 2934.

91. Bollum, F. J. (1974). Terminal deoxynucleotidyl transferase. In *The enzymes* (ed. P. Boyer), Vol. 10, p. 145. Academic Press, New York.

*92. Komori, T., Okada, A., Stewart, V., and Alt, F. W. (1993). Lack of N regions in antigen receptor variable region genes of TdT-deficient lymphocytes. *Science*, **261**, 1171.

*93. Gilfillan, S., Dierich, A., Lemeur, M., Benoist, C., and Mathis, D. (1993). Mice lacking TdT: Mature animals with an immature lymphocyte repertoire. *Science*, **261**, 1175.

94. Lieber, M. R. (1992). The mechanism of V(D)J recombination: a balance of diversity, specificity, and stability. *Cell*, **70**, 873.

95. Lieber, M. R., Hesse, J. E., Mizuuchi, K., and Gellert, M. (1988). Lymphoid V(D)J recombination: nucleotide insertion at signal joints as well as coding joints. *Proc. Natl Acad. Sci. USA*, **85**, 8588.

96. Takeshita, S., Toda, M., and Yamagishi, H. (1989). Excision products of the T cell receptor gene support a progressive rearrangement model of the alpha/delta locus. *EMBO J.*, **8**, 3261.

97. Toda, M., Fujimoto, S., Iwasato, T., Takeshita, S., Tezuka, K., Ohbayashi, T., and Yamagishi, H. (1988). Structure of extrachromosomal circular DNAs excised from T-cell antigen receptor alpha and delta-chain loci. *J. Mol. Biol.*, **202**, 219.

98. Hochtl, J., Muller, C. R., and Zachau, H. G. (1982). Recombined flanks of the variable and joining segments of immunoglobulin genes. *Proc. Natl Acad. Sci. USA*, **79**, 1383.

*99. Lafaille, J. J., DeCloux, A., Bonneville, M., Takagaki, Y., and Tonegawa, S. (1989). Junctional sequences of T cell receptor gamma delta genes: implications for gamma delta T cell lineages and for a novel intermediate of V-(D)-J joining. *Cell*, **59**, 859.

100. Alt, F. W., Reth, M. G., Blackwell, T. K., and Yancopoulos, G. D. (1986). Regulation of immunoglobulin variable-region gene assembly. *Mt Sinai J. Med.*, **53**, 166.

101. Carlsson, L. and Holmberg, D. (1990). Genetic basis of the neonatal antibody repertoire: germline V-gene expression and limited N-region diversity. *Int. Immunol.*, **2**, 639.

102. Bangs, L., Sanz, J. M., and Teale, J. M. (1991). Comparison of D, J_H and junctional diversity in the fetal, adult, and aged B cell repertoires. *J. Immunol.*, **146**, 1996.

103. Carlsson, L., Overmo, D., and Holmberg, D. (1992). Selection against N-region diversity in immunoglobulin heavy chain variable regions during the development of pre-immune B cell repertoires. *Int. Immunol.*, **4**, 549.

104. Feeney, A. J. (1991). Junctional sequences of fetal T cell receptor beta chains have few N regions. *J. Ex. Med.*, **174**, 115.

105. Meek, K., Rathbun, G., Reininger, L., Jaton, J. C., Kofler, R., Tucker, P. W., and Capra, J. D. (1990). Organization of the murine immunoglobulin V_H complex: Placement of two new V_H families (V_H10 and V_H11) and analysis of V_H family clustering and interdigitation. *Mol. Immunol.*, **27**, 1073.

106. Bogue, M., Candeias, S., Benoist, C., and Mathis, D. (1991). A special repertoire of alpha:beta T cells in neonatal mice. *EMBO J.*, **10**, 3647.

107. Schwager, J., Burkert, N., Courtet, M., and Du Pasquier. (1991). The ontogeny of diversification at the immunoglobulin heavy chain locus in *Xenopus*. *EMBO J.*, **10**, 2461.

108. Elliot, J. F., Rock, E. P., Patten, P. A., Davis, M. M., and Chen, Y.-H. (1988). The adult T-cell receptor δ-chain is diverse and distinct from that of fetal thymocytes. *Nature*, **331**, 627.

109. Kyes, S., Pao, W., and Hayday, A. (1991). Influence of site of expression on the fetal gamma delta T-cell receptor repertoire. *Proc. Natl Acad. Sci. USA*, **88**, 7830.

110. McCormack, W. T., Tjoelker, L. W., Carlson, L. M., Petryniak, B., Barth, C. F., Humphries, E. H., and Thompson, C. B. (1989). Chicken IgL gene rearrangement involves deletion of a circular episome and addition of single nonrandom nucleotides to both coding segments. *Cell*, **56**, 785.

111. George, J. F. and Schroeder, H. W. (1992). Developmental regulation of D beta reading frame and junctional diversity in T cell receptor-beta transcripts from human thymus. *J. Immunol.*, **148**, 1230.

*112. Gu, H., Forster, I., and Rajewsky, K. (1990). Sequence homologies, N sequence insertion and J_H gene utilization in $V_H D J_H$ joining: implications for the joining mechanism and the ontogenetic timing of Ly1 B cell and B-CLL progenitor generation. *EMBO J.*, **9**, 2133.

*113. Feeney, A. J. (1990). Lack of N regions in fetal and neonatal mouse immunoglobulin V–D–J junctional sequences. *J. Exp. Med.*, **172**, 1377.

114. Feeney, A. J. (1992). Comparison of junctional diversity in the neonatal and adult immunoglobulin repertoires. *Int. Rev. Immunol.*, **8**, 113.

115. Aguilar, L. K. and Belmont, J. W. (1991). Vγ3 T cell receptor rearrangement and expression in the adult thymus. *J. Immunol.*, **146**, 1348.

116. Bogue, M., Gilfillan, S., Benoist, C., and Mathis, D. (1992). Regulation of N-region diversity in antigen receptors through thymocyte differentiation and thymus ontogeny. *Proc. Natl Acad. Sci. USA*, **89**, 11 011.

*117. Asarnow, D. M., Cado, D., and Raulet, D. H. (1993). Selection is not required to produce invariant T-cell receptor γ-gene junctional sequences. *Nature*, **362**, 158.

118. Itohara, S., Mombaerts, P., Lafaille, J., Iacomini, J., Nelson, A., Clarke, A. R., Hooper, M. L., Farr, A., and Tonegawa, S. (1993). T cell receptor delta gene mutant mice: independent generation of alpha beta T cells and programmed rearrangements of gamma delta TCR genes. *Cell*, **72**, 337.

*119. Allison, J. P. and Havran, W. L. (1991). The immunobiology of T cells with invariant γδ antigen receptors. *Annu. Rev. Immunol.*, **9**, 679.

120. Ichihara, Y., Matsuoka, H., and Kurosawa, Y. (1988). Organization of human immunoglobulin heavy chain diversity gene loci. *EMBO J.*, **7**, 4141.

121. Raulet, D. H., Spencer, D. M., Hsiang, Y.-H., Goldman, J. P., Bix, M., Liao, N.-S., Zulstra, M., Jaenisch, R., and Correa, I. (1991). Control of γδ T-cell development. *Immunol. Rev.*, **120**, 185.

122. Gerstein, R. and Lieber, M. R. (1993). Extent to which homology can constrain coding exon junctional diversity in V(D)J recombination. *Nature*, **363**, 625.

123. Kienker, L. J., Kuziel, W. A., and Tucker, P. W. (1991). T cell receptor γ and δ gene junctional sequences in SCID mice: excessive P nucleotide insertion. *J. Exp. Med.*, **174**, 769.

124. Schuler, W., Ruetsch, N. R., Amsler, M., and Bosma, M. J. (1991). Coding joint formation of endogenous T cell receptor genes in lymphoid cells from scid mice: unusual P-nucleotide additions in VJ-coding joints. *Eur. J. Immunol.*, **21**, 589.

125. Taccioli, G., Cheng, H.-L., Varghese, A. J., Whitmore, G., and Alt, F. W. (1994). A DNA repair defect in Chinese hamster ovary cells affects V(D)J recombination similarly to the murine *scid* mutation. *J. Biol. Chem.*, **269**, 7439.

126. Takeda, S., Zou, Y.-R., Bluethmann, H., Kitamura, D., Muller, U., and Rajewsky, K. (1993). Deletion of the immunoglobulin κ chain intron enhancer abolishes κ chain gene rearrangement in *cis* but not λ chain gene rearrangement in trans. *EMBO J.*, **12**, 2329.

*127. Schatz, D. G., Oettinger, M. A., and Baltimore, D. (1989). The V(D)J recombination activating gene, *Rag-1*. *Cell*, **59**, 1035.

128. Wang, J. C., Caron, P. R., and Kim, R. A. (1990). The role of DNA topoisomerases in recombination and genome stability: a double-edged sword? *Cell*, **62**, 403.

129. Oettinger, M. A. (1992). Activation of V(D)J recombination by *Rag-1* and *Rag-2*. *Trends Genet.*, **8**, 413.

130. Aguilera, A. and Klein, H. L. (1990). *HPR1*, a novel yeast gene that prevents intrachromosomal homology to the *Saccharomyces cerevisiae TOP1* gene. *Mol. Cell. Biol.*, **10**, 1439.

131. Oltz, E. M., Alt, F. W., Lin, W.-C., Chen, J., Taccioli, G., Desiderio, S., and Rathbun, G. (1993). A V(D)J recombinase inducible B cell line: role of transcriptional enhancer elements in directing V(D)J recombination. *Mol. Cell. Biol.*, **13**, 6223.

132. Alt, F. W., Oltz, E. M., Young, F., Gorman, J., Taccioli, G., and Chen, J. (1992). VDJ recombination. *Immunol. Today*, **13**, 306.

133. Wayne, J., Suh, H., Misulovin, Z., Sokol, K. A., Inaba, K., and Nussensweig, M. C. (1994). A regulatory role for recombinase activating genes, *RAG-1* and *RAG-2*, in T cell development. *Immunity*, **1**, 95.

134. Chun, J. J., Schatz, D. G., Oettinger, M. A., Jaenisch, R., and Baltimore, D. (1991). The recombination activating gene-1 (*RAG-1*) transcript is present in the murine central nervous system. *Cell*, **64**, 189.

*135. Shinkai, Y., Rathbun, G., Lam, K. P., Oltz, E. M., Stewart, V., Mendelsohn, M., Charron, J., Datta, M., Young, F., Stall, A. M., and Alt, F. W. (1992). *Rag-2*-deficient mice lack mature lymphocytes owing to inability to initiate V(D)J rearrangement. *Cell*, **68**, 855.

*136. Mombaerts, P., Iacomini, J., Johnson, R. S., Herrup, K., Tonegawa, S., and Papaioannou, V. E. (1992). *Rag-1*-deficient mice have no mature B and T lymphocytes. *Cell*, **68**, 869.

*137. Mombaerts, P., Clarke, A. R., Rudnicki, M. A., Iacomini, J., Itohara, S., Lafaille, J. J., Wang, I., Ichikawa, Y., Jaenisch, R., Hooper, M. I., and Tonegawa, S. (1992). Mutations in T-cell antigen receptor genes α and β block thymocyte development at different stages. *Nature*, **360**, 225.

*138. Shinkai, Y., Koyasu, S., Nakayama, K., Murphy, K. M., Loh, D. Y., Reinherz, E. L., and Alt, F. W. (1993). Restoration of T cell development in *Rag-2*-deficient mice by functional TCR transgenes. *Science*, **259**, 822.

*139. Young, F., Ardman, B., Shinkai, Y., Lansford, R., Blackwell, T. K., Mendelsohn, M., Rolink, A., Melchers, F., and Alt, F. (1994). Influence of immunoglobulin heavy and light chain expression on B cell differentiation. *Genes Dev.*, **8**, 1043.

*140. Spanopoulou, E., Roman, C. A. J., Corcoran, L., Schlissel, M. S., Silver, D. P., Nemazee, D., Nussenzweig, M., Shinton, S. A., Hardy, R., and Baltimore, D. (1994). Functional immunoglobulin transgenes guide ordered B cell differentiation in *Rag-1* deficient mice. *Genes Dev.*, **8**, 1030.

141. Fuschiotti, P., Harindranath, N., Mage, R. G., McCormack, W. T., Dhanarajan, P., and Roux, K. H. (1993). Recombination activating genes-1 and -2 of the rabbit: cloning and characterization of germline and expressed genes. *Mol. Immunol.*, **30**, 1021.

142. Carlson, L. M., Oettinger, M. A., Schatz, D. G., Masteller, E. L., Hurley, E. A.,

McCormack, W. T., Baltimore, D., and Thompson, C. B. (1991). Selective expression of *Rag-2* in chicken B cells undergoing immunoglobulin gene conversion. *Cell*, **64**, 201.

143. Silverstone, A. E., Rosenberg, N., Baltimore, D., Sato, V., Scheid, M. P., and Boyse, E. A. (1978). Correlating terminal deoxynucleotidyl transferase and cell-surface markers in the pathway of lymphocyte ontogeny. In *Differentiation of normal and neoplastic hematopoietic cells.*, p. 433. Cold Spring Harbor Press, Cold Spring Harbor, NY.

144. Desiderio, S. V., Yancopoulos, G. D., Paskind, M., Thomas, E., Boss, M. A., Landau, N., Alt, F. W., and Baltimore, D. (1984). Insertion of N regions into heavy-chain genes is correlated with expression of terminal deoxytransferase in B cells. *Nature*, **311**, 752.

145. Landau, N. R., Schatz, D. G., Rosa, M., and Baltimore, D. (1987). Increased frequency of N-region insertion in a murine pre-B-cell line infected with a terminal deoxynucleotidyl transferase retroviral expression vector. *Mol. Cell. Biol.*, **7**, 3237.

*146. Kallenbach, S., Doyen, N., D'Andon, M. F., and Rougeon, F. (1992). Three lymphoid-specific factors account for all junctional diversity characteristic of somatic assembly of T-cell receptor and immunoglobulin genes. *Proc. Natl Acad. Sci. USA*, **89**, 2799.

147. Bosma, M. J. and Carroll, A. M. (1991). The SCID mouse mutant: definition, characterization, and potential uses. *Annu. Rev. Immunol.*, **9**, 323.

*148. Malynn, B. A., Blackwell, T. K., Fulop, G. M., Rathbun, G. A., Furley, A. J., Ferrier, P., Heinke, L. B., Phillips, R. A., Yancopoulos, G. D., and Alt, F. W. (1988). The scid defect affects the final step of the immunoglobulin VDJ recombinase mechanism. *Cell*, **54**, 453.

149. Blackwell, T. K., Malynn, B. A., Pollock, R. R., Ferrier, P., Covey, L. R., Fulop, G. M., Phillips, R. A., Yancopoulos, G. D., and Alt, F. W. (1989). Isolation of scid pre-B cells that rearrange kappa light chain genes: formation of normal signal and abnormal coding joins. *EMBO J.*, **8**, 735.

150. Carroll, A. M. and Bosma, M. J. (1991). T-lymphocyte development in scid mice is arrested shortly after the initiation of T-cell receptor δ gene recombination. *Genes Dev.*, **5**, 1357.

151. Hendrickson, E. A., Schatz, D. G., and Weaver, D. T. (1988). The scid gene encodes a *trans*-acting factor that mediates the rejoining event of Ig gene rearrangement. *Genes Dev.*, **2**, 817.

*152. Lieber, M. R., Hesse, J. E., Lewis, S., Bosma, G. C., Rosenberg, N., Mizuuchi, K., Bosma, M. J., and Gellert, M. (1988). The defect in murine severe combined immune deficiency: joining of signal sequences but not coding segments in V(D)J recombination. *Cell*, **55**, 7.

153. Ferrier, P., Covey, L. R., Li, S. C., Suh, H., Malynn, B. A., Blackwell, T. K., Morrow, M. A., and Alt, F. W. (1990). Normal recombination substrate V_H to DJ_H rearrangements in pre-B cell lines from scid mice. *J. Exp. Med.*, **171**, 1909.

154. Pennycook, J. L., Chang, Y., Celler, J., Phillips, R. A., and Wu, G. E. (1993). High frequency of normal DJ_H joints in B cell progenitors in severe combined immunodeficiency mice. *J. Exp. Med.*, **178**, 1007.

*155. Fulop, G. M. and Phillips, R. A. (1990). The SCID mutation in mice causes a general defect in DNA repair. *Nature*, **347**, 479.

156. Biedermann, K., Sun, J., Giaccia, A. J., Tosto, L. M. and Brown, J. M. (1991). *scid* Mutation in mice confers hypersensitivity to ionizing radiation and a deficiency in DNA double-strand break repair. *Proc. Natl Acad. Sci. USA*, **88**, 1394.

157. Friedberg, E. C. (1988). Deoxyribonucleic acid repair in the yeast. *Microbiol. Rev.*, **52**, 70.

158. Orr-Weaver, T. L. and Szostak, J. W. (1988). Fungal recombination. *Microbiol. Rev.*, **49**, 33.

*159. Taccioli, G. E., Rathbun, G., Oltz, E., Stamato, T., Jeggo, P. A., and Alt, F. W. (1993). Impairment of V(D)J recombination in double-strand break repair mutants. *Science*, **260**, 207.

160. Pergola, F., Zdzienicka, M. Z., and Lieber, M. R. (1993). V(D)J recombination in mammalian cell mutants defective in DNA double-strand break repair. *Mol. Cell. Biol.*, **13**, 3464.

*161. Taccioli, G. E., Gottleib, T. M., Blunt, T., Priestley, A., Demengoet, J., Mizuta, R., Lehmann, A. R., Alt, F. W., Jackson, S., and Jeggo, P. A. (1994). Ku80 is the product of the *XRCC5* gene: A direct link with DNA repair and V(D)J recombination. *Science*, **265**, 1442.

162. Rathmell, W. K. and Chu, G. (1994). A DNA end-binding factor involved in double-strand break repair and V(D)J recombination *Mol. Cell. Biol.*, **14**, 4741.

163. Getts, R. C. and Stamato, T. D. (1994). Absence of a Ku-like DNA end binding activity in the *xrs* double-strand DNA repair-deficient mutant. *J. Biol. Chem.*, **269**, 15981.

164. Cai, Q.-Q. (1994). Chromosomal location and expression of the genes coding for Ku p70 and p80 in human cell lines and normal tissues. *Cytogenet. Cell Genet.*, **65**, 221.

165. Hafezparast, M., Kaur, G. P., Zdzienicka, M., Athwal, R. S., Lehmann, A. R., and Jeggo, P. A. (1993). Subchromosomal localization of a gene (*XRCC5*) involved in double strand break repair to the region 2q34–36. *Somat. Cell Molec. Genet.*, **19**, 413.

166. Gottlieb, T. M. and Jackson, S. P. (1993). The DNA-dependent protein kinase: requirement for DNA ends and association with Ku antigen. *Cell*, **72**, 131.

167. Dvir, A., Peterson, S. R., Knuth, M. W., Lu, H., and Dynan, W. S. (1992). Ku auto antigen is the regulatory component of a template-associated protein kinase that phosphorylates RNA polymerase II. *Proc. Natl Acad. Sci. USA*, **89**, 11920.

168. Feeney, A. J. and Riblet, R. (1993). DST4: a new, and probably the last, functional D_H gene in the BALB/c mouse. *Immunogenetics*, **37**, 217.

169. Livant, D., Blatt, C., and Hood, L. (1986). One heavy chain variable region gene segment subfamily in the BALB/c mouse contains 500–1000 or more members. *Cell*, **47**, 461.

*170. Broduer, P. H., Osman, G. E., Mackle, J. J., and Lalor, T. M. (1988). The organization of the mouse Igh-V locus. *J. Exp. Med.*, **168**, 2261.

171. Winter, E., Krawinkel, U., and Radbruch, A. (1987). Directed Ig class switch recombination in activated murine B cells. *EMBO J.*, **6**, 1663.

172. Dildrop, R., Krawinkel, U., Winter, E., and Rajewsky, K. (1985). V_H-gene expression in murine lipopolysaccharide blasts distributes over the nine known V_H groups and may be random. *Eur. J. Immunol.*, **15**, 1154.

173. Kofler, R., Geley, S., Kofler, H., and Helmberg, A. (1992). Mouse variable region gene families: complexity, polymorphism and use in non-autoimmune responses. *Immunol. Rev.*, **128**, 5.

174. Tutter, A., Brodeur, P., Shlomchik, M., and Riblet, R. (1991). Structure, map position, and evolution of two newly diverged mouse Ig V_H gene families. *J. Immunol.*, **147**, 3215.

175. Dildrop, R., Gause, A., Muller, W., and Rajewsky, K. (1987). A new V gene expressed in lambda-2 light chains of the mouse. *Eur. J. Immunol.*, **17**, 731.

*176. Rathbun, G. A., Capra, J. D., and Tucker, P. W. (1987). Organization of the murine immunoglobulin V_H complex in the inbred strains. *EMBO J.*, **6**, 2931.

177. Walter, M. A., Dosch, H. M., and Cox, D. W. (1991). A deletion map of the human immunoglobulin heavy chain variable region. *J. Exp. Med.*, **174**, 335.

178. Pennell, C. A., Sheehan, K. M., Brodeur, P. H., and Clarke, S. H. (1989). Organization and expression of VH gene families preferentially expressed by Ly−1+ (CD5) B cells. *Eur. J. Immunol.*, **19**, 2115.

179. Blankenstein, T. and Krawinkel, U. (1987). Immunoglobulin V_H genes of the mouse are organized in overlapping clusters. *Eur. J. Immunol.*, **17**, 1351.

180. Reth, M. G., Jackson, S., and Alt, F. W. (1986). V_HDJ_H formation and DJ_H replacement during pre-B differentiation: Non-random usage of gene segments. *EMBO J.*, **5**, 2131.

181. Rathbun, G., Berman, J., Yancopoulos, G., and Alt, F. W. (1989). Organization and expression of the mammalian heavy-chain variable-region locus. In *Immunoglobulin genes* (ed. T. Honjo, F. W. Alt, and T. H. Rabbits), p. 63. Academic Press, New York.

182. Rathbun, G. A., Otani, F., Milner, E. C. B., Capra, J. D., and Tucker, P. W. (1988). Molecular characterization of the A/J J558 family of heavy chain variable region gene segments. *J. Mol. Biol.*, **202**, 383.

183. Krawinkel, U., Christoph, T., and Blankenstein, T. (1989). Organization of the Ig V_H locus in mice and humans. *Immunol. Today*, **10**, 339.

184. Kemp, D. J., Tyler, B., Bernard, O., Gough, N., Gerondakis, S., Adams, J. M., and Cory, S. (1981). Organization of genes and spacers within the mouse immunoglobulin V_H locus. *J. Mol. Appl. Gen.*, **1**, 245.

185. Blankenstein, T., Bonhomme, F., and Krawinkel, U. (1987). Evolution of pseudogenes in the immunoglobulin V_H-gene family of the mouse. *Immunogenetics*, **26**, 237.

186. Givol, D., Zakut, R., Effron, K., Rechavi, G., Ram, D., and Cohen, J. B. (1981). Diversity of germline immunoglobulin V_H genes. *Nature*, **292**, 426.

187. Cohen, J. B. and Givol, D. (1983). Conservation and divergence of immunoglobulin V_H pseudogenes. *EMBO J.*, **2**, 1795.

188. Schiff, C., Milili, M., and Fougereau, M. (1985). Functional and pseudogenes are similarly organized and may equally contribute to the extensive antibody diversity of the IgV_HII family. *EMBO J.*, **2**, 1795.

189. Loh, D., Bothwell, A. L. M., White-Scharf, M., Imanishi-Kari, T., and Baltimore, D. (1983). Molecular basis of a mouse strain-specific anti-hapten response. *Cell*, **33**, 85.

190. Schreier, P. H., Bothwell, A. L. M., Mueller-Hill, B., and Baltimore, D. (1981). Multiple differences between the nucleic acid sequences of the $IgG2a^a$ and $IgG2a^b$ alleles of the mouse. *Proc Natl Acad. Sci. USA*, **78**, 4495.

191. Clarke, S. J., Claflin, J. L., and Rudikoff, S. (1982). Polymorphisms in immunoglobulin heavy chains suggesting gene conversion. *Proc. Natl Acad. Sci. USA*, **79**, 3280.

192. Bentley, D. L. and Rabbits, T. (1983). Evolution of immunoglobulin V genes: Evidence indicating that recently duplicated human $V_κ$ sequences have diverged by gene conversion. *Cell*, **32**, 181.

193. Rechavi, G., Ram, D., Glazer, L., Zakut, R., and Givol, D. (1983). Evolutionary aspects of immunoglobulin heavy chain variable region (V_H) gene subgroups. *Proc. Natl Acad. Sci. USA*, **80**, 855.

194. Chen, P. P., Kabat, E. A., Wu, T. T., Fong, S., and Carson, D. (1985). Possible involvement of human D minigenes in the first complementarity-determining region of κ light chains. *Proc. Natl Acad. Sci. USA*, **82**, 2125.

195. Aikimenko, M.-A., Marlame, B., and Rougeon, F. (1986). Evolution of the immuno-globulin κ light chain locus in the rabbit: Evidence for differential gene conversion events. *Proc. Natl Acad. Sci. USA*, **83**, 5180.

196. Berman, J. E. and Alt, F. W. (1990). Human heavy chain variable region gene diversity, organization, and expression. *Int. Rev. Immunol.*, **5**, 203.

197. Wu, G. E., Atkinson, M. J., Ramsden, D. A., and Paige, C. J. (1990). V_H gene repertoire. *Sem. Immunol.*, **2**, 207.

198. Siebenlist, U., Ravetch, J. V., Korsmeyer, D., Waldman, T., and Leder, P. (1981). Human immunoglobulin D segments encoded in tandem multigenic families. *Nature*, **294**, 631.

199. Schroeder Jr, H. W., Walter, M. A., Hofker, M. H., Ebens, A., Van Dijk, K. W., Liao, L. C., Cox, D. W., Milner, E. C. B., and Perlmutter, R. M. (1988). Physical linkage of a human immunoglobulin heavy chain variable region gene segment to diversity and joining region elements. *Proc. Natl Acad. Sci. USA*, **85**, 8196.

200. Matsuda, F., Shin, E. K., Hirabayashi, Y., Nagaoka, H., Yoshida, M. C., Zong, S. Q., and Honjo, T. (1990). Organization of variable region segments of the human immunoglobulin heavy chain: duplication of the D5 cluster within the locus and inter-chromosomal translocation of variable region segments. *EMBO J.*, **9**, 2501.

201. Matsuda, F., Lee, K. H., Nakai, S., Sato, T., Kodaira, M., Zong, S. Q., Ohno, H., Fukuhara, S., and Honjo, T. (1988). Dispersed localization of D segments in the human immunoglobulin heavy-chain locus. *EMBO J.*, **7**, 1047.

202. Buluwela, L., Albertson, D. G., Sherrington, P., Rabbitts, P. H., Spurr, N., and Rabbitts, T. H. (1988). The use of chromosomal translocations to study human immunoglobulin gene organization: mapping D_H segments within 35 kb of the C_μ gene and identification of a new D_H locus. *EMBO J.*, **7**, 2003.

203. Meek, K. D., Hasemann, C. A., and Capra, J. D. (1989). Novel rearrangements at the immunoglobulin D locus. Inversions and fusions add to IgH somatic diversity. *J. Exp. Med.*, **170**, 39.

204. Rechavi, G., Bienz, B., Ram, D., Ben-Neriah, Y., Cohen, J. B., Zakut, R., and Givol, D. (1982). Organization and evolution of immunoglobulin V_H gene subgroups. *Proc. Natl Acad. Sci. USA*, **79**, 4405.

205. Kodaira, M., Kinashi, T., Umemura, I., Matsuda, F., Noma, T., Ono, Y., and Honjo, T. (1986). Organization and evolution of variable region genes of the human immunoglobulin heavy chain. *J. Mol. Biol.*, **190**, 529.

206. Lee, K. H., Matsuda, F., Kinashi, T., Kodaira, M., and Honjo, T. (1987). A novel family of variable region genes of the human immunoglobulin heavy chain. *J. Mol. Biol.*, **195**, 761.

207. Shen, A., Humphries, C., Tucker, P., and Blattner, F. (1987). Human heavy-chain variable region gene family nonrandomly rearranged in familial chronic lymphocytic leukemia. *Proc. Natl Acad. Sci. USA*, **84**, 8563.

*208. Berman, J. E., Mellis, S. J., Pollock, R., Smith, C. L., Suh, H., Heinke, B., Kowal, C., Surti, U., Chess, L., Cantor, C. R., and Alt, F. W. (1988). Content and organization of the human Ig V_H locus: definition of three new V_H families and linkage to the IgH locus. *EMBO J.*, **7**, 727.

*209. Matsuda, F., Shin, E. K., Nagoka, H., Matsumara, R., Haino, M., Fukita, Y., Takaishi, S., Imai, T., Riley, J. H., Anand, R., and Honjo, T. (1993). Structure and physical map of 64 variable segments in the 3' 0.8-megabase region of the human immunoglobulin heavy-chain locus. *Nature Genet.*, **3**, 88.

210. Tomlinson, I. M., Walter, G., Marks, J. D., Llewelyn, M. B., and Winter, G. (1992). The repertoire of human germline V_H sequences reveals about fifty groups of V_H segments with different hypervariable loops. *J. Mol. Biol.*, **227**, 776.

211. Walter, M. A., Surti, U., Hofker, M. H., and Cox, D. W. (1990). The physical organization of the human immunoglobulin heavy chain gene complex. *EMBO J.*, **9**, 3303.

212. van Dijk, K. W., Mortari, F., Kirkham, P. M., Schroeder, H. W., and Milner, E. C. (1993). The human immunoglobulin V_H7 segment gene family consists of a small, polymorphic group of six to eight gene segments dispersed throughout the V_H locus. *Eur. J. Immunol.*, **23**, 832.

213. Capra, J. D. and Tucker, P. W. (1989). Human immunoglobulin heavy chain genes. *J. Biol. Chem.*, **264**, 12745.

214. Shin, E. K., Matsuda, F., Nagaoka, H., Fukita, Y., Imai, T., Yokoyama, K., Soeda, E., and Honjo, T. (1991). Physical map of the 3' region of the human immunoglobulin heavy chain locus: clustering of autoantibody-related variable segments in one haplotype. *EMBO J.*, **10**, 3641.

215. Schutte, M. E. M., Ebeling, S. B., Akkermans-Koolhaas, K. E., and Logtenberg, T. (1992). Deletion mapping of IgVH gene segments expressed in human CD5 B cell lines: J_H proximity is not the sole determinant of the restricted fetal V_H repertoire. *J. Immunol.*, **149**, 3963.

216. van Dijk, K. W., Milner, L. A., Sasso, E. H., and Milner, E. C. (1992). Chromosomal organization of the heavy chain variable region gene segments comprising the human fetal antibody repertoire. *Proc. Natl Acad. Sci. USA*, **89**, 10430.

217. van Dijk, K. W., Milner, L. A., and Milner, E. C. (1992). Mapping of human H chain V region genes (V_H4) using deletional analysis and pulsed field gel electrophoresis. *J. Immunol.*, **148**, 2923.

218. van der Maarel, S., van Djik, K. W., Alexander, C. M., Sasso, E. H., Bull, A., and Milner, E. C. B. (1993). Chromosomal organization of the human V_H4 gene family. Location of individual gene segments. *J. Immunol.*, **150**, 2858.

219. Thomlinson, I. M., Walter, G., Marks, J. D., Llewlyn, M. B., and Winter, G. (1992). The repertoire of human V_H sequences reveals about fifty groups of V_H segments with different hypervariable loops. *J. Mol. Biol.*, **227**, 776.

220. Cory, S., Tyler, B. M., and Adams, J. (1981). Sets of immunoglobulin V_κ genes homologous to ten cloned sequences: Implications for the number of germline V_κ genes. *J. Mol. Appl. Genet.*, **1**, 103.

*221. Zachau, H. G. (1989). Immunoglobulin light-chain genes of the κ type in man and mouse. In *Immunoglobulin genes* (ed. T. Honjo, F. W. Alt, and T. H. Rabbitts), p. 91. Academic Press, New York.

222. Heinrich, G., Traunecker, A., and Tonegawa, S. (1984). Somatic mutation created diversity in the major group of mouse immunoglobulin kappa light chains. *J. Exp. Med.*, **159**, 417.

223. Hieter, P., Maizel, J., and Leder, P. (1982). Evolution of human immunoglobulin κ J region genes. *J. Biol. Chem.*, **257**, 1516.

224. Pech, M., Jaenischen, H. R., Pohlenz, H. D., Neumaier, P. S., Klobeck, H. G., and Zachau, H. (1984). Organization and evolution of a gene cluster for human immunoglobulin variable regions of the kappa type. *J. Mol. Biol.*, **146**, 189.

225. Pech, M., Smola, H., Pohlenz, H. D., Straubinger, B., Gerl, R., and Zachau, H. (1985). A large selection of the gene locus encoding human immunoglobulin variable region of the kappa type is duplicated. *J. Mol. Biol.*, **183**, 291.

226. Pargent, W., Meindl, A., Thiebe, R., Mitzel, S., and Zachau, H. G. (1991). The immunoglobulin κ locus. Characterization of the duplicated O regions. *Eur. J. Immunol.*, **21**, 1821.

227. Lautner-Rieske, A., Huber, C., Meindl, A., Pargent, W., Schable, K. F., Thiebe, R., Zocher, I., and Zachau, H. G. (1992). The human immunoglobulin κ locus. Characterization of the duplicated A regions. *Eur. J. Immunol.*, **22**, 1023.

228. Weichhold, G. M., Klobeck, H. G., Ohnheiser, R., Combriato, G., and Zachau, H. G. (1990). Megabase inversions in the human genome as physiological events. *Nature*, **347**, 90.

229. Reilly, E. B., Blomberg, B., Imanishi-Kari, T., Tonegawa, S., and Eisen, H. (1984). Restricted association of V, J and C gene segments for mouse λ light chains. *Proc. Natl Acad. Sci. USA*, **81**, 2484.

230. Sanchez, P. and Cazenave, P. A. (1987). A new variable region in mouse immunoglobulin lambda light chains. *J. Exp. Med.*, **166**, 265.

231. Selsing, E., Durdik, J., Moore, M. W., and Persiani, D. M. (1989). Immunoglobulin λ genes. In *Immunoglobulin genes* (ed. T. Honjo, F. W. Alt, and T. H. Rabbits), p. 111. Academic Press, New York.

232. Scott, C. L., Mushinski, J. F., Huppi, K., Weigert, M., and Potter, M. (1982). Amplification of immunoglobulin λ light chains in populations of wild mice. *Nature*, **300**, 757.

233. Miller, J., Bothwell, A., and Storb, U. (1981). Physical linkage of the constant region genes for immunoglobulin lambda I and lambda III. *Proc. Natl Acad. Sci. USA*, **78**, 3829.

234. Miller, J., Ogden, S., McMullen, M., Andres, H., and Storb, U. (1988). The order and orientation of mouse lambda-genes explain lambda rearrangement patterns. *J. Immunol.*, **141**, 2497.

235. Storb, U., Haasch, D., Arp, B., Cazanave, P. A., and Miller, J. (1989). Physical linkage of mouse λ genes by pulsed-field gel electrophoresis suggests that the rearrangement process favors proximate target sequences. *Mol. Cell. Biol.*, **9**, 711.

236. Persiani, D. M., Durdick, J., and Selsing, E. (1987). Active lambda and kappa antibody gene rearrangement in Abelson murine leukemia virus transformed pre-B cells. *J. Exp. Med.*, **165**, 1655.

237. Bich-Thuy, L. T. and Queen, C. (1989). An enhancer associated with the mouse immunoglobulin lambda λ gene is specific for lambda light chain producing cells. *Nucleic Acids Res.*, **17**, 5307.

238. Eccles, S., Sarner, N., Vidal, M., Cox, A., and Grosveld, F. (1990). Enhancer sequences located 3' of the mouse immunoglobulin λ locus specify high-level expression of an immunoglobulin λ gene in B cells of transgenic mice. *New Biol.*, **2**, 801.

239. Bossy, D., Milili, M., Zucman, J., Thomas, G., Fougereau, M., and Schiff, C. (1991). Organization and expression of the λ-like genes that contribute to the m–x light chain complex in human pre-B cells. *Int. Immunol.*, **3**, 1081.

240. Hieter, P. A., Hollis, G. F., Korsmeyer, S. J., Waldman, T. A., and Leder, P. (1981). Clustered arrangement of immunoglobulin λ constant region genes in man. *Nature*, **294**, 536.

*241. Vasicek, T. J. and Leder, P. (1990). Structure and expression of the human immunoglobulin λ genes. *J. Exp. Med.*, **172**, 609.

242. Combriato, G. and Klobeck, H. G. (1991). V_λ and J_λ–C_λ gene segments of the human immunoglobulin λ light chain locus are separated by 14 kb and rearranged by a deletion mechanism. *Eur. J. Immunol.*, **21**, 1513.

243. Evans, R. J. and Hollis, G. F. (1991). Genomic structure of the human Igλ1 gene

suggests that it may be expressed as an Igλ14.1-like protein or as a canonical B cell Igλ light chain: Implications for Igλ gene evolution. *J. Exp. Med.*, **173**, 305.

244. Guglielmi, P. and Davi, F. (1991). Expression of a novel type of immunoglobulin C lambda transcripts in human mature B lymphocytes producing kappa light chains. *Eur. J. Immunol.*, **21**, 501.

245. Chang, H., Dmitrovsky, E., Hieter, P. A., Mitchell, K., Leder, P., Turoczi, L., Kirsch, I. R., and Hollis, G. F. (1986). Identification of three new Ig λ-like genes in man. *J. Exp. Med.*, **163**, 425.

246. Bauer, S. K., Kudo, A., and Melchers, F. (1988). Structure and pre-B lymphocyte restriction of the VpreB gene in humans and conservation of its structure in other mammalian species. *EMBO J.*, **7**, 111.

247. Hollis, G. F., Evans, R. J., Stafford-Hollis, J. M., Korsmeyer, S. J., and McKearn, J. P. (1989). Immunoglobulin lambda light-chain-related genes 14.1 and 16.1 are expressed in pre-B cells and may encode the human immunoglobulin omega light-chain protein. *Proc. Natl Acad. Sci. USA*, **86**, 5552.

248. Kerr, W. G., Cooper, M. D., Feng, L., Burrows, P. D., and Hendershot, L. M. (1989). Mu heavy chains can associate with a pseudo light chain complex. *Int. Immunol.*, **1**, 355.

249. Kudo, A. and Melchers, F. (1987). A second gene, VpreB in the λ5 locus of the mouse, which appears to be selectively expressed in pre-B lymphocytes. *EMBO J.*, **6**, 2267.

250. Lassoued, K., Nunez, C. A., Billips, L., Kubagawa, H., Montiero, R. C., LeBien, T. W., and Cooper, M. D. (1993). Expression of surrogate light chain receptors is restricted to a late stage in pre-B cell differentiation. *Cell*, **73**, 73.

*251. Kitamura, D., Kudo, A., Schaal, S., Muller, W., Melchers, F., and Rajewsky, K. (1992). A critical role of lambda 5 protein in B cell development. *Cell*, **69**, 823.

*252. Melchers, F., Karasuyama, H., Haasner, D., Bauer, S., Kudo, A., Sakaguchi, N., Jameson, B., and Rolink, A. (1993). The surrogate light chain in B-cell development. *Immunol. Today*, **14**, 60.

253. Schiff, C., Bensmana, M., Guglielmi, P., Milili, M., Lefranc, M. P., and Fougereau, M. (1990). The immunoglobulin lambda-like gene cluster (14.1, 16.1, and F lambda 1) contains gene(s) selectively expressed in pre-B cells and is the human counterpart of the mouse lambda 5 gene. *Int. Immunol.*, **2**, 201.

*254. Sakaguchi, N. and Melchers, F. (1986). Lambda 5, a new light-chain-related locus selectively expressed in pre-B lymphocytes. *Nature*, **324**, 579.

255. Nishimoto, N., Kubagawa, H., Ohno, T., Gartland, G. L., Stankovic, A. K., and Cooper, M. D. (1991). Normal pre-B cells express a receptor complex of mu heavy chains and surrogate light chain proteins. *Proc. Natl Acad. Sci. USA*, **88**, 6284.

256. Kudo, A., Sakaguchi, N., and Melchers, F. (1987). Organization of the murine Ig-related λ5 gene transcribed selectively in pre-B lymphocytes. *EMBO J.*, **6**, 103.

257. Pillai, S. and Baltimore, D. (1987). Formation of disulphide-linked mu2 omega2 tetramers in pre-B cells by the 18k omega immunoglobulin light chain. *Nature*, **329**, 172.

258. Tsubata, T. and Reth, M. (1990). The products of pre-B cell-specific genes (lambda 5 and VpreB) and the immunoglobulin mu chain form a complex that is transported on to the cell-surface. *J. Exp. Med.*, **172**, 973.

259. Karasuyama, H., Kudo, A., and Melchers, F. (1990). The proteins encoded by the VpreB and lambda 5 pre-B cell-specific genes can associate with each other and with mu heavy chain. *J. Exp. Med.*, **172**, 969.

260. Cherayil, B. J. and Pillai, S. (1991). The omega/lambda 5 surrogate immunoglobulin light chain is expressed on the surface of transitional B lymphocytes in murine bone marrow. *J. Exp. Med.*, **173**, 111.

261. Clarke, C., Berenson, J., Goverman, J., Boyer, P. D., Crews, S., Siv, G., and Calame, K. (1982). An immunoglobulin promoter region is unaltered by DNA rearrangement and somatic mutation during B cell development. *Nucleic Acids Res.*, **10**, 7731.

262. Kataoka, T., Nikaido, T., Miyata, T., Morivaki, K., and Honjo, T. (1982). The nucleotide sequence of rearranged and germline immunoglobulin V_H genes of a mouse myeloma MC101 and evolution of V_H genes in mouse. *J. Biol. Chem.*, **257**, 277.

263. Kemp, D. J., Morahan, G., Cowman, A. F., and Harris, A. W. (1983). Production of RNA for secreted immunoglobulin μ chains does not require transcriptional termination 5' to the μ_m exons. *Nature*, **301**, 84.

264. Early, P., Rogers, J., Davis, M., Calame, K., Bond, M., Wall, R., and Hood, L. (1980). Two mRNAs can be produced from a single immunoglobulin μ gene by alternative processing pathways. *Cell*, **20**, 313.

265. Yuan, D. and Tucker, P. W. (1984). Transcriptional regulation of the mu-delta heavy chain locus in normal murine B lymphocytes. *J. Exp. Med.*, **160**, 564.

266. Kuziel, W. A., Word, C. J., Yuan, D., White, M. B., Mushinski, J. F., Blattner, F. R., and Tucker, P. W. (1989). The human immunoglobulin C mu-C delta locus: regulation of mu and delta RNA expression during B cell development. *Int. Immunol.*, **1**, 310.

267. Kelley, D. E., Coleclough, C., and Perry, R. P. (1982). Functional significance and evolutionary development of the 5' terminal regions of immunoglobulin variable-region genes. *Cell*, **29**, 681.

268. Kuehl, W. M., Kaplan, B. A., Scharff, M. D., Nau, M., Honjo, T., and Leder, P. (1975). Characterization of light chain and light chain constant region fragment mRNAs in MPC11 mouse myelomas and variants. *Cell*, **5**, 139.

269. Alt, F. W., Enea, V., Bothwell, A. L. M., and Baltimore, D. (1980). Activity of multiple light chain genes in murine myeloma lines expressing a single, functional light chain. *Cell*, **21**, 1.

270. Alt, F. W., Rosenberg, N., Enea, V., Siden, E., and Baltimore, D. (1982). Multiple immunoglobulin heavy-chain gene transcripts in Abelson murine leukemia virus-transformed lymphoid cell lines. *Mol. Cell. Biol.*, **2**, 386.

271. Baumann, B., Potash, M. J., and Kohler, G. (1985). Consequences of frame-shift mutations at the Ig H chain locus of the mouse. *EMBO J.*, **4**, 351.

272. Jack, H. M., Berg, J., and Wabl, M. (1989). Translation affects immunoglobulin mRNA stability. *Eur. J. Immunol.*, **19**, 843.

273. Kurosawa, Y. and Tonegawa, S. (1982). Organization, structure and assembly of immunoglobulin heavy chain diversity (D) DNA segments. *J. Exp. Med.*, **155**, 201.

274. Coleclough, C. (1983). Chance, necessity, and antibody gene dynamics. *Nature*, **303**, 23.

*275. Alt, F. W., Yancopoulos, G. D., Blackwell, T. K., Wood, C., Thomas, E., Boss, M., Coffman, R., Rosenberg, N., Tonegawa, S., and Baltimore, D. (1984). Ordered re-arrangement of immunoglobulin heavy chain variable region segments. *EMBO J.*, **3**, 1209.

276. Reth, M. G., Ammirati, P., Jackson, S., and Alt, F. W. (1985). Regulated progression of a cultured pre-B cell line to the B cell stage. *Nature*, **317**, 353.

277. Pollock, B. A., Kearney, J. F., Vakail, M., and Perry, R. P. (1984). A biological consequence of variation in the site of D-to-J_H gene rearrangement. *Nature*, **31**, 376.

278. Jeske, D. J., Jarvis, J., Milstein, C., and Capra, D. (1984). Junctional diversity is essential to antibody diversity. *J. Immunol.*, **133**, 1090.

279. Azuma, T., Igras, V., Reilly, E. B., and Eisen, H. (1984). Diversity at the variable joining region boundary of λ light chains has a pronounced effect on immunoglobulin ligand-binding activity. *Proc. Natl Acad. Sci. USA*, **81**, 6139.

280. Sanz, I. and Capra, D. (1987). V$_\kappa$ and J$_\kappa$ gene segments of A/J Ars-A antibodies: Somatic recombination generates the essential arginine at the junction of the variable and joining regions. *Proc. Natl Acad. Sci. USA*, **84**, 1085.

*281. Davis, M. M. and Bjorkman, P. J. (1988). T-cell antigen receptor genes and T-cell recognition. *Nature*, **334**, 395.

282. Manser, T., Wysocki, L. J., Gridley, T., Near, R. I., and Gefter, M. (1985). The evolution of the immune response. *Immunol. Today*, **6**, 94.

283. Gonzalez-Fernandez, A. and Milstein, C. (1993). Analysis of somatic hypermutation in mouse Peyer's patches using immunoglobulin kappa light-chain transgenes. *Proc. Natl Acad. Sci. USA*, **90**, 9862.

284. Rogerson, B., Hackett, J., Peters, A., Haasch, D., and Storb, U. (1991). Mutation pattern of immunoglobulin transgenes is compatible with a model of somatic hypermutation in which targeting of the mutator is linked to the direction of DNA replication. *EMBO J.*, **10**, 4331.

285. Sharpe, M. J., Milstein, C., Jarvis, J. M., and Neuberger, M. S. (1991). Somatic hypermutation of immunoglobulin kappa may depend on sequences 3' of C kappa and occurs on passenger transgenes. *EMBO J.*, **10**, 2139.

286. Sohn, J., Gerstein, R. M., Hsieh, C. L., Lemer, M., and Selsing, E. (1993). Somatic hypermutation of an immunoglobulin mu heavy chain transgene. *J. Exp. Med.*, **177**, 493.

287. Hackett, J., Stebbins, C., Rogerson, B., Davis, M. M., and Storb, U. (1992). Analysis of a T cell receptor gene as a target of the somatic hypermutation mechanism. *J. Exp. Med.*, **176**, 225.

288. Umar, A., Schweitzer, P. A., Levy, N. S., Gearhart, J. D., and Gearhart, P. J. (1991). Mutation in a reporter gene depends on proximity to and transcription of immunoglobulin variable transgenes. *Proc. Natl Acad. Sci. USA*, **88**, 4902.

*289. Betz, A. G., Milstein, C., Gonzalez-Fernandez, A., Pannell, R., Larson, T., and Neuberger, M. S. (1994). Elements regulating somatic mutation: Critical role for the intron enhancer/matrix attachment region. *Cell*, **77**, 239.

*290. Reynaud, C. A., Anquez, V., Dahan, A., and Weill, J. C. (1985). A single rearrangement event generates most of the chicken immunoglobulin light chain diversity. *Cell*, **40**, 283.

291. Thompson, C. B. and Neiman, P. E. (1987). Somatic diversification of the chicken immunoglobulin light chain gene is limited to the rearranged variable gene segment. *Cell*, **38**, 369.

292. Reynaud, C. A., Anquez, V., Grimal, H., and Weill, J. C. (1987). A hyperconversion mechanism generates the chicken light chain preimmune repertoire. *Cell*, **48**, 379.

293. Benatar, T. and Ratcliffe, M. J. (1993). Polymorphism of the functional immunoglobulin variable region genes in the chicken by exchange of sequence with donor pseudogenes. *Eur. J. Immunol.*, **23**, 2448.

294. Egel, R. (1981). Intergenic conversion and reiterated genes. *Nature*, **290**, 191.

295. Baltimore, D. (1981). Gene conversion: some implications for immunoglobulin genes. *Cell*, **24**, 592.

296. Dildrop, R., Bruggemann, M., Radbruch, A. R., Rajewsky, K. and Beyreuther, K. (1982). Immunoglobulin V region variants in hybridoma cells. II. Recombination between V genes. *EMBO J.*, **1**, 635.

297. Krawinkel, U., Zoebelein, G., Bruggemann, M., Radbruch, A., and Rajewsky, K. (1983). Recombination between antibody heavy chain variable-region genes: Evidence for gene conversion. *Proc. Natl Acad. Sci. USA*, **80**, 4997.

298. Maizels, N. (1989). Might gene conversion be the mechanism of somatic hypermutation of mammalian immunoglobulin genes? *Trends Genet.*, **5**, 4.

299. Berek, C. (1993). Somatic mutation and memory. *Curr. Opin. Immunol.*, **5**, 218.

300. Kepler, T. B. and Perelson, A. S. (1993). Cyclic re-entry of germinal center B cells and the efficiency of affinity maturation. *Immunol. Today*, **14**, 412.

301. Thompson, C. B. (1992). Creation of immunoglobulin diversity by intrachromosomal gene conversion. *Trends Genet.*, **8**, 416.

302. Gearhart, P. J. (1993). Somatic mutation and affinity maturation. In *Fundamental immunology* (ed. W. E. Paul), p. 865. Raven Press, New York.

303. Mason, J. O., Williams, G. T., and Neuberger, M. (1985). Transcription cell type specificity is conferred by an immunoglobulin V_H gene promoter that includes a functional consensus sequence. *Cell*, **41**, 479.

304. Grosschedl, R. and Baltimore, D. (1985). Cell-type specificity of immunoglobulin gene expression is regulated by at least three DNA sequence elements. *Cell*, **41**, 885.

*305. Banerji, J., Olson, L., and Schaffner, W. (1983). A lymphocyte-specific cellular enhancer is located downstream of the joining region in immunoglobulin heavy chain genes. *Cell*, **33**, 729.

*306. Gillies, S. D., Morrison, S. L., Oi, V. T., and Tonegawa, S. (1983). A tissue-specific transcription enhancer element is located in the major intron of a rearranged immunoglobulin heavy chain gene. *Cell*, **33**, 717.

307. Neuberger, M. S. (1983). Expression and regulation of immunoglobulin heavy chain gene transfected into lymphoid cells. *EMBO J.*, **2**, 1373.

308. Wang, X.-F. and Calame, K. (1985). The endogenous immunoglobulin heavy chain enhancer can activate tandem V_H promoters separated by a large distance. *Cell*, **43**, 659.

309. Wasylyk, C. and Wasylyk, B. (1986). The immunoglobulin heavy-chain B-lymphocyte enhancer efficiently stimulates transcription in non-lymphoid cells. *EMBO J.*, **5**, 553.

310. Garcia, J. V., Bich-Thuy, L. T., Stafford, J., and Queen, C. (1986). Synergism between immunoglobulin enhancers and promoters. *Nature*, **322**, 383.

311. Sen, R. and Baltimore, D. (1989). Factors regulating immunoglobulin-gene transcription. In *Immunoglobulin genes* (ed. T. Honjo, F. W. Alt and T. H. Rabbits), p. 327. Academic Press, New York.

312. Staudt, L. and Lenardo, M. (1991). Immunoglobulin gene transcription. *Annu. Rev. Immunol.*, **9**, 373.

313. Wabl, M. and Burrows, P. D. (1984). Expression of immunoglobulin heavy chain at a high level in the absence of a proposed immunoglobulin enhancer element in *cis*. *Proc. Natl Acad. Sci. USA*, **81**, 2452.

314. Eckhardt, L. A. and Birshtein, B. K. (1985). Independent immunoglobulin class-switch events occurring in a single myeloma cell line. *Mol. Cell. Biol.*, **5**, 856.

315. Chen, J., Young, F., Bottaro, A., Stewart, V., Smith, R. K., and Alt, F. W. (1993). Mutations of the intronic IgH enhancer and its flanking sequences differentially affect accessibility of the J_H locus. *EMBO J.*, **12**, 4635.

316. Serwe, M. and Sablitzky, F. (1993). Recombination in B cells is impaired but not blocked by targeted deletion of the immunoglobulin heavy chain intron enhancer. *EMBO J.*, **12**, 2321.

317. Grosschedl, R. and Marx, M. (1988). Stable propagation of the active transcriptional state of an immunoglobulin mu gene requires continuous enhancer function. *Cell*, **55**, 645.

318. Yancopoulos, G. D. and Alt, F. W. (1985). Developmentally controlled and tissue-specific expression of unrearranged V_H gene segments. *Cell*, **40**, 271.

*319. Pettersson, S., Cook, G., P., Bruggemann, M., Williams, G. T., and Neuberger, M. S. (1990). A second B cell-specific enhancer 3' of the immunoglobulin heavy-chain locus. *Nature*, **344**, 165.

*320. Cogne, M., Lansford, R., Bottaro, A., Zhang, J., Gorman, J., Young, F., Cheng, H.-L., and Alt, F. W. (1994). A major class switch control region at the 3' end of the IgH locus. *Cell*, **77**, 737.

321. Singh, H., Sen, R., Baltimore, D., and Sharp, P. A. (1986). A nuclear factor that binds to a conserved sequence motif in transcriptional control elements of immunoglobulin genes. *Nature*, **319**, 154.

322. Queen, C. and Baltimore, D. (1983). Immunoglobulin gene transcription is activated by downstream sequence elements. *Cell*, **33**, 741.

323. Picard, D. and Schaffner, W. (1984). A lymphocyte-specific enhancer in the mouse immunoglobulin κ gene. *Nature*, **307**, 80.

*324. Meyer, K. B. and Neuberger, M. S. (1989). The immunoglobulin κ locus contains a second, stronger B-cell-specific enhancer which is located downstream of the constant region. *EMBO J.*, **8**, 1959.

325. Hagman, J., Rudin, C. M., Haasch, D., Chaplin, D., and Storb, U. (1990). A novel enhancer in the immunoglobulin λ locus is duplicated and functionally independent of NF-κB. *Genes Dev.*, **4**, 978.

326. Rudin, C. M. and Storb, U. (1992). Two conserved essential motifs of the murine immunoglobulin λ enhancers bind B-cell-specific factors. *Mol. Cell. Biol.*, **12**, 309.

327. Eisenbeis, C. F., Singh, H., and Storb, U. (1993). PU.1 is a component of a multi-protein complex which binds an essential site in the murine immunoglobulin λ_{2-4} enhancer. *Mol. Cell. Biol.*, **13**, 6452.

328. Kemp, D. J., Harris, A. W., Cory, S., and Adams, J. M. (1980). Expression of the immunoglobulin C_μ gene in mouse T and B lymphoid and myeloid cell lines. *Proc. Natl Acad. Sci. USA*, **77**, 2876.

329. Alessandrini, A. and Desiderio, S. V. (1991). Coordination of immunoglobulin DJ_H transcription and D-to-J_H rearrangement by promoter–enhancer approximation. *Mol. Cell. Biol.*, **11**, 2096.

330. Schlissel, M., Voronova, A., and Baltimore, D. (1991). Helix–loop–helix transcription factor E47 activates germ-line immunoglobulin heavy-chain gene transcription and rearrangement in a pre-T-cell line. *Genes Dev.*, **5**, 1367.

331. Lennon, G. G. and Perry, R. P. (1985). Cμ-containing transcripts initiate heterogeneously within the IgH enhancer region and contain a novel 5'-nontranslatable exon. *Nature*, **318**, 475.

332. Su, L.-K. and Kadesch, T. (1990) The immunoglobulin heavy-chain enhancer functions as the promoter for I_μ sterile transcription. *Mol. Cell. Biol.*, **10**, 2619.

333. Nelson, K. J., Haimovich, J., and Perry, R. P. (1983). Characterization of productive and sterile transcripts from the immunoglobulin heavy chain locus: Processing of μ_m and μ_s mRNA. *Mol. Cell. Biol.*, **3**, 1317.

334. Reth, M. G. and Alt, F. W. (1984). Novel immunoglobulin heavy chains are produced from DJ$_H$ gene segment rearrangement in lymphoid cells. *Nature*, **312**, 418.

335. Gu, H., Kitamura, D., and Rajewsky, K. (1991). B cell development regulated by gene rearrangement: arrest of maturation by membrane-bound D mu protein and selection of D$_H$ element reading frame. *Cell*, **65**, 47.

336. Shimizu, T. and Yamagishi, H. (1992). Biased reading frames of pre-existing D$_H$–J$_H$ coding joints and preferential nucleotide insertions at V$_H$–DJ$_H$ signal joints of excision products of immunoglobulin heavy chain gene rearrangements. *EMBO J.*, **11**, 4869.

337. Tsubata, T., Tsubata, R., and Reth, M. (1991). Cell surface expression of the short immunoglobulin mu chain (D mu protein) in murine pre-B cells is differentially regulated from that of the intact mu chain. *Eur. J. Immunol.*, **21**, 1359.

338. Weinberger, J., Baltimore, D., and Sharp, P. A. (1986). Distinct factors bind to apparently homologous sequences in the immunoglobulin heavy-chain enhancer. *Nature*, **322**, 846.

339. Reth, M. (1992). Antigen receptors on B lymphocytes. *Annu. Rev. Immunol.*, **10**, 97.

340. Nelson, K. J., Mather, E. L., and Perry, R. P. (1984). Lipopolysaccharide-induced transcription of the kappa immunoglobulin locus occurs on both alleles and is independent of methylation status. *Nucleic Acids Res.*, **12**, 1911.

341. Nelson, K. J., Mather, E. L., and Perry, R. P. (1985). Inducible transcription of the unrearranged C$_\kappa$ locus is a common feature of pre-B cells and does not require DNA or protein synthesis. *Proc. Natl Acad. Sci. USA*, **82**, 5305.

342. Van Ness, B. G., Weigert, M., Colecough, C., Mather, E. L., Kelley, D. E., and Perry, R. P. (1981). Transcription of the unrearranged mouse C$_\kappa$ locus: sequence of the initiation region and comparison of activity with a rearranged V$_\kappa$–C$_\kappa$ gene. *Cell*, **27**, 593.

*343. Schlissel, M. S. and Baltimore, D. (1989). Activation of immunoglobulin kappa gene rearrangement correlates with induction of germline kappa gene transcription. *Cell*, **58**, 1001.

344. Mather, E. L. and Perry, R. P. (1981). Transcriptional regulation of immunoglobulin V genes. *Nucleic Acids Res.*, **9**, 6855.

345. Mather, E. L. and Perry, R. P. (1983). Methylation status and DNaseI sensitivity of immunoglobulin genes: Changes associated with rearrangement. *Proc. Natl Acad. Sci. USA*, **80**, 4689.

346. Picard, D. and Schaffner, W. (1984). Unrearranged immunoglobulin L variable region is transcribed in κ-producing myelomas. *EMBO J.*, **3**, 3031.

347. Muller, B. and Reth, M. (1988). Ordered activation of the Igλ locus in Abelson B cell lines. *J. Exp. Med.*, **168**, 2131.

348. Reth, M. and Leclercq, L. (1987). Assembly of variable region gene segments. In *Comprehensive biochemistry* (ed. M. Neuberger).

349. Siden, E. J., Baltimore, D., Clark, D., and Rosenberg, N. (1979). Immunoglobulin synthesis by lymphoid cells transformed *in vitro* by Abelson murine leukemia virus. *Cell*, **16**, 389.

350. Burrows, P., LeJeune, M., and Kearney, J. F. (1979). Evidence that murine pre-B cells synthesize μ heavy chains but not light chains. *Nature*, **280**, 838.

351. Levitt, D. and Cooper, M. D. (1980). Mouse pre-B cells synthesize and secrete μ heavy chains but not light chains. *Cell*, **19**, 617.

352. Siden, E., Alt, F. W., Shinefeld, L., Sato, V. and Baltimore, D. (1981). Synthesis of immunoglobulin μ chain gene products precedes synthesis of light chains during B lymphocyte development. *Proc. Natl Acad. Sci. USA*, **78**, 1823.

*353. Alt, F., Rosenberg, N., Lewis, S., Thomas, E., and Baltimore, D. (1981). Organization and reorganization of immunoglobulin genes in A-MuLV-transformed cells: rearrangement of heavy but not light chain genes. *Cell*, **27**, 381.

354. Perry, R. P., Kelley, D. E., Coleclough, C., and Kearney, J. F. (1981). Organization and expression of immunoglobulin genes in fetal liver hybridomas. *Proc. Natl Acad. Sci. USA*, **78**, 247.

355. Perlmutter, R. M., Kearney, J. F., Chang, S. P., and Hood, L. E. (1985). Developmentally controlled expression of IgVH genes. *Science*, **227**, 1597.

356. Coffman, R. L. and Weissman, I. L. (1983). Immunoglobulin gene rearrangement during pre-B cell differentiation. *J. Mol. Cell. Immunol.*, **1**, 31.

357. Li, Y.-S., Hayakawa, K., and Hardy, R. (1993). The regulated expression of B lineage associated genes during B cell differentiation in bone marrow and fetal liver. *J. Exp. Med.*, **178**, 951.

*358. Hardy, R. R., Carmack, C. E., Shinton, S. A., Kemp, J. D., and Hayakawa, K. (1991). Resolution and characterization of pro-B and pre-pro-B cell stages in normal mouse bone marrow. *J. Exp. Med.*, **173**, 1213.

*359. Kitamura, D., Roes, J., Kuhn, R., and Rajewsky, K. (1991). A B cell-deficient mouse by targeted disruption of the membrane exon of the immunoglobulin μ chain gene. *Nature*, **350**, 423.

360. Chen, J., Trounstine, M., Alt, F. W., Young, F., Kuraha, C., Loring, J. F., and Huzsar, D. (1993). Immunoglobulin gene rearrangement in B cell deficient mice generated by targeted deletion of the J$_H$ locus. *Int. Immunol.*, **5**, 647.

*361. Ehlich, A., Schaal, S., Gu, H., Kitamura, D., Muller, W., and Rajewsky, K. (1993). Immunoglobulin heavy and light chain gene rearrange independently in large B cell precursors. *Cell*, **72**, 659.

362. Yaoita, Y., Matsunami, N., Choi, C. Y., Sugiyama, H., Kishimoto, T., and Honjo, T. (1983). The D–J$_H$ complex is an intermediate to the complete immunoglobulin heavy-chain V region gene. *Nucleic Acids Res.*, **11**, 7303.

363. Sugiyama, H., Akira, S., Kikutani, H., Kishimoto, S., Yamamura, Y., and Kishimoto, T. (1983). Functional V region formation during *in vitro* culture of a murine immature B precursor cell line. *Nature*, **303**, 812.

364. Mizutani, S., Ford, A. M., Weidemann, L. M., Chan, L. C., Furley, A. J., Greaves, M. F., and Molgaard, H. V. (1986). Rearrangement of immunoglobulin heavy chain genes in human T leukaemic cells shows preferential utilization of the D segment (DQ52) nearest to the J region. *EMBO J.*, **5**, 3467.

365. Berman, J., Nickerson, K. G., Pollack, R. R., Barth, J. E., Schuurman, K. B., Knowles, D. M., Chess, L., and Alt, F. W. (1991). V$_H$ gene usage in humans: biased usage of the V$_H$6 gene in immature B lymphoid cells. *Eur. J. Immunol.*, **21**, 1311.

366. Nickerson, K. G., Berman, J., Glickman, E., Chess, L., and Alt, F. W. (1989). Early human IgH gene assembly in Epstein–Barr virus-transformed fetal B cell lines. Preferential utilization of the most J$_H$ proximal D segment (DQ52) and two unusual V$_H$-related rearrangements. *J. Exp. Med.*, **169**, 1391.

367. Chang, Y., Paige, C. J., and Wu, G. E. (1992). Enumeration and characterization of DJ$_H$ structures in mouse fetal liver. *EMBO J.*, **11**, 1891.

368. Shin, E. K., Matsuda, F., Fujikura, J., Akamizu, T., Sugawa, H., Mori, T., and Honjo, T. (1993). Cloning of a human immunoglobulin gene fragment containing both V$_H$–D and D–J$_H$ rearrangements: implication for V$_H$–D as an intermediate to V$_H$–D–J$_H$ formation. *Eur. J. Immunol.*, **23**, 2365.

*369. Nussenzweig, M. C., Shaw, A. C., Sinn, E., Danner, D. B., Holmes, K. I., Morse, H. C., and Leder, P. (1987). Allelic exclusion in transgenic mice that express membrane form of immunoglobulin μ. *Science*, **236**, 816.

370. Ritchie, K. A., Brinster, R. L., and Storb, U. (1984). Allelic exclusion and control of endogenous immunoglobulin gene rearrangement in κ transgenic mice. *Nature*, **312**, 517.

371. Reth, M., Petrac, E., Wiese, P., Lobel, L., and Alt, F. W. (1987). Activation of V kappa gene rearrangement in pre-B cells follows the expression of membrane-bound immunoglobulin heavy chains. *EMBO J.*, **6**, 3299.

372. Iglesias, A., Kopf, M., Williams, G. S., Buhler, B., and Kohler, G. (1991). Molecular requirements for the μ-induced light chain gene rearrangement in pre-B cells. *EMBO J.*, **10**, 2147.

373. Era, T., Ogawa, M., Nishikawa, S.-I., Okamoto, M., Honjo, T., Akagi, K., Miyazaki, J.-I., and Yamamura, K.-I. (1991). Differentiation of growth signal requirement of B lymphocyte precursor is directed by expression of immunoglobulin. *EMBO J.*, **10**, 337.

374. Kubagawa, H., Cooper, M. D., Carroll, A. J., and Burrows, P. D. (1989). Light-chain gene expression before heavy-chain gene rearrangement in pre-B cells transformed by Epstein–Barr virus. *Proc. Natl Acad. Sci. USA*, **86**, 2350.

375. Felsher, D. W., Ando, D. T., and Braun, J. (1991). Independent rearrangement of Igλ genes in tissue culture-derived murine B cell lines. *Int. Immunol.*, **3**, 711.

*376. Kitamura, D., and Rajewsky, K. (1992). Targeted disruption of μ chain membrane exon causes loss of heavy-chain allelic exclusion. *Nature*, **356**, 154.

377. Hieter, P., Korsmeyer, S., Waldmann, T., and Leder, P. (1981). Human immunoglobulin κ light chain genes are deleted or rearranged in λ-producing cells. *Nature*, **290**, 368.

378. Coleclough, C., Perry, R. P., Karjalainen, K., and Weigert, M. (1981). Aberrant rearrangements contribute significantly to the allelic exclusion of immunoglobulin gene expression. *Nature*, **290**, 372.

379. Lewis, S., Rosenberg, N., Alt, F., and Baltimore, D. (1982). Continuing kappa-gene rearrangement in a cell line transformed by Abelson murine leukemia virus. *Cell*, **30**, 807.

380. McIntire, K. R. and Rouse, A. M. (1970). Mouse Ig light chains alteration of κ/λ ratio. *Fed. Proc.*, **29**, 704.

*381. Chen, J., Trounstine, M., Huraha, C., Young, F., Kuo, C. C., Xu, Y., Loring, J. F., Alt, F. W., and Huzsar, D. (1993). B cell development in mice that lack one or both immunoglobulin κ light chain genes. *EMBO J.*, **12**, 821.

*382. Zou, Y.-R., Takeda, S., and Rajewsky, K. (1993). Gene targeting in the Igκ locus: efficient generation of λ expressing B cells independent of gene rearrangement in Igκ. *EMBO J.*, **12**, 811.

383. Jakobovits, A., Vergara, G. J., Kennedy, J. L., Hales, J. F., McGuinness, R. P., Casentini-Borocz, D. E., Brenner, D. G., and Otten, G. R. (1993). Analysis of homozygous mutant chimera mice: Deletion of the immunoglobulin heavy-chain joining region block B-cell development and antibody production. *Proc. Natl Acad. Sci. USA*, **90**, 2551.

*384. Chen, J., Lansford, R., Stewart, V., Young, F., and Alt, F. W. (1993). *Rag-2*-deficient blastocyst complementation: An assay of gene function in lymphocyte development. *Proc. Natl Acad. Sci. USA*, **90**, 4528.

385. Rajewsky, K. (1992). Early and late B-cell development in the mouse. *Curr. Op. Immunol.*, **4**, 171.

386. Chen, J., Shinkai, Y., Young, F., and Alt, F. W. (1994). Probing immune functions in *Rag*-deficient mice. *Curr. Biol.*, **6**, 313.

387. Pillai, S., and Baltimore, D. (1988). The omega and iota surrogate immunoglobulin light chains (psi L) in human pre-B cells. *Curr. Top. Immunol.*, **137**, 136.

388. Sakaguchi, N., Kashiwamura, S. I., Kimoto, M., Thalmann, P., and Melchers, F. (1988). B lymphocyte-lineage-restricted expression of mb-1, a gene with CD3-like structural properties. *EMBO J.*, **7**, 3457.

389. Hombach, J., Leclercq, L., Radbruch, A., Rajewsky, K., and Reth, M. (1988). A novel 34 kd protein co-isolated with the IgM molecule in surface IgM-expressing cells. *EMBO J.*, **7**, 3451.

390. Hermanson, G. G., Eisenberg, D., Kincade, P. W., and Wall, R. (1988). A member of the immunoglobulin gene superfamily exclusively expressed on B-lineage cells. *Proc. Natl Acad. Sci. USA*, **85**, 6890.

*391. Reth, M. (1989). Antigen receptor tail clue. *Nature*, **338**, 383.

392. Karasuyama, H., Rolink, A., Shinkai, Y., Young, F., Alt, F. W., and Melchers, F. (1994). The expression of Vpre-B/λ5 surrogate light chain in early bone marrow precursor B cells of normal and B cell-deficient mutant mice. *Cell*, **77**, 133.

393. Karasuyama, H., Rolink, A., and Melchers, F. (1993). A complex of glycoproteins is associated with VpreB/15 surrogate light chain on the surface of μ heavy chain-negative early precursor B cell lines. *J. Exp. Med.*, **178**, 469.

394. Shinjo, F., Hardy, R. R., and Jongstra, J. (1994). Monoclonal anti-λ5 antibody FS1 identifies a 130 kDa protein associated with λ5 and VpreB on the surface of early pre-B cell lines. *Int. Immunol.*, **6**, 393.

395. Alt, F. W. (1984). Exclusive immunoglobulin genes. *Nature*, **312**, 502.

396. Yancopoulos, G. D. and Alt, F. W. (1986). Regulation of the assembly and expression of variable-region genes. *Annu. Rev. Immunol.*, **4**, 339.

397. Storb, U. (1987). Transgenic mice with immunoglobulin genes. *Annu. Rev. Immunol.*, **5**, 151.

398. Weaver, D., Constantini, F., Imanishi-Kari, T., and Baltimore, D. (1985). A transgenic immunoglobulin mu gene prevents rearrangement of endogenous genes. *Cell*, **42**, 117.

*399. Goodnow, C. C., Crosbie, J., Adelstein, S., Lavoie, T. B., Smith-Gill, S. J., Brink, R. A., Pritchard-Briscoe, H., Wotherspoon, J. S., Loblay, R. H., Raphael, K., and Basten, T. (1988). Altered immunoglobulin expression and functional silencing of self-reactive B lymphocytes in transgenic mice. *Nature*, **334**, 676.

400. Storb, U., Pinkert, C., Arp, B., Engler, P., Gollahon, K., Manz, J., Brady, W., and Brinster, R. (1986). Transgenic mice with μ and κ encoding antiphosphorylcholine antibodies. *J. Exp. Med.*, **164**, 627.

401. Grosschedl, R., Weaver, D., Baltimore, D., and Constantini, F. (1984). Introduction of a μ immunoglobulin gene into the mouse germ line: Specific expression in lymphoid cells and synthesis of a functional antibody. *Cell*, **38**, 647.

402. Bernard, O., Gough, N. M., and Adams, J. M. (1981). Plasmacytomas with more than one Igκ mRNA: Implications for allelic exclusion. *Proc. Natl Acad. Sci. USA*, **78**, 5812.

403. Kwan, S. P., Max, E. E., Seidman, J. G., Leder, P., and Scharff, M. (1981). Two kappa Ig genes are expressed in the myeloma S107. *Cell*, **26**, 57.

404. Storb, U. (1989). Immunoglobulin gene analysis in transgenic mice. In *Immunoglobulin genes* (ed. T. Honjo, F. W. Alt, and T. H. Rabbits) p. 303. Academic Press, New York.

405. Storb, U., Engler, P., Klotz, E., Weng, A., Haasch, D., Pinkert, C., Doglio, L.,

Glymore, M., and Brinster, R. (1992). Rearrangement and expression of immuno-globulin genes in transgenic mice. *Curr. Topic. Microbiol. Immunol.*, **182**, 137.

406. Neuberger, M. S., Caskey, H. M., Pettersson, S., Williams, G. T., and Surani, M. A. (1989). Isotype exclusion and transgene down-regulation in immunoglobulin-lambda transgenic mice. *Nature*, **338**, 350.

407. Ma, A., Fisher, P., Dildrop, R., Oltz, E., Rathbun, G., Achacoso, P., Stall, A., and Alt, F. W. (1992). Surface IgM mediated regulation of *Rag* gene expression in Emu-N-myc B cell lines. *EMBO J.*, **11**, 2727.

408. Turka, L. A., Schartz, D. G., Oettinger, M. A., Chun, J. J. M., Gorka, C., Lee, K., McCormack, W. T., and Thompson, C. B. (1991). Thymocyte expression of the recombination activating *genes RAG-1* and *RAG-2* can be terminated by T-cell receptor cross-linking. *Science*, **253**, 778.

409. Brandle, D., Muller, C., Rulicke, T., Hengartner, H., and Pircher, H. (1992). Engagement of the T-cell receptor during positive selection in the thymus down-regulates *Rag-1* expression. *Proc. Natl Acad. Sci. USA*, **89**, 9529.

410. Gay, D., Saunders, T., Camper, S., and Weigert, M. (1993). Receptor editing: An approach by autoreactive B cells to escape tolerance. *J. Exp. Med.*, **177**, 999.

411. Radic, M. Z., Erikson, J., Litwin, S., and Weigert, M. (1993). B lymphocytes may escape tolerance by revising their antigen receptors. *J. Exp. Med.*, **177**, 1165.

412. Tiegs, S. L., Russell, D. M., and Nemazee, D. (1993). Receptor editing in self-reactive bone marrow B cells. *J. Exp. Med.*, **177**, 1009.

413. Levy, S., Campbell, M. J., and Levy, R. (1989). Functional immunoglobulin light chain genes are replaced by ongoing rearrangements of germline V_κ genes to downstream J_κ segments in a murine B cell line. *J. Exp. Med.*, **170**, 1.

414. Taki, S., Meiering, M., and Rajewsky, K. (1993). Targeted insertion of a variable region gene into the immunoglobulin heavy chain locus. *Science*, **262**, 1268.

415. Malynn, B. A., Berman, J. E., Yancopoulos, G. D., Bona, C. A., and Alt, F. W. (1987). Expression of the immunoglobulin heavy-chain variable gene repertoire. *Curr. Topic. Microbiol. Immunol.*, **135**, 75.

416. Yancopoulos, G. D., Desiderio, S. V., Paskind, M., Kearney, J. F., Baltimore, D., and Alt, F. W. (1984). Preferential utilization of the most J_H-proximal V_H gene segments in pre-B-cell lines. *Nature*, **311**, 727.

417. Blackwell, T. K., Yancopoulos, G. D., and Alt, F. W. (1984). Joining of immunoglobulin heavy-chain variable region gene segments *in vivo* and within a recombination substrate. *UCLA Symposia on Molecular and Cellular Biology*. Liss, A. (ed), Vol. 19, New York.

418. Lawler, A. M., Lin, P. S., and Gearhart, P. (1987). Adult B-cell repertoire is biased toward two heavy-chain variable-region genes that rearrange frequently in fetal pre-B cells. *Proc. Natl Acad. Sci. USA*, **84**, 2454.

419. Freitas, A., Andrade, L., Lembezat, M. P., and Coutinho, A. (1990). Selection of V_H gene repertoires: differentiating B cells of adult bone marrow mimic fetal development. *Int. Immunol.*, **2**, 15.

420. Huetz, F., Carlsson, L., Thornberg, U., and Holmberg, D. (1993). V-region directed selection in differentiating B lymphocytes. *EMBO J.*, **12**, 1819.

421. Atkinson, M. J., Paige, C. J., and Wu, G. E. (1993). Map position and usage of 3' V_H family members: Usage is not position dependent. *Int. Immunol.*, **5**, 1577.

422. Yancopoulos, G. D., Malynn, B. A., and Alt, F. W. (1988). Developmentally regulated and strain-specific expression of murine V_H gene families. *J. Exp. Med.*, **168**, 417.

423. Jeong, H. D. and Teale, J. (1988). Comparison of the foetal and adult functional repertoires by analysis of V_H gene family expression. *J. Exp. Med.*, **168**, 589.

424. Freitas, A., Lembezat, M. P., and Coutinho, A. (1989). Expression of antibody V-regions is genetically and developmentally controlled and modulated by the B lymphocyte environment. *Int. Immunol.*, **1**, 342.

425. Malynn, B., Yancopoulos, G., Barth, J., Bona, C., and Alt, F. (1990). Biased expression of J_H-proximal V_H genes occurs in the newly generated repertoire of neonatal and adult mice. *J. Exp. Med.*, **171**, 843.

426. Gu, H., Tarlinton, D., Muller, W., Rajewsky, K., and Forster, I. (1991). Most peripheral B cells in mice are ligand selected. *J. Exp. Med.*, **173**, 1357.

427. Liao, F., Giannini, S. L., and Birshtein, B. K. (1992). A nuclear DNA-binding protein expressed during early stages of B cell differentiation interacts with diverse segments within and 3′ of the Ig H chain gene cluster. *J. Immunol.*, **148**, 2909.

428. Teale, J. M. and Medina, C. A. (1992). Comparative expression of adult and fetal V gene repertoires. *Int. Rev. Immunol.*, **8**, 95.

429. Kearney, J. F. (1993). Early B-cell repertoires. *Curr. Topic. Microbiol. Immunol.*, **182**, 81.

430. Kearney, J. F., Bartels, J., Hamilton, A. M., Lehuen, A., Solvason, N., and Vakil, M. (1992). Development and function of the early B cell repertoire. *Int. Rev. Immunol.*, **8**, 247.

431. Holmberg, D., Andersson, A., Carlsson, L., and Forsgren, S. (1989). Establishment and functional implications of B-cell connectivity. *Immunol. Rev.*, **110**, 89.

432. Cuisinier, A. M., Guigou, V., Boubli, L., Fougereau, M., and Tonnelle, C. (1989). Preferential expression of V_H5 and V_H6 immunoglobulin genes in early human B cell ontogeny. *Scand. J. Immunol.*, **30**, 493.

433. Schroeder, H. W. J., Hillson, J. L., and Perlmutter, R. M. (1987). Early restriction of the human antibody repertoire. *Science*, **238**, 791.

434. Cuisinier, A. M., Fumoux, F., Moinier, D., Boubli, L., Guigiou, V., Milili, M., Schiff, C., Fougereau, M., and Tonnelle, C. (1990). Rapid expansion of human immunoglobulin repertoire (V_H, V kappa, V lambda) expressed in early fetal bone marrow. *New Biol.*, **2**, 689.

435. Kofler, R., Duchosal, M. A., and Dixon, F. J. (1989). Complexity, polymorphism, and connectivity of mouse V_κ gene families. *Immunogenetics*, **29**, 65.

436. Kofler, R. and Helmberg, A. (1991). A new Ig_κ-V gene family in the mouse (letter). *Immunogenetics*, **34**, 139.

437. Lawler, A. M., Kearney, J. F., Kuehl, M., and Gearhart, P. J. (1989). Early rearrangements of genes encoding murine immunoglobulin kappa chains, unlike genes encoding heavy chains, use variable gene segments dispersed throughout the locus. *Proc. Natl Acad. Sci. USA*, **86**, 6744.

438. Chen, J. and Alt, F. W. (1993). Gene rearrangement and B-cell development. *Curr. Opin. Immunol.*, **5**, 194.

439. Davis, M. M. (1985). Molecular genetics of the T-cell receptor β chain. *Annu. Rev. Immunol.*, **3**, 537.

440. Lieber, M. R., Hesse, J. E., Mizuuchi, K., and Gellert, M. (1987). Developmental stage specificity of the lymphoid V(D)J recombination activity. *Genes Dev.*, **1**, 751.

441. Goodhardt, M., Cavelier, P., Doyen, N., Kallenbach, S., Babinet, C., and Rougeon, F. (1993). Methylation status of immunoglobulin κ gene segments correlates with their recombination potential. *Eur. J. Immunol.*, **23**, 1789.

442. Kelley, D. E., Pollock, B. A., Atchison, M. L., and Perry, R. P. (1988). The coupling

between enhancer activity and hypomethylation of kappa immunoglobulin genes is developmentally regulated. *Mol. Cell. Biol.*, **8**, 930.

443. Storb, U. and Arp, B. (1983). Methylation patterns of immunoglobulin genes in lymphoid cells: correlation of expression and differentiation with undermethylation. *Proc. Natl Acad. Sci. USA*, **80**, 6642.

444. Akira, S., Sugiyama, H., Sakaguchi, N., and Kishimoto, T. (1984). Immunoglobulin gene expression and DNA methylation in murine pre-B cell lines. *EMBO J.*, **3**, 677.

445. Lennon, G. G. and Perry, R. P. (1990). The temporal order of appearance of transcripts from unrearranged and rearranged Ig genes in murine fetal liver. *J. Immunol.*, **144**, 1983.

446. Engler, P., Roth, P., Kim, J. Y., and Storb, U. (1991). Factors affecting the rearrangement efficiency of an Ig test gene. *J. Immunol.*, **146**, 2826.

447. Engler, P., Haasch, D., Pinkert, C. A., Doglio, L., Glymour, M., Brinster, R., and Storb, U. (1991). A strain-specific modifier on mouse chromosome 4 controls the methylation of independent transgene loci. *Cell*, **65**, 939.

448. Ferrier, P., Krippl, B., Blackwell, T. K., Furley, A. J., Suh, H., Winoto, A., Cook, W. D., Hood, L., Costantini, F., and Alt, F. W. (1990). Separate elements control DJ and VDJ rearrangement in a transgenic recombination substrate. *EMBO J.*, **9**, 117.

449. Serwe, M. and Sablitzky, F. (1993). V(D)J recombination in B cells is impaired but not blocked by targeted deletion of the immunoglobulin heavy chain intron enhancer. *EMBO J.*, **12**, 2321.

450. Rathbun, G., Oltz, E. M., and Alt, F. W. (1993). Comparison of *Rag* gene expression in normal and transformed precursor lymphocytes. *Int. Immunol.*, **8**, 997.

451. Schlissel, M. S., Corcoran, L. M., and Baltimore, D. (1991). Virus-transformed pre-B cells show ordered activation but not inactivation of immunoglobulin gene rearrangement and transcription. *J. Exp. Med.*, **173**, 711.

452. Engler, P. and Storb, U. (1987). High-frequency deletional rearrangement of immunoglobulin κ gene segments introduced into a pre-B cell line. *Proc. Natl Acad. Sci. USA*, **84**, 4949.

453. Hsieh, C. L., McCloskey, R. P., and Lieber, M. R. (1992). V(D)J recombination on minichromosomes is not affected by transcription. *J. Biol. Chem.*, **267**, 15613.

454. Fondell, J. D. and Marcu, K. B. (1992). Transcription of germ line V alpha segments correlates with ongoing T-cell receptor alpha-chain rearrangement. *Mol. Cell. Biol.*, **12**, 1480.

455. Goldman, J. P., Spencer, D. M., and Raulet, D. H. (1993). Ordered rearrangement of variable region genes of the T cell receptor gamma locus correlates with transcription of the unrearranged genes. *J. Exp. Med.*, **177**, 729.

456. Holman, P. O., Roth, M. E., Huang, M., and Kranz, D. M. (1993). Characterization of transcripts from unrearranged V alpha 8 genes in the thymus. *J. Immunol.*, **151**, 1959.

457. Okada, A., Mendelsohn, M., and Alt, F. W. (1994). Differential activation of transcription versus recombination of transgenic TCRβ variable region gene segments in B and T lineage cells. *J. Exp. Med.*, **180**, 261.

458. Cedar, H. (1988). DNA methylation and gene activity. *Cell*, **53**, 3.

459. Bird, A. (1992). The essentials of DNA methylation. *Cell*, **70**, 5.

460. Hsieh, C. L. and Lieber, M. R. (1992). CpG methylated minichromosomes become inaccessible for V(D)J recombination after undergoing replication. *EMBO J.*, **11**, 315.

461. Tazi, J. and Bird, A. (1990). Alternative chromatin structure at CpG islands. *Cell*, **60**, 909.

462. Boyes, J. and Bird, A. (1992). Repression of genes by DNA methylation depends on

CpG density and promoter strength: evidence for involvement of methyl-CpG binding protein. *EMBO J.*, **11**, 327.

463. Lichtenstein, M., Keini, G., Cedar, H., and Yehudit, B. (1994). B cell-specific demethylation: A novel role for the intronic κ chain enhancer sequence. *Cell*, **76**, 913.
464. Ferrier, P., Covey, L. R., Suh, H., Winoto, A., Hood, L., and Alt, F. W. (1989). T cell receptor DJ but not VDJ rearrangement within a recombination substrate introduced into a pre-B cell line. *Int. Immunol.*, **1**, 66.
465. Parslow, T. G. and Granner, D. K. (1982). Chromatin changes accompany immunoglobulin κ gene activation. *Nature*, **229**, 449.
466. Jenuwein, T., Forrester, W. C., Qiu, R.-G., and Grosschedl, R. (1993). The immunoglobulin μ enhancer core establishes local factor access in nuclear chromatin independent of transcriptional stimulation. *Genes Dev.*, **7**, 2016.
467. Capone, M., Watrin, F., Fernex, C., Horvat, B., Krippl, B., Wu, L., Scollay, R., and Ferrier, P. (1993). TCRβ and TCRα gene enhancers confer tissue- and stage-specificity on V(D)J recombination events. *EMBO J.*, **12**, 4335.
468. Lauzurica, P. and Krangel, M. S. (1994). Enhancer-dependent and -independent steps in the rearrangement of a human T cell receptor δ transgene. *J. Exp. Med.*, **179**, 43.
469. Dariavach, P., Williams, G. T., Campbell, K., Pettersson, S., and Neuberger, M. S. (1991). The mouse IgH 3' enhancer. *Eur. J. Immunol.*, **21**, 1499.
470. Lieberson, R., Giannini, S. L., Birshtein, B. K., and Eckhardt, L. A. (1991). An enhancer at the 3' end of the mouse immunoglobulin heavy chain locus. *Nucleic Acids Res.*, **19**, 933.
471. Kottman, A. H., Brack, C., Eibel, H., and Kohler, G. (1992). A survey of protein–DNA interaction sites within the murine immunoglobulin heavy chain locus reveals a particularly complex pattern around the DQ52 element. *Eur. J. Immunol.*, **22**, 2113.
472. Kottman, A. H., Zevnick, B., Welte, M., Nielson, P. J., and Kohler, G. (1994). A second promoter and enhancer element within the immunoglobulin heavy chain locus. *Eur. J. Immunol.*, **24**, 817.
473. Anderson, S. J., Abraham, K. M., Nakayama, T., Singer, A., and Perlmutter, R. M. (1992). Inhibition of T-cell receptor beta chain gene rearrangement by overexpression of the non-receptor protein tyrosine kinase p561ck. *EMBO J.*, **11**, 4877.
*474. Anderson, S. J., Levin, S. D., and Perlmutter, R. M. (1993). The protein tyrosine kinase p561ck controls allelic exclusion of T cell receptor b chain genes. *Nature*, **365**, 552.
475. Uematsu, Y., Ryser, S., Dembic, Z., Borgulya, P., Krimpenfort, P., Berns, A., von Boehmer, H., and Steinmetz, M. (1988). In transgenic mice the introduced functional T cell receptor β gene prevents expression of endogenous β genes. *Cell*, **52**, 831.
*476. Bottaro, A., Lansford, R., Xu, L., Zhang, J., Rothman, P., and Alt, F. W. (1993). S region transcription per se promotes basal IgE class switch recombination but additional factors regulate the efficiency of the process. *EMBO J.*, **13**, 665.
*477. Lauster, R., Reynaud, C.-A., Martensson, I.-L., Peter, A., Bucchini, D., Jami, J., and Weill, J.-C. (1993). Promoter, enhancer and silencer elements regulate rearrangement of an immunoglobulin transgene. *EMBO J.*, **12**, 4615.
478. Thomas, B. J. and Rothstein, R. (1989). Elevated recombination rates in transcriptionally active DNA. *Cell*, **56**, 619.
479. Nickoloff, J. A. and Reynolds, R. J. (1990). Transcription stimulates homologous recombination in mammalian cells. *Mol. Cell. Biol.*, **10**, 4837.
480. Voelkel-Meiman, K., Keil, R. L., and Roeder, G. S. (1987). Recombination-stimulating

sequences in yeast ribosomal DNA correspond to sequences regulating transcription by RNA polymerase I. *Cell*, **48**, 1071.

481. Droge, P. and Nordheim, A. (1991). Transcription induced conformational change in a topologically closed DNA domain. *Nucleic Acids. Res.*, **19**, 2941.

482. Brill, S. J. and Sternglanz, R. (1988). Transcription-dependent DNA supercoiling in yeast DNA topoisomerase mutants. *Cell*, **54**, 403.

483. Giaever, G. N. and Wang, J. C. (1988). Supercoiling of intracellular DNA can occur in eukaryotic cells. *Cell*, **55**, 849.

484. Ostrander, E. A., Benedetti, P., and Wang, J. C. (1990). Template supercoiling by a chimera of yeast GAL4 protein and phage T7 RNA polymerase. *Science*, **249**, 1261.

485. Tsao, Y. P., Wu, H. Y., and Liu, L. F. (1989). Transcription-driven supercoiling of DNA: direct biochemical evidence from *in vitro* studies. *Cell*, **56**, 111.

486. Yang, L., Jessee, C. B., Lau, K., Zhang, H., and Liu, L. F. (1989). Template super-coiling during ATP-dependent DNA helix tracking: studies with simian virus 40 large tumor antigen. *Proc. Natl Acad. Sci. USA*, **86**, 6121.

487. Wu, H. Y., Shyy, S. H., Wang, J. C., and Liu, L. F. (1988). Transcription generates positively and negatively supercoiled domains in the template. *Cell*, **53**, 433.

488. Tse, Y., Kirkegaard, K., and Wang, J.-C. (1980). Covalent bonds between protein and DNA. Formation of phosphotyrosine linkage between certain DNA topoisomerases and DNA. *J. Biol. Chem.*, **255**, 5560.

489. Droge, P. (1993). Transcription-driven site-specific DNA recombination *in vitro*. *Proc. Natl Acad. Sci. USA*, **90**, 2759.

490. Giese, K., Cox, J., and Grosschedl, R. (1992). The HMG domain of lymphoid enhancer factor 1 bends DNA and facilitates assembly of functional nucleoprotein structures. *Cell*, **69**, 185.

491. Blasquez, V. C., Xu, M., Moses, S. C., and Garrard, W. T. (1989). Immunoglobulin kappa gene expression after stable integration. I. Role of the intronic MAR and enhancer in plasmacytoma cells. *J. Biol. Chem.*, **264**, 21183.

492. Xu, M., Hammer, R. E., Blasquez, V. C., Jones, S. L., and Garrard, W. T. (1989). Immunoglobulin kappa gene expression after stable integration. II. Role of the intronic MAR and enhancer in transgenic mice. *J. Biol. Chem.*, **264**, 21190.

493. Cockerill, P. N. (1990). Nuclear matrix attachment in several regions of the IgH locus. *Nucleic Acids Res.*, **18**, 2643.

494. Cockerill, P. N., Yuen, M. H., and Garrard, W. T. (1987). The enhancer of the immunoglobulin heavy chain locus is flanked by presumptive chromosomal loop anchorage elements. *J. Biol. Chem.*, **262**, 5394.

495. Denny, C. T., Yoshikai, Y., Mak, T. W., Smith, S. D., Hollis, G. F., and Kirsch, I. R. (1986). A chromosome 14 inversion in a T-cell lymphoma is caused by site-specific recombination between immunoglobulin and T-cell receptor loci. *Nature*, **320**, 549.

496. Baer, R., Forster, A., and Rabbits, T. H. (1987). The mechanism of chromosome 14 inversion in human T-cell lymphoma. *Cell*, **50**, 97.

497. Bakhshi, A., Jensen, J. P., Goldman, P., Wright, J., McBride, O. W., Epstein, A. L., and Korsmeyer, S. (1985). Cloning the chromosomal breakpoint of t(14:18)human lymphomas: clustering around J_H on chromosome 14 and near a transcriptional unit on 18. *Cell*, **41**, 899.

498. Showe, L. C. and Croce, C. M. (1987). The role of chromosomal translocations in B- and T-cell neoplasia. *Annu. Rev. Immunol.*, **5**, 253.

499. Hua, C., Zorn, S., Jensen, J. P., Coupland, R. W., Ko, H. S., Wright, J. J., and

Bakhshi, A. (1988). Consequences of the t(14;18) chromosomal translocation in follicular lymphoma: deregulated expression of a chimeric and mutated BCL-2 gene. *Onc. Res.*, **2**, 263.

500. Cleary, M. L., Smith, S. D., and Sklar, J. (1986). Cloning and structural analysis of cDNAs for bcl-2 and a hybrid bcl-2/immunoglobulin transcript resulting from the t(14:18) translocation. *Cell*, **47**, 19.

501. Strasser, A., Harris, A. W., Bath, M. L., and Cory, S. (1990). Novel primitive lymphoid tumors induced in transgenic mice by cooperation between myc and bcl-2. *Nature*, **348**, 331.

502. Adams, J. M., Harris, A. W., Vaux, D. L., Alexander, W. S., Rosenbaum, H., Klinken, S. P., Strasser, A., Bath, M. L., McNeall, J., and Cory, S. (1989). The transgenic window on lymphoid malignancy. *International Symposium of the Princess Takamatsu Cancer Res Fund*, Vol. 20, 297.

503. Lee, J. T., Innes, J., and Williams, M. E. (1989). Sequential bcl-2 and c-myc oncogene rearrangements associated with the clinical transformation of non-Hodgkin's lymphoma. *J. Clin. Invest.*, **84**, 1454.

504. Adams, J. M. and Cory, S. (1992). Oncogene co-operation in leukaemogenesis. *Cancer Surv.*, **15**, 119.

505. Honjo, T. and Kataoka, T. (1978). Organization of immunoglobulin heavy chain genes and allelic deletion model. *Proc. Natl Acad. Sci. USA*, **75**, 2140.

506. Davis, M. M., Kim, S. K., and Hood, L. E. (1980). DNA sequences mediating class switching in α-immunoglobulins. *Science*, **209**, 1360.

507. Dunnick, W., Rabbitts, T. H., and Milstein, C. (1980). An immunoglobulin deletion mutant with implications for the heavy-chain switch and RNA splicing. *Nature*, **286**, 669.

508. Kataoka, T., Miyata, T., and Honjo, T. (1981). Repetitive sequences in class-switch recombination regions of immunoglobulin heavy chain genes. *Cell*, **23**, 357.

509. Marcu, K., Banerji, J., Penncavage, R., Lang, R., and Arnheim, N. (1980). 5' flanking region of immunoglobulin heavy chain constant region genes display length heterogeneity in germlines of inbred mouse strains. *Cell*, **22**, 187.

510. Nikaido, T., Yamakami-Kataoka, Y., and Honjo, T. (1982). Nucleic acid sequences of switch regions of immunoglobulin C_ϵ and C_γ genes and their comparison. *J. Biol. Chem.*, **257**, 7322.

511. Sakano, H., Maki, R., Kurosawa, Y., Roeder, W., and Tonegawa, S. (1980). Two types of somatic recombination are necessary for the generation of complete immunoglobulin heavy-chain genes. *Nature*, **286**, 676.

512. Marcu, K., Lang, R., Stanton, L., and Harris, L. (1982). A model for the molecular requirements of immunoglobulin heavy chain class switching. *Nature*, **298**, 87.

513. Nikaido, T., Nakai, S., and Honjo, T. (1981). Switch region of immunoglobulin C_μ gene is composed of simple tandem repetitive sequences. *Nature*, **292**, 845.

514. Mowatt, M. and Dunnick, W. (1986). DNA sequence of the murine γ1 switch segment reveals novel structural elements. *J. Immunol.*, **136**, 2674.

515. Stanton, L. W. and Marcu, K. B. (1982). Nucleotide sequence and properties of the murine gamma 3 immunoglobulin heavy chain gene switch region: implications for successive C gamma gene switching. *Nucleic Acids Res.*, **10**, 5993.

516. Szurek, P., Petrini, J., and Dunnick, W. (1985). Complete nucleotide sequence of the murine γ3 switch region and analysis of switch recombination sites in two γ3-expressing hybridomas. *J. Immunol.*, **135**, 620.

517. Scappino, L. A., Chu, C., and Gritzmacher, C. A. (1991). Extended nucleotide sequence of the switch region of the murine gene encoding immunoglobulin E. *Gene*, **99**, 295.

518. Carayannopoulos, L. and Capra, J. D. (1993). Immunoglobulins: Structure and function. In *Fundamental immunology* (ed. W. E. Paul), p. 283. Raven Press, New York.

519. Kenter, A. L., Wuerffel, R., Sen, R., Jamieson, C. E., and Merkulov, G. V. (1993). Switch recombination breakpoints occur at nonrandom positions in the S_γ tandem repeat. *J. Immunol.*, **151**, 4718.

520. Dunnick, W., Hertz, G., Scappino, L., and Gritzmacher, C. (1993). DNA sequences at immunoglobulin switch region recombination sites. *Nucleic Acids Res.*, **21**, 365.

521. Yaoita, Y. and Honjo, T. (1980). Deletion of immunoglobulin heavy chain genes from expressed allelic chromosome. *Nature*, **286**, 850.

522. Cory, S. and Adams, J. M. (1980). Deletions are associated with somatic rearrangement of immunoglobulin heavy chain genes. *Cell*, **19**, 37.

523. Rabbitts, T. H., Forster, A., Dunnick, W., and Bentley, D. L. (1980). The role of gene deletion in the immunoglobulin heavy chain switch. *Nature*, **283**, 351.

524. Coleclough, C., Cooper, D., and Perry, R. P. (1980). Rearrangement of immunoglobulin heavy chain genes during B-lymphocyte development as revealed by studies of mouse plasmacytoma cells. *Proc. Natl Acad. Sci. USA*, **77**, 1422.

*525. Jack, H. M., McDowell, M., Steinberg, C. M., and Wabl, M. (1988). Looping out and deletion mechanism for the immunoglobulin heavy-chain class switch. *Proc. Natl Acad. Sci. USA*, **85**, 1581.

526. Iwasato, T., Shimizu, A., Honjo, T., and Yamagishi, H. (1990). Circular DNA is excised by immunoglobulin class switch recombination. *Cell*, **62**, 143.

527. von Schwedler, U., Jack, H. M., and Wabl, M. (1990). Circular DNA is a product of the immunoglobulin class switch rearrangement. *Nature*, **345**, 452.

528. Yoshida, K., Matsuoka, M., Usuda, S., Mori, A., Ishizaka, K., and Sakano, H. (1990). Immunoglobulin switch circular DNA in the mouse infected with *Nippostrongylus brasieliensis*: Evidence for successive class switching from μ to ε via γ1. *Proc. Natl Acad. Sci. USA*, **87**, 7829.

529. Iwasato, T., Arakawa, H., Shimizu, A., Honjo, T., and Yamagishi, H. (1992). Biased distribution of recombination sites within S regions upon immunoglobulin class switch recombination induced by transforming growth factor beta and lipopolysaccharide. *J. Exp. Med.*, **175**, 1539.

530. DePinho, R., Kruger, K., Andrews, N., Lutzker, S., Baltimore, D., and Alt, F. W. (1984). Molecular basis of heavy-chain class switching and switch region deletion in an Abelson virus-transformed cell line. *Mol. Cell. Biol.*, **4**, 2905.

531. Yancopoulos, G. D., DePinho, R. A., Zimmerman, K. A., Lutzker, S. G., Rosenberg, N., and Alt, F. W. (1986). Secondary genomic rearrangement events in pre-B cells: V_HDJ_H replacement by a LINE-1 sequence and directed class switching. *EMBO J.*, **5**, 3259.

*532. Jung, S., Rajewsky, K., and Radbruch, A. (1993). Shutdown of class switch recombination by deletion of a switch region control element. *Science*, **259**, 984.

533. Calvo, C. F., Giannini, S. L., Martinez, N., and Birshtein, B. (1991). DNA sequences 3' of the IgH chain cluster rearrange in mouse B cell lines. *J. Immunol.*, **146**, 1353.

534. Petrini, J. and Dunnick, W. A. (1989). Products and implied mechanism of H chain switch recombination. *J. Immunol.*, **142**, 2932.

535. Siebenkotten, G., Esser, C., Wabl, M., and Radbruch, A. (1992). The murine IgG1/IgE class switch program. *Eur. J. Immunol.*, **22**, 1827.

536. Snapper, C. M., Finkelman, F. D., Stefany, D., Conrad, D. H., and Paul, W. E. (1988). IL-4 induces co-expression of intrinsic membrane IgG1 and IgE by murine B cells stimulated with lipopolysaccharide. *J. Immunol.*, **141**, 489.

537. Mandler, R., Finkelman, F. D., Levine, A. D., and Snapper, C. M. (1993). Interleukin-4 induction of IgE class switching by lipopolysaccharide-activated murine B cells occurs predominantly through sequential switching. *J. Immunol.*, **150**, 407.

538. Obata, M., Kataoka, T., Nakai, S., Yamagishi, H., Takahashi, N., Yamawaki, K. Y., Nikaido, T., Shimizu, A., and Honjo, T. (1981). Structure of a rearranged gamma 1 chain gene and its implication to immunoglobulin class-switch mechanism. *Proc. Natl Acad. Sci. USA*, **78**, 2437.

539. Petrini, J., Shell, B., Hummel, M., and Dunnick, W. (1987). The immunoglobulin heavy chain switch: structural features of gamma 1 recombinant switch regions. *J. Immunol.*, **138**, 1940.

540. Li, S. C., Rothman, P. B., Chan, C., Hirsch, D., and Alt, F. W. (1994). Expression of I_μ–C_γ hybrid germline transcripts subsequent to Ig heavy chain class-switching. *Int. Immunol.*, **6**, 491.

541. Jung, S., Siebenkotten, G., and Radbruch, A. (1994). Frequency of immunoglobulin E class switching is autonomously determined and independent of prior switching to their classes. *J. Exp. Med.*, **179**, 2023.

542. Wahls, W. P., Wallace, L. J., and Moore, P. D. (1990). Hypervariable minisatellite DNA is a hotspot for homologous recombination in human cells. *Cell*, **60**, 95.

543. Alberts, B., Bray, D., Lewis, J., Raff, M., Roberts, K., and Watson, J. D. (1989). *Molecular biology of the cell*. Benjamin/Cummings Publishing, Menlo Park, CA.

544. Kenter, A. L. and Birshtein, B. K. (1981). Chi, a promoter of generalized recombination on λ phage, is present in immunoglobulin genes. *Nature*, **293**, 402.

545. Lyamichev, V. I., Mirkin, S. M., and Frank-Kamenetskii, M. D. (1986). Structures of homopurine–homopyrimidine tract in superhelical DNA. *J. Biomol. Struct. Dynam.*, **3**, 667.

546. Reaban, M. E. and Griffin, J. A. (1990). Induction of RNA-stabilized DNA conformers by transcription of an immunoglobulin switch region. *Nature*, **348**, 342.

547. Stavnezer, J. (1991). Triple helix stabilization? *Nature*, **351**, 447.

548. Ehrlich, S. D. (1988). Illegitimate recombination in bacteria. In *Mobile DNA* (ed. D. E. Berg and M. M. Howe), p. 799. American Society for Microbiology, Washington, DC.

549. Meuth, M. (1988). Illegitimate recombination in mammalian cells. In *Mobile DNA* (ed. D. E. Berg and M. M. Howe), p. 833. American Society for Microbiology, Washington, DC.

550. Mezard, C., Pompon, D., and Nicolas, A. (1992). Recombination between similar but not identical DNA sequences during yeast transformation occurs within short stretches of identity. *Cell*, **70**, 659.

551. Xu, M. Z. and Stavnezer, J. (1992). Regulation of transcription of immunoglobulin germ-line γ1 RNA: Analysis of the promoter/enhancer. *EMBO J.*, **11**, 145.

552. Wuerffel, R. A., Nathan, A. T., and Kenter, A. L. (1990). Detection of an immunoglobulin switch region-specific DNA-binding protein in mitogen-stimulated mouse splenic B cells. *Mol. Cell. Biol.*, **10**, 1714.

553. Wuerffel, R., Jamieson, C., Morgan, L., Merlulov, G., Sen, R., and Kenter, A. (1992). Switch recombination breakpoints are strictly correlated with DNA recognition motifs for immunoglobulin S gamma-3 DNA-binding proteins. *J. Exp. Med.*, **176**, 339.

554. Williams, M. and Maizels, N. (1991). LR1, a lipopolysaccharide-responsive factor with

binding sites in the immunoglobulin switch regions and heavy chain enhancer. *Genes Dev.*, **5**, 2353.

555. Schultz, C. L., Elenich, L. A., and Dunnick, W. A. (1991). Nuclear protein binding to octamer motifs in the immunoglobulin gamma 1 switch region. *Int. Immunol.*, **3**, 109.

556. Lin, Y. C. and Stavnezer, J. (1992). Regulation of transcription of the germ-line Ig alpha constant region gene by an ATF element and by novel transforming growth factor-beta 1-responsive elements. *J. Immunol.*, **149**, 2914.

557. Waters, S. H., Saikh, K. U., and Stavnezer, J. (1989). A B-cell-specific nuclear protein that binds to DNA sites 5' to immunoglobulin S alpha tandem repeats is regulated during differentiation. *Mol. Cell. Biol.*, **9**, 5594.

558. Severinson, E., Fernandez, C., and Stavnezer, J. (1990). Induction of germ-line immunoglobulin heavy chain transcripts by mitogens and interleukins prior to switch recombination. *Eur. J. Immunol.*, **20**, 1079.

559. Severinson, E., Bergstedt-Lindqvist, S., van der Loo, W., and Fernandez, C. (1982). Characterization of the IgG response induced by polyclonal B cell activators. *Immunol. Rev.*, **67**, 73.

560. Chu, C. C., Paul, W. E., and Max, E. E. (1992). Analysis of DNA synthesis requirement for deletional switching in normal B cells. In *8th International Congress of Immunology Abstracts*, p. 34. Springer-Verlag, Budapest.

561. Kenter, A. L. and Watson, J. V. (1987). Cell cycle kinetics model of LPS-stimulated spleen cells correlates switch region rearrangements with S phase. *J. Immunol. Meth.*, **97**, 111.

562. Dunnick, W., Wilson, M., and Stavnezer, J. (1989). Mutations, duplication, and deletion of recombined switch regions suggest a role for DNA replication in the immunoglobulin heavy-chain switch. *Mol. Cell. Biol.*, **9**, 1850.

563. Dunnick, W. and Stavnezer, J. (1990). Copy choice mechanism of immunoglobulin heavy-chain switch recombination. *Mol. Cell. Biol.*, **10**, 397.

564. Ott, D. E., Alt, F. W., and Marcu, K. B. (1987). Immunoglobulin heavy chain switch region recombination within a retroviral vector in murine pre-B cells. *EMBO J.*, **6**, 577.

565. Ott, D. E. and Marcu, K. B. (1989). Molecular requirements for immunoglobulin heavy chain constant region switch-recombination revealed with switch-substrate retroviruses. *Int. Immunol.*, **1**, 582.

566. Radbruch, A., Liesegang, B., and Rajewsky, K. (1980). Isolation of variants of mouse myeloma X63 that express changed immunoglobulin class. *Proc. Natl Acad. Sci. USA*, **77**, 2909.

567. Alt, F. W., Rosenberg, N., Casanova, R. J., Thomas, E., and Baltimore, D. (1982). Immunoglobulin heavy chain class switching and inducible expression in an Abelson murine leukaemia virus transformed cell line. *Nature*, **296**, 325.

568. Akira, S., Sugiyama, H., Yoshida, N., Kikutani, H., Yamamura, Y., and Kishimoto, T. (1983). Isotype switching in murine pre-B cell lines. *Cell*, **34**, 545.

569. Stavnezer, J., Abbott, J., and Sirlin, S. (1984). Immunoglobulin heavy chain switching in cultured I.29 murine B lymphoma cells: commitment to an IgA or IgE switch. *Curr. Top. Microbiol. Immunol.*, **113**, 109.

570. Chen, Y. W., Word, C. J., Jones, S., Uhr, J. W., Tucker, P. W., and Vitetta, E. S. (1986). Double isotype production by a neoplastic B cell line. I. Cellular and biochemical characterization of a variant of BCL1 that expresses and secretes both IgM and IgG1. *J. Exp. Med.*, **164**, 548.

571. Chen, Y. W., Word, C. J., Dev, V., Uhr, J. W., Vitetta, E. S., and Tucker, P. W. (1986).

Double isotype production by a neoplastic B cell line. II. Allelically excluded production of μ and γ1 heavy chains without C_H gene rearrangement. *J. Exp. Med.*, **164**, 562.

572. Arnold, L. W., Grdina, T. A., Whitmore, A. C., and Haughton, G. (1988). Ig isotype switching in B lymphocytes: Isolation and characterization of clonal variants of the murine Ly-1[+] B cell lymphoma, CH12, expressing isotypes other than IgM. *J. Immunol.*, **140**, 4355.

573. Burrows, P. D., Beck, G. B., and Wabl, M. R. (1981). Expression of μ and γ immunoglobulin heavy chains in different cells of a cloned lymphoid line. *Proc. Natl Acad. Sci. USA*, **78**, 564.

574. Leung, H. and Maizels, N. (1992). Transcriptional regulatory elements stimulate recombination in extrachromosomal substrates carrying immunoglobulin switch-region sequences. *Proc. Natl Acad. Sci. USA*, **89**, 4154.

575. Leung, H. and Maizels, N. (1994). Regulation and targeting of recombination in extrachromosomal substrates carrying immunoglobulin switch region sequences. *Mol. Cell. Biol.*, **14**, 1450.

576. Tuaillon, N., Taylor, L. D., Lonberg, N., Tucker, P. W., and Capra, J. D. (1993). Human immunoglobulin heavy-chain minilocus recombination in transgenic mice: Gene-segment use in μ and γ transcripts. *Proc. Natl Acad. Sci. USA*, **90**, 3720.

577. Taylor, L. D., Carmack, C. E., Huszar, D., Higgins, K. M., Mashayekh, R., Sequar, G., Schramm, S. R., Kuo, C. C., O'Donnell, S. L., Kay, R. M., Woodhouse, C. S., and Lonberg, N. (1994). Human immunoglobulin transgenes undergo rearrangement, somatic mutation, and class switching in mice that lack endogenous IgM. *Int. Immunol.*, **6**, 579.

578. Lafrenz, D., Teale, J. M., Klinman, N. R., and Strober, S. (1986). Surface IgG-bearing cells retain the capacity to secrete IgM. *J. Immunol.*, **136**, 2076.

579. Pernis, B., Forni, L., and Luzzati, A. L. (1977). Synthesis of multiple immunoglobulin classes by single lymphocytes. *Cold Spring Harbor Symp. Quant. Biol.*, **41**, 175.

580. Yuan, D., Vitetta, E. S., and Kettman, J. R. (1977). Cell-surface immunoglobulin. XX. Antibody responsiveness of subpopulations of B lymphocytes bearing different isotypes. *J. Exp. Med.*, **145**, 1421.

581. Teale, J. M., Lafrenz, D., Klinman, N. R., and Strober, S. (1981). Immunoglobulin class commitment exhibited by B lymphocytes separated according to surface isotype. *J. Immunol.*, **126**, 1952.

582. Abney, E. R., Cooper, M. D., Kearney, J. F., Lawton, A. R., and Parkhouse, R. M. E. (1978). Sequential expression of immunoglobulin on developing mouse B lymphocytes: A systematic survey that suggests a model for the generation of immunoglobulin isotype diversity. *J. Immunol.*, **120**, 2041.

583. Wabl, M., Meyer, J., Beck, E. G., Tenkhoff, M., and Burrows, P. D. (1985). Critical test of a sister chromatid exchange model for the immunoglobulin heavy-chain class switch. *Nature*, **313**, 687.

584. Perlmutter, A. P. and Gilbert, W. (1984). Antibodies of the secondary response can be expressed without switch recombination in normal mouse B cells. *Proc. Natl Acad. Sci. USA*, **81**, 7189.

585. Nolan-Willard, M., Berton, M. T., and Tucker, P. (1992). Co-expression of μ and γ1 heavy chains can occur by a discontinuous transcription mechanism from the same unrearranged chromosome. *Proc. Natl Acad. Sci. USA*, **89**, 1234.

586. Yaoita, Y., Kumagai, Y., Okumura, K., and Honjo, T. (1982). Expression of lymphocyte surface IgE does not require switch recombination. *Nature*, **297**, 697.

587. Lutzker, S. G. and Alt, F. W. (1989). Immunoglobulin heavy-chain class switching. In *Mobile DNA* (ed. D. E. Berg and M. M. Howe), p. 693. American Society for Microbiology, Washington, DC.

588. Agabian, N. (1990). *Trans* splicing of nuclear pre-mRNAs. *Cell*, **61**, 1157.

589. Bonen, L. (1993). Trans-splicing of pre-mRNAs in plants, animals, and protists. *FASEB J.*, **7**, 40.

590. Dandeker, T. and Sibbald, P. R. (1990). Trans-splicing of pre-mRNA is predicted to occur in a wide range of organisms including vertebrates. *Nucleic Acids Res.*, **18**, 4719.

591. Shimizu, A., Nussenzweig, M. C., Han, H., Sanchez, M., and Honjo, T. (1991). *Trans*-splicing as a possible molecular mechanism for the multiple isotype expression of the immunoglobulin gene. *J. Exp. Med.*, **173**, 1385.

592. Shimizu, A. and Honjo, T. (1993). Synthesis and regulation of *trans*-mRNA encoding the immunoglobulin epsilon heavy chain. *FASEB J.*, **7**, 149.

593. Han, H., Okamoto, M., Honjo, T., and Shimizu, A. (1991). Regulated expression of immunoglobulin trans-mRNA consisting of the variable region of a transgenic mu chain and constant regions of endogenous isotypes. *Int. Immunol.*, **3**, 1197.

594. Radbruch, A., Muller, W., and Rajewsky, K. (1986). Class switch recombination is IgG1 specific on active and inactive IgH loci of IgG1-secreting B-cell blasts. *Proc. Natl Acad. Sci. USA*, **83**, 3954.

595. Hummel, M., Kaminshka, J., and Dunnick, W. (1987). Switch region content of hybridomas: the two spleen cell IgH loci tend to rearrange to the same isotype. *J. Immunol.*, **138**, 3539.

596. Brunswick, M., June, C. H., Finkelman, F. D., Dintzis, H. M., Inman, J. K., and Mond, J. J. (1989). Dextran conjugated anti-Ig antibody stimulates B cells by repetitive signal transduction: a model for T cell independent B cell activation. *Proc. Natl Acad. Sci. USA*, **86**, 6724.

597. Kearney, J. F. and Lawton, A. R. (1975). B lymphocyte differentiation induced by lipopolysaccharide. I. Generation of cells synthesizing four major immunoglobulin classes. *J. Immunol.*, **115**, 671.

598. Vitetta, E. S., Ohara, J., Myers, C., Layton, J., Krammer, P. H., and Paul, W. E. (1985). Serological, biochemical, and functional identity of B cell-stimulatory factor 1 and B cell differentiation factor for IgG1. *J. Exp. Med.*, **161**, 1726.

599. Coffman, R. L., Ohara, J., Bond, M. W., Carty, J., Zlotnick, A., and Paul, W. E. (1986). B cell stimulatory factor-1 enhances the IgE response of lipopolysaccharide-activated B cells. *J. Immunol.*, **136**, 4538.

600. Snapper, C. M. and Paul, W. E. (1987). Interferon-γ and cell stimulatory factor-1 reciprocally regulate Ig isotype production. *Science*, **236**, 944.

601. Coffman, R. L., Lebman, D. A., and Shrader, B. (1989). Transforming growth factor beta specifically enhances IgA production by lipopolysaccharide-stimulated murine B lymphocytes. *J. Exp. Med.*, **170**, 1039.

602. Finkelman, F. D., Holmes, J., Katona, I. M., Urban, J. J., Beckmann, M. P., Park, L. S., Schooley, K. A., Coffman, R. L., Mosmann, T. R., and Paul, W. E. (1990). Lymphokine control of *in vivo* immunoglobulin isotype selection. *Annu. Rev. Immunol.*, **8**, 303.

603. Noelle, R. J., Ledbetter, J. A., and Aruffo, A. (1992). CD40 and its ligand, an essential ligand–receptor pair for thymus-dependent B-cell activation. *Immunol. Today*, **13**, 431.

604. Jabara, H. H., Fu, S. M., Geha, R. S., and Vercelli, D. (1990). CD40 and IgE: Synergism between anti-CD40 monoclonal antibody and interleukin 4 in the induction of IgE synthesis by highly purified human B cells. *J. Exp. Med.*, **172**, 1861.

605. Armitage, R. J., Fanslow, W. C., Strockbine, L., Sato, T. A., Clifford, K. N., Macduff, B. M., Anderson, D. M., Gimpel, S. D., Davis, S. T., Maliszewski, C. R., *et al.* (1992). Molecular and biological characterization of a murine ligand for CD40. *Nature,* **357**, 80.

606. Spriggs, M. K., Armitage, R. J., Strockbine, L., Clifford, K. N., Macduff, B. M., Sato, T. A., Maliszewski, C. R., and Fanslow, W. C. (1992). Recombinant human CD40 ligand stimulates B cell proliferation and immunoglobulin E secretion. *J. Exp. Med.,* **176**, 1543.

607. Maliszewski, C. R., Grabstein, K., Fanslow, W. C., Armitage, R., Spriggs, M. K., and Sato, T. A. (1993). Recombinant CD40 ligand stimulation of murine B cell growth and differentiation: cooperative effects of cytokines. *Eur. J. Immunol.,* **23**, 1044.

608. Rousset, F., Garcia, E., and Banchereau, J. (1991). Cytokine-induced proliferation and immunoglobulin production of human B lymphocytes triggered through their CD40 antigen. *J. Exp. Med.,* **173**, 705.

609. Defrance, T., Vanbervliet, B., Briere, F., Durand, I., Rousset, F., and Banchereau, J. (1992). Interleukin 10 and transforming growth factor beta cooperate to induce anti-CD40-activated naive human B cells to secrete immunoglobulin A. *J. Exp. Med.,* **175**, 671.

*610. Aruffo, A., Farrington, M., Hollenbaugh, D., Li, X., Milatovich, A., Nonoyama, S., Bajorath, J., Grosmaire, L. S., Stenkamp, R., Beubauer, M., Roberts, R. L., Noelle, R. J., Ledbetter, J. A., Francke, U., and Ochs, H. D. (1993). The CD40 ligand, gp39, is defective in activated T cells from patients with X-linked Hyper-IgM syndrome. *Cell,* **72**, 291.

*611. Korthauer, U., Graf, D., Mages, H. W., Briere, F., Padayachee, M., Malcolm, S., Ugazio, A. G., Notarangelo, L. D., Levinsky, R. J., and Kroczek, R. A. (1993). Defective expression of T-cell CD40 ligand causes X-linked immunodeficiency with hyper-IgM. *Nature,* **361**, 539.

*612. DiSanto, J. P., Bonnefoy, J. Y., Gauchet, J. F., Fischer, A., and de Saint Basile, G. (1993). CD40 ligand mutations in X-linked immunodeficiency with hyper-IgM. *Nature,* **361**, 541.

*613. Kawabe, T., Naka, T., Yoshida, K., Tanaka, T., Fujiwara, H., Suematsu, S., Yoshida, N., Kishimoto, T., and Kikutani, H. (1994). The immune responses of CD40-deficient mice: Impaired immunoglobulin class switching and germinal center formation. *Immunity,* **1**, 167.

*614. Castigli, E., Alt, F. W., Davidson, L., Bottaro, A., Mizoguchi, E., Bhan, A. K., and Geha, R. S. (1994). CD40 deficient mice generated by *Rag-2* deficient blastocyst complementation. *Proc. Natl Acad. Sci. USA,* **91**, 12135.

615. Stavnezer-Nordgren, J. and Sirlin, S. (1986). Specificity of immunoglobulin heavy chain switch correlates with activity of germline heavy chain genes prior to switching. *EMBO J.,* **5**, 95.

616. Lutzker, S., Rothman, P., Pollock, R., Coffman, R., and Alt, F. W. (1988). Mitogen- and IL-4-regulated expression of germ-line Ig gamma 2b transcripts: evidence for directed heavy chain class switching. *Cell,* **53**, 177.

617. Esser, C. and Radbruch, A. (1989). Rapid induction of transcription of unrearranged S gamma 1 switch regions in activated murine B cells by interleukin 4. *EMBO J.,* **8**, 483.

618. Berton, M. T., Uhr, J. W., and Vitetta, E. S. (1989). Induction of germline S gamma 1 transcripts in normal resting B cells by interleukin-4 prior to switch recombination to IgG1 and inhibition by interferon-gamma. *Proc. Natl Acad. Sci. USA,* **86**, 2829.

619. Lebman, D. A., Nomura, D. Y., Coffman, R. L., and Lee, F. D. (1990). Molecular

characterization of germ-line immunoglobulin A transcripts produced during transforming growth factor type beta-induced isotype switching. *Proc. Natl Acad. Sci. USA*, **87**, 3962.

620. Rothman, P., Lutzker, S., Cook, W., Coffman, R., and Alt, F. W. (1988). Mitogen plus interleukin 4 induction of C_ϵ transcripts in B lymphoid cells. *J. Exp. Med.*, **168**, 2385.

621. Rothman, P., Lutzker, S., Gorham, B., Stewart, V., Coffman, R., and Alt, F. W. (1990). Structure and expression of germline immunoglobulin γ3 heavy chain gene transcripts: Implications for mitogen and lymphokine directed class switching. *Int. Immunol.*, **2**, 621.

622. Gerondakis, S., Gaff, C., Goodman, D. J., and Grumont, R. J. (1991). Structure and expression of mouse germline immunoglobulin gamma 3 heavy chain transcripts induced by the mitogen lipopolysaccharide. *Immunogenetics*, **34**, 392.

623. Xu, M. and Stavnezer, J. (1990). Structure of germline immunoglobulin heavy-chain gamma 1 transcripts in interleukin 4 treated mouse spleen cells. *Dev. Immunol.*, **1**, 11.

624. Turaga, P. S. D., Berton, M. T., and Teale, J. M. (1993). Frequency of B cells expressing germ-line γ1 transcripts upon IL-4 induction. *J. Immunol.*, **151**, 1383.

625. Lutzker, S. and Alt, F. W. (1988). Structure and expression of germ line immunoglobulin gamma 2b transcripts [published erratum appears in *Mol. Cell. Biol.* (1988), **8**, 4585]. *Mol. Cell. Biol.*, **8**, 1849.

626. Collins, J. T. and Dunnick, W. A. (1993). Germline transcripts of the γ2a gene: structure and induction by IFN-γ. *Int. Immunol.*, **5**, 885.

627. Rothman, P., Chen, Y. Y., Lutzker, S., Li, S. C., Stewart, V., Coffman, R., and Alt, F. W. (1990). Structure and expression of germ line immunoglobulin heavy-chain epsilon transcripts: interleukin-4 plus lipopolysaccharide-directed switching to C epsilon. *Mol. Cell. Biol.*, **10**, 1672.

628. Gaff, C., Grumont, R. J., and Gerondakis, S. (1992). Transcriptional regulation of the germline immunoglobulin C_α and C_ϵ genes: implications for commitment to an isotype switch. *Int. Immunol.*, **4**, 1145.

629. Gerondakis, S. (1990). Structure and expression of murine germ-line immunoglobulin I_ϵ heavy chain transcripts induced by interleukin 4. *Proc. Natl Acad. Sci. USA*, **87**, 1581.

630. Rothman, P., Li, S. C., Gorham, B., Glimcher, L., Alt, F., and Boothby, M. (1991). Identification of a conserved lipopolysaccharide-plus-interleukin-4-responsive element located at the promoter of germ line ε transcripts. *Mol. Cell. Biol.*, **11**, 5551.

*631. Zhang, J., Bottaro, A., Li, S., Stewart, V., and Alt, F. W. (1993). A selective defect in IgG2b switching as a result of targeted mutation of the Iγ2b promoter and exon. *EMBO J.*, **2**, 3529.

632. Xu, L., Gorham, B., Li, S. C., Bottaro, A., Alt, F. W., and Rothman, P. B. (1993). Replacement of germ-line epsilon promoter by gene targeting alters control of immunoglobulin heavy chain class switching. *Proc. Natl Acad. Sci. USA*, **90**, 3705.

633. Gilmour, D. S., Pflugfelder, G., Wang, J. C., and Lis, J. T. (1986). Topoisomerase I interacts with transcribed regions in *Drosophila* cells. *Cell*, **44**, 401.

634. Tanaka, T., Chu, C. C., and Paul, W. E. (1993). An anitsense oligonucleotide complementary to a sequence in I gamma 2b increases gamma 2b germline transcripts, stimulates B cell DNA synthesis, and inhibits immunoglobulin secretion. *J. Exp. Med.*, **175**, 597.

635. Wakatsuki, Y. and Strober, W. (1993). Effect of downregulation of germline transcripts on immunoglobulin A isotype differentiation. *J. Exp. Med.*, **178**, 129.

*636. Gu, H., Zou, Y. R., and Rajewsky, K. (1993). Independent control of immunoglobulin switch recombination at individual switch regions evidenced through Cre-*loxP*-mediated gene targeting. *Cell*, **73**, 1155.

637. Klein, S., Sablitzky, F., and Radbruch, A. (1984). Deletion of the IgH enhancer does not reduce immunoglobulin heavy chain production of a hybridoma IgD class switch variant. *EMBO J.*, **3**, 2473.

638. Aguilera, R. J., Hope, T. J., and Sakano, H. (1985). Characterization of immunoglobulin enhancer deletions in murine plasmacytomas. *EMBO J.*, **4**, 3689.

639. Klein, S., Gerster, T., Picard, D., Radbruch, A., and Schaffner, W. (1985). Evidence for transient requirement of the IgH enhancer. *Nucleic Acids Res.*, **13**, 8901.

640. Schaffner, W. (1988). Gene regulation. A hit-and-run mechanism for transcriptional activation? *Nature*, **336**, 427.

641. Neuberger, M. S. and Calabi, F. (1983). Reciprocal chromosome translocation between c-myc and immunoglobulin g2b genes. *Nature*, **305**, 240.

642. Dillon, N. and Grosveld, F. (1993). Transcriptional regulation of multigene loci: multi-level control. *Trends Genet.*, **9**, 134.

643. Fiering, S., Kim, C. G., Epner, E. M., and Groudine, M. (1993). An 'in-out' strategy using gene targeting and FLP recombinase for the functional dissection of complex DNA regulatory elements: Analysis of the β-globin locus control region. *Proc. Natl Acad. Sci. USA*, **90**, 8469.

644. Gregor, P. D. and Morrison, S. L. (1986). Myeloma mutant with a novel 3' flanking region: Loss of normal sequence and insertion of repetitive elements leads to decreased transcription of normal processing of the alpha heavy-chain gene products. *Mol. Cell. Biol.*, **6**, 1903.

645. Matthias, P. and Baltimore, D. (1993). The immunoglobulin heavy chain locus contains another B-cell specific 3' enhancer close to the α constant region. *Mol. Cell. Biol.*, **13**, 1547.

646. Giannini, S. L., Singh, M., Calvo, C.-F., Ding, G., and Birshtein, B. K. (1993). DNA regions flanking the mouse Ig 3' a enhancer are differentially methylated and DNase I hypersensitive during B cell differentiation. *J. Immunol.*, **150**, 1772.

647. Blunt, T., Finnie, N. J., Taccioli, G. E., Smith, G. C. M., Demengeot, J., Gottlieb, T. M., Mizuta, R., Varghese, A. J., Alt, F. W., Jeggo, P. A., and Jackson, S. P. (1995). Defective DNA-dependent protein kinase activity is linked to V(D)J recombination and DNA repair defects associated with the murine *scid* mutation. *Cell*, **80**, 813

648. Kirchgessner, C. U., Patil, C. K., Evans, J. W., Cumo, C. A., Fried, L. M., Carter, T., Oettinger, M. A., and Brown, J. (1995). DNA-dependent kinase (p350) as a candidate gene for the murine Scid defect. *Science*, **267**, 1178.

2 | T cell antigen receptor genes

MARK M. DAVIS and YUEH-HSIU CHIEN

1 Introduction

The characteristics of T cell recognition and the nature of their antigen receptors has long been a difficult and controversial area for immunologists. However, the development of cloned T cell lines and hybrids (as reviewed in ref. 1), the generation of antisera (2) and monoclonals (3–7) enabling the identification of polymorphic T cell receptor (TCR)–heterodimer–CD3 complexes (8–11) and the isolation of four different *TCR* genes (α, β, γ, δ) (12–18) has generated a large amount of information in the past decade. It has also shifted the focus of this area from the question of 'What molecules mediate T cell recognition?' to 'How do they work?' and 'How are they selected in the thymus?' Although antibodies and their genes (see Chapter 1) are a natural touchstone for understanding some aspects of TCRs, the emphasis in this chapter will be on the unique features of the different T cell antigen receptors and their possible biological significance.

That T cell recognition itself operates differently from that of antibody binding to antigen was first clearly suggested by the experiments of Zinkernagel and Doherty, who showed that viral antigen recognition by cytotoxic T cells was compatible only with a certain major histocompatibility complex (MHC) haplotype on the infected cell (19, 20). Work in other laboratories also suggested a similar form of MHC-restricted recognition occurs in other types of T cells (21, 22). We now know that, at least for the majority of T lymphocytes, this involves the recognition of fragments of antigens (for example peptides) bound to specific MHC molecules (see Chapters 4 and 5). As all proteins must be degraded and hence at least in principle available for detection in this way, it may be much harder for pathogens to escape recognition by T cells than by B cells.

The proteins responsible for this recognition on the surfaces of T cells are referred to as 'T cell antigen receptors' or just 'T cell receptors'. These occur as either of two distinct heterodimers, $\alpha\beta$ or $\gamma\delta$, both of which are expressed with the non-polymorphic CD3 γ, δ, ϵ, ζ and in some cases its RNA splicing variant η or $F_c\epsilon$ chains. The CD3 polypeptides, especially ζ and its variants, are critical for intracellular signalling. These are covered in Chapter 3 and therefore will not be discussed further here. The $\alpha\beta$ TCR heterodimer is the predominant form in most lymphoid compartments (90–95 per cent of the cells) of humans and mice and is responsible for most classical helper or cytotoxic T cell responses. In most or all cases, its ligand is a peptide antigen bound to a class I or class II MHC molecule. T cells bearing $\gamma\delta$ TCR are less numerous than the $\alpha\beta$ type in most cellular compartments of these organisms. However, they make up the majority of T lymphocytes in cows (23), for example, and a substantial fraction in chickens (24). The role or roles that $\gamma\delta$ T cells play in the immune system is unknown but is a very active area of research.

2. TCR polypeptides

The search for the molecules responsible for T cell recognition first focused on deriving antisera or monoclonal antibodies specific for molecules on T cell-surfaces. Ultimately, a number of groups identified 'clonotypic' sera (2) or mono-clonal antibodies (3–7), many of the latter precipitating polymorphic $\alpha\beta$ hetero-dimers (8–11). Parallel work exploited the small differences (2 per cent) observed between B and T cell gene expression (25) and isolated both a mouse (12, 13) and a human (14) T cell specific gene which had antibody-like V, J, and C region sequences and could rearrange in T lymphocytes (12, 13). This was identified as *TCRβ* by partial sequence analysis of immunoprecipitated materials (26). Sub-sequent subtractive cloning work rapidly identified two other candidate T cell receptors cDNAs identified as TCRα (16, 17) and TCRγ (15). The role of TCRγ was a puzzle until work by Brenner *et al*. (27) showed that it was expressed on a small (5–10 per cent) subset of peripheral T cells together with another polypeptide, TCRδ. The structure of *TCRδ* remained uncharacterized until the work of Chien *et al*. (18) who discovered it located in the *TCRα* locus, between V_α and J_α.

As shown in Fig. 1, all TCR polypeptides have a similar primary structure, with distinct variable (V), diversity (D) in the case of TCR β and δ, joining (J), and constant (C) regions exactly analogous to their immunoglobulin counterparts, sharing many of the amino acids thought to be important for the characteristic variable and constant domains of immunoglobulins (13–18, 28–31). The mouse C_β region, for example, shares 40 per cent of its amino acid sequences with C_κ and C_λ (13, 14). Characteristic of TCR polypeptides are a single C region domain (versus the two to four for immunoglobulins) followed by a connecting peptide or hinge region, usually containing the cysteine for the disulfide linkage found joining the two chains of the heterodimer (some human TCR $\gamma\delta$ isoforms lack this cysteine and consequently are not disulfide-linked; refs 32, 33). N-linked glycosylation sites vary from two to four for each polypeptide, with no indications of O-linked sugar addition. C-terminal to the connecting peptide sequences are the transmembrane regions, which have no similarity to those of *IgH* genes, but instead have one or two positively-charged residues which appear important for CD3 interaction and signalling (34), perhaps because of some necessary interaction with the acidic residues found in all CD3 transmembrane regions.

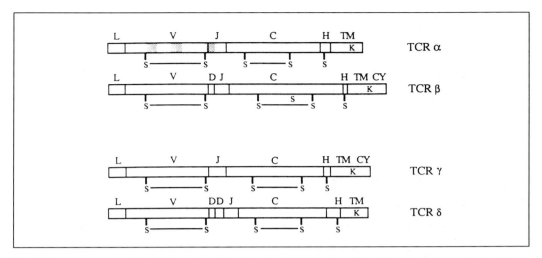

Fig. 1 Primary structure of the TCR polypeptides. Leader (L), V, D, J, and C region elements are indicated, as are cysteine residues (5) and disulfide bonds. The approximate locations of the CDR1, CDR2, and CDR3 loops are shown by shading. The constant regions are divided into a canonical Ig-like domain, a hinge-like region (H), an approximate transmembrane (TM) region, and a cytoplasmic portion (CY). The cytoplasmic portion is only appreciably hydrophilic in β and γ (18) and thus it is only indicated in those polypeptides. K indicates a conserved lysine in the transmembrane region.

3. *TCR* genes

As shown in Fig. 2, *TCR* gene segments are organized similarly to those of immunoglobulins and the same recombination machinery seems responsible for joining separate *V* and *D* segments to a particular *J* and *C*. This was initially indicated by the fact that the characteristic seven and nine nucleotide conserved sequences adjacent to the V, D, and J regions with the 12 or 23 nucleotide spacing between them, first described in immunoglobulin genes, are also present in *TCRs* (29, 30). As discussed in Chapter 1, these sequences appeared to be essential for DNA binding by the recombination machinery and if they are significantly altered, it cannot proceed. The most conclusive evidence of this common rearrangement mechanism has been shown recently by the fact that both a naturally occurring recombination-deficient mouse strain (severe combined immune deficiency, SCID; ref. 35) and mice engineered to lack recombinase activating genes 1 (36) or 2 (37) are unable to rearrange either *TCR* or immunoglobulin gene segments properly. As with immunoglobulins, if the V regions and the J regions are in the same transcriptional direction, the intervening DNA is deleted during recombination. DNA circles of this material can be observed in the thymus (38, 39), the principal site of TCR recombination (see below). In the case of *TCRβ* (40) and *TCRδ* (41), there is a single V region 3′ to the C in the opposite transcriptional orientation to J and C (Fig 2). Thus rearrangement to these gene segments occurs via an inversion. Variable points of joining are seen along the *V*, *D*, and *J* gene segments as well as random nucleotide addition (N regions) in postnatal TCRs (42) as with

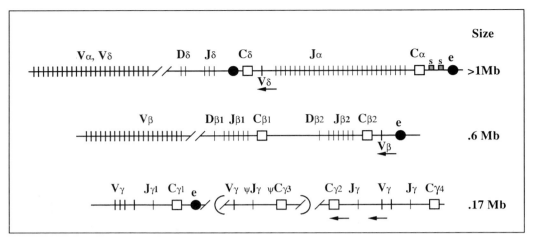

Fig. 2 Genomic organization of *TCR* genes in mice showing approximate location of V, D, J, and C regions as well as enhancer (e) and silencer (s) regions where these have been localized. In the case of $C_\gamma 1$, the existence of a silencer has been postulated but its location is not known. The approximate size of the loci in megabases (Mb; 10^6 bp) is also given. Except where indicated by arrows, the transcriptional orientation of the gene segments is left to right.

IgH genes. The addition of several nucleotides together in an inverted repeat pattern, referred to as P element insertion, at the V-J junction of TCR γ chains has also been observed (43).

Transcriptional regulation has been studied extensively, with enhancer sequences having been identified for all four genes (as reviewed in ref. 45, and indicated in Fig. 2). These TCR enhancers all share sequences with each other and some of the nuclear factors which bind to them are also shared with Ig genes. As with immunoglobulins, promoter sequences are located 5' to the *V* gene segments. In general, $D \rightarrow J_\beta$ rearrangement and transcription occurs fairly often in B cells (45), particularly in tumour lines, but V_β rearrangement or transcription appears highly specific to T cells. Particularly of interest with respect to the question of what determines whether a T cell expresses TCRαβ versus γδ are the observations of 'silencer' sequences 3' of C_α (46) and in the $C_\gamma 1$ locus (47). It has been suggested that these 'repressor sites' could turn off the expression of either of these genes, promoting a progression towards the other lineage.

The chromosomal locations of the different *TCR* loci have been delineated in both mouse and humans and the results are summarized in Table 1 (as reviewed in refs 29, 48). As with immunoglobulin genes, these *TCR* genes are the site of oncogene translocations in human neoplasms (48), particularly the α/δ locus, perhaps because it spans the longest developmental window in terms of gene expression, since *TCRδ* is the first gene to rearrange in ontogeny and *TCRα* is the last (as discussed in more detail below). Alternatively, this may just be a function of target size, since α/δ is in excess of 1 Mb (as shown in Fig. 2). Interestingly, in humans *TCRα/δ* is on the same chromosome as the *IgH* locus and $V_H \rightarrow J_\alpha$

Table 1 Chromosomal locations of the α/δ, β, and γ loci [a]

Gene	Mouse	Human
TCRα/δ	14C–D	14q11–12
TCRβ	6B	7q35
TCRγ	13A2–3	7p15

[a] As reviewed in refs 29 and 48.

rearrangements (by inversion) have been observed in some human tumour material (49, 50). The functional significance of this, if any, is not known.

Despite the many similarities with Ig genes, there are some unique features of the *TCR* loci and their organization which are worthy of comment. The most striking of these is the location of the *TCRδ* locus with the *TCRα* such that any V_α to J_α rearrangement must necessarily delete the *TCRδ* locus from that chromosome (18). This has in fact happened in most peripheral $\alpha\beta$ cells and seems designed to prevent the co-expression of a $\gamma\delta$ heterodimer (in addition to the possible role of silencer sequences discussed above). This arrangement would also allow for the use of common V regions for *TCRα* and *TCRδ*. While this has been observed (51), it seems rare and generally V_α and V_δ gene segments are distinct.

Another interesting feature of *TCR* gene structure is the tandem nature of J_β–C_β in the *TCRβ* locus. This arrangement is preserved in all species that have been characterized thus far (mouse, human, chicken, frog) but its purpose remains obscure. The two C_β coding sequences are identical in the mouse and nearly so in humans and other species (30) so are unlikely to represent two functionally distinct forms of C_β. The J_β clusters have relatively unique sequences, however, and data concerning mice which have deleted $J_\beta 2$ suggest that they have impaired responses (52). Therefore it may be that this J_β C_β duplication is a convenient way of increasing the number of J_β gene segments. This is a reasonably attractive explanation considering that a third distinguishing feature of *TCR* genes is the very large number (approximately 50) of J_α gene segments (29, 30, 53), far more than has been described for any Ig locus (see Section 4). The weakness of this argument is that it fails to explain why C_α is not duplicated.

Several important genetic mechanisms that are well known in antibody genes do not appear to occur in *TCR* genes, particularly somatic hypermutation, any secreted form or any form of C-region switching. Considering the nature of TCR function however, it does not appear that any of the above mechanisms would be useful, in that cell-surface receptors which have cell-surface ligands generally seem to have lower affinity than molecules which have to bind ligands in solution (54), yet still retain a high degree of specificity (55). Hence there may be no need for the increase in affinity that immunoglobulins achieve through somatic hypermutation. In addition, such a mutator mechanism operating in the periphery might negate the effects of thymic selection (see below). With respect to a secreted form, one could argue that there is no need for a TCR version because that is what antibodies are for. In addition, the relatively low affinities reported for TCRs thus far (5×10^{-5} M $-$ 1×10^{-7} M) (56–58) would mean that the concentration of a secreted form would have to be very high to have any effect (see Section 7). As for C_H switching, the different Ig types seem designed such that a given V region specificity can be associated with distinct constant regions which have different properties in solution (such as complement fixation, basophil binding, etc.). In lacking a secreted form therefore, TCRs would have no need to 'switch' their C regions.

4. Rearrangement and expression of T cell receptor genes in the thymus

As with B lymphocytes, T cells progress through a series of distinct developmental stages, associated with specific surface markers and *TCR* gene rearrangements (as reviewed in refs 59–61). Of great experimental convenience is the fact most of this occurs in the thymus (see Fig. 3), the lobes of which are easily removed and can yield up to 10^8 T cells at different stages of development in young adult mice. T cell development appears to start with the migration of stem cells from either the liver in a fetus or from the bone marrow of more mature organisms to the thymus. Early cells are largely negative for CD4 and CD8 expression and are thus easily distinguished from the later stages. Soon afterwards, the first *TCR* rearrangements begin, with *TCRδ* first, followed by β and γ, all being detectable in early fetal

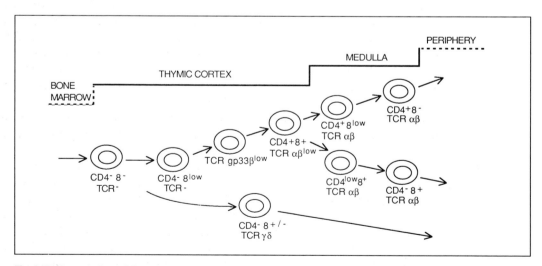

Fig. 3 Differentiation of T cells in the thymus. T cells progress through various modalities of CD4 and CD8 expression as they rearrange their *TCR* genes and become either αβ or γδ TCR bearing cells. The mature cells accumulate in the medulla where they then migrate to the periphery (e.g. spleen, lymph nodes, etc.).

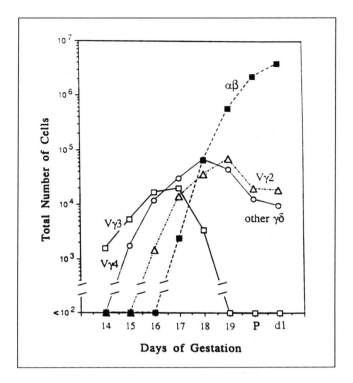

Fig. 4 Specific utilization of V_γ gene segments during fetal thymic development (from ref. 69). Particular V_γ gene segments are rearranged and expressed in an ordered fashion during fetal thymic development. The following symbols correspond to the adjacent V_γ region: $V_\gamma 3$ (–□–), $V_\gamma 4$(–○–), $V_\gamma 2$(–△–), and $\alpha\beta$ T cells are represented by –■–.

thymocytes of the mouse (day 14–15). $\gamma\delta$ TCR bearing cells are first detectable at this stage, and only on day 16 do *TCRα* rearrangements begin and $\alpha\beta$ TCRδ bearing T cells first appear (Fig. 4). For thymocytes destined to be $\alpha\beta$ cells, cell transfer studies (62, 63) show that these cells first express low levels of CD8, then increase the density of this marker as well as turning on the expression of CD4. This CD4$^+$8$^+$ cell type makes up the majority (80 per cent) of thymocytes. As discussed in more detail below, TCR rearrangements and the decision to become an $\alpha\beta$ or $\gamma\delta$ T cell is made just prior to this stage.

In these CD4$^+$8$^+$ cells, TCR expression is low and many of the cells have reduced amounts of cytoplasm similar to 'resting' T and B cells in the periphery. These cells are very sensitive to a variety of shocks, such as corticosteroids or irradiation by programmed cell death (apoptosis), and most will die due to a variety of selective processes (see Section 5). All the cells mentioned up until now are located in the cortical epithelium region of the thymic lobes. A fraction of thymocytes can be observed maturing along at least two distinct pathways, one which will become the CD4$^+$ $\alpha\beta$ T cells and the other will become CD8$^+$ $\alpha\beta$ T cells. The former cells recognize class II MHC-restricted antigens and will be mostly of the helper pheno-type, that is they will augment B cell and other responses. The later subset of T cells will recognize class I MHC antigens and will predominantly be cytotoxic T cells. Thus, they are able to kill cells bearing the appropriate peptide/MHC complex directly. This progression to either of these two end points is characterized by a gradual loss of either CD8 or CD4 expression, an increase in TCR expression to about 25–30 times more per cell, and also a somewhat increased cell size (63). These are the mature thymocytes which have survived selection and they congregate in the medullar (inner) region of the thymus where they will be exported to peripheral lymphoid organs. Recent work with *TCR* gene-targeted 'knock-out' mice has been very useful in clearing up some points related to thymic differentiation and has revealed others (64–66). One of these is the fact that, if *TCRα* is inactivated, the lack of an αβ lineage has no apparent effect on γδ cells and vice versa, largely putting to rest speculation that the two cell types might collaborate in some way. *TCRβ* inactivation has also revealed that functional *TCRβ* expression is required for the progression into the CD4$^+$8$^+$ stage (64). That is, when *TCRβ* is not present, no cells appear in this category. When *TCRβ* is restored by the introduction of a transgenic B, this pathway resumes (64) and some TCRβ is expressed on the cell-surface. Further biochemical evidence shows that this is due to TCRβ being co-expressed as a heterodimer with a non-TCRα polypeptide, gp33 (67). Although the gp33 polypeptide has not yet been sequenced, this expression of a surrogate α chain prior to *TCRα* rearrangement is strikingly similar to the case of pre-B cell differentiation where the *IgH* gene is rearranged and expressed on the cell-surface with a surrogate light chain (λ5) prior to Igκ or λ (see Chapter 1). Thus there appear to be carefully delineated checkpoints at critical stages of both T and B cell differentiation, possibly designed to reduce the amount of cell wastage due to faulty V(D)J rearrangement (as discussed in more detail below).

4.1 Programmed γδ *TCR* rearrangements

Another interesting aspect of *TCR* gene rearrangements in thymus is the sequential rearrangement and expression of specific V_γ gene segments at distinct times during fetal and neonatal life. As shown in Fig. 4, these particular sequences appear to be activated at set times and produce distinct types of γδ T cells (68, 69), perhaps most notably the dendritic epidermal type of γδ T cells, which appear first in fetal thymic development (day 14 of gestation in the mouse) and have identical V(D)J sequences for both TCRγ and TCRδ. Other waves of γδ T cells have a more typical sequence diversity in the junctional region, but still seem to follow this rigid pattern of expression. The purpose of this pattern or of these cells is not known, but the specific rearrangements appear to be 'hard-wired' and not based on thymic selection as animals with a disabled *TCRδ* gene still produce the characteristic TCR_γ rearrangements at the usual times (66). Studies of Weissman and colleagues have further showed that these specific rearrangements are induced by the thymus (70).

4.2 Allelic exclusion

As with immunoglobulins, TCRs generally exhibit allelic exclusion, that is, only one specific α or β chain (or γ or δ) is expressed by a given cell. This can also be seen by the fact that in *TCR* transgenic animals with the appropriate genetic background, the transgene(s)-encoded TCR polypeptide(s) is expressed on most T cells, blocking endogenous rearrangement (71). While this rule seems very strictly true for TCRβ, it seems much less rigid in the case of TCRα, where fully 10–30 per cent of T cell lines or hybridoma appear to have two potentially functional α chains (72). Nevertheless, in cases where it has been analysed further, often one of the two possible transcripts appears on the cell-surface. Therefore, possibly only certain α chains are able to pair with a particular β chain and a failure of the first α chain to do so triggers a second rearrangement.

4.3 Choice of γδ versus αβ lineage

As discussed previously, several observations have suggested mechanisms whereby a cell bears an αβ versus a γδ TCR. Firstly, the 'silencers' observed in TCRα (46) and TCRγ (47) suggest a genetic switching off of one or the other. Secondly the location of *TCRδ* with the α locus (18) suggests that a failure to achieve a functional *TCRδ* gene could trigger a default pathway rearranging *TCRα* and deleting the *TCRδ* locus. In fact, most peripheral αβ T cells have deleted *TCRδ* from both chromosomes (18). Also interesting is the observation in the human δ locus of characteristic rearrangements called TEA sites which seem designed to inactivate *TCRδ* by deletion (73). The existence of such a rearrangement would argue for a lineage decision independent of rearrangement outcomes (for example the 'default model'). Experiments designed to distinguish between these possibilities have produced mixed results. Winoto *et al.* (74) isolated circular DNAs excised from thymocytes and found that *TCRδ* was not rearranged, but Takeshita *et al.* found defective rearrangements (75). Although not resolved, the weight of current opinion favours the predetermined lineage option since the *TCR* gene knock-out experiments discussed previously show no apparent effect on the absence of one lineage over another (65, 66), although this could be due to a homeostatic effect.

5. TCR repertoire selection in the thymus

Perhaps the most interesting aspect of thymic differentiation is the massive amount of T cell death that occurs there. From thymocyte labelling experiments, Scollay *et al.* (76) have estimated that only 1–5 per cent of all thymocytes ever migrate to the periphery, while the rest apparently die in the thymus, seemingly by apoptosis. There appear to be at least three general categories underlying this selection. These are: failure to rearrange functional TCR heterodimers; 'negative selection', for example the removal of self-reactive cells; and lastly, 'positive' selection in which TCRs are apparently screened for some weak complementarity to self-peptide/MHC complexes. While the contributions of these processes to thymic attrition are not known, a discussion of each is outlined below.

5.1 Failure to rearrange an expressible TCR heterodimer

Two-thirds of each V(D)J rearrangement should be out-of-frame. If cells are rigidly separated into $\alpha\beta$ and $\gamma\delta$ pools, then a T cell would have a one-third chance of successfully rearranging one chromosome's TCR and then the remaining two-thirds would have an additional chance (one-third) to succeed on the second chromosome. Thus the fraction of T cells that would be expected to be successful in expressing one TCR polypeptide would be

$$1/3 + \frac{2/3}{3} = 0.55.$$

As each TCR is a heterodimer, the odds of getting expressible chains of each would be 0.55×0.55 or 0.3. Thus 70 per cent of T cells could be expected to die due to failure to properly rearrange one or both chains of their TCRs. The actual number is likely to be lower as there are a number of third and fourth chances that seem to be available, at least to some extent. One of these involves the *TCRβ* locus where, at least *in vitro*, a non-productive VDJ $C_\beta 1$ rearrangement was replaced by a new VDJ $C_\beta 2$ rearrangement (77). There is a noticeable bias towards $C_\beta 2$ rearrangements in antigen-specific T cell lines and hybrids (70 per cent actual versus 50 per cent predicted), consistent with this being a way in which T cells can try again to obtain a productive *TCRβ* (78). Similarly, the large number of J_α elements in the *TCRα* locus make possible additional $V_\alpha \rightarrow J_\alpha$ rearrangements on the same chromosome that could bypass a non-productive α rearrangement, as observed by Livak *et al.* (79). Therefore the 70 per cent number calculated above is likely to represent an upper limit. The existence of these apparent 'salvage' pathways is particularly relevant to the discovery of the gp33 protein as a surrogate α chain (67). Having such a protein, which presumably does not rearrange (by analogy to $\lambda 5$), would allow cells to 'try out' their *TCRβ* rearrangements prior to negative and positive selection (see Sections 5.2 and 5.3).

5.2 Negative selection

If a T cell reacts with a self-peptide/MHC complex or a self-superantigen/MHC in the thymus, it generally dies very quickly. This was first seen directly by Kappler *et al.* (80, 81) who observed that particular V_β bearing T cells disappeared during the $CD4^+8^+$ stage in certain strains of mice. This case and others like it now seem attributable to a unique class of genes, termed self-superantigens, expressed on cells in the thymus (especially B cells) and resulting in their death by apoptosis. These have now been found to be a series of endogenous retroviral alleles of a previously unknown type II membrane protein derived from mouse mammary tumour virus (MMTV) as discussed in Section 8. Soon afterwards, *TCR* transgenic mice were made which were reactive to self- or introducible foreign antigens, and these were found to delete either early in the $CD4^+8^+$ cells or somewhat later (82–84). The advantage of *TCR* transgenic mice is that in the right selective environment, more than 50 per cent of all T cells will express the transgenic receptor, making it very easy to follow the fate of those cells. This was particularly useful in experiments which showed that direct injection of the appropriate peptide in *TCR* transgenic mice results in massive cell death by apoptosis within hours (85). There remains some controversy about the exact stage at which death by negative selection occurs, since the extensive work of Guidos *et al.* (63) in a non-transgene dependent system shows that at least one form of Mls/MMTV-mediated deletion occurs immediately after the CD4$^+$8$^+$ stage. In contrast, the *TCR* transgenic data suggest that this form of selection can happen as early as the beginning of the CD4$^+$8$^+$ stage (82, 83). However, this may be misleading since the presence of $\alpha\beta$ transgenes accentuates negative selection, probably because TCRs are expressed somewhat earlier and at a slightly higher density than normal (84).

In any case, the penalty for self-reactivity in the thymus is generally death, although a variety of half measures have also been observed to a lesser extent, such as lower CD8 density, lower TCR density and paralysis (anergy), so death may not be the only possible result (59–61). It may be, however, that the *TCR* transgenic systems so overload the animal with T cells of a given specificity that the leakiness and half measures that are seen are not likely to be encountered normally. What about T cells that recognize self-peptides that are not usually found in the thymus? It seems that these usually escape. There are a number of systems which indicate that targeting antigens to a specific organ often hides them from the view of T cells and has no effect on thymic development. Whether there is then a T cell response in the periphery or not seems variable. In one very clear case of viral antigen expression in pancreatic islet cells, reactive T cells were clearly present in the periphery, but did no harm unless they were first 'primed' with a viral infection (86). Consistent with this finding are results (87) showing that stimulation *in vitro* to activate resting T cells is much more stringent (requiring more than one signal) than re-stimulation of cells that had been previously activated (requiring only a signal through the TCR).

5.3 Positive selection

Perhaps the most mysterious form of thymic TCR repertoire selection is that which seems to pre-screen TCRs for some 'goodness-of-fit' with self-MHC/peptide complexes, referred to as 'positive selection' because TCRs are selected for their promising qualities as opposed to being eliminated due to their deleterious ones (for example 'negative selection'). This was first suggested by the experiments of Bevan (88) and Zinkernagel (89) in which MHC F$_1$ bone marrow cells of mice show preferential MHC restriction depending on which parental thymus they were exposed to. Thus A × B cells would recognize antigen with A type MHC molecules when allowed to develop in an A thymus and B type MHC restriction when exposed to a B thymus. These experiments were later confirmed by transgenic studies in which specific MHC molecules were shown to be crucial to the progression to mature CD8$^+$ or CD4$^+$ T cells (90–93). Cortical epithelial thymic stromal cells appear to be particularly important in this process (93). It is also clear from these experiments that T cells make an all-or-none decision to become CD4$^+$ or CD8$^+$, which can be predicted based on their original MHC class II or class I preference, respectively (90–94). The way in which this choice is made is still being debated (61, 95).

What could be the mechanism of positive selection? Even before the structure of MHC molecules was known it was suggested that cross-reactive self-antigens were important (96), since the goal seems to be to select TCRs that can recognize self-antigens to some extent but not too well (or they become a liability). That such self-peptides bound to MHC molecules might be characterized by a weak affinity for a given TCR (97) is an extension of this idea. Evidence showing that mutations in the binding groove of various MHC molecules could affect the outcome of positive selection were the first real physical evidence that specific self-peptides as well as MHC molecules are important (98–101). Later experiments showed that peptide mixtures exogenously added to cultures of fetal thymocytes could promote positive selection (102, 103). Since it is also clear that the peptides occupying a given MHC molecule can vary from 0.005 per cent to 20 per cent of the total, Janeway et al. first recognized that avidity considerations (that is the affinity of a given peptide in an MHC for a TCR, times its density) could be a critical parameter for selection (104). This was also suggested by the data of Berg et al. (99) who showed that even 2–3 fold differences in the density of an MHC molecule could dramatically affect the efficiency of positive selection. In addition to showing the importance of ligand density, this latter result also suggests that positive selection in the thymus is an inefficient process, at least in a *TCR* transgenic model. Similarly, data of Roby et al. and Lee et al. showed that increasing the density of CD8 molecules or thymocytes could shift the outcome for particular T cells from positive to negative selection (105, 106).

Very recent experiments of Ashton–Richardt et al. (107) and Hogquist et al. (108)

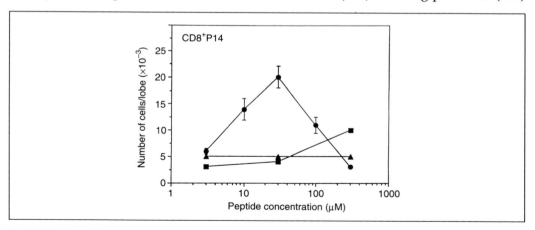

Fig. 5 Induction of either positive or negative selection by the same peptide in fetal thymic organ culture (from ref. 107). Shown are the effects of a range of concentrations of an influenza haemagglutinin peptide (P14) on the number of mature T cells. The CD8+P14 T cells derive from a TCR transgenic animal specific for the influenza peptide and thus express the same TCR together with the CD8 marker. As seen in the figure, intermediate concentrations of peptide promote differentiation (in Class I MHC processing deficient mice) but at higher levels trigger negative selection.

strongly support the differential avidity model of Janeway and colleagues (104). Thus, as shown in Fig. 5, low doses of a foreign peptide infused into fetal thymic organ cultures are able to trigger positive selection of *TCR* transgenic T cells specific for that peptide plus a self-MHC molecule. Higher ($10 \times$) concentrations of peptide trigger negative selection. Thus even what is probably the highest affinity peptide/ MHC complex can trigger positive selection when present at low enough concentrations (occupancy). Thus while affinity or differential activation properties can be a factor ((especially from the data of Hogquist *et al.* (108)) they need not be. Therefore, the phenomenon of positive and negative selection appears to be a continuum, in which limited stimulation through the TCR with a self-peptide MHC ligand in the thymus results in positive selection and cell survival whereas too low a level of stimulation results in cell death, as does too high a level of stimulation (triggering negative selection). In this way, the thymus seems to select the most useful (and least harmful) TCR repertoire.

6. αβ TCR recognition of peptide/MHC complexes

Overall, a clear difference that emerges between *TCR* and immunoglobulin genes is that the number of *TCR* gene segments and their pattern of rearrangement seems heavily weighted towards diversifying the V–(D)–J junctional region, equivalent to the CDR3 loop of immunoglobulins (30). This analysis is summarized in Table 2 and shows that the large number of J regions in TCRα and -β, the use of two or even three D regions simultaneously in TCRδ, as well as the ubiquity of N-region addition and the small number of V region segments all serve to skew TCR diversity into this one region and to be lacking in others. Thus *TCRαβ* and γδ genes are inherently capable of 10^4–10^7 times more diversity in their CDR3 regions than are mouse immunoglobulins (excluding the effects of somatic mutation).

Table 2 Sequence diversity in *TCR* and Ig genes [a]

	Ig		TCRαβ		TCRγδ	
	H	κ	α	β	γ	δ
V segments	250–1000	250	75	25	7	10
D segments	10	0	0	2	0	2 [b]
Ds read in all frames	rarely	—	—	often	—	often
N-region addition	V–D, D–J	none	V–J	V–D, D–J	V–J	V–D1, D1–D2, D1–J
J segments	4	4	50	12	2	2
V region combinations	62 500		1875		50	
Junctional combinations	10^{11}		10^{16}		10^{18}	

[a] Adapted from ref. 30.
[b] Used simultaneously.

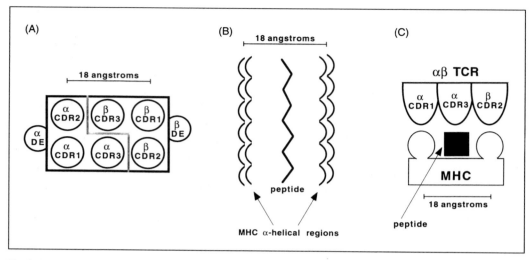

Fig. 6 Model of αβ TCR recognition of peptide/MHC complexes (from ref. 109). As discussed in the text, the dimensions and placement of CDR loops on antibodies (A) and of the peptide binding site of MHC molecules (B), as well as concentrations of TCR sequence diversity in CDR3 loops suggested the model shown in part (C).

Why might this be? Looking at immunoglobulin structures one sees that the CDR3 loops from the heavy and light chain V regions line up in the middle of the binding site (Fig. 6A) flanked on either side by the CDR1 and CDR2 loops. The general structure and dimensions of an Ig binding site are likely to be very similar to those of a TCR, and such a structure fits very well to that of the peptide-binding groove of MHC molecules such that the CDR3 loops could scan the peptide (Fig. 6B and C) and the CDR1 and 2 loops could contact the α helices of the MHC (30, 31, 110). This could explain the heavy concentration of diversity in the CDR3 loops of TCRs as being the most efficient way to recognize a very large number of possible peptides. Such a model suggests a unique specialization of TCRs for recognizing their 'hybrid' ligand, and also has suggested a simple evolutionary model in which a self-MHC receptor could be converted into an MHC-peptide receptor (for example a TCR) by a single transposon insertion event (30).

The above model is supported by the characterization of many TCRα and -β sequences which often show conserved residues in their CDR3 regions (when the ligand is identical, refs 111–116). In several cases, either fortuitous (112, 113, 116) or intentional changes (117) in a CDR3 sequence alter or abolish antigen recognition. CDR3 looping 'swapping' studies, however, have thus far only resulted in the complete loss of reactivity (118) rather than any transfer of peptide specificity from one heterodimer to another. Recently, Jorgensen *et al.* (119, 120) used a novel strategy to test this model in a comprehensive way. This method uses peptides altered at residues important for T cell recognition to immunize mice and then examines the resulting TCR repertoire. To keep the TCRs as related as possible, the mice immunized were transgenic for the α or β chain of the original TCR. This essentially holds one chain or the other 'constant' in the animal's T cell population and yields immediate information, since one or the other usually gives a better response. The TCR chain giving the weakest response is likely to be the one responsible for recognition at that residue. One of the most striking findings in this analysis was that a Lys→Glu change on the peptide triggered a Glu→Lys change in the V_α CDR3 of the responding T cells while keeping the same V_α and $V(D)J_\beta$. This and other data all point to a 'salt bridge' between this position on the peptide and the opposite charge on the TCR. Changes at a second T cell-sensitive residue on the peptide, three amino acids C-terminal to the first, provoked changes in both V_β segment usage and V_β CDR3 sequences. Again, the CDR3 changes often included an opposite charge motif. The V_β changes are more difficult to rationalize, but seem more to do with some overall structural imperatives than to necessarily imply direct V_β→peptide contact. This is also seen in the several situations where the V_βs are identical but the CDR3s are different, and the TCRs can distinguish between Lys or Glu at that position in the peptide. A third residue on this model (the most C-terminal) has now been analysed and appears to be a V_α CDR3 determinant (120). With three points of contact now established and the orientation of peptides within MHC molecules seemingly invariant, we have proposed a topology (Fig. 7) for this particular complex which would place the V_α over the E_α helix and V_β over the E_β helix. This is consistent with the conclusions of Taylor *et al.* regarding influenza hemagglutinin-reactive T cells (121) and is in the opposite orientation from that proposed by Hong *et al.* (122), although in both of these cases the data are less direct. Thus it remains unclear as to whether TCR orientation with respect to peptide/MHC complexes will be uniform.

The expected dimensions of a TCR binding could mean that direct TCR contact with the α helices of MHC molecules would occur at the same time as TCR–peptide contact. The original models (30, 31, 110) suggested that this would be governed by the CDR1 and CDR2 loops of V_α and V_β. That this is no uniform 'docking site' for TCRs on a given MHC was first shown by Ajitkumar (123) and Peccoud (124) who provided evidence that different TCRs are affected differently by amino acid

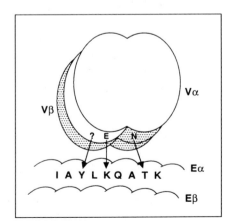

Fig. 7 Mapping peptide–TCR contacts (from ref. 109). A summary of the work of Jorgensen *et al.* (119, 120) showing the various amino acid residues (shown using a one-letter code) of the insect cytochrome c peptide which interact with the CDR3 loops (internal circles) of Vα and Vβ. Eα and Eβ represent the α helices of these class II MHC polypeptides.

substitutions (either naturally occurring or introduced) on the α helices of class I or class II MHC molecules, respectively. Ehrich *et al.* (125) have refined this type of analysis significantly by looking at a broad panel of class II MHC mutants (in I-E^k) and their effects on T cell reactivity of very similar TCRs. Surprisingly, they found that V_β usage does not predict which MHC mutants will affect reactivity, but that even very slight (single amino acid) changes in V_α or V_β CDR3 regions (and nowhere else) had a marked affect.

The MHC mutants affected are too broadly spaced to be explained by local effects. Thus the most likely explanation is that changes in TCR–peptide interactions have very widespread consequences for TCR–MHC contact. Together with the lack of a V_β correlation, this also suggests that TCR–MHC contacts are not as specific as those observed for TCR–peptide (although they are still necessary, as even very slight changes such as Ala→Val on the MHC can completely abrogate T cell reactivity). The most striking evidence for this view is supplied by Ehrich *et al.* using cross-reactive TCRs (125). These are single T cell hybrids which can recognize either of two different forms of the antigenic peptide, moth cytochrome c (99K) or (99E). Using the panel of MHC mutants, it was shown that the same TCR appears to contact the MHC very differently, depending on which peptide is being recognized (125). This variation in mutant sensitivity is equivalent to that seen with completely different T cell receptors recognizing different peptides. These data show that TCR–MHC contact is very 'fluid', in that it can be made in a variety of ways even with the same TCR and MHC, with the final configuration dominated by the peptide.

7. Affinity of αβ TCRs for peptide/MHC ligands

Measuring the affinity of αβ TCR for peptide/MHC ligands has been very difficult, partly because of the membrane-bound nature of each component and partly due to what (in most cases) seems to be a very low affinity. The first of these problems necessitated developing means of producing soluble forms of MHC or TCR or both. This has now been accomplished by a variety of methods including replacing the normal transmembrane regions with signal sequences for GPI linkage, Ig chimeras, expression and refolding from *Escherichia coli*, and other methods (as reviewed in ref. 54). The first measurements of affinity of αβ TCR for defined peptide/MHC ligands were obtained by Matsui *et al.* (56) and Weber *et al.* (57). In the former case, shown in Fig. 8, Matsui and co-workers were able to compete away the binding of a labelled anti-TCR V_β antibody with high concentrations of a soluble peptide/MHC complex, deriving a K_D for TCR affinity of 50 μM. This interaction is highly specific as a variant peptide (MCC99E) bound to the MHC does not compete as shown in Fig. 8A. When a second T cell is used, this time specific for MCC(99E), the situation is reversed (Fig. 8B) and the 99E/MHC complex now competes (again giving a 50 μM affinity).

Weber *et al.* (57) measured the ability of a soluble TCR to block specific T cell stimulation. Comparing the dose–response curve of this inhibition with that of an

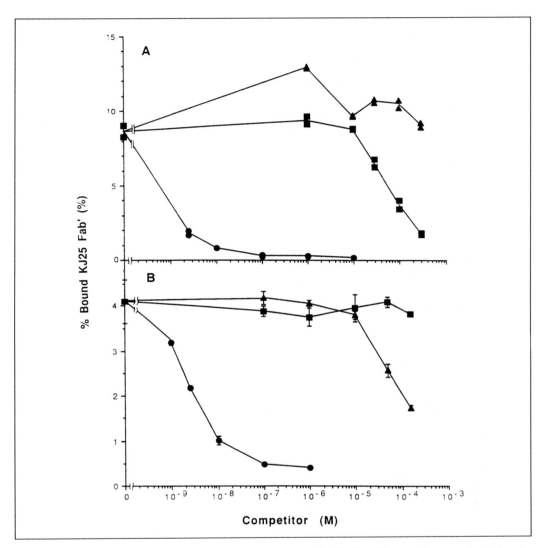

Fig. 8 Low affinity of αβ TCR binding to peptide/MHC (from ref. 56). Specific peptides were loaded on to soluble class II MHC protein and the resultant complexes were used to compete for binding with a labelled anti-TCR antibody Fab fragment. Panel A shows competition with a cytochrome c (MCC) peptide/MHC complex (—■—) but not an MCC (99E) complex (▲) with an MCC specific T cell. The filled circles, ●, show unlabelled antibody competing with itself. Panel B shows that the MCC (99E) complexes but not the MCC complexes are able to compete with the antibody for an MCC (99E) specific T cell. In both cases, a K_D of ~50 μM was calculated.

antibody to the same MHC, they estimated a TCR K_D of 10 μM. While these two numbers are in reasonably close agreement, recent data of Sykulev *et al.* (58) on an alloreactive cytotoxic T cell-derived TCR utilizing the competition method suggest a much higher affinity (~0.1 μM). Various related peptides loaded on to the MHC gave decreasing affinities, and these relative affinities correlated with the efficiency of cytotoxicity, even down to a K_D of 10^{-3} M (58). Which result is more typical is not yet clear. If the low affinity measurements of Matsui and Weber are more representative, TCRs would rank rather low on the affinity scale (54) even compared with other cell surface molecules on T cells (such as CD28). Even with higher affinities, the fact that T cells need relatively few ligands (50–1000) (126–128 and Reay *et al.*, unpublished observation) for activation or killing versus the more numerous (by 50–1000 times) adhesion molecule ligands would suggest that TCR molecules are not driving T cell interactions with other cells. That is, it seems likely that they are able to bind only after the (in some cases, higher affinity) adhesion molecules had already engaged (54, 56, 129). Such multistep recognition processes are employed in other biological systems, such as sequence-specific DNA recognition where the protein molecules often have an intrinsic affinity for any DNA, or in lymphocyte homing and extravasation, where three distinct steps have been proposed in which different surface molecules are successively engaged (130). Adhesion molecule engagement as the first step in T cell recognition has been proposed previously (131), based on experiments showing that antibodies to the T cell adhesion molecule LFA-1 block T cell aggregation, whereas anti-CD3 does not, in a cytotoxic T cell and transformed target model and in another T cell model where dendritic cells are the antigen presenters (132). The biological effect of such a progression would be to keep T cells away from cells and tissues which lack adhesion molecule ligands and to focus them on 'approved' antigen presenting cells. This would serve two functions, namely suppressing autoimmunity by keeping T cells away from most tissues and organs and secondly, keeping T cells from 'wasting their time' surveying such cells. It is interesting to note that one consequence of inflammation is the expression of T cell adhesion molecule ligands such as ICAM-1 and V-CAM-1 on tissues that do not normally express these molecules (133). In the context of the model discussed above, this would be a way in which a damaged or infected tissue could 'draw the attention' of T lymphocytes.

8. Superantigens

One of the most surprising findings that has come out of the study of TCRs is the discovery of a completely unexpected ligand for $\alpha\beta$ T cells, referred to as 'superantigens' because while they also bind to MHC molecules, they are much less sensitive to polymorphism than are peptide antigens. This is due to the fact that they bind outside of the peptide binding groove of class II MHC molecules, whereas most of the MHC polymorphism in focused within this site. Also, unlike peptide/MHC ligands, they bind to the 'sides' and CDR1 + 2 of TCR V_β as shown in Fig. 9 and do not appear to involve the V_β CDR3 or V_α except indirectly. This class of antigen was first observed as endogenous loci in particular mouse strains by Festenstein (134), which could weakly stimulate heterogeneous T cells and could cause the elimination of specific V_β bearing T cells (see Section 5). Thus they were referred to as the *Mls* (murine lymphocyte stimulatory) loci. They are now

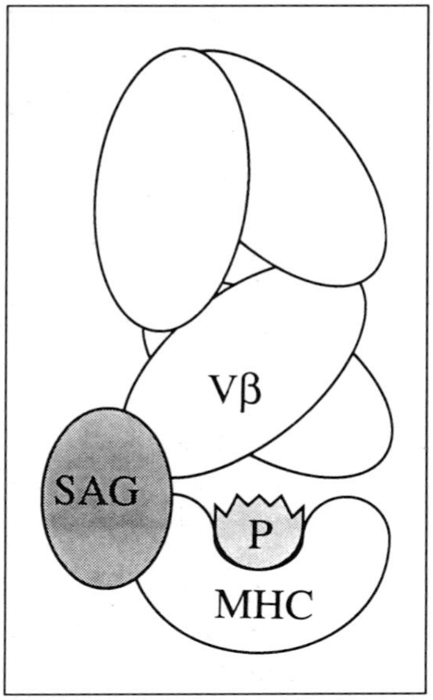

Fig. 9 Superantigen (SAG) ligation of TCR and class MHC molecules (adapted from ref. 138). P indicates the peptide antigen. Unlike peptides, superantigens do not bind in the MHC groove nor do they contact CDR3 loops. Instead they appear to bind to the solvent exposed portion of a Vβ (including CDR1, 2, and D→E loops (see Fig. 6).

known to be derived from a type II membrane-bound retroviral protein derived from mouse mammary tumour virus (MMTV). There are now at least ten such *MMTV/Mls* loci that have been defined and mapped to various locations in the mouse genome (135–137). They each have particular V_βs which they react to. A second class of 'superantigens' were first observed by Janeway *et al.* who showed that bacterial toxins could also stimulate T cells in an MHC-unrestricted fashion and did not require protein processing (138). There were also shown to be specific for particular TCR V_βs (135–137). As shown in Fig. 9, both MHC and TCR V_β epitopes are bound simultaneously, thus ligating TCR + MHC much as the correct peptide would and triggering T cell activation in many cases (but other effects such as 'anergy', for example paralysis of the T cell, have also been described; refs 135–137)). Why organisms as diverse as bacteria and retroviruses have such molecules is extremely interesting. There must be advantages, perhaps in occupying/diverting/exhausting the immune system in order to weaken a specific (and pathogen toxic) response.

9. γδ T cell recognition
9.1 Specificity of diverse γδ T cells

The question of what γδ T cells recognize and what function, or functions, they serve in the immune system has been a mystery ever since their discovery in 1984 by Saito *et al.* (15) studying a cDNA sequence that rearranged in T cells but was neither TCRα nor -β. There are reports of γδ T cells recognizing self-antigens, mycobacteria and heat-shock proteins, bacteria superantigens, carbohydrates, and classical as well as non-classical MHC molecules, but very few of these reactivities have been extensively characterized (139). In particular, it has been clear for some time that γδ T cells do not have the same propensity for recognizing classical MHC molecules that αβ T cells do. In fact, most antigen-specific γδ T cells show no demonstrable MHC restriction (68, 140). Furthermore, whereas a large fraction (approximately 10 per cent) of αβ T cells derived from a mixed lymphocyte reaction (MLR) show MHC allele specific or alloreactivity, very few γδ T cells do so, although a number of laboratories have identified clones that recognize classical or non-classical MHC molecules (such as TL, Qa, and CD1; refs 68, 139). In many cases, these γδ T cell clones also show a broad cross-reactivity (140, 141) that is not seen for αβ alloreactive T cells. These data all suggest that there may be fundamental differences in antigen recognition and/or functional consequences between γδ T cells and αβ T cells.

Because, in most cases, classical MHC molecules are clearly not involved in γδ T cell recognition, it has been suggested that non-classical MHC molecules might serve as antigen-presenting molecules or that as-yet-unidentified surface molecules on γδ T cells themselves do so. As these possibilities are very difficult to test experimentally, the requirements for antigen processing and presentation of γδ T cell recognition have been difficult to assess. Recently, Schild et al. (142) have taken a different approach to this question by carrying out experiments to ask when γδ T cell recognition does involve MHC molecules, what kind of antigen processing is required? Is any of the specificity conferred by bound peptide and is the MHC molecule recognized similarly by γδ and αβ TCRs?

The results are very striking. They show that peptides (bound to the MHC) do not confer specificity and that no conventional antigen processing pathways are required in the MHC recognition by these γδ T cells. All variations in the ability of different stimulator cells to activate these γδ T cells can be attributed solely to the level of their surface MHC expression, regardless of the species origin (mouse, hamster, human) and cell type (B cells, T cells, fibroblasts) of the stimulator cell. Manipulations that influence the repertoire of peptides loaded on to MHC molecules also show no effect. In addition, the data show intriguing indications that the epitope of the γδ TCRs is also distinct from that of αβ T cell recognition. That is, the MHC molecules are not recognized from the 'top' of the peptide–MHC complex, but from the 'side,' away from the peptide binding groove.

What could be the molecular basis for these surprising differences in γδ versus αβ T cell recognition? Recently, Rock et al. (143) characterized the length distribution of CDR3 regions in different immune receptor chains (Fig. 10). It was found that the CDR3 lengths of both α and β TCR polypeptides are nearly identical and have very constrained distributions. In contrast, CDR3 lengths of Ig heavy chains are long and variable, while those of light chains are much shorter and less variable. Surprisingly, the δ chain CDR3 lengths are long and variable, but γ chains are much shorter and constrained. In this respect, γδ TCR CDR3 length distributions are similar to those of Ig and distinct from those of αβ TCR. Thus, γδ TCRs

Fig. 10 CDR3 length distributions shows a similarity between human immunoglobulin H and L chains (panel A) and γδ TCRs (panel C) (from ref. 143). CDR3 length distributions show marked differences between TCRα, β and TCRαβ (panel B) (see text). Mouse sequences show a similar pattern.

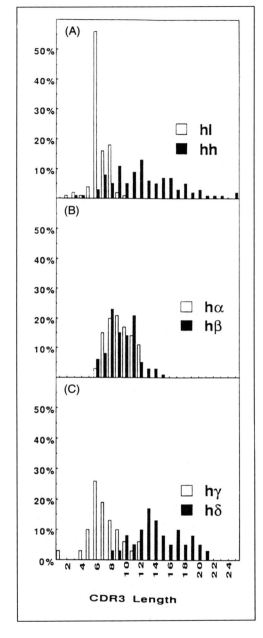

as a group may be more 'immunoglobulin-like' in their antigen recognition properties.

Therefore, MHC recognition by γδ T cells may represent a fortuitous cross-reaction with other non-MHC structures which are the physiologically relevant ligands for γδ T cells. The low frequency and broad cross-reactivity of MHC reactive γδ T cells would then reflect a recognition that is immunoglobulin-like, focusing on common 'features' shared by MHC molecules. This is in contrast to αβ T cell allorecognition which appears to represent cross-reactivity by cells selected to recognize peptides on self-MHC molecules, with the high frequency of such clones reflecting the diversity of different peptides bound to MHC molecules. In this context, it is interesting that some other γδ T cell ligands, such as heat-shock protein molecules, may have an MHC class I-like structure (144).

γδ T cells share many cell-surface proteins with αβ T cells and are able to secrete lymphokines and express cytolytic activities in response to antigenic stimulation. The experimental results presented here suggest that γδ T cells can mediate cellular immune functions without a requirement for antigen processing. Thus pathogens, damaged tissues, or even T cells can be recognized directly, and cellular immune responses can be initiated, without a requirement for antigen degradation and specialized antigen presenting cells, such as B cells, macrophages, or dendritic cells. This would allow greater flexibility than classical αβ T cell responses. As shown in Table 2, the estimated sequence variability of the CDR3 region of γδ TCRs is if anything greater than that of αβ, TCR and immunoglobulins, suggesting that γδ T cells might recognize a wide variety of antigens. This may include carbohydrate antigens as suggested by the reports of Pfeffer et al. (145, 146) which show that the major γδ T cell stimulatory components of mycobacteria are of low molecular weight (2–10 kDa), resistant to proteases, and able to bind lectins. Such unique properties may allow γδ T cells to effectively complement the more 'classical' type of αβ T cell-mediated cellular immunity.

9.2 Function of diverse γδ T cells

The function of γδ T cells in the immune system is an area of continual interest but little hard data are currently available. While bacterial and other infections can trigger γδ T cell proliferation and accumulation, these largely appear to be secondary effects and directed more at self-components than specific for the foreign antigen (139). Some promising recent results have been reported however. Mombaerts et al. (147) using TCR 'knock-out' mice, recently found significant differences in how mice deficient in γδ T cells were able to cope with a Listeria infection. Using these same mice, Tsuji et al. were able to show a clear γδ T cell driven response to malaria in infected mice (148). These systems are providing some of the first clues as to what role γδ T cells might play in an organism's defence.

9.3 Monomorphic γδ T cells

In some ways, even more remarkable than the discovery of γδ TCRs and the cells that express them, was the finding, first by Asarnow (149), that there were whole classes of γδ TCR bearing cells which showed identical sequences for both VJ$_\gamma$ and VDDJ$_\delta$. This was initially shown for a type of γδ T cell found embedded in the skin of mice referred to as dendritic epidermal cells (or dEC). These express Thy1, are negative for CD4 and CD8 and derive from the first 'wave' of fetal γδ T cells. They are immobile in the epidermis and spaced in regular arrays. It has been proposed that they serve some signalling or 'alarm' function (69, 149) and results show that they can be stimulated to secrete cytokines with keratinocyte extracts (150). γδ T cells with invariant TCR sequences have also been described in mucous membranes of the mouse (151, 152) although so far not in humans or other species. They appear to represent either a fascinating form of retrogression in which a polymorphic receptor can be converted into a monomorphic one to fulfil some specific receptor function, or perhaps a relic of the original antigen receptor precursor.

References

References marked * are recommended for further reading.

1. Fathman, C. G. and Frelinger, J. G. (1983). T-lymphocyte clones. *Annu. Rev. Immunol.*, **1**, 633.
2. Infante, A. J., Infante, P. D., Gillis, S., and Fathman, C. G. (1982). Definition of T cell idiotypes using anti-idiotype antisera produced by immunization with T cell clones. *J. Exp. Med.*, **155**, 1100.
3. Allison, J., McIntrye, B., and Bloch, D. (1982). Tumor-specific antigen of murine T-lymphoma defined with monoclonal antibody. *J. Immunol.*, **129**, 2293.
4. Haskins, K., Hannum, C., White, J., Roehm, N., Kubo, R., Kappler, J., and Marrack, P. (1984). The major histocompatibility complex-restricted antigen receptor on T cells. I. Isolation with a monoclonal antibody. *J. Exp. Med.*, **160**, 452.
5. Meuer, S. C., Acuto, O., Hussey, R. E., Hodgdon, J. C., Fitzgerald, K. A., Schlossman, S. F., and Reinherz, E. L. (1983). Evidence for the T3-associated 90k heterodimer as the T cell antigen receptor. *Nature*, **303**, 808.
6. Samelson, L. E., Germain, R. N., and Schwartz, R. H. (1983). Monoclonal antibodies against the antigen receptor on a cloned T cell hybrid. *Proc. Natl Acad. Sci. USA*, **80**, 6972.
7. Kaye, J., Porcelli, S., Tite, J., Jones, B., and Janeway, C. A. (1983). Both a monoclonal antibody and antisera specific for determinants unique to individual cloned helper T cell lines can substitute for antigen and antigen-presenting cells in the activation of T cells. *J. Exp. Med.*, **158**, 836.
8. Kappler, J., Kubo, R., Haskins, K., Hannum, C., Marrack, P., Pigeon, M., McIntyre, B., and Allison, J. (1983). The major histocompatibility complex-restricted antigen receptor on T cells in mouse and man. V. Identification of constant and variable peptides. *Cell*, **35**, 295.
9. Kappler, J. (1983). The mouse T cell receptor: Comparison of MHC-restricted receptors on two T cell hybridomas. *Cell*, **34**, 727.

10. McIntyre, B. and Allison, J. (1983). The mouse T cell receptor: Structural heterogeneity of molecules of normal T cells defined by xenoantiserum. *Cell*, **34**, 739.

11. Meuer, S. C., Fitzgerald, K. A., Hussey, R. E., Hodgdon, J. C. Schlossman, S. F., and Reinherz, E. L. (1983). Clonotypic structures involved in antigen-specific human T cell function. Relationship to the T3 molecular complex. *J. Exp. Med.*, **157**, 705.

12. Hedrick, S., Cohen, D. I., Nielsen, E. A., and Davis, M. M. (1984). Isolation of cDNA clones encoding T cell-specific membrane-associated proteins. *Nature*, **308**, 149.

13. Hedrick, S., Nielsen, E. A., Kavaler, J., Cohen, D. I., and Davis, M. M. (1984). Sequence relationships between putative T cell receptor polypeptides and immunoglobulins. *Nature*, **308**, 153.

14. Yanagi, Y., Yoshikai, Y., Leggett, K., Clark, S. P., Aleksander, I., and Mak, T. W. (1984). A human T cell-specific cDNA clone encodes a protein having extensive homology to immunoglobulin chains. *Nature*, **308**, 145.

15. Saito, H., Kranz, D. M., Takagaki, Y., Hayday, A. C., Eisen, H. N., and Tonegawa, S. (1984). Complete primary structure of a heterodimeric T cell receptor deduced from cDNA sequences. *Nature*, **309**, 757.

16. Chien, Y., Becker, D. M., Lindsten, T., Okamura, M., Cohen, D. I., and Davis, M. M. (1984). A third type of murine T-cell receptor gene. *Nature*, **312**, 31.

17. Saito, H., Kranz, D. M., Takagaki, Y., Hayday, A. C., Eisen, H. N., and Tonegawa, S. (1984). A third rearranged and expressed gene in a clone of cytotoxic T lymphocytes. *Nature*, **312**, 36.

18. Chien, Y., Iwashima, M., Kaplan, K., Elliott, J., and Davis, M. M. (1987). A new T cell receptor gene located within the α locus and rearranged early in T cell differentiation. *Nature*, **327**, 677.

19. Zingernagel, R. and Doherty, P. (1974). Restriction of *in vitro* T cell-mediated cytotoxicity in lymphocytic choriomeningitis within a syngeneic or semiallogeneic system. *Nature*, **248**, 701.

20. Zinkernagel, R. and Doherty, P. (1975). H-2 compatibility requirement for T cell-mediated lysis of target cells infected with lymphocytic choriomeningitis virus: Different cytotoxic T cell specificities are associated with structures from H-2K or H-2D. *J. Exp. Med.*, **141**, 1427.

21. Katz, D. H., *et al.* (1973). Cell interactions between histocompatible T and B lymphocytes. II. Failure of physiologic cooperative interactions between T and B lymphocytes from allogenic donor strains in humoral response to hapten-protein conjugates. *J. Exp. Med.*, **137**, 1405.

22. Rosenthal, A. S. and Shevach, E. M. (1973). The function of macrophages in antigen recognition by guinea pig lymphocytes. I. Requirement for histocompatible macrophages and lymphocytes. *J. Exp. Med.*, **138**, 1194.

23. Hein, W. R. and Mackay, C. R. (1991). Prominence of γδ T cells in the ruminant immune system. *Immunol. Today*, 12, 30.

24. Sowder, J. T., Chen, C. L., Ager, L. L., Chan, M. M., and Cooper, M. D. (1988). A large subpopulation of avian T cells express a homologue of the mammalian T γ/δ receptor. *J. Exp. Med.*, **167**, 315.

25. Davis, M. M., Cohen, D. I., Nielsen, E. A., DeFranco, A. L., and Paul, W. E. (1982). The isolation of B and T cell-specific genes. In *B and T cell tumors*, UCLA Symposia on Molecular and Cellular Biology, ed. E. Vitteta, Vol. 24, pp. 215–20. Academic Press, New York.

26. Acuto, O., Fabbi, M., Smart, J., Poole, C. B., Protentis, J., Royer, H-D., Schlossman,

S., and Reinherz, E. L. (1984). Purification and N-terminal amino acid sequence of the beta subunit of a human T cell antigen receptor. *Proc. Natl Acad. Sci. USA*, **81**, 3855.

27. Brenner, M. B., Strominger, J. L., and Krangel, M. S. (1986). Identification of a putative second T cell receptor. *Nature*, **322**, 145.

28. Novotny, J., Tonegawa, S., Saito, H., Kranz, D. M., and Eisen, H. N. (1986). Secondary, tertiary, and quaternary structure of T-cell-specific immunoglobulin-like polypeptide chains. *Proc. Natl Acad. Sci. USA*, **83**, 742.

29. Kronenberg, M., Siu, G., Hood, L. E., and Shastri, N. (1986). The molecular genetics of the T-cell antigen receptor and T-cell antigen recognition. *Annu. Rev. Immunol.*, **4**, 529.

*30. Davis, M. M. and Bjorkman, P. J. (1988). T-cell antigen receptor genes and T-cell recognition. *Nature*, **334**, 395.

*31. Chothia, C., Boswell, D. R., and Lesk, A. M. (1988). The outline structure of the T-cell alpha beta receptor. *EMBO J.*, **7**, 3745.

32. Brenner, M. B. *et al.* (1987). Two forms of the T cell receptor gamma protein found on peripheral blood cytotoxic T lymphocytes. *Nature*, **325**, 6891.

33. Brenner, M. B. (1988). The γδ T cell receptor. In *Advances in Immunology* (ed.), Vol. 43, pp. 133. Academic Press, New York.

34. Tan, L., Turner, J., and Weiss, T. M. (1991). Regions of the T cell receptor alpha and beta chains that are responsible for interactions with CD3. *J. Exp. Med.*, **173**, 1247.

35. Bosma, M. J. (1992). B and T cell leakiness in the scid mouse mutant. *Immunodefic. Rev. (England)*, **3**, 261.

36. Mombaerts, P., Lacomini, J., Johnson, R. S., Herrup, K., Tonegawa, S., and Papaioannou, V. E. (1992). *Rag-1* deficient mice have no mature B and T lymphocytes. *Cell*, **68**, 869.

37. Shinkai, Y., Rothbun, G., Lam, K.-P., Oltz, E. M., Stewart, V., Mendelsohn, M., Charron, J., Datta, M., Young, F., Stall, A. M., and Alt, F. W. (1992). *Rag-2*-Deficient mice lack mature lymphocytes owing to inability to initiate V(D)J rearrangement. *Cell*, **68**, 855.

38. Fujimoto, S. and Yamagishi, H. (1987). Isolation of an excision product of T-cell receptor alpha-chain gene rearrangements. *Nature*, **327**, 242.

39. Okazaki, K., Davis, D. D., and Sakano, H. (1987). T cell receptor beta gene sequences in the circular DNA of thymocyte nuclei: direct evidence for intramolecular DNA deletion in V–D–J joining. *Cell*, **49**, 477.

40. Malissen, M. C., McCoy, D., Blanc, J., Trucy, C., Devaux, A.-M., Schmitt-Verhulst, F., Fitch, L., Hood, L. E., and Malissen, B. (1986). *Nature*, **319**, 28.

41. Iwashima, M., Green, A., Davis, M. M., and Chien, Y.-h. (1988). Variable region (Vδ) gene segment most frequently utilized in adult thymocytes is 3' of the constant (Cδ) region. *Proc. Natl Acad. Sci. USA*, **85**, 8161.

42. Elliott, J. F., Rock, E. P., Patten, P. A., Davis, M. M., and Chien, Y. (1988). The adult T-cell receptor δ-chain is diverse and distinct from that of fetal thymocytes. *Nature*, **331**, 627.

43. Lafaille, J. J., De Cloux, A., Bonneville, M., Takagaki, Y., and Tonegawa, S. (1989). Junctional sequences of T cell receptor gamma delta genes; implications for gamma delta T cell lineages and for a novel intermediate of V–(D)–J joining. *Cell*, **59**, 859.

44. Leiden, J. M. (1993). Transcriptional regulation of T cell receptor genes. *Annu. Rev. Immunol.*, **11**, 539.

45. Waldmann, T. A., Davis, M. M., Bongiovanni, K. F., and Korsmeyer, S. J. (1985).

Rearrangements of genes for the antigen receptor on T cells as markers of lineage and clonality in human lymphoid neoplasms. *New Engl. J. Med.*, **313**, 776.

46. Winoto, A. and Baltimore, D. (1989). αβ lineage-specific expression of the T cell receptor α gene by nearby silencers. *Cell*, **59**, 649.

47. Ishida, I., Verbeek, S., Bonneville, M., Itohara, S., Berns, A., and Tonegawa, S. (1990). T-cell receptor γδ and γ transgenic mice suggest a role of a γ gene silencer in the generation of αβ T cells. *Proc. Natl Acad. Sci. USA*, **87**, 3067.

*48. Hedrick, S. M. and Eidelman, F. J. (1993). T lymphocyte antigen receptors. In *Fundamental immunology* (ed. William E. Paul) 3rd edn, pp. 383. Raven Press, Ltd, New York.

49. Baer, R., Chen, K.-C., Smith, S. D., and Rabbitts, T. H. (1986). Fusion of an immuno-globulin variable gene and a T cell receptor constant gene in the chromosome 14 inversion associated with T cell tumors. *Cell*, **43**, 705.

50. Denny, C. T., Yoshikai, Y., Mak, T. W., Smith, S. D., Hollis, G. F., and Kirsch, I. R. (1986). A chromosome 14 inversion in a T cell lymphoma is caused by site-specific recombination between immunoglobulin and T cell receptor loci. *Nature*, **320**, 549.

51. Guglielmi, P., Davi, F., D'Auriol, L., Bories, J.-C., Dausset, J., and Bensussan, A. (1988). Use of variable α region to create functional T-cell receptor δ chain. *Proc. Natl Acad. Sci. USA*, **85**, 5634.

52. Woodland, D. L., Kotzin, B., and Palmer, E. (1990). Functional consequences of a T cell receptor Dβ2 and Jβ2 gene segment deletion. *J. Immunol.*, **14**, 379.

*53. Koop, B. F., Wilson, R. K., Wang, K., Vernooij, B., Zallwer, D., Kuo, C. L., Seto, D., Toda, M., and Hood, L. (1992). Organization, structure, and function of 95 kb of DNA spanning the murine T-cell receptor C alpha/C delta region. *Genomics*, **13**, 1209.

*54. Davis, M. M. and Chien, Y.-h. (1993). Topology and affinity of T-cell receptor mediated recognition of peptide–MHC complexes. *Curr. Opin. Immunol.*, **5**, 45.

55. Reay, P. A., Kantor, R. M., and Davis, M. M. (1994). Use of global amino acid replacements to define the requirements for MHC binding and T cell recognition of moth cytochrome c (93–103). *J. Immunol.*, **152**, 3946.

*56. Matsui, K., Boniface, J. J., Reay, P. A., Schild, H., Fazekas de St Groth, B., and Davis, M. M. (1991). Low affinity interaction of peptide–MHC complexes with T cell receptors. *Science*, **254**, 793.

*57. Weber, S., Traunecker, A., Oliveri, F., Gerhard, W., and Karjalainen, K. (1992). Specific low-affinity recognition of major histocompatibility complex plus peptide by soluble T-cell receptor. *Nature*, **356**, 793.

58. Sykulev, Y., Brunmark, A., Jackson, M., Cohen, R. J., Peterson, P. A., and Eisne, H. N. (1994). Kinetics and affinity of reactions between an antigen-specific T-cell receptor and peptide–MHC complexes. *Immunity*, **1**, 15.

59. Fowlkes, B. J. and Pardoll, D. M. (1989). Molecular and cellular events of T cell development. In *Advances in immunology* (ed.), Vol. 44, p. 207. Academic Press, New York.

*60. Shortman, K. (1992). Cellular aspects of early T-cell development. *Curr. Opin. Immunol.*, **4**, 140.

*61. Robey, E. and Fowlkes, B. J. (1994). Selective events in T cell development. *Annu. Rev. Immunol.*, **12**, 675.

62. Guidos, C. J., Weissman, I. L., and Adkins, B. (1989). Intrathymic maturation of murine T lymphocytes from CD8[+] precursors. *Proc. Natl Acad. Sci. USA*, **86**, 7542.

63. Guidos, C. J., Danska, J. S., Fathman, C. G., and Weissman, I. L. (1990). T cell receptor-mediated negative selection of autoreactive T lymphocyte precursors occurs after commitment to the CD4 or CD8 lineages. *J. Exp. Med.*, **172**, 835.

*64. Mombaerts, P., Clarke, A. R., Rudnicki, M. A., Iacomini, J., Shigeyoshi, I., Lafaille, J. J., Wang, L., Ishikawa, Y., Jaenisch, R., Hooper, M. L., and Tonegawa, S. (1992). Mutations in T-cell receptor genes α and β block thymocyte development at different stages. *Nature*, **360**, 225.

65. Philpott, K. L., Viney, J. L., Kay, G., Rastan, S., Gardiner, E. M., Chae, S., Hayday, A. C., and Owen, M. J. (1992). Lymphoid development in mice congenitally lacking T cell receptor αβ-expressing cells. *Science*, **256**, 1448.

66. Itohara, S., Mombaerts, P., Lafaille, J., Iacomini, J., Nelson, A., Clarke, A. R., Hooper, M. L., Farr, A., and Tonegawa, S. (1993). T cell receptor delta gene mutant mice: independent generation of alpha beta T cells and programmed rearrangements of gamma delta TCR genes. *Cell*, **72**, 337.

*67. Groettrup, M., Ungewiss, K., Azogui, O., Palacios, R., Owen, J. J., Hayday, A. C., and von Boehmer, H. (1994). A novel disulfide-linked heterodimer on pre-T cells consists of the T cell receptor β chain and a 33 kd glycoprotein. *Cell*, **75**, 283.

*68. Raulet, D. H. (1989). The structure, function and molecular genetics of the γ/δ T cell receptor. *Annu. Rev. Immunol.*, **7**, 175.

*69. Allison, J. P. and Havran, W. L. (1991). The immunobiology of T cells with invariant γδ antigen receptors. *Annu. Rev. Immunol.*, **9**, 679.

70. Ikuta, K., Kina, T., MacNeil, I., Uchida, N., Peault, B., Chien, Y., and Weissman, I. L. (1990). A developmental switch in thymic lymphocyte maturation potential occurs at the level of hematopoietic stem cells. *Cell*, **62**, 863.

71. Blüthmann, H., Kisielow, P., Uematsu, Y., Malissen, M., Krimpenfort, P., Berns, A., von Boehmer, H., and Steinmetz, M. (1988). T cell specific deletion of T cell receptor transgenes allows functional rearrangement of endogenous α and β genes. *Nature*, **334**, 156.

72. Malissen, M., Trucy, J., Jouvin-Marche, E., Cazenave, P. A., Scollay, R., and Malissen, B. (1992). Regulation of TCR alpha and beta gene allelic exclusion during T-cell development. *Immunol. Today* (England), **13**, 315.

73. De Villartay, J. P. and Cohen, D. L. (1990). Gene regulation within the TCR-γδ locus by specific deletion of the TCR-δ cluster. *Res. Immunol.*, **141**, 618.

74. Winoto, A. and Baltimore, D. (1989). Separate lineages of T cells expressing the alpha beta and gamma delta receptors. *Nature*, **338**, 430.

75. Takeshita, S., Toda, M., and Yamagishi, H. (1989). Excision products of T cell receptor gene support a progressive rearrangement model of the alpha/delta locus. *EMBO J.*, **8**, 3261.

76. Scollay, R., Butcher, E. C., and Weissman, I. L. (1980). Thymus cell migration. Quantitative aspects of cellular traffic from the thymus to the periphery in mice. *Eur. J. Immunol.*, **10**, 210.

77. Kaye, J. and Hedrick, S. M. (1988). Analysis of specificity for antigen, Mls, and allogeneic MHC by transfer of T-cell receptor α- and β-chain genes. *Nature*, **336**, 580.

78. Lee, N. E. and Davis, M. M. (1988). T cell receptor β-chain genes in BW5147 and other AKR tumors: Deletion order of murine V_β gene segments and possible 5' regulatory regions. *J. Immunol.*, **140**, 1695.

79. Livak, F., Schatz, D. G., Strasser, A., Crispe, I. N., and Shortman, K. (1993). Multiple rearrangements in T cell receptor alpha chain genes maximize the production of useful thymocytes. *J. Exp. Med.*, **178**, 615.

*80. Kappler, J. W., Wade, T., White, J., Kushnir, E., Blackman, M., Bill, J., Roehm, N., and Marrack, P. (1987). A T cell receptor Vβ segment that imparts reactivity to a Class II major histocompatibility complex product. *Cell*, **49**, 263.

*81. Kappler, J. W., Roehm, N., and Marrack, P. (1987). T cell tolerance by clonal elimination in the thymus. *Cell*, **49**, 273.

82. Kisielow, P., Blüthmann, H., Staerz, U. D., Steinmetz, M., and von Boehmer, H. (1988). Tolerance in T cell receptor transgenic mice involves deletion of nonmature CD4$^+$8$^+$ thymocytes. *Nature*, **333**, 742.

83. Sha, W. C., Nelson, C. A., Newberry, R. D., Kranz, D. M., Russell, J. H., and Loh, D. Y. (1988). Positive and negative selection of an antigen receptor on T cells in transgenic mice. *Nature*, **336**, 73.

84. Berg, L. J., Fazekas de St. Groth, B., Pullen, A. M., and Davis, M. M. (1989). Phenotypic differences between αβ versus β T cell receptor transgenic mice undergoing negative selection. *Nature*, **340**, 559.

85. Murphy, K. M., Heimberger, A. B., and Loh, D. Y. (1990). Induction by antigen of intrathymic apoptosis of CD4$^+$8$^+$ TCR^{lo} thymocytes *in vivo*. *Science*, **250**, 1720.

*86. Ohashi, P. S., Oehen, S., Buerki, K., Pircher, H., Ohashi, C. T., Odermatt, B., Malissen, B., Zinkernagel, R. M., and Hengartner, H. (1991). Ablation of 'tolerance' and induction of diabetes by virus infection in viral antigen transgenice mice. *Cell*, **65**, 305.

87. Sagerström, C., Kerr, E. M., Allison, J. P., and Davis, M. M. (1993). Activation and differentiation requirements for primary T cells *in vitro*. *Proc. Natl Acad. Sci. USA*, **90**, 8987.

88. Bevan, M. J. (1977). In a radiation chimaera, host H-2 antigens determine the immune responsiveness of donor cytotoxic cells. *Nature*, **269**, 417.

89. Zinkernagel, R. M., Callahan, G., Althage, A., Cooper, S., Klein, P., and Klein, J. (1978). On the thymus in the differentiation of H-2 self-recognition by T cells: evidence for dual recognition. *J. Exp. Med.*, **147**, 882.

*90. Kisielow, P., Teh, H.S., Blüthmann, H., and von Boehmer, H. (1988). Positive selection of antigen specific T cell in thymus by restricting MHC molecules. *Nature*, **335**, 730.

91. Teh, H.-S., Kisielow, P., Scott, B., Kishi, H., Uematsu, Y., Blüthmann, H., and von Boehmer, H. (1988). Thymic MHC antigens and the αβ T cell receptor determine the CD4/CD8 phenotype of T cells. *Nature*, **335**, 229.

92. Sha, W. C., Nelson, C. A., Newberry, R. D., Kranz, D. M., Russell, J. H., and Loh, D. Y. (1988). Selective expression of an antigen receptor on CD8-bearing T lymphocytes in transgenic mice. *Nature*, **335**, 271.

*93. Berg, L. J., Pullen, A. M., Fazekas de St Groth, B., Mathis, D., Benoist, C., and Davis, M. M. (1989). Antigen/MHC-specific T cells are preferentially exported from the thymus in the presence of their MHC ligands. *Cell*, **58**, 1035.

94. Kaye, J., Hsu, M. L., Sauron, M. E., Jameson, S. C., Gascoigne, N. R., and Hedrick, S. N. (1989). Selective development of CD4$^+$ T cells in transgenic mice expressing a Class II MHC restricted antigen receptor. *Nature*, **341**, 746.

*95. von Boehmer, H. (1994). Positive selection of lymphocytes. *Cell*, **76**, 219.

96. Singer, A., Mizuochi, T., Munitz, T. I., and Gress, R. E. (1986). Role of self-antigens in the selection of the developing T cell repertoire. In *Progress in immunology*, Vol. VI, p. 60. Academic Press, New York.

97. Schwartz, R. H. (1982). Functional properties of I region gene products and theories of immune response (Ir) gene function. In *Ia Antigens*, Vol. 1, Mice (ed. S. Ferrone and C. S. David), pp. 199. CRC Press, Baton Rouge, Florida.

98. Nikolic–Zugic, J. and Bevan, M. (1990). Role of self-peptides in positively selecting the T cell repertoire. *Nature*, **344**, 65.

99. Berg, L. J., Frank, G. D., and Davis, M. M. (1990). The effects of MHC gene dosage and allelic variation on T cell receptor selection. *Cell*, **60**, 1043.

100. Sha, W. C., Nelson, C. A., Newberry, R. D., Pullen, J. K., Pease, L. R., Russell, J. H., and Loh, D. Y. (1990). Positive selection of transgenic receptor bearing thymocytes by K^b antigen is altered by K^b mutations that involve peptide binding. *Proc. Natl Acad. Sci. USA*, **87**, 6186.

101. Jacobs, H., von Boehmer, H., Melief, C.J., and Berns, A. (1990). Mutations in the MHC Class I antigen presenting groove affect both negative and positive selection of T cells. *Eur. J. Immunol.*, **20**, 2333.

102. Hogquist, K. A., Gavin, M. A., and Bevan, M. J. (1993). Positive selection of $CD8^+$ T cells induced by major histocompatibility complex binding peptides in fetal thymus organ culture. *J. Exp. Med.*, **177**, 1464.

103. Ashton-Rickhardt, Van Kaer, P. G., Schumacher, T. N. M., Ploegh, H. L., and Tonegawa, S. (1993). Peptide contributes to the specificity of positive selection of $CD8^+$ T cells in the thymus. *Cell*, **73**, 1041.

*104. Janeway, C., Rudensky, S., Rath, S., and Murphy, D. (1992). It is easier for a camel to pass the needle's eye. *Curr. Biol.*, **2**, 26.

105. Roby, E. A., Ramsdell, F., Kioussis, D., Sha, W. C., Loh, D. Y., Axel, R., and Fowlkes, B. J. (1992). The level of CD8 expression can determine the outcome of thymic selection. *Cell*, **64**, 1084.

106. Lee, N. A., Loh, D. Y., and Lacy, E. (1992). CD8 surface expression alters the fate of α/β T cell receptor-expressing thymocytes in transgenic mice. *J. Exp. Med.*, **175**, 1013.

*107. Ashton-Richardt, P. G., Bandeira, A., Delaney, J. R., Kaer, L. V., Pircher, H-P., Zinkernagel, R. M., and Tonegawa, S. (1994). Evidence for a differential avidity model of T cell selection in the thymus. *Cell*, **76**, 65.

*108. Hogquist, K. A., Jameson, S. C., Heath, W. R., Howard, J. L., Bevan, M. J., and Carbone, F. R. (1994). T cell receptor antagonist peptides induced positive selection. *Cell*, **76**, 17.

109. Rock, E. P. and Davis, M. M. (1993). Structural aspects of T cell receptor recognition of antigen–MHC complexes. *Accts. Chem. Res.*, **26**, 435.

110. Claverie, J. M., Prochnicka, C. A., and Bouguelert, L. (1989). Implications of a Fab-like structure for the T-cell receptor. *Immunol. Today.*, **10**, 10.

111. Hedrick, S., Engel, I., McElligott, D. L., Fink, P. J., Hsu, M-L., Hansburg, D., and Matis, L. A. (1988). Selection of amino acid sequences in the beta chain of the T cell antigen receptor. *Science*, **239**, 1541.

112. Lai, M. -Z., Jang, Y. J., Chen, L.-K., and Gefter, M. J. (1990). Restricted V–(D)–J junctional regions in the T cell response to lambda-repressor. Identification of residues critical for antigen recognition. *J. Immunol.*, **144**, 4851.

113. Wither, J., Pawling, J., Phillips, L., Delovitch, T., and Hozumi, N. (1991). Amino acid residues in the T cell receptor CDR3 determine the antigenic reactivity patterns of insulin-reactive hybridomas. *J. Immunol.* **146**, 3513.

114. Acha-Orbea, H., Mitchell, D. J., Timmerman, L., Wraith, D. C., Tausch, G. S., Waldor, M. K., Zamvil, S. S., McDevitt, H. O., and Steinmon, L. (1988). Limited heterogeneity of T cell receptors from lymphocytes mediating autoimmune encephalo-myelitis allows specific immune intervention. *Cell*, **54**, 263.

115. Urban, J. L., Kumar, V., Kono, D. H., Gomez, C., Horvath, S. J., Clayton, J., Ando,

D. G., Sercare, E., and Hood, L. (1988). Restricted use of T cell receptor V genes in murine autoimmune encephalomyelitis raises possibilities for antibody therapy. *Cell*, **54**, 577.

116. Danska, J. S., Livingstone, A. M., Paragas, V., Ishihara, T., and Fathmon, C. G. (1990). The presumptive CDR3 regions of both T cell receptor alpha and beta chains determine T cell specificity for myoglobin peptides. *J. Exp. Med.*, **172**, 27.

117. Engel, I. and Hedrick, S. M. (1988). Site-directed mutations in the VDJ junctional region of T cell receptor beta chain cause changes in antigenic peptide recognition. *Cell*, **54**, 473.

118. Patten, P., Yokota, T., Rothbard, J., Chien, Y., Arai, K. and Davis, M. M. (1984). Structure, expression and divergence of T-cell receptor β-chain variable regions. *Nature*, **312**, 40.

*119. Jorgensen, J. L., Esser, U., Fazekas de St Groth, B., Reay, P. A., and Davis, M. M. (1992). Mapping T cell receptor/peptide contacts by variant peptide immunization of single-chain transgenics. *Nature*, **355**, 224.

120. Jorgensen, J. L. and Davis, M. M. (1995). (In preparation).

121. Taylor, A. H., Haberman, A. M., Gerhard, W., and Caton, A. J. (1990). Structure–function relationships among highly diverse T cells that recognize a determinant from influenza virus hemagglutinin. *J. Exp. Med.*, **172**, 1643.

122. Hong, S. C., Chelouche, A., Lin, R. H., Shaywitz, D., Braunstein, N. S., Glimcher, L., and Janeway, C. A., Jr. (1992). An MHC interaction site maps to the amino-terminal half of the T cell receptor alpha chain variable domain. *Cell*, **69**, 999.

123. Ajitkumar, P., Geier, S. S., Kesari, K. V., Borriello, F., Nakagawa, M., Bluestone, J. A., Saper, M. A., Wiley, D. C., and Nathenson, S. G. (1988). Evidence that multiple residues on both the alpha-helices of the Class I MHC molecule are simultaneously recognized by the T cell receptor. *Cell*, **54**, 47.

124. Peccoud, J., Dellabona, P., Allen, P., Benoist, C., and Mathis, D. (1990). Delineation of antigen contact residues on an MHC Class II molecule. *EMBO J.*, **9**, 4215.

*125. Ehrich, E. W., Devaux, B., Rock, E. P., Jorgensen, J. L., Davis, M. M., and Chien, Y-h. (1993). T cell receptor interaction with peptide/MHC and superantigen/MHC ligands is dominated by antigen. *J. Exp. Med.*, **178**, 713.

126. Demotz, S., Grey H. M., and Sette A. (1990). The minimal number of class II MHC–antigen complexes needed for T cell activation. *Science*, **249**, 1028.

127. Harding, C. V. and Unanue, E. R. (1990). Quantitation of antigen-presenting cell MHC class II/peptide complexes necessary for T-cell stimulation. *Nature*, **346**, 574.

128. Christinck, E. R., Luscher, M. A., Barber, B. H., and Williams, D. B. (1991). Peptide binding to Class I MHC on living cells and quantification of complexes regained for CTL lysis. *Nature*, **352**, 67.

*129. Williams, A. F. and Beyers, A. D. (1992). T-cell receptors: at grips with interactions. *Nature*, **356**, 746.

130. Butcher, E. C. (1992). Leukocyte–endothelial cell recognition: Three (or more) steps to specificity and diversity. *Cell*, **67**, 1033.

131. Spits, H., van Schooten, W., Keizer, H., van Seventer, G., Van de Rijn, M., Terhorst, C., and DeVries, J. E. (1986). Alloantigen recognition is preceded by non-specific adhesion of cytotoxic T cells and target cells. *Science*, **232**, 403.

132. Inaba, K., Romani, N., and Steinman, R. M. (1989). An antigen-independent contact mechanism as an early step in T cell-proliferative responses to dendritic cells. *J. Exp. Med.*, **170**, 527.

133. Springer, T. A. (1994). Traffic signals for lymphocyte recirculation and leukocyte emigration: the multistep paradigm. *Cell*, **76**, 301.

134. Festenstein, H. (1973). Immunogenetic and biological aspects of *in vitro* lymphocyte allotransformation (MLR) in the mouse. *Transplant. Rev.*, **15**, 62.

135. Herman, A., Kappler, J. W., and Marrack, P. (1991). Superantigens: Mechanism of T-cell stimulation and role in immune responses. *Annu. Rev. Immunol.*, **9**, 745.

136. Scherer, M. T., Ignatowicz, L., Winslow, G. M., Kappler, J. W., and Marrack, P. (1983). Superantigens: Bacterial and viral proteins which manipulate the immune system. *Annu. Rev. Cell Biol.*, **9**, 101.

*137. Acha-Orbea, H. and Palmer, E. (1992). Mls—a retrovirus exploits the immune system. *Immunol. Today*, **13**, 77.

138. Janeway, C. A. Jr, Yagi, J., Conrad, P. J., Katz, M. E., Jones, B., Vroegrop, S., and Buxser, S. (1989). T-cell responses to Mls and to bacterial proteins that mimic its behavior. *Immunol. Rev.*, **107**, 61.

*139. Haas, W., Pereira, P., and Tonegawa, S. (1993). Gamma/Delta cells. *Annu. Rev. Immunol.*, **11**, 637.

140. Matis, L. A., Fry, A. M., Crow, R. Q., Cotterman, M. M., Dick, R. F., and Bluestone, J. A. (1989). Structure and specificity of a Class II MHC alloreactive γδ T cell receptor heterodimer. *Science*, **245**, 746.

141. Kabelitz, D., Bender, A., Schondelmaier, S., Da Silva Lobo, M. L., and Janssen, O. J. (1990). Human cytotoxic lymphocytes. V. Frequency and specificity of gamma/delta cytotoxic lymphocyte precursors activated by allogeneic or autologous stimulator cells. *J. Immunol.*, **145**, 2827.

*142. Schild, H., Mavaddat, N., Litzenberger, C., Ehrich, E. W., Davis, M. M., Bluestone, J. A., Matis, L., Draper, R. K., and Chien, Y-h. (1994). The nature of major histocompatibility complex recognition by γδ T cells. *Cell*, **76**, 27.

143. Rock, E. P., Sibbald, P. R., Davis, M. M., and Chien, Y-h. (1994). CDR3 length in antigen-specific immune receptors. *J. Exp. Med.*, **179**, 323.

144. Gething, M. J. and Sambrook, J. (1992). Protein folding in the cell. *Nature*, **355**, 33.

145. Pfeffer, K., Schoel, B., Gulle, H., Kaufmann, S. H. E., and Wagner, H. (1990). Primary responses of human T cells to mycobacteria: a frequent set of γ/δ T cells are stimulated by protease-resistent ligands. *Eur. J. Immunol.*, **20**, 1175.

146. Pfeffer, J., Schoel, B., Plesnila, N., Lipford, G. B., Kromer, S., Deusch, K., and Wagner, H. (1992). A lectin binding, protease-resistent mycobacterial ligand specifically activates Vg9⁺ human γδ T cells. *J. Immunol.*, **148**, 575.

*147. Mombaerts, P., Arnoldi, J., Russ, F., Tonegawa, S., and Kaufmann, S. H. E. (1993). Different roles of αβ and γδ T cells in immunity against an intracellular bacterial pathogen. *Nature*, **365**, 53.

*148. Tsuji, M., Mombaerts, P., LeFrancois, L., Nussenzweig, R. S., Zavala, F., and Tonegawa, S. (1994). γδ T cells contribute to immunity against the liver stages of malaria in αβ T-cell-deficient mice. *Proc. Natl Acad. Sci. USA*, **91**, 345.

*149. Asarnow, D. M., Kuziel, W. A., Bonyhadi, M., Tigelaar, R. E., Tucker, P. W., and Allison, J. P. (1988). Limited diversity of γ/δ antigen receptor genes of Thy-1⁺ dendritic epidermal cells. *Cell*, **55**, 837.

*150. Havran, W. L., Chien, Y. H., and Allison, J. P. (1991). Recognition of self-antigens by skin-derived T cells with invariant γδ antigen receptors. *Science*, **252**, 1430.

151. Sim, G. K. and Augustin, A. (1990). Dominantly inherited expression of BID, and invariant undiversified T cell receptor δ chain. *Cell*, **61**, 397.

152. Itohara, S., Farr, A. G., Lafaille, J. J., Bonneville, M., Takagaki, Y., Haas, W., and Tonegawa, S. (1990). Homing of a γ/δ thymocyte subset with homogenous T-cell receptors to mucosal epithelial. *Nature*, **343**, 754.

3 | T lymphocyte signal transduction

COX TERHORST, HERGEN SPITS, FRANK STAAL,
and MARK EXLEY

1. Introduction

That the immune system combines an extraordinary specificity of recognition with extremely complex control mechanisms has been amply demonstrated by the studies of antigen recognition by T lymphocytes. In response to antigen recognition, resting T lymphocytes are required to undergo a complex series of events collectively known as T cell activation. Although much has been learned about T cell activation in the last years, an orderly array of the molecular events involved in the whole process remains to be established. There is heterogeneity in the effector cells within the T cell compartment comprising T helper (T_H) and cytotoxic T lymphocyte (CTL) subsets. These two subsets have been further functionally and phenotypically subdivided (for example 'naive' $CD4^+T_H0$, 'memory' $CD4^+T_H1$, and $CD4^+T_H2$ cells; $CD8^+CTL$, and $CD4^-/CD8^-$ T cells). Another complicating factor in attempts to understand the network of signal transduction events is an apparent redundancy in signal transduction pathways utilized in T cell activation, which can prevent unequivocal interpretations of the data. In addition, certain antigenic epitopes might induce the production of a different set of cytokines than others. Our most important task is to understand which signal transduction mechanisms govern the decision to initiate activation or to cause antigen-specific 'anergy' of T cells, the latter leading to a more or less prolonged state of tolerance. This review will highlight the early pathways of antigen-driven activation as currently thought to take place in T lymphocytes.

2. Antigen recognition by T cell receptors

The T cell repertoire is characterized by an extremely diverse population of distinct T cell receptor (TCR) V-regions which facilitate highly specific responses to antigens. The prototype T lymphocytes ($\alpha\beta$ T cells) respond to non-self only when their T cell antigen receptor (TCR-$\alpha\beta$) engages antigen-derived peptides complexed with cell surface molecules encoded by the polymorphic genes of the major histocompatibility complex (MHC) (1–3). MHC restriction of the T cell repertoire develops as a consequence of selective processes acting on immature T lymphocytes as they contact antigen presenting cells expressing MHC molecules complexed with specific peptides during early development within the thymus gland. T cell development within the thymus involves both positive and negative selection (4,5). A variable fraction of peripheral T cells have been found to display an alternative clonotypic structure termed the TCR-γδ (6–9). In contrast to αβ T cells, no common restriction element is presently known for γδ T cells. However, both forms of TCR are non-covalently associated with the CD3 complex of molecules, which are essential for transduction of the antigen-binding signal. The components of the TCR/CD3 complex are summarized in Table 1 and discussed in detail below (Section 3.1).

Recognition of MHC/peptide antigen by an αβ T lymphocyte is a dynamic process involving complex cell-to-cell interactions before T cell activation takes place (Fig. 1). This process is necessary because it incorporates two fundamental properties of the T lymphocyte. First, T lymphocytes migrate throughout the body

Table 1 Components of the TCR/CD3 complex

Subunit	Size (kDa)	Primary Structure [a]			Glycosylation	Phosphorylation	Chromosomal location
		Ex.	TM	Cyt.			
TCR							
α_H	45–60 (32)	227	27	4	+	−	14q11.2
α_M	45–55 (29)	227	27	4	+	−	14C–D
β_H	40–50 (32)	220	23	3	+	−	7q35
β_M	40–55 (32)	220	23	3	+	−	6B
γ_H	45–60 (30)	218	25	4	+	−	7p14–p15
γ_M	45–60 (30)	218	25	4	+	−	13A2–3
δ_H	40–60 (37)	223	27	5	+	−	14q11.2
δ_M	40–60 (31)	223	27	5	+	−	14C–D
CD3							
γ_H	25–28 (16)	89	27	44	+	P-Ser/P-Tyr	11q23
γ_M	21 (16)	89	27	44	+	P-Ser/P-Tyr	9
δ_H	20 (14)	79	27	44	+	P-Tyr	11q23
δ_M	28 (16)	79	27	46	+	P-Tyr	9
ε_H	20 (14)	104	26	55	−	P-Tyr	11q23
ε_M	25 (18)	87	26	55	−	P-Tyr	9
ζ_H	16	9	21	112	−	P-Tyr/P-Ser	1q22–q25
ζ_M	16	9	21	113	−	P-Tyr/P-Ser	1
η_M	20	9	21	155	−	P-Try?	1
ω_H	28	?	?	?	−	?	?
ω_M	28	?	?	?	−	?	?
FcεR Iγ_H	9	6	20	42	−	P-Tyr	1q23
FcεR Iγ_M	9	6	20	42	−	P-Tyr	1q23

[a] This lists the numbers of amino acids which are extracellular (Ex.), transmembrane (TM), or located on the cytoplasmic side of the plasma membrane (cyt.).

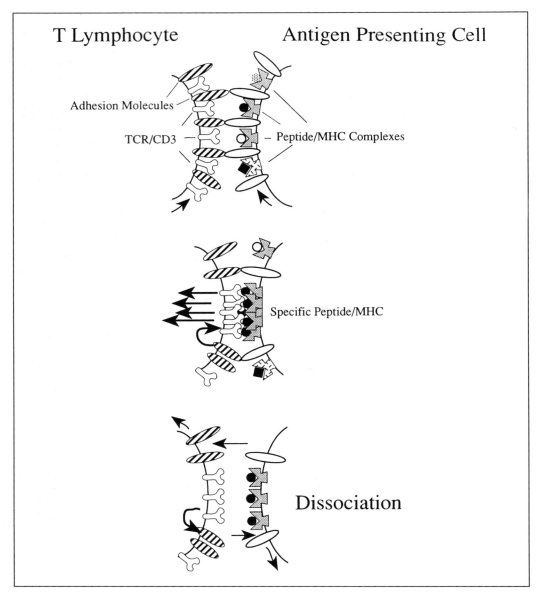

Fig. 1 Initial interactions between a T cell and antigen presenting cell, intracellular signalling, and subsequent cell dissociation.

and have the ability to adhere to many cell types. Second, an antigen-presenting cell, which simultaneously processes many protein–antigens, expresses on its cell-surface a large number of different MHC/peptide–antigen complexes. Consequently, only a small number of these complexes can be recognized by a given αβ T cell. For these two reasons, T cell antigen recognition proceeds in two distinct stages: a non-specific adhesion step without antigen recognition followed by the specific interaction between the TCR and antigen/MHC (10–12). This cell–cell adhesion allows T lymphocytes to screen potential antigen-presenting cells. Adhesion is aborted in the absence of specific antigen recognition by the T cell receptor and is intensified when the correct TCR/MHC/peptide contact is made.

Primary adhesion molecules which contribute to the binding avidity between a T lymphocyte and an antigen-presenting cell are CD2 and CD11a/CD18 (LFA-1). Their natural cell surface bound ligands (CD58 or LFA-3) and (CD54 or ICAM-1), respectively (13) are expressed on all potential antigen-presenting cells (Table 2). CD4 and CD8 constitute a distinct category in the adhesion step because, like the T cell receptor, their natural ligands are the class I or class II MHC surface glycoproteins. Several experiments have demonstrated that direct interactions exist between CD8/class I and CD4/class II. These interactions play a major role in positive selection during intrathymic maturation (4, 5, 14). Indeed, the CD4/CD8 phenotype of a given T lymphocyte largely determines its effector function: T cells recognizing class II MHC/antigen normally display CD4, whereas those

Table 2 Major T cell co-stimulatory structures

Antigen CD	Old name	M_r Human[a]	(kDa) Mouse[a]	Oligomeric state	Phosphorylation state	Function/ligand
CD2	LFA-2	50 (38)	55 (38)	heteroligomer?	Tyr	CD58 (LFA-3) receptor
CD4		55 (51)	55 (51)	?	Ser/Thr	MHC class II receptor and *lck* anchor
CD5		67	67	heteroligomer?	Tyr	CD72 receptor
CD6		130	130	homodimer?	Tyr	?
CD7		38	38	homodimer?	Tyr	?
CD8 α		34 (24)	38 (25)	α/α	Ser/Thr	MHC class I receptor and *lck* anchor
α'		—	34 (22)	α-CD1/α/α'		
β		—	30 (22)	α/β		
CD27		55	55	homodimer	Tyr?	CD70 receptor
CD28		44 (23)	45	homodimer	Tyr (Ser/Thr?)	CD80 B7-1 receptor B7-2 B70 receptor
	CTLA-4	—	45	homodimer	Tyr Ser/Thr?	B7 family receptor and T cell co-stimulation
—	gp39 CD40L	39 (38)	39	homotrimer	Tyr?	CD40 ligand/receptor and T cell co-stimulation
CD45 (RA/RB/RO)	LCA, T200	170–240	170–240	?	Tyr	Tyr-phosphatase CD22 receptor?
CD11a	LFA-1 α	180 (130)	180 (130)	α/β	- (through LFA-1)	ICAM-1 receptor
CD18	β	95 (82)	95 (77)	α/β	- (through LFA-1)	
CD54	ICAM-1	90	90	?	—	CD11a/CD18 receptor
CD95	Fas/Apo-1	35	35	?	?	FasL receptor: either apoptosis or co-stimulation

[a] Numbers in brackets are the M_r values for the unglycosylated polypeptides.

with affinity for class I MHC molecules express CD8. All of these adhesion molecules (for example CD2, CD4, CD8, and CD11a/CD18) also play a role in T cell–target cell interactions, signal transduction, and T cell activation (13). Reciprocally, TCR signal transduction increases affinity between the adhesion molecules LFA-1 and I-CAM (13).

The order of events in screening T cell/antigen presenting cells was first studied for the effector function of CTL, because of the ability to isolate CTL-target cell pairs (10). The Mg^{2+}-dependent CTL/target cell adhesion phase is followed by a Ca^{2+}-dependent phase that leads to the delivery of the lethal hit. Next, the CTL dissociates from the target cell, recycles to attack another target cell, and target cell death proceeds in the absence of the CTL. If the CTL adheres to a cell which does not carry the correct MHC/antigen peptide combination, no lethal hit occurs and dissociation takes place more rapidly. CTL activation by target cells is accompanied by an extensive rearrangement of the cytoplasm of the CTL. The microtubule organizing centre and the Golgi complex re-orient towards the contact site between the CTL and target cell. Furthermore, vesicular components are accumulated in the region in close proximity to the contact site with the target cell. Monoclonal antibodies and antigen specific T cell clones were instrumental in demonstrating that the initial contact between CTL and target cell does not require the interaction between the TCR and its target antigen. Antibodies directed at LFA-1 and CD2 inhibit both non-specific and specific adhesions between CTL and target cells. Monoclonal antibodies directed at the adhesion partners (anti-LFA-3 and I-CAM) similarly block adhesion at the level of the target cells (13). In contrast, anti-clonotypic TCR and anti-CD3 reagents do not inhibit adhesion but do block the lytic phase in soft agar assays. Anti-CD4 and anti-CD8 inhibit both adhesion and the lytic phase (10, 15).

Interactions between T_H cells and antigen-presenting cells follow the same general pattern. Recent experiments by Mark Davis and his colleagues (11) using solubilized class II MHC/peptide complexes allowed for an estimate of the binding constant and indicated low affinities. This led to a more refined model (Fig. 1). Only a very small number of MHC/peptide complexes would be engaged with a specific TCR in a given area of cell–cell contact established by adhesion (16, 17). Next, TCR/CD3 and MHC molecules complexed with the correct peptide are thought to migrate into the interface between antigen-presenting cell and T cell. This establishes a high local density of TCR/CD3 complexes, thus promoting a cooperativity in antigen binding, which in turn leads to T cell activation. Activation results not only in recruitment of intracellular transduction molecules but also in a transient increase in adhesion (13). Down-regulation of adhesion molecules would subsequently permit cell detachment. In case TCR/CD3 complexes are unable to cluster because of the absence of recognition, detachment occurs immediately. Several recent observations fit into this cooperativity model. TCR/CD3 complexes themselves have a propensity to form higher order structures on the T cell surface (18) and MHC class II are thought to form dimers on the cell surface (19). Previously it was shown that the CD8 co-receptors are found on the surface of human T cells and thymocytes as dimers and multimers (20). Thus, receptor ligand/engagement on the interface of the antigen-presenting cell is likely to result in receptor aggregation. Therefore, the initiation of signal transduction involves the T cell receptor for antigen, adhesion molecules, and other co-activators.

3. Structure of the T cell receptor for antigen

T cell receptors for antigen themselves are multimeric membrane protein complexes (1–3) which include the clonotypically variable polypeptide chains (TCR$\alpha\beta$ or -$\gamma\delta$) and the invariant CD3-γ, -δ, ε, and -ζ proteins (Table 1). Whilst the TCR$\alpha\beta$ or -$\gamma\delta$ glycoproteins interact with antigen, the CD3 proteins are involved in regulation of assembly and signal transduction of the receptor complex. Unlike other cell surface receptors, which display specialized segments in their cytoplasmic tails, the TCR utilizes the CD3 proteins for specialized signal transduction functions. The precise arrangements of the members of the TCR/CD3 complex are not known in the absence of X-ray crystallographic data. At the moment, we favour the divalent model of the TCR/CD3 complex (21) as shown in Fig. 2, because it incorporates the following observations:

1. In a T–T hybridoma, specific monoclonal antibodies against one TCR$\alpha\beta$ pair could co-modulate the second TCR$\alpha\beta$ pair (18).
2. Experiments involving sucrose gradient centrifugation in the presence of mild non-denaturing detergents identified a TCR/CD3 complex of approximately 300 kDa (18). This cell-surface receptor, which is more than 100 kDa larger than expected from the minimal 8-subunit complex ($\alpha\beta\gamma\delta\varepsilon_2\zeta_2$), migrated with a higher sedimentation rate than the B cell receptor (250 kDa).
3. An ($\alpha\beta$)$_2$, γ, δ, ε_2, ζ_2 model accommodates:
   - the results of pair formation studies in COS and CHO cells (22) and in T cells (for instance TCRβ/CD3-γ) (21, 23);
   - 'charge' bridges between positively and negatively charged residues within the trans-membrane regions (24–28);
   - CD3-ε–ε homodimers found in cell-surface receptors (28, 29);
   - the so-called 'half receptors' ($\alpha\beta\gamma\varepsilon\zeta$ and $\alpha\beta\delta\varepsilon\zeta$), which have been identified as intermediates in the formation of complete TCR/CD3 complexes (30, 31);
   - a 1:1 ratio of $\alpha\beta$-TCR and CD3-ε molecules as detected by quantitative antibody measurements (32, 33).

It is likely that a divalent model also applies to $\gamma\delta$ T cells, since the sedimentation behaviour of $\gamma\delta$ TCR in sucrose velocity gradients is identical to that of $\alpha\beta$ TCR and since quantitative binding studies with monoclonal antibodies suggest a divalent receptor (18, 33).

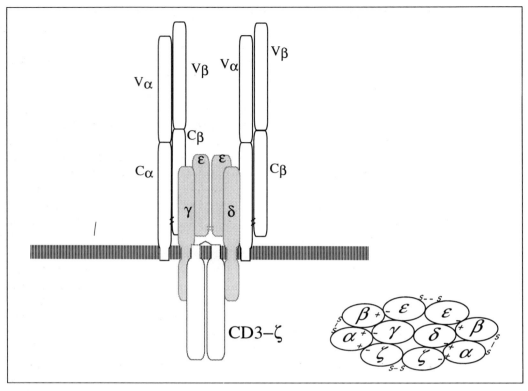

Fig. 2 Model for the structure of the TCRαβ/CD3 complex.

3.1 The components of the TCR/CD3 complex

TCRα, -β, -γ, and -δ are membrane glycoproteins each consisting of a leader sequence, an externally-disposed N-terminal extracellular domain, a single transmembrane domain, and a cytoplasmic tail (1, 2, 21) (Table 1). Most T cell receptor αβ heterodimers are covalently linked through disulfide bonds, whilst TCRγδ receptors often associate with one another non-covalently. The αβ and γδ TCR glycoproteins belong to immunoglobulin superfamily and resemble immunoglobulins in that they contain variable and constant domains. The constant regions of the TCR extracellular domains are anchored in the plasma membrane by an α-helical transmembrane region. The cytoplasmic domain consists of only a few amino acids. The charged residues in the transmembrane regions are of importance for the assembly of the receptor, as discussed in Section 3.2. In one study, utilizing transfection of chimeric proteins into non-T cells, the transmembrane domain of TCRα was shown to be unable to function as an anchor for a heterologous ectodomain in the lipid bilayer of the endoplasmic reticulum (ER). Instead, this CD4-α_{tm} chimeric protein was degraded in the lumen of the ER or was secreted (34). One interpretation of these data, namely that TCRα does not contain a bona fide transmembrane domain, was favoured by the authors. However, since secretion of TCR α-chain by T cells has not been reported, the result could reflect the properties of the chimeric protein. Regardless of the behaviour of the transmembrane region of the independent TCRα polypeptide chain, its incorporation into the different TCR/CD3 complexes is well established (1, 2, 21).

CD3-γ and -δ are glycoproteins which are usually found in single copies in the mature TCR/CD3 complex (Fig. 3). In contrast, two copies of the non-glycosylated CD3-ε (Fig. 3) are detected in the cell surface TCR/CD3 complex (29) and 20–30 per cent of these CD3-ε molecules are found to be disulfide cross-linked (35). In non-T cells, for example in transfected COS or CHO cells or in natural killer cells, disulfide cross-linked CD3-ε homodimers are the predominant species. Dimer formation involves solely the CD3-ε ectodomains, as determined by transfection and expression of truncated cDNAs into COS cells (B. Mueller and C. Terhorst, unpublished data). It is likely that in the mature TCR/CD3 complex the dimeric form is sustained by other protein–protein interactions even in the absence of interchain disulfide bridges. Since similarities have been found between the CD3-γ and -ε amino acid sequences and members of the immunoglobulin supergene family (1, 36; and Fig. 4A–C), intrachain disulfide bridges have been shown in Fig. 3. There is good evidence for intrachain disulfide bonds within CD3-ε based on the difference in mobility in SDS-PAGE with or without reducing agents (37, 38). The transmembrane regions of CD3-γ, -δ, and -ε contain Asp or Glu, which play a dominant role in interchain contacts with TCRα or -β (Fig. 4A–C). The cytoplasmic tails of the three related CD3 molecules from several species can be readily aligned (Fig. 4A), from which the closer relationship of γ and δ can be seen (39).

Unlike the other CD3 proteins, CD3-ζ has a very short ectodomain of nine amino acids and a long cytoplasmic tail of 112(H) or 113(M) amino acids (40, 41; Fig. 4D) and does not belong to the Ig supergene family (Fig. 3). A substantially truncated form of ζ (deletion Δ66–114) can still assemble into the TCR/CD3 complex and reach the cell surface, but is functionally inert (42). A considerable degree of flexibility in T cell signalling is demonstrated by the use of alternative signalling components in different TCR/CD3 complexes. CD3-η is identical to CD3-ζ with the exception of the C-terminal region as it is the product of an alternative splicing event (Fig. 3, Fig. 4D). CD3-η, which is expressed in a small fraction of peripheral T lymphocytes, is functionally rather similar to CD3-ζ. In a subset of αβ and γδ T cells, namely mucosal intestinal epithelial lymphocytes (IEL), the IgE Fc receptor (FcεR 1γ) chain can replace CD3-ζ (43, 44). The transmembrane segment of CD3-ζ is homologous to the FcεR1γ chain (Figs 3 and 4E) and permits its expression in the TCR cell surface complex. TCR/FcεR1γ$^+$ cells might utilize alternative T cell activation pathways (see Section 4.2.1).

3.2 Assembly in the endoplasmic reticulum

Assembly of a complete TCR/CD3 complex containing all six polypeptide chains (TCRα,-β and CD3-γ, -δ, -ε, -ζ) within the ER is an obligate requirement for efficient

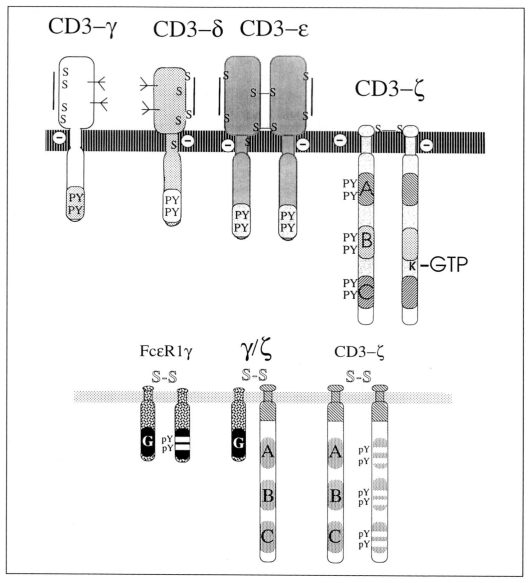

Fig. 3 Diagrams of the CD3 polypeptides. Disulfide bonds, oligosaccharide structures, transmembrane charge residues, location of GTP$_{oxi}$ labelling site and ITAMs are shown.

cell surface expression (1, 28, 45, 46). Single subunits and partial complexes are, in principle, unable to leave the ER and are actively retained within the ER (2, 21). To avoid accumulation of single chains or incomplete TCR/CD3 complexes, they are degraded at a rate that is determined by their subunit composition. Whereas the TCRα, -β, and CD3-δ polypeptide chains are degraded rapidly with half-lives of 90 minutes, CD3-γ, -ε, and -ζ, and all subcomplexes that contain them, are relatively long-lived (6–8 hours) (47–53). This selective degradation of TCR proteins implies that proteolysis of newly-synthesized proteins is carefully regulated within the ER (54, 55). Survival of the TCR within the ER and consequent cell surface expression thus result from a careful balance between receptor assembly and receptor degradation. This is reflected in the sequence of events followed during the building of the TCR/CD3 complex.

During the initial stages of TCR/CD3 complex biosynthesis in T cells, the first structures that can be detected by immunoprecipitation are associated CD3 core subcomplexes: CD3-γεε, CD3-δεε and/or CD3-δεεγ (37, 56). T cell receptor chains are then added to the CD3 core (Fig. 5) one at a time in either order, since both TCRα/CD3 and TCRβ/CD3 subcomplexes can be found. Ample evidence exists that in T cells or T cell mutants, the stable CD3-γ and CD3-ε subunits of the CD3 complex prevent rapid degradation of the newly synthesized TCRα, -β, and CD3-δ chains (37, 47, 56). Assembly around CD3-ε and -γ continues until an αβγδε complex is completed (Fig. 5A). The final subunit to assemble with the complex is CD3-ζ or other members of the ζ family (28, 45, 46, 57) (Fig. 5A). The ζ family members are always isolated as dimers and dimerization occurs prior to assembly with the αβγδε subcomplex (Fig. 5). Interestingly, even after addition of the CD3-ζ chain dimer, the receptor is not yet ready to leave the ER. Although addition of CD3-ζ to the receptor is the last assembly event that we can readily detect, it does not appear that this is the rate-limiting step in receptor transport. Addition of CD3-ζ to the pre-assembled αβγδε subcomplex is rapid, yet it takes substantially longer for pulse-labelled human T cell receptors to reach the *medial*-Golgi. This is possibly due to further conformational maturation of the receptor which could be rate-limiting for subsequent transport to the cell surface.

The charged residues found within the transmembrane anchors of the TCR/CD3 components have long been implicated in receptor assembly. Replacement of the 'forbidden' negatively-charged amino acid of the CD3 subunits with a neutral alanine abolished interactions between TCRα and CD3-δ, and TCRβ and CD3-ε (28). Whereas the transmembrane regions of TCRα and TCR-δ each contain two positively-charged amino acids (α_H^{Arg239} and α_H^{Lys244}; δ_H^{Arg250} and δ_H^{Lys255}, respectively), the TCRβ and -γ transmembrane regions each contain only one ($\beta_H^{Lys\ 271}$ and γ_H^{Lys296}). Removal of charged residues from the membrane anchors of the TCRα and -β chains also prevents pairing with CD3 components (24–27). The position of the charged residues plays an important role: these interactions are most favoured when the charged residues are located at the same level within the membrane (58, 59). Reconstitution experiments of a TCRα Jurkat cell line showed that both positively-charged residues of the transmembrane region of TCRα (i.e. Arg^{239} and Lys^{244}) needed to be modified to prevent incorporation into the complex (25). The negatively-charged transmembrane residues in the CD3 subunits do not affect formation of εγ, εε, and εδ dimers since these associations also take place between the charge-deleted CD3 mutants (22, 28). Surprisingly, the presence of identical (and therefore mutually repulsive) charges in the ζ transmembrane was

(A)   **CD3-γ Amino acid sequences**

Ectodomain

```
                 1                    20                      40
CD3-γH    QSIKGNHLVKVYDYQE*DGSVLLTCDAEAKNITWFKDGKMIGFLTEDKKTLNLGK
CD3-γM    -TN-AKN--Q-DGSRG*-------C---D-T-K-L---SI-SP-NAT-HKW---S
CD3-γR    QQKEEK--F--D-S-QG-------C-FNE-T---L---HR-SPPNAT-S-W---N
CD3-γS    ********---D-N--*----I-ICVTDE-K---L--M-E-SSGDTN-L-WD--S
CD3γ/δC       -V-*GLSMSVK-VS-K-F-QCQESKDLNTNYLWK-GKEE-GNMRQ*-D-GA
```

```
                 60                    80
CD3-γH    NAKDPRGMYQCKGSQNKSKPLQVYYRMCQNCIELNAAT
CD3-γM    -------T--CQ-AKET-N-------CE-C----IG-
CD3-γR    G---------Cr-AKK--QL------LCE-C----MG-
CD3-γS    ST-----I-EC---S-E--S--I---CQ-C----LA-
CD3γ/δC   IYD----T-TCQRDE-VNST-H-H---C--C--VDAP-
```

Transmembrane Domain

```
                        100
CD3-γH    ISGFLFAEIVSIFVLAVGVYFIA
CD3-γM    ----I-EVI---F--L------
CD3-γR    V---I--E-I---F---------
CD3-γS    VA--I-tE-----L--------
CD3γ/δC   ---IVV-Dv-ATVL--IA-C-T
```

Cytoplasmic Domain

```
                 120            140                   160
CD3-γH    GQDGVRQSRASDKQTLLPNDQLYQPLKDREDDQYSHLQGNQLRRN
CD3-γM    -----------------Q-E--Y--L----y--Y--L------KK
CD3-γR    ----------------Q-E-VY--L----yE-Y-rL----V-KK
CD3-γS    --E------------N----Y--L-E-----Y--Lrkk
CD3γ/δC   ---KGLM---R-N-IA----Y--LGE-N-G-Y-qLATAKARK
```

(B)   **CD3-δ Amino acid sequences**

Ectodomain

```
             1                    20                      40
CD3-δH    FKIPIEELEDRVFVNCNTSITWVEGTVGTLLSD1TRLDLGKRIL****DPRGIYR*CNGTD1YK
CD3-δM    ---QVT-Y--K---TCN--VMHLD--EGWFAKNK------GV-****------L*CN--EQLA
CD3-δR    ---EVV-Y--K---NCN--IRHLD-S-ERWLTKNKS-I-*-G-****----MGYMCN--EELA
CD3-δS    RALEVL-A--K-ILKCN---LLQ--A-QEVS-NKT-N----E****----M-Q*CGenaksf
CD3γ/δC   GVHGLSMSVK-VSGK--LQCQE-KDLNTNYLWKKGKEELGNMRQLDLGAIYD----T-T*CQRDENVN
```

```
          60                80
CD3-δH    KDESTVQVHYRMCQSCVELDPAT
CD3-δM    -VV-S-------C-nC----SG-
CD3-δR    -EV-----Y---C-nC----s--
CD3-δS    ****-L--Y---C-nC----s--
CD3γ/δC   ***-LH-----C-nCI-v-AP-
```

Transmembrane Domain

```
                     100
CD3-δH    VAGIIVTDVIATLLLALGVFCFA
CD3-δM    M--V-FIDL----------yC--
CD3-δR    L--V-I-DL----------yC--
CD3-δS    L--L-I-DI----------yC--
CD3γ/δC   IS--V-aD-V--v---IA-yCIT
```

Cytoplasmic Domain

```
                 120                 140
CD3-δH    GHETGRLSGAADTQALLRNDQVYQPLRDRDDAQYSHLGGNWARNK
CD3-δM    ------P----EV----K-E-LY--L---E-T-Y-rL----P---KS
CD3-δR    ----------V---V--K-E-LY--L-------Y-rL----P---KS
CD3-δS    ------F-R-----V-MG---LY--L-E-N---Y-rL-DK-----
CD3γ/δC   -QDK-LM-R-S-R-N-IA---LY--LGE-N-G-Y-qLATAKA-K
```

Fig. 4 (A–D) Sequences of the CD3 polypeptides of human, mouse, rat, sheep, and the consensus sequence. (E) Several CD3-ζ and FcεR-γ chains aligned.

(C)   **CD3-ε Amino acid sequences**

Ectodomain

```
                1              20                    40
CD3-εH    DGNEEMGGITQTPYKVSISGTTVILTCPQYP*GSEILWQHNDKNIGGDEDDKNIGSDED
CD3-εM         DDAENIE-------S-E--C-LDS*DENLK-EK-GQELP*******QKHDK
CD3-εS         DDTE-N--E-----NS-E--C-KDF*ENG-Q-KRNNEQMK*******-HN-K
CD3-εD    DEDFKASDDLTSISPEKRF-------E-VV-C-DVFGYDN-K-EK--NLVE*******-ASNR
```

```
         60             80                   100
CD3-εH    HLSLKEFSELEQSGYYVCYPRGSKPEDANFYLYLRARVCENCMEMDVMS
CD3-εM    --V-QD---V-D-----C-TPA-***NK-T----K---C-YCV-L-LTA
CD3-εS    Y-L-DQ---M-S-----QCLATEGNT-A-HT*---K---CK-C--VNLLE
CD3-εD    E-SQ-----VDD----AC-ADS***IKEKS--------Ca-Ci-VNL-A
```

Transmembrane Domain

```
                  120
CD3-εH    VATIVIVDICITGGLLLLVYYWS
CD3-εM    --I-I--D-C--L---MVI----
CD3-εS    ----IV-D-Cv-L----
CD3-εD    -V--IVaD-CL-L----M-----
```

Cytoplasmic Domain

```
         140               160                   180
CD3-εH    KNRKAKAKPVTRGAGAGGRQRGQNKERPPPVPNPDYEPIRKGQRDLYSGLNQRRI
CD3-εM    -------------T---S-P--------Y--I-------Y--L---AV
CD3-εS    -S-----T-M-------P----R--------Y--I-------Y--L---GV
CD3-εD    -T---N----M--T---S-P----K-K--------Y--I----Q--Y--L---G-
```

(D)   **Amino acid sequences of CD3-ζ/η**

	Ectodomain	*Transmembrane Domain*	*Cytoplasmic Domain*

```
          1                  20                    40
CD3-ζH    QSFGLLDPKLCYLLDGILFIYGVILTALFLRVKFSRSAEPPAYQQGQ
CD3-ζM    ----------C---D---------I---Y--A-------TA-NL-DP
CD3-ηM    ----------C---D---------I---Y--A-------TA-NL-DP
CD3-ζR    ----------C-M-D---------V---Y--A-------DAA--L-DP
CD3-ζS    ----------C---D---------V---F--A-------DA----H--
                        _____ζA_____
```

```
                    60                    80
CD3-ζH    NQLYNELNLGR*REEYDVLDKRRGRDPEMGGKP*RRKNPQ
CD3-ζM    ---Y--L----*---Y--LE---A--------QQ------
CD3-ηM    ---Y--L----*---Y--LE---A--------QQ------
CD3-ζR    ---Y--L----*---Y--L--K-P--------QQ--R---
CD3-ζS    -PVY--L-v--*---Y--L-R-G-F--------Q-K---H
                        _____ζB_____
```

```
                  100
CD3-ζH    EGLYNELQKDKMAEAYSEIGMKG**ERRRGKGH
CD3-ζM    --vY-aL--------Y--I-tK-**--------
CD3-ηM    --vY-aL--------Y--I-tK-**--------
CD3-ζR    --vY-aL--------Y--I--K-**--------
CD3-ζS    -vvY--LR-------Y--I--K-DNQ-------
                   _____ζC_____
```

```
         120                140
CD3-ζH    DGLYQGLSTAT*KDTYDALHMQALPPR
CD3-ζM    ---Y--L----*---Y--L---T-A--
CD3-ηM    ---Y-DSHFQAVQFGNRREREGSELTRTLGLRARPKGESTQQSSQSCASVFSIPTLWSPWPPSSSSQL
CD3-ζR    ---Y--L----*---Y--L---T----
CD3-ζS    --vY--L----*---Y--L--------
```

(E)   **Comparison of FcεR1-γ and CD3-ζ amino acid sequences**

	Ectodomain	*Transmembrane Domain*	*Cytoplasmic Domain*

```
          1                  20                    40
FcεR1γH   LGEPQLCYILDAILFLYGIVLTLLYCRLKIQVRKAAITSYE
FcεR1γM   ------C---D-v--------------------A-R-
CD3-ζH    QSFG-LD-K-C-L-DG---I--VI--A-F--V-FSRSAEPP-YQQ
```

```
              50           60
FcεR1γH   KSDGVYTGLSTRN*QETYETLKHEKPPQ
FcεR1γM   -A-A-Y--Ln--s*---Y--L-------
CD3-ζH    GQNQLYNELNLGR*r-eYDvLDKRRGRDPEMGGKPRRKN
```

```
CD3-ζH    PQEGLYNELQKDKMAEAYSEIGMKGERRRGK
CD3-ζH    GHDGLYQGLSTAT*KDTYDALHMQALPPR
```

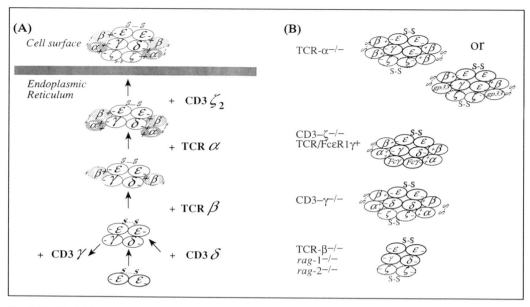

Fig. 5 (A). Model for assembly of the TCR/CD3 complex, showing transmembrane interactions. (B). Model for structures of partial receptors formed in deficient mice as shown.

found to be essential for dimerization and (less surprisingly) subsequent assembly with the TCR/CD3 complex (57).

The stability of CD3-γ and -ε in the ER plays a dominant role in the first phases of receptor assembly. Within the ER of doubly-transfected COS and CHO cells, a large number of polypeptide pairs can be observed (22, 28, 48, 52). Specifically, association of the stable CD3-γ and -ε polypeptide chains with labile TCRα or -β or CD3-δ subunits prevents degradation. In contrast, dimers composed of intrinsically labile subunits (for instance TCRα/CD3-δ or TCRβ/CD3-δ) are degraded rapidly (22, 28, 48). Thus, incorporation of stable chains into complexes appears to be an obligate requirement for receptor survival.

Studies of polypeptide chain pairing have been refined by site-directed mutagenesis experiments and by the use of chimeric proteins where single domains are coupled to reporter proteins (for example the human or murine IL-2Rα ectodomain) (22, 27, 59–62). Physiological control of degradation occurs via ER Ca^{2+} levels (54, 62).

Specific amino acid sequences within the membrane spanning domains of the TCRα and -β chains have been shown to cause rapid degradation of these polypeptides within the ER (27, 54, 60, 61). The TCRα and TCRβ ectodomains are by themselves more stable to degradation in the ER than the complete TCRα or -β. $TCRα_{tm}$ or $TCRβ_{tm}$ joined to another ectodomain (for example IL-2Rα) causes the IL-2Rα ectodomain to be degraded (22, 27, 54, 60). In addition to amino acid sequence features, the conformation adopted by the newly synthesized TCRα, -β, and CD3-δ proteins will also determine their sensitivity to ER proteases. Moreover, the location of the TCRα, -β, and CD3-δ proteins within the ER and further along the exocytic apparatus may contribute to their rapid degradation.

Interactions between membrane spanning domains is not sufficient for the protection of the TCRβ chain from ER proteolysis. When sites of binding between TCRβ and CD3-γ, -δ, or -ε were restricted to the membrane anchor of the TCRβ chain, stabilization by CD3 subunits was markedly reduced. The presence of the $C_β$ domain containing the first 150 amino acids of the TCR ectodomain greatly increases the stability of all pairs formed in the ER. For assembly with CD3-ε, stability is further enhanced by the $V_β$ amino acid sequences (22). Thus, the efficient neutralization of transmembrane proteolytic targeting information requires associations between membrane spanning domains and the presence of receptor ectodomains. Interactions between receptor ectodomains may slow the dissociation of CD3 subunits from the TCRβ chain and prolong the masking of transmembrane targeting information. In similar experiments, after the chimeric $TCRα_{tm}$–IL-2Rα protein paired with CD3-δ, degradation of each chain was reduced (26). However, this is in contrast with the observation that intact TCRα/CD3-δ pairs are degraded rapidly and so some caution in interpreting chimera experiments may be warranted.

4. Signal transduction through the TCR/CD3 complex

In addition to serving a role in the regulation of TCR expression, the CD3 proteins initiate signal transduction pathways leading to antigen-specific T cell proliferation, cytokine production, or cytolysis (1, 2, 63, 64). Several biochemical events immediately following T cell triggering have been defined. Cell surface activating stimuli share the common feature of antigen receptor aggregation and include antigen (Fig. 6A), superantigens, anti-TCR, and anti-CD3 antibodies, and other mitogens including lectins. Since the CD3 proteins have relatively short cytoplasmic tails, they do not display any enzymatic functions by themselves, but rather recruit signal transduction molecules (Fig. 6B). The best studied signal transduction elements of the TCR/CD3 complex are CD3-εε and CD3-ζζ, but CD3-γ and -δ also play a distinct role. For instance, the earliest detectable biochemical event following TCR engagement is tyrosine phosphorylation of CD3-ζ (65), CD3-ε, and to a somewhat lesser extent -γ and -δ (35, 66). A significant degree of autonomy in the mode of action of CD3-ε and -ζ has been demonstrated using CD3-ζ⁻ T cell lines (42) and in T cells transfected with the cytoplasmic tails of CD3-ε and -ζ linked to several different ectodomains (42, 67–69). Cross-linking of the heterologous ectodomains was sufficient to cause both early (tyrosine phosphorylation/Ca⁺ mobilization) and late (IL-2 secretion/CTL activity) activation events.

There are several protein tyrosine kinases (PTKs) involved in CD3 tyrosine phosphorylation: the src-family PTKs-lck (70) and fyn (71), syk (72), and ZAP 70 (64, 72–74), and possibly tsk (75). The order of recruitment of these PTKs in T cell

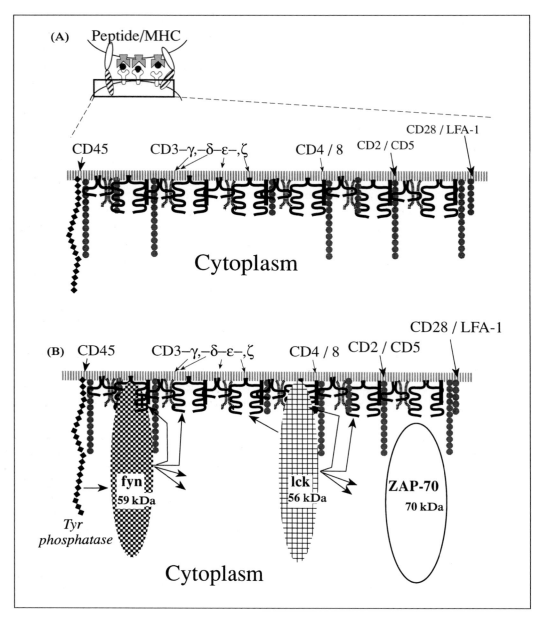

Fig. 6 T cell activation by receptor aggregation (A) leads to recruitment of multiple signalling molecules (B).

activation is not completely understood, but it would appear that the fyn and lck act upstream of ZAP-70 and syk (64, 72–74). To emphasize the importance of PTK in this regard, tyrosine kinase inhibitors, unlike the depletion of protein kinase C (PKC), block the very earliest stages of T cell activation (76–78). The requirement for a cycle of tyrosine phosphorylation–dephosphorylation in T cell activation is also underscored by the involvement of CD45, a membrane-bound tyrosine phosphatase. Absence of CD45 in some T cell lines inhibits the activation pathway to varying degrees, and reconstitution with CD45 cDNA restores full activity (64, 79). Conversely, phosphatase inhibitors block the positive effects of CD45 transfection upon activation (64; see also Section 5.4).

Phosphorylation of CD3-ζ upon engagement of the TCR/CD3 complex by antigen occurs on the six tyrosine residues of the three so-called immune receptor tyrosine-based activation motifs (ITAMs), YxxI/L$\{x\}_{10}$YxxI/L (Figs 3, 4, 7) (refs 80, 81). Similarly, the two ITAMs of the CD3-$\varepsilon\varepsilon$ pair and the single motifs of CD3-γ and -δ (Fig. 7) are phosphorylated upon activation (35, 66). Since ten copies of these ITAMs exist in the complete TCR/CD3 complex (six in ζ, two in ε, one each in γ and δ), a model in which the cumulative involvement of the receptor transduction modules regulates the subsequent events has been proposed (64). However, the individual ITAMs (ζA, ζB, and ζC and those of CD3-γ, -δ, and -ε) may play distinct roles in signal transduction required both during thymic development and subsequently in the periphery. Interestingly, *in vivo* much of surface TCR associated ε chain appears to be constitutively tyrosine phosphorylated. So regulation of signalling may be complex (81). Studies of mutant mice (*fyn*$^{-/-}$, *lck*$^{-)-}$, and *CD3-ζ*$^{-/-}$) and immunodeficiency patients (*CD3-γ*$^{-/-}$, -$\varepsilon$$^{-/-}$, and *ZAP70*$^{-/-}$) support this notion (see Sections 4.1 and 4.2).

Increased turnover of phospholipids by phosphorylated phospholipase Cγ1 (82, 83), a rapid and transient rise in intracellular free Ca^{2+} (84), and activation of phosphatidyl-inositol 3' kinase (PI 3-kinase; refs 85, 86) can be measured within seconds to a few minutes of activation. The role of G-protein(s) in coupling TCR activation to phospholipid metabolism has been inferred, but evidence for such a role is ambiguous (87). These events (Fig. 7), which are coupled to and immediately downstream of receptor activation, will be discussed in detail. Events that are more distal to the TCR/CD3 complex (that is to say, Ser/Thr phosphorylation of PKC, raf-1, and MAP2 kinases) have been reviewed elsewhere (63, 64, 88). Further events result in transduction of the activation signal from the inner surface of the plasma membrane through the cytoplasm to the nucleus, culminating in transcriptional changes, effector functions, and in many cases cell division.

4.1 Signal transduction molecules associated with CD3-ζ and CD3-ε

4.1.1 src-family tyrosine kinases

Members of the src family of non-receptor PTKs initiate phosphorylation of the Tyr residues in the CD3 ITAMs (89). T lymphocytes contain transcripts encoded by at least three members of the *src* gene family; *fyn*, *lck*, and *yes*. All *src*-related genes encode closely related products of molecular masses varying between 55 and 72 kDa (89). Due to myristylation of a conserved amino-terminal glycine residue

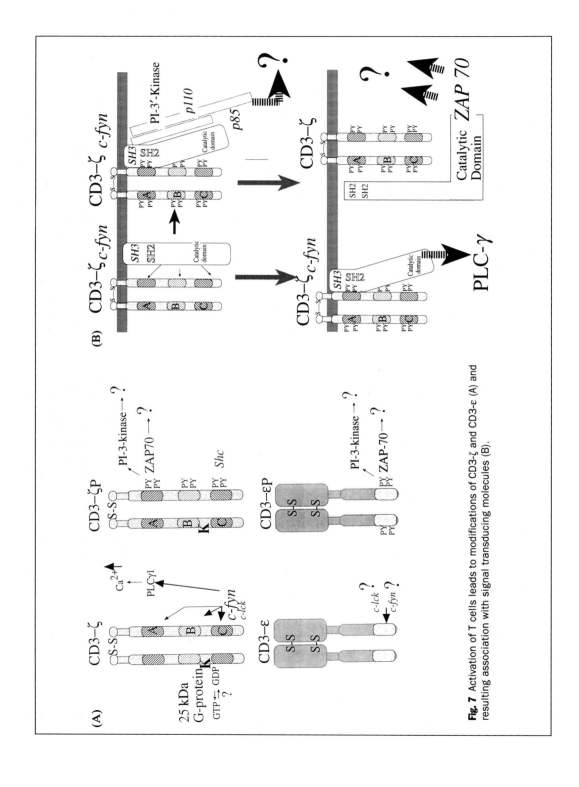

Fig. 7 Activation of T cells leads to modifications of CD3-ζ and CD3-ε (A) and resulting association with signal transducing molecules (B).

(Gly2), src-related enzymes are tightly membrane associated, in particular with the plasma membrane (90). The tyrosine kinase domain (src homology domain 1, SH1) is the most highly conserved portion of all receptor tyrosine kinase molecules and contains a consensus sequence, GlyYGlyXXGlyX(15–20)Lys that functions as part of the binding site for ATP (90). src-Related non-receptor PTKs contain an adjacent non-catalytic domain of approximately 100 amino acids, the src homology domain 2 (SH2) which interacts with phosphorylated Tyr residues (89, 91, 92). A third domain, SH3, is found to mediate interactions with proline-rich sequences. It has been implicated in protein–protein interaction that involves protein kinases and their substrates (92). SH2 domains compliment the action of the catalytic kinase activity by communicating the phosphorylation states of signal transduction proteins to elements of the signalling pathway. SH2 domains are also found in other cytosolic signalling molecules including, phospholipase C (PLC) PLC-γ1, the p85 subunit of PI-3-kinase, ras GTPase-activating protein, and tyrosine phosphatases. In each of these proteins, the sequence of the SH2 domain is thought to be important in either localizing the protein to tyrosine-phosphorylated proteins or controlling its activity (91). Crystal structures of SH2 domains of src and lck (93, 94) provide the first view of the mechanism by which phosphorylated tyrosine residues are recognized by SH2 domains, confirming the predictions and binding data (91, 95).

p56^{lck}, a src family tyrosine kinase, is a lymphocyte-specific kinase non-covalently associated with CD4 and CD8 (70, 89). Increasing evidence demonstrates that a large part of the accessory function provided by CD4 and CD8 during T cell activation is mediated through p56^{lck} (88; Section 5.1). However, in co-transfection experiment's lck has been found to interact with CD3-ζ and -ε (Metzger, Terhorst et al. in preparation). The structure of p56^{lck} is typical of the members of the src family. It contains an amino-terminal glycine, a unique NH2 domain, SH2 and SH3 domains, and a highly conserved catalytic carboxy-terminal domain, with autophosphorylatable positive and negative regulatory tyrosines (Y394 and Y505, respectively) (70, 89).

p59^{fyn} which is physically associated with CD3-ζ (71) exists as two isoforms differing within a stretch of 51 amino acids spanning the end of the src homology domain and the beginning of an otherwise classical tyrosine kinase domain (89). Activation of the TCR/CD3 complex activates associated fyn (96). Recently, it was shown that elements of the TCR/ζ induced signal transduction pathways can be reconstituted in Cos cells (97):

- a Vβ-specific monoclonal antibody triggers cell-surface $(\beta-\zeta)_2$ chimeras expressed in Cos cells;

- co-transfection of c-fyn results in phosphorylation of all three ITAMs in $(\beta-\zeta)_2$;

- $(\beta-\zeta)_2$/c-fyn phosphorylates Tyr residues of endogenous PLC-γ1. Phosphorylated PLC-γ1 catalyses the conversion of phosphatidylinositol 4,5,bisphosphate to inositol (1,4,5)-triphosphate (IP$_3$) and diacylglycerol;

- triggering $(\beta-\zeta)_2$ results in IP_3-dependent release of Ca^{2+} from intracellular stores.

In this assay, c-lck could only phosphorylate CD3-ζ weakly. Whether c-fyn is involved in phosphorylation of CD3-ε is uncertain. In general, it is thought that these two tyrosine kinases are involved in the first steps of the tyrosine phosphorylation cascade initiated by the TCR/CD3 complex (89). The critical importance of both fyn and lck in thymic development and T cell activation are emphasized by the respective kinase deficient mice (98–100), which fail to generate normal mature thymocytes (see also Sections 4.3 and 5.1).

4.1.2 The protein tyrosine kinases ZAP-70, syk, and other kinases involved in T cell activation

A 70 kDa tyrosine phosphoprotein (ZAP-70 for ζ-associated protein) associates with CD3-ζ within 1 minute of TCR stimulation and has been characterized as a novel tyrosine kinase (73, 74). The role of the 72 kDa tyrosine kinase syk in T cells is less understood. However, syk is expressed in thymocytes and some T cells as well as in B cells and mast cells (64, 72). These closely related receptor-associated kinases have two SH2 domains and a single kinase domain with relatively long and characteristic variable linking regions (64). The phosphorylation of ZAP-70 and its association with CD3-ζ is independent of the other TCR/CD3 chains since stimulation of a CD8/CD3-ζ chimeric receptor in a TCR-negative cell leads to co-precipitation of ZAP-70 with the chimeric protein. The association and tyrosine phosphorylation of ZAP-70 occur exclusively with either activated receptor complex. One likely function of ZAP-70 is to couple the TCR through CD3-ζ to more distal signalling molecules. Tyrosine phosphorylation of CD3-ζ allows its interaction with the two SH2 domains of ZAP-70 (74). Similarly, the ZAP-70 SH2 domains can interact with the phosphotyrosine residues in CD3-ε (101). Interestingly, the SH2 domain of lck binds ZAP-70, whilst the phosphotyrosine residues in ZAP-70 are not positioned in a consensus sequence known to be recognized by lck (94, 102). Therefore, the interaction of ZAP-70 with phosphorylated ζ and ε and possibly all CD3 cytoplasmic tail ITAMs is probably via the tandem SH2 domains of ZAP-70. Lck may then bind to phosphotyrosines elsewhere within ZAP-70, rather than directly to ζ and ε. Fyn, however, appears to interact in a stable constitutive manner with at least ζ and, therefore, may interact with downstream signalling molecules such as ZAP-70 in an entirely distinct and complementary way from that of lck (Fig. 7; see also Sections 4.3 and 5.1). An important observation stressing the role of ZAP-70 in T cell activation is the finding that certain T cell immunodeficiencies are due to failure to make a stable ZAP-70 molecule. In these patients, vulnerable to AIDS-like opportunistic infections, $CD4^+$ T cells are found in the periphery, but these cells do not respond to activation stimuli (103).

Although less information is currently available, the cytoplasmic tail of CD3-ε is also able to function in T cell activation in the absence of other CD3 components.

The cytoplasmic tail of CD3-ε contains a single ITAM, whereas CD3-ζ has three similar motifs (Fig. 4). Ligation of chimeras with either CD3-ε or CD3-ζ cytoplasmic tails results in only partly identical patterns of cellular protein tyrosine phosphorylation in T cells (69). It is now known that ZAP-70 can interact with both CD3-ζ and CD3-ε (101, 104). It seems likely, however, that the two separate structures within the TCR/CD3 complex function in parallel, distinct, but partially overlapping pathways (Fig. 7), which together provide the most sensitive TCR/CD3 signalling moiety and enabling only a few hundred MHC–antigen peptide complexes to activate a given T cell. A precedent for this can be found in the high-affinity Fcε receptor γ chain. A chimera encoding this cytoplasmic tail functions optimally in basophil-like cell lines, whereas the CD3-ζ and CD3-ε tails function most efficiently in T cells (69).

Most recently, a different potential function has been identified for the remaining two CD3 tails. CD3-γ and CD3-δ cytoplasmic tails can associate constitutively with the ser kinase raf-1 (105), previously positioned downstream of ras in the signal transduction cascade (106).

4.1.3 PI-3-kinase

Phosphatidylinositol 3-kinase (PI-3-kinase), a key enzyme in the signalling pathways of both receptor and non-receptor tyrosine kinases, has two unique properties: phosphorylation of the D-3 position of the inositol moiety of phosphatidylinositol (PI) to produce phosphatidyl inositol 3-phosphate (PI-3P) and specific association with many activated protein kinases (91). PI-3-kinase has been shown to physically associate and to form a complex with PDGF-, EGF-, CSF-1-, and insulin-receptors, with oncogene products of polyoma middle T, and v-*src*, v-*fms*, v-*abl*, and v-*yes*. Mutants of receptors and oncoproteins that fail to associate with PI-3-kinase fail to mediate growth signals and transformation. Historically, three forms of PI kinase have been distinguished; type I phosphorylates the D-3 position of the inositol ring, whereas type II and III enzymes phosphorylate the D-4 position (91).

PI-3-kinase has been purified to homogeneity and phosphorylates PI at the D-3 position of the inositol ring irrespective of phosphorylation at positions 4 and 5. Purified bovine brain PI-3-kinase is composed of 85 kDa and 110 kDa subunits. The 85 kDa subunit lacks PI-3-kinase activity and acts as an adapter, coupling to the 110 kDa subunit. The PI-3-kinase regulatory subunit, p85, is one of the family of SH2-containing proteins whose interactions are mediated by binding to phosphorylated tyrosine residues within distinct motifs (91, 107). Cloning revealed the existence of two 85 kDa subunits (p85α and p85β), each possessing one SH3 domain and two SH2 domains (reviewed in ref. 108). p110 cloning and expression in Cos cells showed that p110 is catalytically active only when complexed with p85 (108, 109). It is still unknown how PI-3-kinase products might act as second messengers, although two clues do exist. Firstly, PI-3-kinase products specifically activate the Ca^{2+}-independent protein kinase C-ζ isoform (110). PKC is known to be downstream of the TCR/CD3 complex in transducing the activation signal to the nucleus. However, p21*ras* appears to specifically activate PKC in T cells (ref. 106; see Section 4.1.4). Different PKC isoforms may be involved in these different signalling events. Very recently, the expression of p110 cDNA has demonstrated a second activity of PI-3-kinase potentially involved in signal transduction; PI-3-kinase has an intrinsic protein ser kinase activity. Autophosphorylation by this ser kinase activity inhibited PI-3-kinase activity (111). This observation presumably provides a second possibly inhibitory component to PI-3-kinase signalling.

T cell activation results in association of PI-3-kinase with the TCR (85) and an increase in levels of 3'-phosphorylated polyphosphoinositides (86). We found an increase in phosphotyrosine-associated PI-3-kinase activity in response to TCR/CD3 activation, and that PI-3-kinase associated with the CD3-ζ chain specifically through the ζA motif, and not ζB or ζC (112). Association was maximal with double-phosphorylated ζA and correlated with phosphorylation of ζA tyrosine residues in activated T cells and simultaneous association of a 70 kDa phospho-protein, presumably ZAP-70. Future experiments will address whether ZAP-70 can act as an adaptor by interacting simultaneously with CD3-ζ and PI-3-kinase, or whether PI-3-kinase can interact with CD3-ζ independently. A third attractive hypothesis is that a src kinase could also act as an adaptor for PI-3-kinase. Support for this suggestion comes from the detection of complexes between PI-3-kinase and both p56lck and p59fyn (113, 114). These complexes associate via the src kinase family member SH3 domains probably via a proline-rich sequence on the PI-3-kinase regulatory subunit p85. How activation increases these interactions is unclear, since only SH2 domains respond to receptor activation via association with Tyr-phosphorylated sequences (91). Thus it may be that src kinases can act as adaptor molecules for association with downstream signalling molecules. Since fyn associates constitutively with the TCR/CD3 complex via CD3-ζ at low stoichiometry (71) and lck associates with CD4/CD8 (70), these two players are already available to participate in TCR/CD3 signalling.

In summary, it remains to be shown whether PI-3-kinase association with the activated TCR/CD3 complex is direct or indirect. It is possible that either src family kinases lck and fyn or syk/ZAP-70 kinases act as adaptors in this interaction. Furthermore, it has not yet been shown that T cell PI-3-kinase is directly coupled to mitogenic or other signalling pathways (for instance cytotoxicity). However, the association of PI-3-kinase with mitogenic and other signalling pathways of many receptors and the activation-dependent association of PI-3-kinase with the TCR strongly implicate this enzyme in TCR signalling.

4.1.4 GTP binding by CD3-ζ

A novel and highly efficient *in situ* protein labelling technique allowed the detection of a GTP/GDP binding site in the CD3-ζ chain. This method differed from classical affinity labelling techniques in that the native ligand was directly introduced into permeabilized cells, thus allowing binding or exchange of native GTP to a GTP-binding protein. A reactive nucleotide analogue (GTP_{oxi}) was then generated transiently by *in situ* oxidation with periodate. Using this method, more than 50 per cent of $p21^{H-ras}$ could be cross-linked to its endogenous bound nucleotide. In the same way, the CD3-ζ chain was specifically cross-linked to GTP but not ATP or CTP. This reaction was inhibited by the addition of dGTP or dGDP but not by dATP, dGMP, or pyrimidine nucleotides (87). Based on these data, the CD3-ζ protein appears to represent a novel guanosine nucleotide binding structure or is very closely associated with a GTP-binding protein.

Identification of lysine 108 as the primary GTP_{oxi} modification site in human ζ implied that, unlike classical GTP binding proteins such as $p21^{H-ras}$ or elongation factor Tu, the ribose moiety was in close proximity to the putative phosphoryl binding site, potentially included in the glycine-rich sequence K*(108)G XXXXGKGXXGXXG (87). Since the GTP binding of CD3-ζ was found in functionally active T cell receptor/CD3 complexes as well as in intracellular ζζ homodimers, an energetically driven conformational change of CD3-ζ could play a role in cell signal transduction. Two experimental results suggest that this is indeed the case. First, in the murine system, a ζ protein with the mutation G115V (equivalent to human G114V) has been tested functionally (115, 116). These authors found that both antigen-triggered IL-2 production and tyrosine phosphorylation of CD3-ζ were abrogated in a transfected variant of the murine T–T hybridoma 2B4 expressing the mutated ζ protein. Mutant K130R, corresponding to the human CD3-ζ mutant K129R, did not impair antigen-stimulated T cell activation. Since in our assay mutant G114V did not label with GTP_{oxi} whereas K129R did, the functional defect in the murine G115V mutant might result from an altered GTP/GDP binding capacity (87). Second, the absence of GTP_{oxi} cross-linking in a set of TCR/CD3$^+$ Jurkat mutants correlated with the complete impairment of TCR mediated Ca^{2+} mobilization, IL-2 production, and CD3-ζ tyrosine phosphorylation (117). These two observations lead to the conclusion that antigen-triggered tyrosine phosphorylation and GTP/GDP binding of CD3-ζ might be linked. This model implies that interaction of CD3-ζ with a specific tyrosine kinase, tyrosine phosphatase, or other unidentified macromolecules may be regulated by its GTP/GDP binding function.

The physiological role of guanine nucleotide binding by CD3-ζ requires further investigation. The putative nucleotide binding domain of CD3-ζ is not classical in structure, and the possibility that CD3-ζ binds to a small G-protein needs to be considered. Three experiments support this notion:

- Truncated murine CD3-ζ can still be labelled by GTP_{oxi} (87).

- Preliminary studies have identified a 28 kDa GTP_{oxi} labelled protein associated with CD3-ζ (Jkang and Terhorst, unpublished).

- Recently, it has been found that the Tyr-phosphorylated CD3-ζ cytoplasmic tail ζC motif can associate with the SHC:Grb-2:mSOS complex upon T cell activation (118). This would indirectly link CD3-ζ to the small G-proteins.

The small G proteins are ubiquitously expressed and are required for control of cell-cycle progression and intracellular trafficking. Downward and co-workers (reviewed in ref. 106) demonstrated that ras activation occurs in primary T cells within minutes of TCR stimulation and that this activation is apparently dependent upon PKC activation (119). It remains to be determined whether PKC is responsible for the ras activation. Alternative possibilities are that other serine/threonine kinases such as raf-1 or MAP-2 kinase subserve this role. ras-GAP involvement in T cell activation appears to take place downstream of the TCR/CD3 transduction modules (106).

4.1.5 Specificity of interaction of signalling molecules with CD3 cytoplasmic tail motifs

Specific interaction of each ITAM with cytoplasmic signalling molecules might occur. These motifs can be subdivided into families based upon sequence differences (Fig. 4.D) which could be due to evolutionary relationships or to subdivision of functions to different motifs. Why does the CD3-ζ chain need three distinct motifs, and why should the TCR complex contain multiple copies of six different motifs, probably ten in all (six in CD3-ζ_2, two in each CD3-ε pair, and one each in the CD3-γ–δ subcomplex)? In chimera experiments, any of the three ζ motifs can partially replace the intact cytoplasmic tail in both early and late activation assays when transiently expressed at high levels (42, 102, 120). Furthermore, the three small ITAMs of the ζ tail are able to function independently of the remaining amino acids, although with reduced efficiency (102, 120). A single synthetic molecule containing three ζA motifs can elicit the response seen with the complete cytoplasmic tail chimera (102). The ZAP-70 tyrosine kinase has also been shown to interact with other CD3 tails (101). This interaction depends upon tyrosine phosphorylation following TCR/CD3 activation and the individual ITAM closest to the membrane, ζA, is able to reconstitute this interaction (102). It is not yet clear whether each of the three ITAMs can interact with ZAP-70, or if there is preferential association with ζA and the single motifs of other CD3 tails. However, the individual ζC motif is most active in promoting IL-2 secretion following activation, although at only 10 per cent of the level of the complete ζ cytoplasmic tail (120). Furthermore, CD3-ε and CD3-ζ cytoplasmic tails stimulate distinct patterns of T cell protein tyrosine phosphorylation (69). Activation of a second pathway involving GTP-binding proteins was suggested by the finding that ζ itself is a GTP-binding protein (87) and that the ability of ζ to bind GTP correlates with activation (117). GTP can be chemically cross-linked to a site between ζB and ζC, suggesting that GTP could specifically interact with either ζB or ζC during activation. Finally, the phosphorylated ζC motif has recently been shown to interact with the SHC:Grb-2:mSOS signalling complex (118). Although other motifs were not tested, these observations and the association of PI-3-kinase with the phosphorylated ζA motif (112) suggest distinct functional roles for the six different ITAMs of the TCR/CD3 complex (Fig. 7).

4.2 On the role of CD3-ζ and -ε in transgenic and mutant mice

4.2.1 The CD3-$\zeta^{-/-}$ mouse

The development of T cells within the thymus is a complex and selective process (Fig. 8) recently discussed by von Boehmer and Kisielow (4) and Janeway (5).

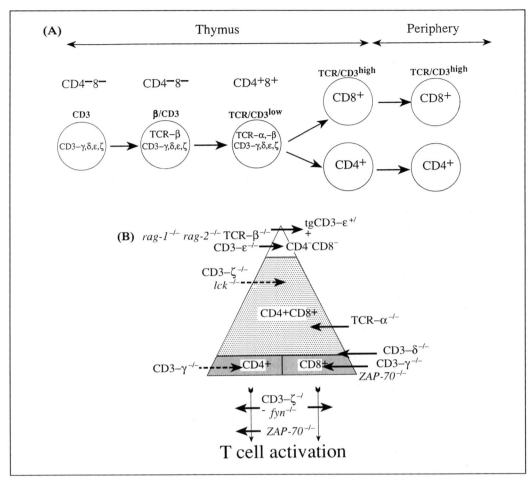

Fig. 8 (A) Development of T cells within the thymus. (B) Maturation of T cells within the thymus and expression of mature phenotypes in the periphery are blocked by mutation of the genes shown.

Elimination of the ζ gene from the murine genome by homologous recombination results in an abnormally developed thymus (10–15 per cent of wild type size) in which primarily TCRαβ⁻, CD4⁻8⁻, and TCRαβ very low CD4⁺8⁺ are found (121–123; and see Fig. 8). Our preliminary biochemical data show that, as predicted from studies with cell lines, a minute amount of an αβγδε-containing complex is expressed on the surface of the $CD3\text{-}\zeta/\eta^{-/-}$ thymocytes. Analysis of the D–J joining segments of $V_\beta 4$, $V_\beta 8$, and $V_\beta 12$ in mRNA isolated from CD4⁺CD8⁺ thymocytes and from splenic T cells confirms that cells with productively rearranged V_β regions are selected. In the $CD3\text{-}\zeta^{-/-}$ mouse very few if any 'single positive' thymocytes are detected (121–123). This could be due to a partial block in the transition from CD4⁺8⁺ to single positive cells or could be a result of the very small pool of CD4⁺8⁺ cells.

In contrast to the thymus, spleen and lymph nodes contain large numbers of single positive TCR/CD3$^{\text{very low}}$CD4$^+$ and TCR/CD3$^{\text{very low}}$CD8$^+$ T lymphocytes. This could be explained by a model in which small numbers of CD3-ζ$^-$ thymocytes are positively selected by interacting with MHC class I and class II proteins. A second possible interpretation is that the TCR/CD3$^{\text{very low}}$ single positive cells may have been generated through an alternative pathway not involving positive selection for MHC. Studies with *CD3-ζ/η$^{-/-}$* × MHC *class I$^{-/-}$* mice reveal that CD8$^+$ T lymphocytes are absent in the spleen and lymph nodes, suggesting that positive selection takes place in *CD3-ζ/η$^{-/-}$* mice. Similarly, CD4$^+$ T cells are not found in *CD3-ζ/η$^{-/-}$* × MHC *class II$^{-/-}$* mice, demonstrating that positive selection occurs even when very low levels of the *CD3-ζ$^-$* TCR are present on the cell surface (Fig. 5B). Surprisingly, the total number of CD4$^+$ T cells is drastically lower in some of the *CD3-ζ/η$^{-/-}$* × MHC *class I$^{-/-}$* mice when compared to the *CD3-ζ$^{-/-}$* animals. In some *CD3-ζ/η$^{-/-}$* × MHC *class II$^{-/-}$* animals, the number of CD8$^+$ T lymphocytes is reduced compared to CD3-ζ$^-$ mice (124). This may reflect the stochastic processes involved in thymic selection. Thus, a model is arising in which the TCR/CD3$^{\text{very low}}$ complexes can still engage in positive selection resulting in the export to the periphery of small numbers of mature thymocytes. How these single positive T cells expand in the peripheral organs is uncertain. Careful studies of the TCR V-region repertoire need to be conducted to examine whether a relatively small number of T cell clones have expanded. Pauciclonality will have to be studied at the mRNA level because V-region specific antibody reagents do not stain the TCR/CD3-ζ$^-$ complex on the surface of spleen cells. Although splenic *CD3-ζ$^{-/-}$* T lymphocytes express several cell surface markers that are connected with T cell activation, the parameters of proliferation are unknown. For example, these cells do not respond to mitogens (concanavalin-A, anti-CD3) or in mixed lymphocyte cultures (She and Terhorst, in preparation).

4.2.2 TCR/FcεR1γ$^+$ receptors

Recently, the significance of a novel type of T cell of the murine intestinal epithelium (Fig. 9) was uncovered by studying *CD3-ζ$^{-/-}$* mutant mice (121). In these animals high levels of αβ or γδ TCR/CD3 complexes are invariably found on the surface of intraepithelial (T) lymphocytes (IELs) isolated from the small intestine or the colon (Fig. 9), but not dendritic epithelial T cells in the skin. This is due to association of TCRα, -β/CD3-γ, δ, ε with the FcεRIγ polypeptide chain (125). Intraepithelial T cells expressing TCR/FcεR1γ$^+$ complexes are functionally indolent, as judged by their poor responses to anti-CD3, anti-TCR, concanavalin-A, and mixed lymphocyte culture, but are able to respond to stimulation with phorbol myristate acetate (PMA) and ionomycin (121 and in preparation). In redirected cytotoxic T lymphocyte (CTL) assays with ^{51}Cr-labelled P815 cells and anti-CD3 antibodies, IEL from the small intestine do not cause chromium release, in contrast to the wild-type IEL (124). However, *in vitro* stimulation of these cells may give rise to the production of certain cytokines.

The majority of the CD3$^+$ IEL are TCRαβ$^+$CD8$^{\alpha\alpha+}$ cells. A smaller percentage is

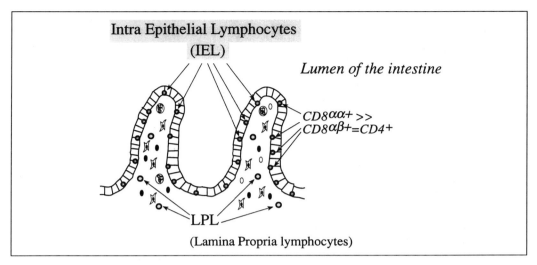

Fig. 9 Tissue distribution of intra-epithelial lymphocytes.

TCRαβ$^+$, CD8$^{αβ+}$, or CD4$^+$. In young animals, a variable fraction of these TCR/FcεR1γ$^+$ IEL are γδ T cells and the percentage of γδ T cells increases with age or with an immunological challenge (She, in preparation). The phenotype of TCR/FcεR1γ$^+$ in the IEL of wild-type mice is similar to that in the *CD3-ζ$^{-/-}$* mice. Interestingly, in γδ TCR transgenic mice a high percentage of these γδ T cells in spleen and lymph nodes express TCR/FcεR1γ$^+$ complexes (44).

Further analyses of the T cells in the intestinal epithelium of *CD3-ζ/η$^{-/-}$* × MHC *Class I$^{-/-}$* mice reveal that only the CD8αβ$^+$ IELs and lamina propria lymphocytes (LPL) of both the small and large intestine, which are predominantly TCR/CD3$^{very\,low}$, can be positively selected for class I MHC (that is to say CD8$^+$). Similarly, the CD4$^+$ lymphocytes, which are absent in the intestinal epithelium of *CD3-ζ/η$^{-/-}$* × MHC class II$^{-/-}$ mice are TCR/CD3$^{very\,low}$. By contrast, the majority of TCR/FcεR1γ$^+$ IEL which are CD8$^{αα+}$ and to a lesser extent CD4$^+$ or CD8$^{αβ+}$, are not selected by either class I or class II MHC antigens. These observations support the concept of extrathymic development of TCR/FcεR1γ$^+$CD8$^{αα+}$ IEL (Fig. 10A), using alternative restriction elements or without restriction, rather than conventionally in the thymus (Fig. 10B).

Preferential expression of the *FcεR1γ* gene in intestinal epithelial T cells might be caused by induction through environmental factors or be due to an independent programming of precursor lymphocytes involved in extrathymic maturation. At the moment we favour an induction model because of an observation by Mizoguchi *et al.* (126), who describe an *in vivo* effect of a murine colon carcinoma cell line MC38 which induces an impaired cytotoxic function of splenic CD8$^+$ T cells. Interestingly, the splenic T cells completely lack ζ and expressed the TCR/FcεR1γ$^+$ complexes on the cell surface (Fig. 5B). In agreement with the data of the colon cancer model is our observation that the re-directed cytotoxic activity of ζ$^{-/-}$ IELs

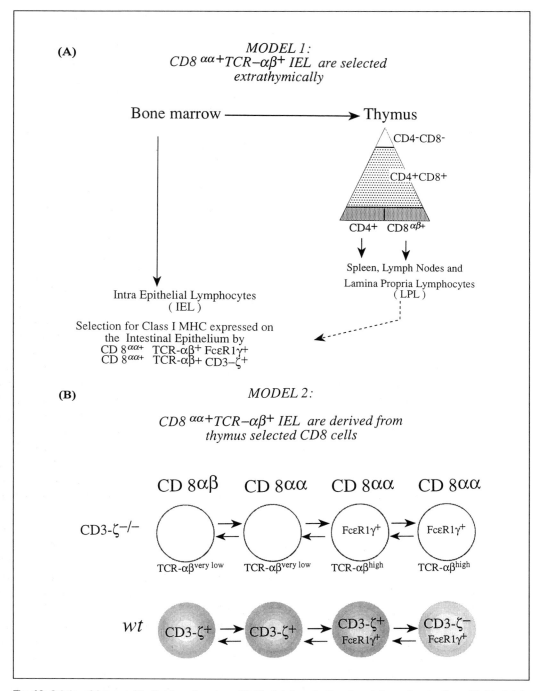

Fig. 10 Origin of intra-epithelial lymphocytes. (A) Model 1: extrathymic origin and selection. (B) Model 2: intra-epithelial lymphocytes undergo thymic development and post-thymic maturation.

are completely abrogated when compared to wild type. However, in contrast to Mizoguchi *et al.* (126) we have detected a significant level of the src-like tyrosine kinases, c-fyn and c-lck in the $\zeta^{-/-}$ IEL. Thus, the colon cancer-induced TCR/ FcεR1γ$^+$ cells appear similar, but not identical, to the CD3-$\zeta^{-/-}$ IELs. Nevertheless the experiments suggest that a non-responsive TCR/FcεR1γ$^+$ cell can be induced by environmental factors, which may in part be provided by intestinal epithelium. Replacement of CD3-ζ by the FcεR1γ chain in the TCR/CD3 complex resulting in a relatively non-responsive cell type may be a physiological correlate of the observation that FcεR1γ chain chimeras function less effectively than CD3-ζ chimeras in T cells and optimally in a mast cell line (69).

4.2.3 Overexpression of the CD3-ε signal transduction module in transgenic mice abrogates development of T cells and natural killer cells

To study the role of the signal transduction module of the T cell antigen receptor associated CD3-ε protein in the development of T lymphocytes and natural killer (NK) cells, a series of transgenic mice were generated with the human *CD3-ε* gene. In transgenic animals containing more than 30 copies of the *CD3-ε^H* gene, both T cell and NK cells were completely absent. The CD3-ε^H protein appeared to be indispensable for establishing the *T cell$^-$/NK cell$^-$* phenotype, because deletion of the exons coding for the ecto- and the transmembrane domains or replacement by the murine gene resulted in a wild-type phenotype. *T cell$^-$* mice could also be generated by using a gene fragment coding for the 55 amino acid non-enzymatic CD3-ε cytoplasmic tail (127). Taken together these results demonstrated that overexpression of the cytoplasmic tail of the human CD3-ε protein affected both thymic maturation and NK cell function. The most plausible explanation for these observations was that aberrant signal transduction by the CD3-ε^H module led to abrogation of T cell and NK cell development. These observations not only strongly supported the concept that NK cells and T cells stem from a common precursor cell, but also that signal transduction via CD3-ε plays an important role in the development of both cell lineages. The influence of the CD3-ε protein was exerted in a very early thymocyte population (Thy-1$^+$Pgp-1$^+$) or in precursors thereof and in precursor NK cells, and before expression of endogenous *CD3-γ*, -δ, and -ε genes or rearrangement of the *TCRα* and -β receptor genes. Therefore, the possibility of an intracellular signalling event specifically dependent upon CD3-ε must be considered.

Transgene coded mRNAs were found in fetal thymuses on day 13 of gestation, whereas on day 14 of gestation, transgenic mRNA levels were dramatically higher than those of the endogenous CD3-ε transcripts. Analysis of a series of transgenic lines generated with different genomic CD3-ε constructs revealed that the extent of the immunodeficiency correlated with the number of copies of the transgenes and with the enhancer species used. Taken together, the data demonstrated that high levels of expression of CD3-ε caused the block in early thymocyte development. Since CD3-ε has been found in activated mature human natural killer cells and in human fetal NK cells (128–130), the influence of the *CD3-ε* transgene on natural killer cell development was studied. Surprisingly, no NK cells could be found in transgenic mice carrying high copy numbers of *CD3-ε* gene in their genome. These observations strongly indicate that recruitment of signal transduction molecules by the cytoplasmic tail of CD3-ε played an important role in abrogating development of both cell types, and lend support to the theory that NK cells and T cells stem from a common precursor cell.

The following conclusions can be drawn from these studies:

1. In transgenic animals containing more than 30 copies of the $CD3-ε^H$ gene (but not $CD3-δ^H$), both peripheral T lymphocytes and NK cells were absent.

2. High levels of expression of the CD3-ε protein is indispensable for this block in T cell$^-$ and NK$^-$ cell development. High levels of transcripts and protein were found if the human enhancer (E^H) was present, but not when the murine enhancer (E^M) was used. In fact, 35 copies of the complete murine gene resulted in a wild-type phenotype.

3. A truncated $CD3-ε^H$ (transmembrane and cytoplasmic domains only) or human/murine chimeric proteins also caused the *T cell$^-$/NK$^-$* phenotype (i.e. the $CD3-ε^H$ cytoplasmic tail can cause the block in development). Deletion of most of the protein resulted in a wild-type phenotype.

4. The most plausible explanation for these observations is that aberrant (intracellular) signal transduction through the CD3-ε cytoplasmic tail in Thy-1$^-$ CD3$^-$4$^-$8$^-$, Pgp1$^+$, IL-2Rα-thymocytes leads to cell death.

5. Aberrant intracellular signal transduction via CD3-ε blocked NK cell development.

6. These observations not only strongly supported the concept that NK cells and T cells stem from a common precursor cell, but also that signal transduction via CD3-ε plays an important role in the development of both cell lineages.

Since overexpression of CD3-δ (127) and CD3-η (131) do not cause this effect, it is likely that CD3-ε plays a distinct role in early T lymphocyte and NK cell development. Whether CD3-ε recruits unique signal transduction molecules in the different cell types (that is to say, pre-thymocytes, mature T cells, and NK cells) will need to be investigated.

5. The primary co-stimulatory events

Compared to other mitogenic signalling receptors, activation of the T cell antigen receptor is very tightly controlled. Aberrant activation of the TCR can potentially initiate disastrous immune responses, such as those seen in autoimmune diseases and other immunopathologies. As a consequence, T cell activation requires the strict co-ordinate and kinetically regulated ligation of the TCR and at least one other receptor or 'co-stimulatory molecule' on T cells and antigen presenting cells. In some cases, induction of co-stimulatory molecules or ligands on either T cells or antigen presenting cells occurs following initial cell–cell interaction by cytokine exchange and further signalling. Some co-stimulatory molecules are constitutively expressed, but may be up-regulated during activation. T cell co-stimulation involves a variety of different molecules falling into several receptor families, of which immunoglobulin superfamily members are the most prominent (132). Other families include the TNF (133) and TNF receptor superfamilies (134) and the integrins and selectins (13). T cell co-stimulatory molecules and molecules whose ligation can in some cases also negatively effect T cell activation include CD2, CD4, CD5, CD6, CD7, CD8, CD27, CD28, and CTLA-4, gp39 (CD40L), CD45, CD11–CD18 complex (LFA-1), CD54 (ICAM-1), and CD95 (Fas) and are summarized in Table 2. Excellent reviews have discussed the role of CD2, CD4, and CD8 (12, 13, 135), CD28, CTLA-4, and gp39/CD40L (136, 137), CD45 (138), CD11–CD18, and CD54 (13), CD27 and CD95 (134). Recent summaries of known data on T cell–APC interaction have been published (64, 137). We shall concentrate on recent findings on several key co-stimulatory molecules.

5.1 The role of CD4 and CD8

CD4 and CD8 are co-receptors that interact with peptides bound to class II and class I MHC antigens on antigen presenting cells (135, 139). Both antigens belong to the Ig supergene family indicating a common ancestry, but they are structurally different since CD4 is a monomer of 55 kDa consisting of four Ig-like domains, while there are two CD8 chains, α and β, which are encoded by different genes (140, 141). CD8β is expressed only in conjunction with CD8α, but CD8α can be present as a homodimer. CD8$^+$ TCR$\alpha\beta$ cells express the CD8$\alpha\beta$ heterodimer, but most CD8$^+$TCR $\gamma\delta^+$ T cells (8, 9) and some NK cells express a homodimer of CD8α (142). Furthermore it has been demonstrated in the mouse that CD8$^+$TCR $\alpha\beta$ cells developed in the intestinal epithelium express CD8$\alpha\alpha$ homodimers. Each of these CD8 chains has an N-terminal Ig-like domain followed by an extended heavily glycosylated region.

Expression of CD4 and CD8 correlates with specificity of T cells (143) for antigenic peptides bound to class II and class I MHC molecules, respectively. Consistent with the correlation of CD4 and CD8 expression with MHC specificity are findings that the extracellular domains of CD4 and CD8 bind with low affinity to class II and class I MHC antigens, indicating that they assist the binding of the TCR to these antigens (144, 145).

Crystal structures of parts of CD4 (the two amino-terminal Ig domains) and of CD8 (a dimer of the amino-terminal Ig domain of the α chain; ref. 146) that are involved in binding to MHC antigens have been published. The CD8α dimer has a structure similar to Ig variable structures. Analysis using space-filling models of CD8 and HLA-A2 suggests that CD8/class I MHC molecule binding involves loops corresponding to complement-determining regions (CDR) like regions of the CD8 Ig domain (146). Indeed mutations in the Ig-CDR like regions of CD8α interfere with binding of class I MHC antigens. The proposed CD8 binding site in the class I MHC molecule is located in the α3 domain. Three segments seem to be involved in binding CD8α to class I MHC encompassing residues 222–229, and containing four highly conserved acidic residues, which makes this region strongly negatively charged. It is, therefore, plausible that electrostatic forces are involved in CD8/class I MHC binding. In contrast to the box-like structure of the crystallized CD8α fragment, the two Ig domain CD4 fragments look like rod-shaped molecules (147–149). The contact between CD4 and class II MHC molecules probably involves a long stretch of the CD4 molecule. Using cell adhesion assays and IL-2 production as a read out, regions in the first three Ig-like domains of CD4 have been implicated in MHC class II binding. Whether these regions are directly involved in binding or whether mutations have some indirect effect which diminishes CD4–class II MHC binding, remains to be determined. The CD4 binding site on class II MHC molecules is located on a stretch consisting of residues 137–143 of the β chain (150, 151). This region is homologous to that of class I MHC that interacts with CD8α.

Transfection experiments have shown that CD4 and CD8 are indispensable for most antigen-specific T cells. For example, transfection of cDNAs encoding a class I MHC specific TCR expressed in a class II MHC specific hybridoma conferred only class I MHC specific reactivity upon co-transfection with CD8 cDNA (152, 153). Interestingly, ligation of the TCR/CD3 complex increases the affinity of CD8 for class I MHC molecules (154, 155). This would provide a CD8[+] CTL with a mechanism to amplify avidity for the target cell. The mechanism of this amplification of CD8-mediated adhesion by TCR ligation is not known. A similar increase in CD4-mediated adhesion by triggering the TCR has not been reported.

Monoclonal antibodies against CD4 and CD8 block the function of T cells expressing these antigens. Experiments with CD4 and CD8 CTL have shown that these antibodies have an effect on adhesion of T cells and target cells, not only by the obvious interference of CD4/8–MHC binding, but also by affecting signal transduction through the TCR (15, 156). Evidence has been presented that CD4 and the TCR can associate with each other (135, 157), indicating that CD4 is not merely an adhesion molecule, but can assist in signalling through the TCR.

Clustering of the TCR and CD8 during activation of CD8[+] T cells has not been demonstrated. However, other evidence supports the notion that CD8 also assists in TCR-mediated signal transduction. Eichman and colleagues have demonstrated that antibody-mediated, simultaneous cross-linking of CD4 or CD8 to the TCR resulted in a stronger activation compared to cross-linking of the TCR alone (158). Therefore, it is now generally accepted that CD4 and CD8 are co-receptors for the TCR. In view of the findings that CD4–CD8 and TCR contact class II–I MHC antigens at different sites, it is plausible that a CD4–CD8 and TCR complex are brought in close contact with each other, when they interact with the same MHC molecule. It should, however, be noted that the ways CD4 and CD8 assist in TCR-mediated signal transduction may not be entirely comparable (159) (for instance through different clustering with the TCR, as described above).

The idea that CD4 and CD8 are involved in signal transduction is underscored by the finding that these antigens associate with the cytoplasmic tyrosine kinase p56[lck] (70, 89, 160, 161), since protein tyrosine kinases are involved in control of cell growth and differentiation (90). P56[lck] has one SH2 and one SH3 domain, and a highly conserved C-terminal kinase domain, which are characteristics of the cytoplasmic src-like kinases (162). Like other members of this family, p56^{lck} has a unique N-terminal sequence as well as the above mentioned structural motifs of this family. The p56^{lck} enzyme is specifically expressed in T cells and NK cells and is membrane-associated. The CD4–CD8 binding site of p56^{lck} is localized in the unique N-terminal 32 residues (163). Site-directed mutagenesis revealed that two cysteine residues in critical regions in CD4, CD8α, and p56^{lck} are required for association of p56^{lck} to CD4 and CD8α. The p56^{lck} binding region in CD4 and CD8α that contains these two cysteine residues is highly conserved between species and contains the only homologies between the cytoplasmic tails of these antigens (163). In addition to the two cysteine residues, the p56^{lck} binding motif has two basic residues and two non-conserved residues in this region. The six-residue motif of CD8α is sufficient to confer the ability of p56^{lck} binding to heterologous proteins. In contrast, the six-residue motif of CD4 is not sufficient to confer binding, and other sequences more membrane-proximal (in the CD4 cytoplasmic tail) are required for optimal association with p56^{lck}.

CD8 cDNA constructs that lack a cytoplasmic tail have a limited capacity to assist class I MHC specific TCRs. Since tail-less CD8α cannot interact with p56^{lck}, those findings suggest that association with p56^{lck} is essential for appropriate functioning of CD8. More direct proof for the importance of p56^{lck} association with co-receptors for T cell activation was provided by Glaichenhaus et al. (164), who demonstrated that changing the two cysteine residues in the p56^{lck} binding motif of CD4 to alanine abrogated the ability of CD4 to support antigenic peptide and superantigen specific stimulation of a class II specific T cell hybridoma (164). CD8α was not able to substitute for CD4 despite the fact that the APC used to stimulate the T cell hybridoma expressed class I MHC antigens. Together, these findings support the notion that co-receptor function of CD4 and CD8 requires interaction of these molecules and TCR with the same MHC antigen.

Whether interaction of CD8α with p56^{lck} is required for activation and cytotoxic activity of CD8$^+$ T cells was investigated using mice in which the endogenous CD8 was replaced by a mutated form that fails to associate with p56^{lck} (14). A sensitive in vitro kinase assay was used to exclude the presence of any p56^{lck} associated with CD8 in these mice. CD8$^+$ CTL could be generated in vitro from spleen cells of such mice. These CTL, which have no CD8-associated p56^{lck}, still lysed the stimulator cells in a CD8 dependent way (14), indicating that p56^{lck}/CD8 association is dispensable for CD8-dependent cytotoxic activity.

It has been reported that anti-CD4 can inhibit T cell responses in the absence of an interaction of the TCR with MHC antigens (165), indicating that CD4 can transmit a 'negative' signal. Consistent with this notion, Haughn et al. (166) have found that in some cases association of p56^{lck} with CD4 inhibits activation through the TCR/CD3 complex. A class II MHC specific hybridoma was used for these studies that did not need CD4 for antigen recognition. A CD4$^-$ mutant of this line was transfected with wild-type CD4 or 'tail-less' CD4 mutants. It was shown that activation of the transfected hybridoma by anti-TCR monoclonal antibodies (mAbs) was blocked in cells transfected with wild-type but not in those with mutant CD4.

To explain these findings it has been proposed that CD4 molecules that do not participate in TCR-mediated activation, may inhibit this activation by sequestration of p56[lck] (166).

5.1.1 The role of CD4 and CD8 in T cell development: induction of CD4 and CD8 in the thymus

T cell development in the thymus proceeds through a series of phenotypically distinct stages. In the human, CD4 and CD8 negative precursor cells seed the thymus and first acquire CD4 followed by CD8α and CD8β. Differential up-regulation of CD8α and β has never been observed in the murine thymus. In some mouse strains, CD8 is up-regulated before CD4, whereas in others the opposite has been observed (167, 168). The functional role of CD4 or CD8 on immature thymocytes is unclear. Mice lacking β2 microglobulin have almost no CD8[+]CD3[+] single positive (SP) thymocytes, but have normal proportions of CD8[+]CD3[+] immature thymocytes and their CD4[+]CD8[+] double positive (DP) progeny (169), indicating that a CD8/class I MHC interaction is not involved in the transition of CD8[+] immature to DP thymocytes. Recent findings have provided evidence that expression of the TCR β chain is required for transition of CD4[-]CD8[-] to DP thymocytes. Mice with a *TCRβ* gene disrupted by gene targeting lack CD4[+]CD8[+] thymocytes and have only a few thymocytes compared to wild-type mice. In contrast, TCRα negative mice have a normal level of cellularity in the thymus and CD4[+]CD8[+] thymocytes. In addition, immunodeficient mice that lack T and B cells because of a defect in DNA repair (scid mice) or because they lack recombination activation genes (*RAG*) critical for V(D)J recombination of the *TCR* genes, do not have CD4[+]CD8[+] thymocytes, but introduction of a *TCRβ* transgene in the genome of these mice results not only in the appearance of CD4[+]CD8[+] cells but also in a numerical increase of cells in the thymus (170–172). P56[lck] plays a pivotal role in the transition of double-negative (DN) to DP thymocytes and the accompanying expansion of the thymocytes, since p56[lck] negative mice (98) and mice with a dominant negative p56[lck] transgene also lack DP thymocytes. Introduction of a *TCRβ* transgene in these latter mice cannot correct the p56[lck] defect in inducing the appearance of CD4[+]CD8[+] thymocytes (173). The appearance of DP thymocytes does not require the *TCRα*, because *TCRα* 'knock-out mice' have normal numbers of DP thymocytes (172). In conclusion, both the β-chain and p56[lck] are necessary for transition of DN thymocytes to the DP-stage. The *in vivo* role of CD4 and CD8 as determined by analysis of CD4 or CD8 'knock-out' mice will be described further below (Section 5.1.2).

It was demonstrated recently that immature T cell progenitors in the thymus express low levels of TCRβ associated with CD3 components and another protein with a molecular weight of 33 kDa, named gp33 (174–176). These findings together indicate that a signal through the TCRβ chain/X/CD3 complex, involving p56[lck], is required for expansion of thymocytes and the appearance of CD4[+]CD8[+] thymocytes. In addition to induction and expansion of CD4[+]CD8[+] thymocytes, p56[lck] is required for inhibition for further rearrangement of the *TCRβ* gene, thereby controlling allelic exclusion.

5.1.2 The role of CD4 and CD8 in positive selection in the thymus: mechanism of down-regulation of CD4 and CD8

Development of DP thymocytes into CD4$^+$ and CD8$^+$ SP CD3^{high} cells requires positive selection involving engagement of the TCR with MHC antigens. It has been shown that CD4 and CD8 play a direct role in T cell positive selection in the thymus (177–181). CD8 negative mice do not possess cytotoxic cells (182), whereas CD4 mutant mice lack helper cells (183). In neither CD4 mutant mice nor CD8 mutant mice was the lack of CD4$^+$ and CD8$^+$ cells compensated for by the reciprocal subset, and the number of CD4$^-$CD8$^-$ T cells was increased, demonstrating that CD4 and CD8 are required for the generation of functional CD4$^+$ or CD8$^+$ T cells. The interaction of CD4 and CD8 with MHC molecules is instrumental in generating CD4$^+$ and CD8$^+$ T cells, since class I negative mutant mice obtained by targeted disruption of the β2 microglobulin gene (169) or the gene encoding the peptide transporter TAP-1 (184), have only a minimal number of CD8$^+$ single positive (SP) thymocytes compared to normal mice. Similarly, class II MHC negative mice lack CD4$^+$ SP thymocytes (185, 186).

It is not yet clear how CD4 and CD8 are down-regulated from DP thymocytes. One view contends that down-regulation of CD4 and CD8 is not dependent on the MHC specificity of the TCR, but occurs randomly. Once the accessory molecules are down-regulated, interaction of the TCR with the correct MHC molecule rescues the SP thymocytes from programmed death. Because down-regulation occurs randomly, it can be expected that some SP thymocytes are present expressing the 'wrong' CD4 or CD8. For example, some CD4$^+$ SP thymocytes should then be present in the thymus of class II MHC negative mice. CD4$^+$ cells with a fairly high CD3/TCR expression were indeed observed in class II MHC negative mice, although these cells exhibited some immature features. These cells were present in the cortex (185, 186), expressed low levels of CD4, and the level of CD3 expression was slightly lower than in mature CD4$^+$ T cells. It was concluded that those cells were already committed to the CD4$^+$ SP mature lineage in the absence of class II MHC antigens. Furthermore, forced expression of a CD4 transgene in all T lineage cells rescued class II MHC specific T cells bearing endogenous CD8. These data together support the concept that CD4 and CD8 are randomly down-regulated (this is referred to as the 'stochastic' model). However, another prediction of that hypothesis is that expression of a CD8 transgene should rescue class I MHC specific CD4$^+$ T cells, because the CD8 transgene would replace the down-regulated endogenous CD8 in supporting the class I MHC specific TCR. This prediction was not supported by the experimental data (178, 187, 188). The alternative hypothesis is that down-regulation of CD4 and CD8 is dependent on the specificity of the TCR, interaction of the TCR with class I or class II MHC results in down-regulation of CD4 and CD8, respectively. This hypothesis (the so-called 'inductive' model) explains the data of Robey *et al.* (178) and Borgulya *et al.* (187), but fails to take into account the presence of thymocytes committed to the CD4+ SP T cell lineage in class II negative mice (185, 186). Thus, neither theory alone can adequately explain all the data and more information will be needed for an understanding of the mechanisms that underlie down-regulation of CD4 and CD8 and thymic selection of single positive T cells (5).

The mechanisms of signal transduction that result in selection of CD4+ and CD8+ thymocytes are unclear. As p56^{lck}, this tyrosine kinase was also found to be associated with CD4 and CD8 in the thymus (189), the possibility was raised that p56^{lck} is involved in the selection of SP thymocytes. Analysis of *p56^{lck}* mutant mice were not informative in this respect, since the mutation results in a profound block at an earlier stage, before the appearance of CD4 and CD8 (98). However, if p56^{lck} is involved in positive selection, association of this PTK with CD4 and CD8 is not critical, since CD4+ T cells can develop in the thymus of mice with CD4 lacking a cytoplasmic tail, rendering it unable to associate with CD4 (190). Similarly, Loh and co-workers demonstrated that association of p56^{lck} is dispensable for the development of CD8+ T cells (14).

5.2 Signal transduction events induced by triggering of CD28 and CTLA-4

CD28 is a disulfide linked homodimer of two 44 kDa glycoprotein chains. The extracellular domain comprises 202 amino acids. The intracellular domain is 41 amino acids long and is likely to be involved in CD28-mediated signal transduction (191). The human CD28 protein shows 67 per cent identity with the mouse form in the extracellular domain (136, 192, 193) and 77 per cent in the cytoplasmic tail. CD28 is highly homologous to another T cell antigen CTLA-4 (192). These molecules are both members of the Ig supergene family, in particular of the subfamily of molecules with a single V domain (136). The genes for CD28 and CTLA-4 are tightly linked, since they map on the same regions on chromosome 1 in mouse and chromosome 2 in human (192), suggesting that they arose from a relatively recent gene duplication event. The overall homology between mouse and human CTLA-4 is 76 per cent. Interestingly, the cytoplasmic tails of mouse and human CTLA-4 are identical (192). In contrast to CD28 (see below), CTLA-4 transcripts are not detectable on resting T cells but are essentially restricted to activated T cells. However, expression of CTLA-4 on activated T cells is 20-fold lower than that of CD28 (194).

CD28 is expressed predominantly but not exclusively on cells of the T cell lineage. Early T cell progenitors in the thymus that are CD34+, but are negative for CD3, CD4, and CD8, express CD28 (136, 193, 195). In human thymus it is present at low levels on CD34+ T cell precursors and on immature CD4+CD8+ CD3^{low} thymocytes. Essentially all CD3/TCR+ thymocytes express CD28 in mouse (196) and man (197, 198). In addition, all neonatal cord blood T cells, exemplary of recent thymic emigrants, are CD28 positive (199). In contrast, a variable proportion of the $CD8^+$ peripheral blood T cells (199) and, in some donors, a small proportion of the $CD4^+$ cells (200) do not express CD28. $CD28^-$ T cells express CD11b, which is usually not expressed on $CD28^+$ cells (200, 201). The origin and the function of $CD28^-$ T cells are unclear. It has been proposed that $CD4^+CD28^-$ T cells have an extrathymic origin (200). On the other hand, since expression of CD28 can be modulated by activation signals and by IL-2 (197), it may be possible that $CD28^-$ and $CD28^+$ cells represent different maturation stages and not different subsets. There is some evidence that $CD28^+$ and $CD28^-$ T cells have different functions. $CD28^+CD8^+$ and $CD28^-CD8^+$ cells may contain percursors of cytotoxic cells (202) and suppressor cells (203–205), respectively.

Expression of CD28 is not restricted to $TCR\alpha\beta^+$ cells since it is present on 50 per cent of the $CD4^-CD8^-$ $TCR\gamma\delta^+$ cells (206) and on all $CD4^+CD8^-$ $TCR\gamma\delta$ cells (8). CD28 is not exclusively expressed on T cells, since fetal NK cells also carry CD28 (129, 207). In contrast, adult peripheral blood NK cells are CD28 negative (142). The function of CD28 expressed on fetal NK cells is unclear. An NK leukaemia, YT, has been described that mediates its cytotoxic activity against B cell lines through CD28 (207), but a signalling role of CD28 in the activity of fetal NK cells has not been demonstrated. Resting B cells do not express CD28, but some plasma B cells express CD28 (208).

Monoclonal antibodies specific for CD28 have strong co-stimulatory activities (136, 193, 209, 210). Stimulation through CD28 is generally considered to represent a 'secondary signal'. This is based on findings that anti-CD28 mAbs alone do not induce IL-2 production and proliferation of T cells, whereas these antibodies efficiently stimulate T cells in the presence of mitogens, anti-CD3 and anti-CD2 mAbs. The pathway of the CD28-mediated signal transduction is different from and may be independent of that mediated by the TCR/CD3 complex. In contrast to TCR/CD3 signalling, CD28-mediated signal transduction is not blocked by cyclosporin (Cs) A and agents that raise intracellular cAMP concentration. Cyclosporin binds to cyclophilin, inhibiting the activity of the serine threonine phosphatase calcineurin. Calcineurin is a Ca^{2+}/calmodulin-dependent enzyme and is involved in induction of the transcription factor NF-AT, required for IL-2 transcription (211–213). Thus the CD28 signal transduction pathway may circumvent the Ca^{2+} dependent phosphatase calcineurin, supporting the notion that this pathway is Ca^{2+} independent. However, there is also evidence for Ca^{2+} dependent signal transduction through CD28, since co-stimulation through CD28 can increase the Ca^{2+} flux induced by mitogens and anti-CD3 mAbs. Moreover, cross-linking of CD28 on the cell surface of T cells with anti-CD28 mAbs followed by secondary anti-Ig antibodies results in an increase of cytoplasmic Ca^{2+} (214). These findings raise the possibility that signal transduction through CD28 can deliver a Ca^{2+} independent and a Ca^{2+} dependent signal.

Cross-linking of CD28 with antibody results in a small increase in tyrosine phosphorylation in Jurkat cells (215–217). Maximal tyrosine phosphorylation of predominantly a 100 kDa substrate follows pre-treatment of the cells with phorbol ester and anti-CD3 mAb, following cross-linking of CD28 with antibodies, or with the cell-bound counterstructure of CD28, B7 (see also below) (215–217). This suggests that tyrosine phosphorylation is involved in CD28-mediated signal transduction. Consistent with this notion are the observations that an inhibitor of src family protein tyrosine kinases, Herbimycin-A, blocks IL-2 production by PBMC T cells induced by phorbol ester and anti-CD28, while no effect of this inhibitor was observed on IL-2 production induced by a combination of the Ca^{2+} ionophore, ionomycin, and phorbol ester. These latter activations also do not induce tyrosine phosphorylation in T cells. Ligation of CD28 also results in tyrosine phosphorylation of CD28 itself (217) and a small set of phosphoproteins mostly distinct from those produced by TCR stimulation (216). In particular, the pp100 may also be the pp100 seen following CTL triggering leading to degranulation (218). This suggests a synergistic signalling via two parallel tyrosine kinase pathways. Activation of CD28 also leads to association of CD28 with PI-3-kinase (219) and this may be tyrosine kinase dependent. Therefore, activation of a parallel PTK-dependent pathway and simultaneous association of PI-3-kinase with CD28 and with the TCR/CD3 complex (see Section 4.1) may provide the dual signalling required for activation of mature resting T cells.

Co-stimulation of T cells with anti-CD28 mAb increases mRNA expression and secretion of cytokines including IL-2, IL-3, GM-CSF, IFNγ, TNFα, and lymphotoxin. Signal transduction via CD28 results in an increased stability of mRNA for these cytokines (220). In addition, it has been shown that CD28 signal transduction enhances the transcription rate of IL-2 by activating a CD28-specific element of the IL-2 promoter (221, 222). The CD28 responsive element in the 5' end of the IL-2 promoter has similarities with the NF-κB binding motif (223). Very similar binding motifs have been observed in the promoters of the IL-3 and GM-CSF genes (N. Arai, personal communication), suggesting that CD28 signalling may be involved in induction of these cytokine genes as well (136, 193). In contrast, evidence is beginning to suggest that CTLA-4 may induce a negative signal after induction during activation, providing a potentially complementary feedback inhibition loop (224).

5.2.1 B7/BB1 and B70/B7-2 are counterstructures of CD28 and CTLA-4: biological consequences of CD28-B7 interactions

Recently it has been reported that a B cell activation antigen, characterized by the antibodies BB1 (225) and B7 (226), is a natural ligand for CD28 (227–229). B7, now also called CD80, is expressed in very low densities on resting B cells and can be up-regulated by activation (225, 227, 230–232). B7 expression on B cells was further enhanced by stimulation in the presence of IL-2 and IL-4 (231). B7 is an activation antigen on many leucocytes. T cells, NK cells, monocytes, and freshly isolated dendritic cells fail to express B7. IFN-γ induces B7 on monocytes (230). Freshly isolated human PBMC dendritic cells do not express B7 but this is up-regulated after 1 day of *in vitro* culture in monocyte-conditioned medium (233). Binding of T cells further enhances B7 expression on dendritic cells (234). B7 is not expressed on resting T cells, but is induced following activation of T cells with mitogens (235). Interestingly, anti-B7 mAbs inhibited the activation of resting T cells by activated alloantigenic T cell clones, suggesting that B7 expressed on activated T cells may play a role in T–T cell interactions (235). More recently Yssel *et al.* (236) reported that the cytokine IL-7 induces B7 expression on neonatal cord blood cells, possibly suggesting that IL-7 promotes T–T cell interactions.

Binding of CD28 to B7 is of moderate affinity with a K_D of about 200 nM. A fusion protein of an immunoglobulin Cγ chain and CTLA-4 also binds to B7 (229, 237, 238). The affinity of binding of CTLA-4-Ig for B7 was approximately 20-fold higher than that of a CD28-Ig protein (237). CTLA-4-Ig inhibited T cell proliferation much more efficiently than CD28-Ig, consistent with the difference in binding affinities. The CD28/B7 association mediates heterotypic cell–cell interactions. Interaction of CD28 with B7 expressed on transfected CHO cells or with immobilized B7-Ig fusion protein, results in augmentation of T cell proliferation and cytokine production after stimulation with low doses of phytohaemagglutinin, phorbol ester, or anti-CD3 mAbs. This augmentation is specific since it can be blocked by anti-B7 mAbs (136).

Anti-CD28 antibodies are known to inhibit MLC responses. An important role of the CD28/B7 in primary MLCs has now been established in two recent studies. Koulova *et al.* reported that anti-B7 and anti-CD28 mAbs block allogeneic MLR (239). The role of CD28/B7 interactions in MLR has been rigorously established. A Burkitt lymphoma cell line, Ramos, which does not express B7, only weakly stimulates T cells in a primary MLR. After transfection of Ramos cells with a B7 cDNA, the cells stimulated a vigorous allogeneic T cell response. The same group also reported that B7 expression strongly augmented proliferation of resting T cells to anti-CD3 mAbs (199).

The B7–CD28 interaction plays an important role in maturation of CTL precursors. Resting T cells can kill target cells that express B7 in redirected anti-CD3 induced killing assays (199). While expression of B7 is required for cytotoxicity, it is not sufficient because target cells are efficiently lysed by resting T cells only when both B7 and ICAM-1 are expressed on the target cells (199). The capacity to lyse B7 and ICAM-1 expressing target cells in a short-term (4 hour) ^{51}Cr release assay resided predominantly in the CD45RO$^+$ 'memory' T cells. Both CD4 and CD8$^+$ T cells could mediate killing activity. However, CD45RO$^-$ 'naive' T cells can lyse these target cells as efficiently as CD45RO$^+$ memory cells in a long-term (8 hour) assay, suggesting that cytotoxic cells can be recruited from 'naive' precursors. The finding that the activity mediated by the 'naive' cells is completely blocked by the protein synthesis inhibitor cycloheximide, while this inhibitor does not affect activities mediated by activated CTL clones (199), is consistent with the notion that CTL effectors are generated from precursor CTL in the co-culture with B7/ICAM-1$^+$ target cells. B7 expression on target cells is not essential for lysis by CTL in anti-CD3 redirected killing assays, once the CTL are activated; expression of either ICAM-1 or LFA-3 on the target cells is required and is sufficient to support lysis of activated CTL (240). The finding that influenza virus specific CTL clones lyse HLA-A2$^+$ mouse L cells co-transfected with ICAM-1 or LFA-3, but not HLA-A2$^+$ L cells co-transfected with B7 (240) indicates that B7 cannot replace ICAM-1 or LFA-3 to induce lysis by activated effector CTL. In conclusion, these results strongly suggest that the B7–CD28 interaction is required for the recruitment of effector CTL from naive resting T cells, but once effector CTL are generated these cells can lyse target cells that do not express B7. The importance of interactions with B7 with its counterstructures for primary immune responses *in vivo* has been demonstrated by studies which showed that interference of CD28/B7 interactions through soluble CTLA-4-Ig can prevent allograft rejection (241) and inhibit primary antibody responses (136, 193, 242).

The functional role B7 plays in T cell activation has prompted investigations into the expression and the possible function of this molecule in the thymus. It was found that the anti-BB-1 antibody reacts with non-T cells in the interlobular thymic tissue. However, anti-B7 antibody does not react with interlobular thymic tissue and cultures of human thymic stromal cells do not express B7 (A. Galy and H. Spits, unpublished observations). Since B7 is expressed in the thymus, presumably on interdigitating dendritic cells and on B cells, there has been speculation that B7 is important for negative selection in the thymus. However, soluble CTLA-4-Ig that efficiently blocks antigen- and superantigen-mediated responses of peripheral T cells does not affect positive or negative selection in the mouse thymus (243).

Consistent with the importance of CD28 ligands in T cell activation, CD28 deficient mice have reduced T-dependent immune responses (244). However, the partial nature of this block has implied the presence of at least one more co-stimulatory pathway. Recently, a second counterstructure for CD28/CTLA-4 was described both in mouse, termed B7-2 (245–247), and in human, coined B-70 (248). B7-2 and B-70 show only 30 per cent homology with the mouse and human B7-1, respectively. The expression profile of B-70 is similar to that of B7-1 with the important exception that B-70 is expressed on resting monocytes and dendritic cells. Antibodies against B-70 strongly block primary MLR, in contrast to anti-B7 antibodies that only partially block MLR. These findings seem to suggest that B70/B7-2 is critical for the initiation of an immune response that can be enhanced further by B7-1 once it is induced (248).

5.3 Alternative T cell co-stimulatory structures

Evidence for yet more co-stimulatory molecules on T cells suggests that the immune response is flexible in the use of multiple alternative pathways. However, T cell activation is also stringently controlled, particularly at the point of initial activation of the immune response at the naive T cell level. Simultaneous ligation of two molecules (typically TCR and CD28) are minimal requirements for activation of memory T cells. Full activation of naive T cells also requires the correct ligation kinetics of the first two molecules and a third initial activation-induced molecule (gp39 or CTLA-4). The incomplete nature of the CD28-deficient mouse T cell activation defect (244) and data showing co-stimulatory activity other than via

CD28 (249) suggested that the initiation of the immune response via naive T cells can involve other co-stimulatory activities (250). Like the B7:CD28/CTLA-4 pathway, the alternative co-stimulatory pathways fall into two categories. The first category involves molecules constitutively expressed by T cells. Some of these molecules are constitutively expressed even by naive T cells, and are thus available for involvement in initiation of the T cell-dependent immune response. The prototype constitutively expressed co-stimulatory receptor on T cells now appears to be B7-2 (see Section 5.2). The second group are induced following initial T cell receptor ligation. CTLA-4 may represent an inducible T cell negative regulatory molecule (224, 251). It now seems that the CD40 ligand on T cells, gp39 (CD40L, see below), represents the most important inducible co-stimulatory molecule whose ligation stimulates both the T cell and the antigen-presenting cell. Representatives of both categories of co-stimulatory pathway can act synergistically with other pathways, and naive T cells appear to require the ligation of the greatest number of such pathways for maximal activation (250). Memory T cells appear to have reduced stringency requirements for complete activation, reflecting the increased competence of the immune system to subsequent or crossreactive challenge. The integration of multiple signals from the antigen receptor and the array of potential co-stimulatory molecules presumably reflects the importance of T cell regulation in the immune response and subsequent health of the organism.

5.3.1 Co-stimulation and the T cell receptor for CD40–gp39 (CD40L)

Recently, the importance of the CD40–gp39 interaction in the immune response has become clear. The CD40 ligand is a T cell surface type 2 glycoprotein related to TNF (252, 253). TCR ligation induces gp39 expression. One effective way to induce gp39 expression is ligation of B7 molecules with CD28 or CTLA-4 (254). In turn, ligation of gp39 results in T cell stimulation of B cell immunoglobulin class switching and secretion. This has been shown by elucidation of the molecular defect in a genetic immunodeficiency. Mutation of the *gp39* gene is responsible for the defective antibody class switching of hyper-IgM syndrome (137, 255). T cell-independent B cell responses and responses induced by bypassing gp39 (anti-CD40) are unaffected by this mutation. Thus, T cell help for B cell activation is primarily directed through the CD40–gp39 co-stimulatory pathway. Hyper-IgM patients are also susceptible to pathogens normally dealt with by T cells, producing AIDS-like opportunistic infections, underlining the role of gp39 co-stimulation in normal T cell activation (255). To date, the molecular mechanism of gp39 signalling remains unknown. Strikingly, however, the symptoms of hyper-IgM syndrome are also similar to those of $ZAP70^{-/-}$ immunodeficient patients (103) (see Section 4.1), which may reflect so far undetected similarities in mechanism.

5.3.2 Other co-stimulatory molecules

Evidence for the activity of other candidate T cell co-stimulatory molecules includes that for CD7 (256) and CD27. Like gp39, CD27 is induced upon T cell activation and acts synergistically with T cell receptor-associated activation to induce T cell proliferation (257). In contrast to CD40 and gp39, CD27 on T cells is a TNF receptor family member and the corresponding structure on activated B cells. CD70 is a TNF superfamily gene (258). The signalling pathways in T cells for gp39 and CD27 and in B cells for CD40 and CD70 are likely to reflect this converse relationship. Another TNF receptor family member, CD95 (Fas), has been implicated in both positive and negative signalling events. The ligation of the Fas antigen on T cells typically causes a programmed cell death (apoptosis) response (259). However, ligation with an anti-Fas mAb has recently been shown to enable Fas to function as a co-stimulatory molecule for TCR/CD3 activation (259). It should be remembered that high-affinity ligation of the TCR/CD3 complex of thymocytes also results in apoptosis, a model for thymic negative selection (reviewed in ref. 5). Thus a single molecule can have different signalling outcomes at different stages during T cell development. In conclusion, several co-stimulatory signalling molecules from distinct gene families appear to function in particular T cell subsets and at certain points of T cell differentiation.

5.4 The phosphotyrosine phosphatase CD45

The CD45 common leucocyte antigen family comprises a group of membrane glycoproteins that are expressed by cells of lymphoid and myeloid origin. CD45 (also called leucocyte common antigen, L-CA, or in the mouse Ly-5) can be expressed in different isoforms that are generated by alternative splicing of the exons encoding a portion of the extracellular domain (reviewed in ref. 138). The *CD45* gene spans approximately 120 kb containing 34 (mouse, ref. 260) or 33 (human, ref. 261) exons. Exons 4, 5, and 6 are optional, because they can either all be omitted (180 kDa form, CD45RO), or all included (220 kDa form, CD45RA), or used in combinations to generate additional isoforms (262, 263). Only six of the eight total possible mRNAs have been isolated as cDNAs (264).

The differential expression of the CD45RO and CD45RA isoforms by human T cells have been used to identify so-called 'memory' and 'naive' phenotypes (265, 266). CD45RA$^+$ cells do not respond very well to recall antigens and are poor in providing help for antibody production (266). These cells produce high levels of IL-2 but no IL-4. CD45RO$^+$ T cells, in contrast, produce IL-4 and IL-6 (267), which stimulate B cell proliferation and antibody production.

Besides the extracellular domain, all CD45 molecules have a common, large intracellular domain, which contains two subdomains that have phosphotyrosine phosphatase (PTPase) activity. Originally, the PTPase activity of CD45 was discovered after the primary sequence of the first PTPase, human placental PTPase 1B, had been determined (268). Based on sequence homology, it was inferred that CD45 must have intrinsic PTPase activity; which was confirmed shortly thereafter, through purification of human spleen CD45 (269).

The finding that CD45, a major lymphocyte surface antigen (> 300 000 molecules/cell), encodes a PTPase stimulated a number of studies regarding its function.

First, studies using antibodies to cross-link CD45 with other surface markers (CD3, CD2) have, in general, shown to be inhibitory for early activation signals through the TCR. Engagement of CD45 in TCR-mediated activation inhibited tyrosine phosphorylation of certain substrates, inositol phosphate production, and the generation of a calcium flux (270–272). Some investigators could not confirm these findings, possibly because of different conditions used to cross-link CD45 (273).

More direct evidence for the involvement of CD45 in T cell activation has been obtained using cell lines that lack CD45. Mutant CD45$^-$ T cell clones proliferate poorly in response to antigen or anti-CD3, compared to the wild-type clone and a CD45$^+$ revertant (79). A CD45$^-$ variant of the human HPB-ALL T cell line did not mobilize calcium or generate inositol phosphates after stimulation with anti-CD3. However, full responsiveness was restored by reconstitution of the cells with CD45 (274). Moreover, CD45$^-$ mutants of the Jurkat leukaemia cell line have demonstrated that tyrosine phosphorylation patterns are abnormal and IL-2 production is deficient compared to wild-type Jurkat (79, 274). However, others could not show strong defects in proximal events after TCR-stimulation in CD45$^-$ Jurkat cells, but more pronounced defects in later T cell responses (i.e. cytokine production) (275) were observed. Taken together, these findings imply that CD45 regulates TCR/CD3-mediated T cell activation.

The importance of CD45 for T cell responses is underscored by studies using mice that are made deficient for CD45 through homologous recombination (276). In these CD45-exon 6 deficient mice cytotoxic T cell responses to LCMV are absent. Moreover, CD45 apparently also plays an important role in T cell maturation, because T cell development is (incompletely) blocked at the CD4$^+$CD8$^+$ stage (Kishihara *et al.* 1993) (276).

The PTPase CD45 probably regulates T cell activation through dephosphorylation of its substrates. Although *in vivo* substrates have not been identified conclusively, it seems likely that p56^{lck} and p59^{fyn} are major targets. CD45 dephosphorylates the C-terminal negative-regulatory tyrosine residue of p56^{lck} and p59^{fyn} *in vitro*, thereby activating these kinases (77, 277). Support for regulation by CD45 of the tyrosine phosphorylation status of p56^{lck} comes again from CD45$^-$ mutants, in which p56^{lck} is hyperphosphorylated on C-terminal Tyr505 (278). Other signalling molecules such as the CD3-ζ chain, PLCγ1 (279) or even the tyrosine kinase ZAP-70 could be substrates for CD45 as well. If CD45 indeed acts on these molecules (or their substrates) its effect would be to inhibit T cell responses. If, however, the src-like kinases are the sole substrates, the effect would be to increase activation. The nature of the complex in which CD45 participates, as well as the number of CD45 molecules within the complex, is probably critical to determining a negative or positive outcome of the response. This may explain why cross-linking CD45 with TCR components is inhibitory, whereas CD45 deficient cells are not hyperresponsive but fail to be activated after triggering of the same receptor (270–272).

How CD45 activity is regulated remains unclear. The different CD45 isoforms that exist and its receptor-like structure suggest that CD45 activity may be regulated through binding of its ligands. Indeed, one group has identified the B cell surface antigen CD22 as a ligand for CD45 (280). In contrast, the conserved cytoplasmic domain, containing the PTPase activity, may be sufficient for signalling as has been shown recently (281, 282). However, although the extracellular domain is not required for signalling via the TCR, it may well participate in regulation of the magnitude of response.

Several reports have dealt with phosphorylation of CD45 itself as a way to modify its activity. Treating T cells with phorbol esters or IL-2 leads to phosphorylation of CD45 on serine residues but with no effect on PTPase activity (283). Tyrosine phosphorylation of CD45 occurs after treatment of Jurkat cells with anti-CD3 together with the potent oxidant and phosphatase inhibitor phenylarsine oxide (284, 285). Because of the strong redox-sensitivity of PTPases (285; and Staal and Herzenberg, personal communication), it is unclear if CD45 is tyrosine phosphorylated under physiological conditions and whether this phosphorylation regulates CD45 activity.

In addition to CD45, several other PTPases are expressed in T cells. For instance, the cell membrane PTPase LAR is present in many cell types including lymphocytes (138). In addition, a T cell specific cytoplasmic PTPase (TPTP), distinct from PTPase 1B, has been identified (reviewed in ref. 286). The functions of these and other PTPases will undoubtedly be identified in the near future.

6. Conclusions

Many signal transduction molecules are involved in T cell activation, but few of the pathways are completely understood. Since antigen recognition takes place on the interface of antigen presenting cells and T lymphocytes, the multitude of events accompanying ligand receptor interaction needs to be taken into account. The hierarchy of events is still to be established. Given the large number of different peptide–MHC complexes on the surface of antigen presenting cells and because of the relatively low affinities between TCR-αβ and MHC/peptide, aggregation of the receptors is a prerequisite for activation. Thus, on the cytoplasmic face of the plasma membrane, the relatively short cytoplasmic tails of the CD3 polypeptide chains and the adhesion and co-stimulatory molecules recruit a large number of signal transduction molecules. What kind of interplay occurs between the protein tyrosine kinases, G-proteins, and enzymes of lipid metabolism? How is this interplay terminated leading to the dissociation of the antigen presenting cell and the T cell? How is peripheral tolerance induced and can this be exploited to modulate immunological diseases? How does HIV interfere in signal transduction events? Also, how similar are the rules of positive and negative selection of thymocytes to those of mature T cells? Before a rational drug therapy leading to specific control of early signal transduction events in T cells can be developed, a better understanding of some of these ground rules will be required. None the less, the many contact points that are already established should allow for the development of novel assays which can be used in rapid drug screening. In summary, although enormous steps forward have been made in understanding

T cell activation, much needs to be learned about the molecular underpinnings and the regulation of this event which are at the centre of the immune response.

Acknowledgements

We would like to thank Karen Moore, Millenium Inc., for information on TCR/CD3 chromosomal locations, Alexis Bywater and Tina Myers for patient help with the manuscript.

References

References marked * are recommended for further reading

1. Clevers, H. *et al.* (1988). The T cell receptor/CD3 complex: a dynamic protein ensemble. *Annu. Rev. Immunol.*, **6**, 629.
2. Ashwell, J. and Klausner, R. (1990). Genetic and mutational analysis of the T cell antigen receptor. *Annu. Rev. Immunol.*, **8**, 139.
3. Weiss, A. (1991). Molecular and genetic insights into T cell antigen receptor structure and function. *Annu. Rev. Genet.*, **25**, 487.
*4. Von Boehmer, H. and Kisielow, P. (1993). Lymphocyte lineage commitment: instruction versus selection. *Cell*, **73**, 207.
*5. Janeway, C. (1994). Thymic selection: two pathways to life and two to death. *Immunity*, **1**, 3.
6. Borst *et al.* (1987). A T cell receptor gamma/CD3 complex found on cloned functional lymphocytes. *Nature*, **325**, 688.
7. Brenner *et al.* (1988). The gamma delta T cell receptor. *Adv. Immunol.*, **43**, 133.
8. Spits, H. (1991). Human T cell receptor γδ+ T cells. *Sem. Immunol.*, **3**, 119.
9. Porcelli, S., Brenner, M., and Band, H. (1991). Biology of the human γδ T cell receptor. *Immunol. Rev.*, **120**, 137.
10. Spits, H. *et al.* (1986). Alloantigen recognition is preceded by nonspecific adhesion of cytotoxic T cells and target cells. *Science*, **232**, 403.
11. Matsui, K. *et al.* (1992). Low affinity interaction of peptide–MHC complexes with T cell receptors. *Science*, **254**, 1788.
*12. Bierer, B. *et al.* (1989). The biologic roles of CD2, CD4, and CD8 in T cell activation. *Annu. Rev. Immunol.*, **7**, 579.
*13. Springer, T. (1990). Adhesion receptors of the immune system. *Nature*, **346**, 425.
14. Chan, I. and Loh, D. (1993). Thymic selection of cytotoxic T cells independent of CD8α-lck association. *Science*, **261**, 1581.
15. Blanchard, D. *et al.* (1987). The role of the T cell receptor CD8 and LFA-1 in different stages of the cytolytic reaction mediated by alloreactive T lymphocyte clones. *J. Immunol.*, **138**, 2417.
16. Kanagawa, O. and Ahlem, C. (1989). Requirement of the T cell antigen receptor occupancy for the target cell lysis by cytolytic T lymphocytes. *Int. Immunol.*, **1**, 8178.
*17. Davis, M. and Chien, Y. (1993). Topology and affinity of T-cell receptor mediated recognition of peptide–MHC complexes. *Curr. Opin. Immunol.*, **5**, 45.
18. Exley, M. *et al.* (1995). Evidence for multivalent structure of T cell antigen receptor complex. *Mol. Immunol.*, **32**, 829.

19. Brown, J. *et al.* (1993). Three-dimensional structure of the human class II histocompatibility antigen HLA-DR1. *Nature*, **364**, 33.

20. Snow, P. and Terhorst, C. (1983). The T8 antigen is a multimeric complex of two distinct subunits on human thymocytes but consists of homomultimeric forms on peripheral blood T lymphocytes. *J. Biol. Chem.*, **258**, 14675.

21. Exley, M., Terhorst, C., and Wileman, T. (1991). Structure, assembly and intracellular transport of the T cell receptor for antigen. *Semin Immunol.*, **3**, 283.

22. Wileman, T. *et al.* (1993). Associations between subunit ectodomains promote T cell antigen receptor assembly and protect against degradation in the ER. *J. Cell Biol.*, **122**, 67.

23. Brenner, M., Trowbridge, I., and Strominger, J. (1985). Cross-linking of human T cell receptor proteins: association between the T cell idiotype beta subunit and the T3 glycoprotein heavy subunit. *Cell*, **40**, 183.

24. Morley, B. *et al.* (1988). The lysine residue in the membrane-spanning domain of the β chain is necessary for cell-surface expression of the T cell antigen receptor. *J. Exp. Med.*, **168**, 1971.

25. Blumberg, R. *et al.* (1990). Assembly and function of the T cell antigen receptor: requirement of either the lysine or arginine residues in the transmembrane region of the α chain. *J. Biol. Chem.*, **265**, 14036.

26. Bonifacino, J., Cosson, P., and Klausner, R. (1990). Co-localized transmembrane determinants for ER degradation and subunit assembly explain the intracellular fate of TCR chains. *Cell*, **63**, 503.

27. Cosson, P. *et al.* (1991). Membrane protein association by potential intramembrane charge pairs. *Nature*, **351**, 414.

28. Hall, C. *et al.* (1991). Requirements for cell surface expression of the human TCR/CD3 complex in non-T cells. *Int. Immunol.*, **3**, 359.

29. Blumberg, R. *et al.* (1990). Structure of the T cell antigen receptor: evidence for two CD3 ε subunits in the T cell receptor/CD3 complex. *Proc. Natl Acad. Sci. USA*, **87**, 7220.

30. Alarcon, B. *et al.* (1991). The CD3-γ and CD3-δ subunits of the T cell antigen receptor can be expressed within distinct functional TCR/CD3 complexes. *EMBO J.*, **10**, 903.

31. Kappes, D. and Tonegawa, S. (1991). Surface expression of alternative forms of the TCR/CD3 complex. *Proc. Natl Acad. Sci. USA*, **88**, 337.

32. Meuer, S. *et al.* (1984). The human T cell receptor. *Annu. Rev. Immunol.*, **2**, 23.

33. Rao, P. *et al.* Differential expression of CD3 epitopes in α/β and γ/δ TCR-bearing T cells: evidence for a dimeric receptor. (In preparation.)

34. Shin, J., Lee, S., and Strominger, J. (1993). Translocation of TCRα chains into the lumen of the endoplasmic reticulum and their degradation. *Science*, **259**, 1901.

35. Sancho, J. *et al.* (1993). The T cell receptor associated CD3-ε protein is phosphorylated upon T cell activation in the two tyrosine residues of a conserved signal transduction motif. *Eur. J Immunol.*, **23**, 1636.

36. Gold, D. *et al.* (1986). Isolation of cDNA clones encoding the 20K non-glycosylated polypeptide chain of the human T cell receptor/T3 complex. *Nature*, **321**, 431.

37. Alarcon, B. *et al.* (1988). Assembly of the human T cell receptor–CD3 complex takes place in the endoplasmic reticulum and involves intermediary complexes between the CD3-γ,δ,ε core and single T cell receptor α or β chains. *J. Biol. Chem.*, **263**, 2953.

38. Ley, S. *et al.* (1989). Surface expression of CD3 in the absence of T cell receptor. *Eur. J. Immunol.*, **19**, 2309.

39. Bernot, A. and Auffray, C. (1991). Primary structure and ontogeny of an avian CD3 transcript. *Proc. Natl Acad. Sci. USA*, **38**, 2550.

40. Weissman, A. *et al.* (1988). Molecular cloning and chromosomal localization of the human T-cell receptor ζ chain: Distinction from the molecular CD3 complex. *Proc. Natl Acad. Sci. USA*, **85**, 9709.

41. Baniyash, M. *et al.* (1989). The isolation and characterization of the murine T cell antigen receptor ζ chain gene. *J. Biol. Chem.*, **264**, 13252.

42. Wegener, A.-M. *et al.* (1992). The T cell receptor/CD3 complex is composed of at least two autonomous transduction modules. *Cell*, **68**, 83.

43. Orloff, D. *et al.* (1990). Family of disulphide-linked dimers containing the ζ and η chains of the T-cell receptor and the γ chain of Fc receptors. *Nature*, **347**, 189.

44. Qian, D. *et al.* (1993). The γ chain of the high-affinity receptor for IgE is a major functional subunit of the T cell antigen receptor complex in γδ T lymphocytes. *Proc. Natl Acad. Sci. USA*, **90**, 11875.

45. Weissman, A. *et al.* (1989). Role of the ζ chain in the expression of T cell antigen receptor: genetic reconstitution studies. *EMBO J.*, **8**, 3651.

46. Carson, G. *et al.* (1991). Six chains of the human T cell antigen–receptor CD3 complex are necessary and sufficient for processing the receptor heterodimer to the cell surface. *J. Biol. Chem.*, **266**, 7883.

47. Minami, Y. *et al.* (1987). Building a multichain receptor: synthesis, degradation and assembly of the T cell antigen receptor. *Proc. Natl Acad. Sci. USA*, **84**, 291.

48. Berkhout, B., Alarcon, B., and Terhorst, C. (1989). Transfection of genes encoding the T cell receptor-associated CD3 complex into COS cells results in assembly of the macromolecular structure. *J. Biol. Chem.*, **263**, 8628.

49. Chen, C. *et al.* (1988). Selective degradation of T cell antigen receptors retained in a pre-Golgi compartment. *J. Cell Biol.*, **107**, 2149.

50. Lippincott-Schwartz, J. *et al.* (1988). Degradation from the endoplasmic reticulum: disposing of newly synthesized proteins. *Cell*, **54**, 209.

51. Bonifacino, J. *et al.* (1989). Pre-golgi degradation of newly-synthesized T cell antigen receptor chains: intrinsic sensitivity and role of subunit assembly. *J. Biol. Chem.*, **109**, 73.

52. Wileman, T. *et al.* (1990). The transmembrane anchor of the T cell antigen receptor β chain contains a structural determinant of pre-Golgi proteolysis. *Cell Regul.*, **1**, 907.

53. Wileman, T., Pettey, C., and Terhorst, C. (1990). Recognition for degradation in the endoplasmic reticulum and lysosomes prevents the transport of single TCRβ and CD3δ subunits of the T cell antigen receptor to the surface of cells. *Int. Immunol.*, **2**, 743.

*54. Bonifacino, J. and Lippincott-Schwartz, J. (1991). Degradation of proteins within the endoplasmic reticulum. *Curr. Opin. Cell Biol.*, **3**, 592.

55. Young, J. *et al.* (1993). Regulation of selective protein degradation in the endoplasmic reticulum. *J. Biol. Chem.*, **268**, 19810.

56. de Waal Malefijt, R. *et al.* (1989). Introduction of T cell receptor (TCR)-α cDNA has differential effects on TCR-γδ/CD3 expression by Peer and Lyon-1 cells. *J. Immunol.*, **142**, 3634.

57. Rutledge, T. *et al.* (1992). Transmembrane helical interactions: zeta chain dimerization and functional association with the T cell antigen receptor. *EMBO J.*, **11**, 3245.

58. Manolios, N., Bonifacino, J., and Klausner, R. (1990). Transmembrane helical interactions and the assembly of the T cell receptor complex. *Science*, **249**, 274.

59. Letourneur, F. and Klausner, R. (1992). A novel di-leucine motif and a tyrosine-based

motif independently mediate lysosomal targeting and endocytosis of CD3 chains. *Cell*, **69**, 1143.

60. Bonifacino, J., Suzuki, C., and Klausner, R. (1990). A peptide sequence confers retention and rapid degradation in the endoplasmic reticulum. *Science*, **247**, 79.

61. Wileman, T. *et al.* (1990). The γ and ε subunits of the CD3 complex inhibit pre-Golgi degradation of newly synthesized T cell antigen receptors. *J. Cell Biol.*, **110**, 973.

62. Wileman, T. *et al.* (1991). Depletion of cellular calcium accelerates protein degradation in the endoplasmic reticulum. *J. Biol. Chem.*, **266**, 4500.

63. Weiss, A. (1993). T cell antigen receptor signal transduction: A tale of tails and cytoplasmic protein-tyrosine kinases. *Cell*, **73**, 209.

*64. Weiss, A. and Littman, D. (1994). Signal transduction by lymphocyte antigen receptors. *Cell*, **76**, 263.

65. Baniyash, M. *et al.* (1988). Tyrosine phosphorylation of TCR ζ chain. *J. Biol. Chem.*, **263**, 18225.

66. Qian, D. *et al.* (1993). Multiple components of the T cell antigen receptor complex become tyrosine-phosphorylated upon activation. *J. Biol. Chem.*, **268**, 4488.

67. Irving, B. and Weiss, A. (1991). The cytoplasmic domain of the T cell receptor ζ chain is sufficient to couple to receptor-associated signal transduction pathways. *Cell*, **64**, 891.

68. Romeo, C. and Seed, B. (1991). Cellular immunity to HIV activated by CD4 fused to T cell or Fc receptor polypeptides. *Cell*, **64**, 1037.

69. Letourneur, F. and Klausner, R. (1992). Activation of T cells by a tyrosine kinase activation domain in the cytoplasmic tail of CD3. *Science*, **255**, 79.

70. Rudd, C. (1990). CD4, CD8 and the TCR–CD3 complex: a novel class of protein-tyrosine kinase receptor. *Immunol. Today*, **11**, 400.

71. Samelson, L. *et al.* (1990). Association of the fyn protein tyrosine kinase with the T cell antigen receptor. *Proc. Natl Acad. Sci. USA*, **87**, 4358.

72. Kolanus, W., Romeo, C., and Seed, B. (1993). T cell activation by clustered tyrosine kinases. *Cell*, **74**, 171.

73. Chan, A. *et al.* (1991). The ζ chain is associated with a tyrosine kinase and upon T-cell antigen receptor stimulation associates with ZAP-70, a 70kDa tyrosine phosphoprotein. *Proc. Natl Acad. Sci. USA*, **88**, 9166.

*74. Chan, A. *et al.* (1992). ZAP-70: A 70 kd protein-tyrosine kinase that associates with the TCR ζ chain. *Cell*, **71**, 649.

75. Heyeck, S. and Berg, L. (1993). Developmental regulation of a murine T-cell-specific tyrosine kinase gene, *Tsk*. *Proc. Natl Acad. Sci. USA*, **90**, 669.

76. June, C. *et al.* (1990). Inhibition of tyrosine phosphorylation prevents T cell receptor-mediated signal transduction. *Proc. Natl Acad. Sci. USA*, **87**, 7722.

77. Mustelin, T., *et al.* (1990). T cell antigen receptor-mediated activation of phospholipase C requires tyrosine phosphorylation. *Science*, **247**, 1584.

78. Graber, M. *et al.* (1992). The protein-tyrosine kinase inhibitor herbimycin A, but no genestein, specifically inhibits signal transduction by the T cell antigen receptor. *Int. Immunol.*, **4**, 1201.

79. Pingel, J. and Thomas, M. (1989). Evidence that the leukocyte-common antigen is required for antigen-induced T lymphocyte proliferation. *Cell*, **58**, 1055.

80. Koyasu, S. *et al.* (1992). Phosphorylation of multiple CD3ζ tyrosine residues leads to formation of pp21 *in vitro* and *in vivo*. *J. Biol. Chem.*, **267**, 3375.

81. van Oers, N. *et al.* (1993). Constitutive tyrosine phosphorylation of the T cell receptor

(TCR) ζ subunit: Regulation of TCR-associated protein tyrosine kinase activity by TCR ζ. *Mol. Cell. Biol.*, **13**, 5771.

82. Park, D., Rho, H., and Rhee, S. (1991). CD3 stimulation causes phosphorylation of phospholipase C-γ1 on serine and tyrosine residues in a human T cell line. *Proc. Natl Acad. Sci. USA*, **88**, 5453.

83. Weiss, A. *et al.* (1991). Stimulation of the T cell antigen receptor induces tyrosine phosphorylation of phospolipase C-γ1. *Proc. Natl Acad. Sci. USA*, **88**, 5484.

84. Gardner, P. (1989). Calcium and T lymphocyte activation. *Cell*, **59**, 15.

85. Thompson, P. *et al.* (1992). Association of PI 3-kinase with the TCR/CD3 complex upon T cell activation. *Oncogene*, **7**, 719.

86. Ward, S. *et al.* (1992). Regulation of D-3 phosphoinositides during T cell activation via the T cell antigen receptor/CD3 complex and CD2 antigens. *Eur. J. Immunol.*, **22**, 45.

87. Peter, M. *et al.* (1992). The T cell receptor ζ chain contains a GTP/GDP binding site. *EMBO J.*, **11**, 933.

88. Crabtree, G. (1989) Contingent genetic regulatory events in T lymphocyte activation. *Science*, **243**, 355.

89. Mustelin, T. and Burn, P. (1993). Regulation of src family tyrosine kinases in lymphocytes. *Immunol. Today*, **14**, 215.

90. Hunter, T. and Cooper, J. (1985). Protein tyrosine kinases. *Annu. Rev. Biochem.*, **54**, 897.

91. Cantley, L. *et al.* (1991). Oncogenes and signal transduction. *Cell*, **64**, 281.

*92. Pawson, T. and Gish G. (1992). SH2 and SH3 domains: From structure to function. *Cell*, **71**, 359.

93. Waksman, G. *et al.* (1992) Crystal structure of the phosphotyrosine recognition domain SH2 of v-src complexed with tyrosine-phosphorylated peptides. *Nature*, **358**, 646.

94. Eck, M., Shoelson, S., and Harrison, S. (1993). Structure of the Lck SH2 domain complexed to high-affinity tyrosine phosphopeptide substrate. *Nature*, **362**, 87.

95. Songyang, Z. *et al.* (1993). SH2 domains recognize specific phosphopeptide sequences. *Cell*, **72**, 767.

96. Tsygankov, A. *et al.* (1992). Activation of tyrosine kinase p60^fyn following T cell antigen receptor cross-linking. *J. Biol. Chem.*, **267**, 18259.

97. Hall, C., Sancho, J., and Terhorst, C. (1993). Reconstitution of T cell receptor ζ-mediated calcium mobilization in nonlymphoid cells. *Science*, **261**, 915.

98. Molina, T. *et al.* (1992). Profound block in thymocyte development in mice lacking p56lck. *Nature*, **357**, 161.

99. Stein, P. *et al.* (1992). *pp59fyn* mutant mice display differential signaling in thymocytes and peripheral T cells. *Cell*, **70**, 741.

100. Appleby, M. *et al.* (1992). Defective T cell receptor signaling in mice lacking the thymic isoform of *p59fyn*. *Cell*, **70**, 751.

101. Straus, D. and Weiss, A. (1993). The CD3 chains of the T cell antigen receptor associate with the ZAP-70 tyrosine kinase and are tyrosine phosphorylated after receptor stimulation. *J. Exp. Med.*, **178**, 1523.

102. Irving, B., Chan, A., and Weiss, A. (1993). Functional characterization of a signal transducing motif present in the T cell receptor ζ chain. *J. Exp. Med.*, **177**, 1093.

*103. Arpaia, E. *et al.* (1994). Defective T cell receptor signaling and CD8^+ thymic selection in humans lacking ZAP-70 kinase. *Cell*, **76**, 947.

104. Wange, R. *et al.* (1993). Tandem SH2 domains of ZAP-70 binds to T cell antigen receptor and CD3ε from activated Jurkat T cells. *J. Biol. Chem.*, **268**, 19797.

105. Loh, C. *et al.* (1994). Association of Raf with the CD3 δ and γ chains of the T cell receptor–CD3 Complex. *J. Biol. Chem.*, **269**, 8817.

106. Downward, J., Graves, J., and Cantrell, D. (1992). The regulation and function of p21ras in T cells. *Immunol. Today*, **13**, 89.

107. Pawson, A. and Gish, G. (1992). Role of SH2 domains in signal transduction. *Cell*, **71**, 359.

108. Fry, M. *et al.* (1993). New insights into protein tyrosine kinase receptor signalling complexes. *Protein Sci.*, **2**, 1785.

109. Hiles, I. *et al.* (1992). Phosphatidylinositol 3-kinase: structure and expression of the 110 kd catalytic subunit. *Cell*, **70**, 419.

110. Nakanishi, H., Brewer, K., and Exton, J. (1993). Activation of the ζ isozyme of protein kinase C by phosphatidylinositol 3,4,5-triphosphate. *J. Biol. Chem.*, **268**, 13.

*111. Dhand, R. *et al.* (1994). PL 3-kinase is a dual specificity enzyme: autoregulation by an intrinsic protein-serine kinase activity. *EMBO J.*, **13**, 522.

112. Exley, M. *et al.* (1994). Association of phosphatidylinositol 3-kinase associates with a specific sequence in the T cell receptor ζ chain upon T cell activation. *J. Biol. Chem.*, **269**, 15140.

113. Prasad, K. *et al.* (1993). Phosphatidylinositol (PL) 3-kinase and PL 4-kinase binding to the CD4–p56lck complex: SH3 domain binds to PL 3-kinase but not PL 4-kinase. *Mol. Cell. Biol.*, **13**, 7708.

114. Prasad, K. *et al.* (1993). Src-homology 3 domain of protein kinase p59fyn mediates binding phosphatidylinositol 3-kinase in T cells. *Proc. Natl Acad. Sci. USA*, **90**, 7366.

115. Frank, S. *et al.* (1990). Structural mutations of the T cell receptor ζ chain and its role in T cell activation. *Science*, **249**, 174.

116. Frank, S. *et al.* (1992). Mutagenesis of T cell antigen receptor ζ chain tyrosine residues. *J. Biol. Chem.*, **267**, 13656.

117. Sancho, J. *et al.* (1993). Coupling of GTP binding to the T cell receptor TCR ζ-chain with TCR-mediated signal transduction. *J. Immunol.*, **150**, 3230.

118. Ravichandran, K. *et al.* (1993). Interaction of Shc with the ζ chain of the T cell receptor upon T cell activation. *Science*, **262**, 902.

119. Izquierdo, M. *et al.* (1992). Role of protein kinase C in T-cell antigen receptor regulation of p21ras: Evidence that two p21ras regulatory pathways coexist in T cells. *Mol. Cell. Biol.*, **12**, 3305.

120. Romeo, C., Amiot, M., and Seed, B. (1992). Sequence requirements for induction of cytolysis by the T cell antigen/Fc receptor ζ chain. *Cell*, **68**, 889.

121. Liu, C.-P. *et al.* (1993). Abnormal T cell development in CD3ζ−/− mutant mice and identification of a novel T cell population in the intestine. *EMBO J.*, **12**, 4863.

122. Love, P. *et al.* (1993). T cell development in mice that lack the ζ chain of the T cell antigen receptor complex. *Science*, **261**, 918.

123. Ohno, H. *et al.* (1993). Developmental and functional impairment of T cells in mice lacking CD3ζ chains. *EMBO J.*, **12**, 4357.

124. Simpson, S. *et al.* (1995). Selection of peripheral and intestinal T lymphocytes lacking CD3-ζ. *Int. Immunol.*, **7**, 287.

125. Ohno, H. *et al.* (1994). Preferential usage of the Fc receptor γ chain in the T cell antigen receptor complex by γδ T cells localized in epithelia. *J. Exp. Med.*, **179**, 365.

126. Mizoguchi, H. *et al.* (1992). Alternations in signal transduction molecules in T lymphocytes from tumor-bearing mice. *Science*, **258**, 1795.

127. Wang, B. *et al.* (1994). A block in both early T lymphocyte and natural killer cell development in transgenic mice with high copy numbers of the CD3-ε gene. *Proc. Natl Acad. Sci. USA*, **91**, 9402.

128. Lanier, L. *et al.* (1992). Expression of cytoplasmic CD3ε proteins in activated human adult natural killer (NK) cells and CD3γ,δ,ε complexes in fetal NK cells. *J. Immunol.*, **149**, 1876.

129. Phillips, J. *et al.* (1992). Ontogeny of human natural killer (NK) cells: fetal NK cells mediate cytolytic function and express cytoplasmic CD3 εδ proteins. *J. Exp. Med.*, **175**, 1055.

130. Lanier, L., Spits, H., and Phillips, J. (1992). Developmental relationship of natural killer and T lymphocytes. *Immunol. Today*, **13**, 392.

131. Hussey, R. *et al.* (1993). Overexpression of CD3η during thymic development does not alter the negative selection process. *J. Immunol.*, **150**, 1183.

132. Williams, A. and Barclay, A. (1988). The immunoglobulin superfamily—domains for cell surface recognition. *Annu. Rev. Immunol.*, **6**, 381.

133. Suda, T. *et al.* (1993). Molecular cloning and expression of the Fas ligand, a novel member of the tumor necrosis factor family. *Cell*, **75**, 1169.

*134. Smith, C., Farrah, T., and Goodwin, R. (1994). The TNF receptor superfamily of cellular and viral proteins: Activation, costimulation, and death. *Cell*, **76**, 959.

*135. Janeway, C. (1992). The T cell receptor as a multicomponent signalling machine: CD4/CD8 coreceptors and CD45 in T cell activation. *Annu. Rev. Immunol.*, **10**, 645.

*136. Linsley, P. and Ledbetter, J. (1993). The role of the CD28 receptor during T cell responses to antigen. *Annu. Rev. Immunol.*, **11**, 191.

*137. Clark, E., and Ledbetter, J. (1994). How B and T cells talk to each other. *Nature*, **367**, 425.

*138. Trowbridge, I. and Thomas, M. (1994). CD45: an emerging role as a protein tyrosine phosphatase required for lymphocyte activation and development. *Annu. Rev. Immunol.*, **12**, 85.

*139. Miceli, M. and Parnes, J. (1993). Role of CD4 and CD8 in T cell activation and differentiation. *Adv. Immunol.*, **53**, 59.

140. DiSanto, J., Knowles, R., and Flomenberg, N. (1988). The human Lyt-3 molecule requires CD8 for cell surface expression. *EMBO J.*, **7**, 3465.

141. Norment, A. and Littman, D. (1988). A second subunit of CD8 is expressed in human T cells. *EMBO J.*, **7**, 3433.

142. Nagler, A. *et al.* (1989). Comparative studies of human FcRIII-positive and negative NK cells. *J. Immunol.*, **143**, 3183.

143. Swain, S. (1983). T cell subsets and the recognition of MHC class. *Immunol. Rev.*, **74**, 129.

144. Doyle, C. and Strominger, J. (1987). Interaction between CD4 and class II MHC molecules mediates cell adhesion. *Nature*, **330**, 256.

145. Norment, A. *et al.* (1988). Cell adhesion mediated by CD8 and MHC class I molecules. *Nature*, **336**, 79.

146. Leahy, D., Axel, R., and Hendrickson, W. (1992). Crystal structure of a soluble form of the human T cell coreceptor CD8 at 2.6 Å resolution. *Cell*, **68**, 1145.

147. Wang, J. *et al.* (1990). Atomic structure of a fragment of human CD4 containing two immunoglobulin-like domains. *Nature*, **348**, 411.

*148. Parham, P. (1992). CD8 and CD4: The box and the rod. *Nature*, **357**, 538.

149. Ryu, S.-E. *et al.* (1990). Crystal structure of an HIV-binding recombinant fragment of human CD4. *Nature*, **348**, 411

150. Cammarota, G. *et al.* (1992). Identification of a CD4 binding site on the β2 domain of HLA-DR molecules. *Nature*, **356**, 799.

151. Konig, R., Huang, L., and Germain, R. (1992). MHC class II interaction with CD4 mediated by a region analogous to the MHC class I binding site for CD8. *Nature*, **356**, 796.

152. Dembic, Z. *et al.* (1987). Transfection of the CD8 gene enhances T-cell recognition. *Nature*, **326**, 510.

153. Gabert, J. *et al.* (1987). Reconstitution of MHC Class I specificity by transfer of the T cell receptor and Ly-2 genes. *Cell*, **50**, 545.

154. O'Rourke, A. and Mescher, M. (1992). Cytotoxic T-lymphocyte activation involves a cascade of signalling and adhesion events. *Nature*, **358**, 253.

*155. O'Rourke, A. and Mescher, M. (1993). The roles of CD8 in cytotoxic T lymphocyte function. *Immunol. Today*, **14**, 183.

156. Blanchard, D. *et al.* (1988). CD4 is involved in a post-binding event in the cytolytic reaction mediated by human CD4$^+$ cytotoxic T lymphocyte clones. *J. Immunol.*, **140**, 1745.

157. Kupfer, A. *et al.* (1987). Coclustering of CD4 (L3T4) molecule with the T cell receptor in induced by specific direct interaction of helper T cells and antigen-presenting cells. *Proc. Natl Acad. Sci. USA*, **84**, 5888.

158. Eichmann, K. *et al.* (1987). Synergy between T cell signalling triggered through the TCR and CD4 or CD8. *Eur. J Immunol.*, **17**, 643.

*159. Julius, M., Maroun, C., and Haughn, L. (1993). Distinct roles for CD4 and CD8 as co-receptors in antigen receptor signalling. *Immunol. Today*, **14**, 177.

160. Rudd, C. *et al.* (1988). The CD4 receptor is complexed in detergent lysates to a protein-tyrosine kinase (pp56) from human T lymphocytes. *Proc. Natl Acad. Sci. USA*, **85**, 5190.

161. Veillette, A. *et al.* (1988). The CD4 and CD8 T cell surface antigens are associated with the internal membrane tyrosine kinase p56lck. *Cell*, **44**, 301.

*162. Perlmutter, R. *et al.* (1993). Regulation of lymphocyte function by protein phosphory-lation. *Annu. Rev. Immunol.*, **11**, 451.

163. Turner, J. *et al.* (1990). Interaction of the unique N-terminal region of tyrosine kinase p56lck with cytoplasmic domains of CD4 and CD8 is mediated by cysteine motifs. *Cell*, **60**, 755.

164. Glaichenhaus, N. *et al.* (1991). Requirement of association of p56lck with CD4 in antigen-specific signal transduction in T cells. *Cell*, **64**, 511.

165. Bank, I. and Chess, L. (1985). Perturbation of the T4 molecule transmits a negative signal to T cells. *J. Exp. Med.*, **162**, 1294.

166. Haughn, L. *et al.* (1992). Association of tyrosine kinase p56lck with CD4 inhibits the induction of growth through the alpha beta T cell receptor. *Nature*, **358**, 328.

167. Hugo, P. *et al.* (1990). Ontogeny of a novel CD4$^+$CD8$^-$CD3$^-$ thymocyte sub-population: a comparison with CD4$^-$CD8$^+$CD3$^-$ thymocytes. *Int. Immunol.*, **2**, 209.

168. MacDonald, H. *et al.* (1988). Positive selection of CD4$^+$ thymocytes controlled by MHC class II gene products. *Nature*, **336**, 47.

169. Ziljstra, M. *et al.* (1990). β2 microglobulin deficient mice lack CD4$^-$CD8$^+$ cytotoxic T cells. *Nature*, **344**, 742.

170. Shinkai, Y. *et al.* (1993). Restoration of T cell development in *Rag-2* deficient mice by functional TCR transgenes. *Science*, **259**, 822.

* 171. von Boehmer, H. (1991). Positive and negative selection of the αβ T-cell repertoire *in vivo*. *Curr. Opin. Immunol.*, **3**, 210.

172. Mombaerts, P. *et al.* (1992). Mutations in cell antigen receptor genes α and β block thymocyte development at different stages. *Nature*, **360**, 225.

173. Anderson, S. *et al.* (1993). Protein kinase p56 lck controls allelic exclusion of TCR-β genes. *Nature*, **356**, 552.

174. Groettrup, M. and von Boehmer, H. (1993). T cell receptor β chain dimers on immature thymocytes from normal mice. *Eur. J. Immunol.*, **23**, 1393.

175. Groettrup, M. *et al.* (1993). A novel disulphide-linked heterodimer on pre-T cells consist of the T cell receptor β chain and a 33kd glycoprotein. *Cell*, **75**, 283.

176. Groettrup, M. and von Boehmer, H. (1993). A role for a pre-T-cell receptor in T-cell development. *Immunol. Today*, **14**, 610.

177. MacDonald, H., Budd, R., and Howe, R. (1988). A CD3⁻ subset of CD4⁻CD8⁺ thymocytes: a rapidly cycling intermediate in the generation of CD4⁺CD8⁺ cells. *Eur. J. Immunol.*, **18**, 519.

178. Robey, E. *et al.* (1992). The level of CD8 expression can determine the outcome of thymic selection. *Cell*, **69**, 1089.

179. Scott, B. *et al.* (1989). The generation of mature T cells requires interaction of the αβ T cell receptor with major histocompatibility antigens. *Nature*, **338**, 591.

180. Teh, H. *et al.* (1991). Participation of CD4 coreceptor molecules in T cell repertoire selection. *Nature*, **349**, 241.

181. Zuniga-Pflucker, J., Longo, D. and Kruisbeek, A. (1989). Positive selection of CD4⁻CD8⁺ T cells in the thymus of normal mice. *Nature*, **338**, 76.

182. Fung-Leung, W.-P. *et al.* (1991). CD8 is needed for development of cytotoxic T cells but not helper T cells. *Cell*, **65**, 443.

183. Rahemtulla, A. *et al.* (1991). Normal development and function of CD8⁺ cells but markedly decreased helper cell activity in mice lacking CD4. *Nature*, **353**, 180.

184. Van Kaer, L. *et al.* (1992). TAP1 mutant mice are deficient in antigen presentation, surface Class I molecules, and CD4⁻–8⁺ T cells. *Cell*, **71**, 1205.

185. Cosgrove, D. *et al.* (1991). Mice lacking MHC Class II molecules. *Cell*, **66**, 1051.

186. Grusby, M. *et al.* (1991). Depletion of CD4⁺ T cells in major histocompatibility complex Class II-deficient mice. *Science*, **253**, 1417.

187. Borgulya, P. *et al.* (1991). Development of the CD4 and CD8 lineages of T cells: instruction versus selection. *EMBO J.*, **10**, 913.

188. Seong, R., Chamberlain, J., and Parnes, J. (1992). Signal for T cell differentiation to a CD4⁺ cell lineage is delivered by a CD4 transmembrane and/or cytoplasmic tail. *Nature*, **356**, 718.

189. Veillette, A. *et al.* (1988). Engagement of CD4 and CD8 expressed on immature thymocytes induces activation of intracellular tyrosine phosphorylation pathways. *J. Exp. Med.*, **170**, 1671.

190. Killeen, N. and Littman, D. (1993). Helper T-cell development in the absence of CD4–p56lck association. *Nature*, **364**, 729.

191. Aruffo, A. and Seed, B. (1987). Molecular cloning of a CD28 cDNA by a high efficiency COS cell expression system. *Proc. Natl Acad. Sci. USA*, **84**, 8573.

192. Harper, K. *et al.* (1991). CTLA-4 and CD28 activated lymphocyte molecules are closely related in both mouse and human as to sequence, message expression, gene structure, and chromosomal location. *J. Immunol.*, **147**, 1037.

* 193. Schwartz, R. (1992). Costimulation of T lymphocytes: The role of CD28, CTLA-4, and

B7/BB1 in interleukin-2 production and immunotherapy. *Cell*, **71**, 1065.

194. Linsley, P. *et al.* (1992). Co-expression and functional cooperation of CTLA-4 and CD28 on activated T lymphocytes. *J. Exp. Med.*, **176**, 1595.

195. Barcena, A. *et al.* (1993). Phenotypic and functional analysis of T cell precursors in the human fetal liver and thymus. CD7 expression in the earliest stages of T and myeloid cell development. *Blood*, **82**, 3401.

196. Gross, J., Callas, E., and Allison, J. (1992). Identification and distribution of the costimulatory receptor CD28 in the mouse. *J. Immunol.*, **149**, 380.

197. Turka, L. *et al.* (1991). Signal transduction via CD4, CD8 and CD28 in mature and immature thymocytes. *J. Immunol.*, **146**, 1428.

198. Yang, S. *et al.* (1988). A novel activation pathway for mature thymocytes. Costimulation of CD2 (T, p50) and CD28 (T, p44) induces autocrine interleukin-2/interleukin 2 receptor-mediated cell proliferation. *J. Exp. Med.*, **168**, 1457.

199. Azuma, M. *et al.* (1992). CD28 interaction with B7 costimulates primary allogeneic proliferative responses and cytotoxicity mediated by small, resting T lymphocytes. *J. Exp. Med.*, **175**, 353.

200. Morishita, Y. *et al.* (1989). A distinct subset of human CD4$^+$ cells with a limited alloreactive T cell receptor repertoire. *J. Immunol.*, **143**, 2783.

201. Yamada, H. *et al.* (1985). Monoclonal antibody 9.3 and anti-CD11 antibodies define reciprocal subsets of lymphocytes. *Eur. J. Immunol.*, **15**, 1164.

202. Damle, N. *et al.* (1983). CD28-positive T cells contain cytotoxic T cell precursors. *J. Immunol.*, **131**, 2296.

203. Koide, J. and Engleman, E. (1990). Differences in surface phenotype and mechanism of action between alloantigen-specific CD8$^+$ cytotoxic and suppressor T cell clones. *J. Immunol.*, **144**, 32.

204. Li, S. *et al.* (1990). Human suppressor T cell clones lack CD28. *Eur. J. Immunol.*, **20**, 1281.

205. Lum, L. *et al.* (1982). CD28-T cells are suppressor cell precursors. *Cell. Immunol.*, **72**, 122.

206. Testi, R. and Lanier, L. (1989). Functional expression of CD28 on T cell antigen receptor γδ-bearing T lymphocytes. *Eur. J. Immunol.*, **19**, 185.

207. Azuma, M. *et al.* (1992). Involvement of CD28 in major histocompatibility complex-unrestricted cytotoxicity mediated by a human NK leukemia line. *J. Immunol.*, **149**, 1115.

208. Kozbor, D. *et al.* (1987). Tp44 molecules involved in antigen-independent T cell activation are expressed on human plasma cells. *J. Immunol.*, **138**, 4128.

209. Hara, T., Fu, S., and Hansen, J. (1985). Human T cell activation II: A new activation pathway used by a major T cell population via a disulfide-bonded dimer of a 44 kilodalton polypeptide (9.3 antigen). *J. Exp. Med.*, **161**, 1513.

210. Ledbetter, J. *et al.* (1985). Antibodies to Tp67 and Tp44 augment and sustain proliferative responses of activated T cells. *J. Immunol.*, **135**, 2331.

211. Clipstone, N. and Crabtree, G. (1992). Identification of calcineurin as a key signaling enzyme in T lymphocyte activation. *Nature*, **357**, 695.

212. O'Keefe, S. *et al.* (1992). FK-506 and CsA-sensitive activation of the interleukin-2 promoter by calcineurin. *Nature*, **357**, 692.

*213. Schreiber, S. and Crabtree, G. (1992). The mechanism of action of cyclosporin A and FK506. *Immunol. Today*, **73**, 136.

214. Ledbetter, J. *et al.* (1987). Cross-linking of surface antigens causes mobilization of intracellular ionized calcium in T lymphocytes. *Proc. Natl Acad. Sci. USA*, **84**, 1384.

215. Vandenberghe, P. *et al.* (1992). Antibody and B7/BB1-mediated ligation of the CD28 receptor induces tyrosine phosphorylation in human T cells. *J. Exp. Med.*, **175**, 951.

216. Lu, Y. *et al.* (1992). CD28-induced T cell activation: Evidence for a protein-tyrosine kinase signal transduction pathway. *J. Immunol.*, **149**, 24.

217. Hutchcroft, J. and Bierer, B. (1994). Activation-dependent phosphorylation of the T-lymphocyte surface receptor CD28 and associated proteins. *Proc. Natl Acad. Sci. USA*, **91**, 3260.

218. Anel, A., Richieri, G., and Kleinfeld, A. (1994). A tyrosine phosphorylation requirement for cytotoxic T lymphocyte degranulation. *J. Immunol.*, **269**, 9506.

219. Truitt, K., Hicks, C., and Imboden, J. (1993). Stimulation of CD28 triggers an association between CD28 and phosphatidylinositol 3-kinase in Jurkat T cells. *J. Exp. Med.*, **179**, 1071.

220. Lindstein, T. *et al.* (1989). Regulation of lymphokine messenger RNA stability by a surface-mediated T cell activation pathway. *Science*, **244**, 339.

221. Fraser, J. *et al.* (1991). Regulation of interleukin-2 gene enhancer activity by the T cell accessory molecule CD28. *Science*, **251**, 313.

222. Fraser, J., Newton, M., and Weiss, A. (1992). CD28 and T cell antigen receptor signal transduction co-ordinately regulate interleukin 2 gene expression in response to superantigen stimulation. *J. Exp. Med.*, **175**, 1131.

223. Verweij, C., Geerts, M., and Aarden, L. (1991). Activation of interleukin-2 gene transcription via the T cell surface molecule CD28 is mediated through an NF-κb-like response element. *J. Biol. Chem.*, **266**, 14179.

224. Walunas, T. *et al.* (1994). CTLA-4 can function as a negative regulator of T cell activation. *Immunity*, **1**, 405.

225. Yokochi, T., Holly, R., and Clark, E. (1982). B lymphoblast antigen (BB-1) expressed on Epstein–Barr virus-activated B cell blasts, B cell lymphoblastoid cell lines, and Burkitt's lymphomas. *J. Immunol.*, **128**, 823.

226. Freedman, A. *et al.* (1987). B7, a B cell-restricted antigen that identifies preactivated B cells. *J. Immunol.*, **139**, 3260.

227. Freeman, G. *et al.* (1989). B7, a new member of the Ig superfamily with unique expression on activated and neoplastic B cells. *J. Immunol.*, **143**, 2714.

228. Linsley, P., Clark, E., and Ledbetter, J. (1990). T cell antigen CD28 mediates adhesion with B cells by interacting with activation antigen B7/BB-1. *Proc. Natl Acad. Sci. USA*, **87**, 5031.

229. Freeman, G. *et al.* (1991). Structure, expression and T cell costimulatory activity of the murine homologue of the human B lymphocyte activation antigen B7. *J. Exp. Med.*, **174**, 625.

230. Freedman, A. *et al.* (1991). Selective induction of B7/BB-1 on interferon-gamma stimulated monocytes: A potential mechanism for amplification of T cell activation through the CD28 pathway. *Cell. Immunol.*, **137**, 429.

231. Valle, A. *et al.* (1991). IL-4 and IL-2 upregulate the expression of antigen B7, the B cell counterstructure to T cell CD28: an amplification mechanism for T–B cell interactions. *Int. Immunol.*, **3**, 229.

232. Valle, A. *et al.* (1992). MAb 104 a murine monoclonal antibody, recognizes the B7 antigen which is expressed on activated B cells and HTLV-1 transformed T cells. *Immunology*, **69**, 531.

233. O'Doherty, U. *et al.* (1993). Dendritic cells freshly isolated from human blood express

CD4 and mature in typical immunostimulatory dendritic cells after culture in monocyte-conditioned medium. *J. Exp. Med.*, **178**, 1067.

234. Young, J. *et al.* (1992). The B7/BB-1 antigen provides one of several costimulatory signals for the activation of CD4$^+$ T lymphocytes by human blood dendritic cells *in vitro*. *J. Clin. Invest.*, **90**, 229.

235. Azuma, M. *et al.* (1993). Functional expression of B7/BB-1 on activated T lymphocytes. *J. Exp. Med.*, **177**, 845.

236. Yssel, H., Schneider, P., and Lanier, L. (1993). Interleukin-7 specifically induces the B7/BB-1 antigen on human cord blood and peripheral blood T cells and T cell clones. *Int. Immunol.*, **5**, 753.

237. Linsley, P. *et al.* (1991). Binding of the B cell activation antigen B7 to CD28 costimulates T cell proliferation and interleukin 2 mRNA accumulation. *J. Exp. Med.*, **173**, 721.

238. Linsley, P. *et al.* (1991). CTLA-4 is a second receptor for the B cell activation antigen B7. *J. Exp. Med.*, **174**, 561.

239. Koulova, L. *et al.* (1991). The CD28 ligand B7/BB1 provides costimulatory signal for alloactivation of CD4$^+$ T cells. *J. Exp. Med.*, **173**, 759.

240. de Waal Malefyt, R. *et al.* (1993). CD2/LFA-3 or LFA-1/ICAM-1 but not CD28/B7 interactions can augment cytotoxicity by virus specific CD8$^+$ CTL. *Eur. J. Immunol.*, **23**, 418.

241. Lenschow, D. *et al.* (1992). Long-term survival of xenogeneic pancreatic islet grafts induced by CTLA-4Ig. *Science*, **257**, 789.

242. Linsley, P. *et al.* (1992). Immunosuppression *in vivo* by a soluble form of the CTLA-4 T cell activation molecule. *Science*, **257**, 792.

243. Jones, L. *et al.* (1993). No role for CD28/B cell interactions in negative selection of T cells. *Int. Immunol.*, **5**, 503.

244. Shahinian, A. *et al.* (1993). Differential T cell costimulatory requirements in CD28-deificient mice. *Science*, **261**, 609.

245. Hathcock, K. *et al.* (1993). Identification of an alternative CTLA-4 ligand costimulatory for T cell activation. *Science*, **262**, 905.

246. Freeman, G. *et al.* (1993). Cloning of B7-2: A CTLA-4 counter-receptor that co-stimulates human T cell proliferation. *Science*, **262**, 909.

247. Lenschow, D. *et al.* (1993). Expression and functional significance of an additional ligand for CTLA-4. *Proc. Natl Acad. Sci. USA*, **90**, 11054.

248. Azuma, M. *et al.* (1993). B70 antigen is a second ligand for CTLA-4 and CD28. *Nature*, **366**, 76.

249. Johnson, J. and Jenkins, M. (1994). Monocytes provide a novel costimulatory signal to T cells that is not mediated by the CD28/B7 interaction. *J. Immunol.*, **152**, 429.

*250. Janeway, C. and Bottomly, K. (1994). Signals and signs for lymphocyte responses. *Cell*, **76**, 275.

*251. Jenkins, M. (1994). The ups and downs of T cell costimulation. *Immunity*, **1**, 443.

252. Armitage, R. *et al.* (1992). Molecular and biological characterization of a murine ligand for CD40. *Nature*, **357**, 80.

253. Hollenbaugh, D. *et al.* (1992). The human T cell antigen gp39, a member of the TNF gene family, is a ligand for the CD40 receptor: expression of a soluble form of gp39 with B cell co-stimulatory activity. *EMBO J.*, **11**, 4313.

254. deBoer, M. *et al.* (1993). Ligation of B7 with CD28/CTLA-4 on T cells results in CD40 ligand expression, interleukin-4 secretion and efficient help for antibody production by B cells. *Eur. J. Immunol.*, **23**, 3120.

*255. Hill, A. and Chapel, H. (1994). The fruits of cooperation. *Nature*, **361**, 494.

256. Haynes, B., Eisenbarth, G., and Fauci, A. (1979). Human lymphocyte antigens: Production of a monoclonal antibody that defines functional thymus-derived lymphocyte subsets. *Proc. Natl Acad. Sci. USA*, **76**, 5829.

257. Hintzen, R. *et al.* (1994). Characterization of the human CD27 ligand, a novel member of the TNF gene family. *J. Immunol.*, **152**, 1762.

258. Bowman, M. *et al.* (1994). The cloning of CD70 and its identification as the ligand for CD27. *J. Immunol.*, **152**, 1756.

259. Alderson, M. *et al.* (1993). Fas transduces activation signals in normal human T lymphocytes. *J. Exp. Med.*, **178**, 2231.

260. Saga, Y. *et al.* (1988). Organization of the ly-5 gene. *Mol. Cell. Biol.*, **8**, 4889.

261. Hall, L. *et al.* (1988). Complete exon–intron organization of the human leukocyte common antigen. *J. Immunol.*, **141**, 2781.

262. Barclay, A. *et al.* (1987). Lymphocyte-specific heterogeneity in the rat leukocyte common antigen (T200) is due to differences in polypeptide sequences near the NH2-terminus. *EMBO J.*, **6**, 1259.

263. Ralph, S. *et al.* (1987). Structural variants of the human T200 glycoprotein (leukocyte common antigen). *EMBO J.*, **6**, 1251.

264. Streuli, M. *et al.* (1987). Differential usages of three exons generates at least five different mRNAs encoding human leukocyte common antigen. *J. Exp. Med.*, **166**, 1548.

265. Byrne, J., Butler, J., and Cooper, M. (1988). Differential activation requirements for virgin and memory T cells. *J. Immunol.*, **141**, 3334.

266. Merkenschlager, M. *et al.* (1988). Limiting dilution analysis of proliferative responses in human lymphocyte populations defined by the monoclonal antibody UCHL1: implications for differential CD45 expression in T cell memory formation. *Eur. J. Immunol.*, **18**, 1653.

267. Akbar, A., Salmon, M., and Jannosy, G. (1991). The synergy between naive and memory T cells during activation. *Immunol. Today*, **12**, 184.

268. Charbonneau, H. *et al.* (1988). The leukocyte common antigen (CD45): a putative receptor-linked tyrosine phosphatase. *Proc. Natl Acad. Sci. USA*, **85**, 7182.

269. Tonks, N. *et al.* (1988). Demonstration that the leukocyte common antigen CD45 is a protein tyrosine phosphatase. *Biochemistry*, **27**, 8695.

270. Kiener, P. and Mittler, R. (1989). CD45 protein tyrosine phosphatase cross-linking inhibits T cell receptor/CD3 mediated activation in human T cells. *J. Immunol.*, **143**, 23.

271. Marvel, J. *et al.* (1991). Evidence that CD45 regulates the activity of phospholipase C in mouse T lymphocytes. *Eur. J. Immunol.*, **21**, 195.

272. Samelson, L. *et al.* (1990). Activation of tyrosine phosphorylation in human T cells via the CD2 pathway: regulation by the CD45 tyrosine phosphatase. *J. Immunol.*, **145**, 2448.

*273. Alexander, D. *et al.* (1992). The role of CD45 in T cell activation: resolving paradoxes. *Immunol. Today*, **13**, 477.

274. Koretzky, G. *et al.* (1990). Tyrosine phosphatase CD45 is essential for coupling of T cell antigen receptor to the phosphatidylinositol pathway. *Nature*, **346**, 66.

275. Peyron, J. *et al.* (1991). The CD45 protein tyrosine phosphatase is required for the completion of the activation program leading to lymphokine production in the Jurkat human T cell line. *Int. Immunol.*, **3**, 1357.

276. Kishihara, K. *et al.* (1993). Normal B lymphocyte development but impaired T cell maturation in CD45-Exon6 protein tyrosine phosphatase-deficient mice. *Cell*, **74**, 143.

277. Mustelin, T., Coggeshall, K., and Altman, A. (1989). Rapid activation of the T-cell tyrosine protein kinase pp58[lck]. *Proc. Natl Acad. Sci. USA*, **86**, 6302.

278. Ostergaard, H. *et al.* (1989). Expression of CD45 alters phosphorylation of the lck-encoded tyrosine protein kinase in murine lymphoma T cell lines. *Proc. Natl Acad. Sci. USA*, **86**, 8959.

279. Kanner, S., Deans, J., and Ledbetter, J. (1992). Regulation of CD3-induced phospholipase-C-γ1 tyrosine phosphorylation by CD4 and CD45 receptors. *Immunology*, **75**, 441.

280. Stamenkovic, I. *et al.* (1991). The B lymphocyte adhesion molecule CD22 interacts with leukocyte common antigen CD45RO on T cells and α2–6 sialotransferase. *Cell*, **66**, 1133.

281. Hovis, R. *et al.* (1993). Rescue of signaling by a chimeric protein containing the cytoplasmic domain of CD45. *Science*, **260**, 544.

282. Desai, D. *et al.* (1993). Ligand-mediated negative regulation of a chimeric transmembrane receptor tyrosine phosphatase. *Cell*, **73**, 541.

283. Valentine, M. *et al.* (1991). Interleukin-2 stimulates serine phosphorylation of CD45 in CTLL-2,4 cells. *Eur. J. Immunol.*, **21**, 913.

284. Stover, D. *et al.* (1991). Protein tyrosine phosphatase CD45 is phosphorylated transiently on tyrosine upon activation of Jurkat T cells. *Proc. Natl Acad. Sci. USA*, **88**, 7704.

285. Garcia-Morales, P. *et al.* (1990). Tyrosine phosphorylation in T cells is regulated by phosphatase activity: Studies with phenyl arsine oxide. *Proc. Natl Acad. Sci. USA*, **87**, 9255.

286. Hunter, T. (1989). Protein tyrosine phosphatases: the other side of the coin. *Cell*, **58**, 1013.

4 | Structure and function of MHC class I and class II antigens

DAVID A. JEWELL and IAN A. WILSON

1. Introduction

The major histocompatibility complex (MHC) is a contiguous set of genes that code for polymorphic cell surface receptors involved in antigenic peptide presentation (reviewed in ref. 1). Although originally observed in transplant rejection (2), it is now clear that the biological role of MHC proteins is to initiate cell-mediated immune responses. The co-ordinated activities of MHC proteins act to rid the body of xenobiotic infections (3). In order to combat a wide spectrum of pathogens, the class I and class II antigens have evolved to act as indicators of foreign intrusion at the surface of the host cell. The MHC antigens bind to peptide fragments, derived from the pathogen's proteins, while within the cell, and thereby direct peptides to the cell surface for presentation to immune cells. Antigenic peptide fragments (4, 5) tend to bind with high-affinity to the MHC molecules by virtue of their 'bound' shape and chemical complementary to the MHCs peptide binding groove. The lifetime of these MHC–peptide complexes on the cell surface is an important parameter of MHC restriction. Stable or long-lived peptide–MHC complexes help to increase the probability that the immune system will detect the presence of foreign antigens. T cells are immune cells that have been selected in the thymus to distinguish the host's 'self' peptides from foreign peptides, or 'non-self' antigens. These specialized lymphocytes express MHC-restricted $\alpha\beta$ T cell receptors (TCR) on their cell surface and are able to recognize specific MHC–peptide complexes (reviewed in Chapter 2). Although not absolutely required (6), co-expression of certain accessory molecules, CD8 (for class I) and CD4 (for class II), facilitate the efficient engagement of MHC–peptide complexes by TCR (6–8).

The first crystal structure (9, 10) of an MHC molecule, the class I human leucocyte antigen, HLA-A2, generated considerable excitement as it provided a structural basis for the interpretation of MHC alloreactivity (cytolytic activity with different or allele-specific T cell receptors) (9, 10). Isolation of a soluble HLA-A2 fragment for crystallization required papain cleavage of its natural membrane-bound form. Because this purification procedure did not impose any selection on the bound peptides, it yielded molecular complexes containing a mixture of bound peptides. As a result, only an average image, representing the mixture of bound peptides, could be fitted into the electron density that was present in a large groove on the surface of the class I molecule (9–11). Consequently, the interpretation of this structure was restricted to a description of the more general features of the ligand–MHC class I interaction (10, 11).

An alternative solution to this problem, which ensured the homogeneity of the bound peptide, came from the cloning and expression of the extracellular domains for the class I proteins in *Drosophila melanogaster* cells (12). The *Drosophila* system, which lacks a peptide transport system, simplified both the crystallization and binding studies of mouse MHC class I (murine H2–K^b) peptide complexes by allowing post-purification loading of class I molecules with single peptides (12–17). Other methods, in which single peptides were exchanged into purified MHC antigen proteins, were also successful but proved to be more arduous ways of making individual peptide–MHC class I complexes (18–23). These breakthroughs in methods for obtaining homogenous MHC–peptide complexes have enabled us to obtain the structural and biochemical data that provide a more detailed picture of how class I and class II molecules bind and present peptides and, in particular, how MHC class I molecules can accommodate a variety of peptide sequences in their binding site and yet still bind them with high affinity (13–23). A recent and novel approach in making class II MHC–peptide complexes has been to couple the peptide antigen covalently to the class II molecule so as to stabilize the expressed soluble form of the class II molecule (24). This method will, no doubt, prove a useful approach in guaranteeing *in vitro* homogeneity of bound peptides as well as helping to ensure *in vivo* homogeneity of the MHC antigenic determinant, making interpretation of cytolytic assays more straightforward (25).

Although the general principles governing MHC–peptide recognition have been established, some questions that accompanied the now eight-year-old report of the first class I structure still remain (10): How does the T cell receptor simultaneously engage the MHC molecule and yet discriminate among the different bound peptides? Does alloreactive recognition of MHC class I and class II antigens involve interaction with the bound peptide?

The focus of this chapter is to review the structures of MHC class I molecules and compare these with the recently reported structures for the class II MHC molecule, HLA-DR1 (26, 27). Some of the structural features of these immunologically important molecules will be discussed in the context of how they might help to provide a molecular basis for understanding MHC-restricted cell-mediated immunity. Similarly, much progress is currently being made in understanding T cell recognition of these MHC complexes through the structural studies of TCR, CD4, CD8, and other accessory molecules, such as superantigens (28–32). These studies are expected to add greatly to our understanding of the molecular events that bring about cell-mediated immunity.

2. Gene arrangement of MHC

In many ways, the study of the approximately 4 Mb long region of DNA making up the MHC has contributed general insights into gene function, DNA polymorphism, linkage, and recombination (for reviews see refs 1 and 33). Three unique loci of the MHC have been identified as encoding three groups of proteins: those involved in antigen processing and transport (proteasome complexes and peptide transporters; reviewed in Chapter 5), in antigen presentation on the cell surface (for example MHC class I and class II antigens; this chapter) and in protein stabilization and complement activation (class III region encodes heat-shock proteins and components of the complement system, for example C2, Bf, C4A and C4B; see Chapter 8. A linkage map showing the relationship of the human class I, class II, and the class III regions is outlined in Fig. 1.

The three independent segments of the class I loci, occupying different chromosomal positions, are designated *A*, *B*, and *C* in human and *K*, *D*, and *L* for mouse. These highly polymorphic genes encode the 'classical' class I antigens, HLA in human and H2 in mouse, and are co-dominantly expressed in almost every nucleated cell of vertebrates. Unlike class I proteins, class II (human HLA-DR, -DQ, and -DP) molecules are found primarily on the surface of B cells in complex with peptide antigens derived from proteins that are predominantly extracellular in origin (3). Found within the class II gene segments are the coding regions for peptide transporters (of the ABC transporter superfamily), TAP-1 and TAP-2 (34, 35), and subunits of the proteasome complex, LMP2 and LMP7 in mouse (36, 37) and RING10 and RING12 in humans (38, 39). It has been hypothesized that these MHC-linked components can be activated and controlled in concert so as to increase antigen processing and display during the course of viral or microbial infections (34).

3. Antigen presentation

The discovery that MHC class I molecules can restrict the peptide sequences displayed on the antigen presenting cell (APC) was made in observing allele-

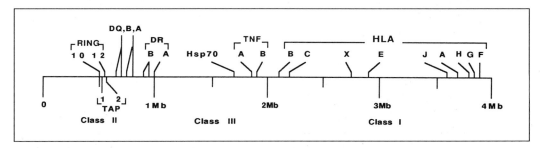

Fig. 1 The genetic map for the human MHC segments found on the short arm of chromosome 6. The centromeric end is on the left and the telomeric end is to the right. RING, proteasome proteins; TAP, peptide transporters; DQ (-B, -A) and DR, class II antigens; Hsp 70, heat-shock proteins; TNF, tumour necrosis factor; HLA, class I antigens.

specific presentation of virus-derived peptides (4, 40). Subsequent extraction of peptides from purified MHC molecules further allowed the identification of a range of 'self-peptide' sequences that were also determined to have allele-specific motifs (41; for review see ref. 5). These experiments helped to establish that the primary sources for many of the peptides bound to MHC class I molecules are proteins found in the cytoplasm and the cell nucleus. The putative pathways used by MHC class I and MHC class II antigens in presenting peptide on the surface of the cell are represented in Fig. 2.

In the course of viral infection, it is hypothesized that virus-derived peptides can become preferentially bound over 'self' peptides when they are at sufficiently high concentration and show suitable complementarity to a particular MHC antigen's peptide binding groove. Some autoimmune diseases are thought to occur as a result of 'self' peptides being mistaken as foreign antigens by activated T cells. These cytotoxic cells are believed to have been reactivated to kill uninfected cells by pathogen-derived peptide homologues of self-peptides (42–44).

For class I, peptides in the size range of octameric to nonameric lengths are most common. For each allele, at least two positions in the peptide are thought to serve as anchor residues (Table 1). These positions correspond (in a nonamer) to the C-terminal residue and usually the second and/or sixth residue of the peptide, with the second position predominating as the other anchor position. An additional requirement appears to be that the anchor positions are separated by two or more residues. The C-terminal residues are restricted to either non-polar or, in HLA, to large positively-charged amino acids.

The preference for anchor amino acids to be hydrophobic and/or have large side chains suggests that they are used in providing a substantial portion of the peptide binding energy. Like some proteases (for example trypsin), full thermodynamic advantage is taken by the MHC molecules in recognizing large basic residues (Arg or Lys, for example P2 of HLA-B27 and P9 of Aw68) by making both intimate van der Waals contacts with the aliphatic portion of the side chain and hydrogen bonds with the buried charge group (45). Interestingly, although Trp could also potentially fulfil some of these requirements, it has yet to be identified as a C-terminal, or other anchor, residue in any of the 'classical' class I or class II allele-specific peptides. A consequence of identifying allele-specific motifs is that they

---

Fig. 2 (A) The proposed pathway for class I presentation of peptide fragments by antigen-presenting cells (APC). The antigenic peptides are produced as cleavage products of the proteasome complex (LMP2 and LMP7). These fragments are then transported into the endoplasmic reticulum (ER) by the transport molecules, TAP1 and TAP2. Expressed heavy chains of the class I molecule are loaded with peptides while associating with the light chain, β_2-microglobulin (β_2m), and the chaperone protein, calnexin, in order to form the stable peptide-MHC class I complexes. Once on the cell surface, the complex may be engaged by the T cell receptor and CD8 from cytotoxic T cells. Binding of the APC through the MHC–peptide–TCR complex initiates a cell-killing signal in the T cell which specifically targets the APC for cell death. (B) The proposed pathway for class II presentation of endocytosed protein fragments. Exogenous proteins are brought into the APC by way of endocytic routes. The low pH of the endosome is thought to be crucial for peptide loading of the class II molecule which can dissociate from its invariant chain and refold under these conditions (59). Nuc, nucleus; ER, endoplasmic reticulum; TGN, *trans* Golgi network; Inv chain, invariant chain.

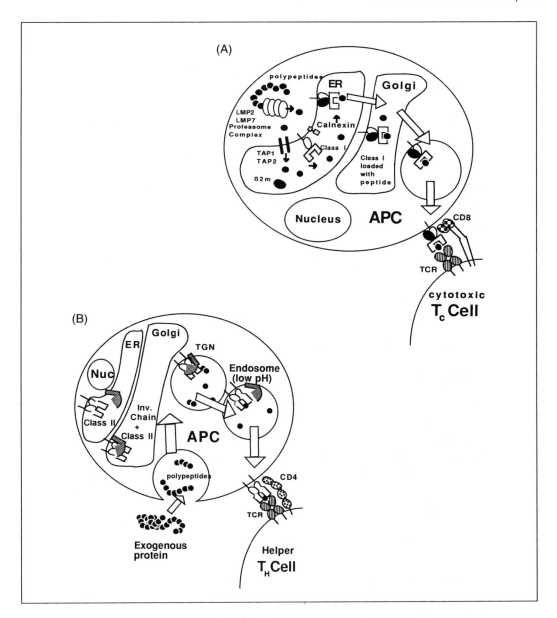

may be used, with relatively good success, in predicting potential MHC-restricted peptide ligands contained within a given protein's primary structure. This has, in turn, permitted the identification of potential T cell epitopes prior to their actual isolation from MHC–peptide complexes (5, 46, 47), although their actual presentation requires both correct processing and peptide transport.

Quantitation of the peptide affinities for MHC class I molecules (48) has only

Table 1 Peptide sequence motifs for MHC class I antigens

	Allele	Consensus[a]	C-terminal amino acid
Mouse	H2–K^d	1Y3456789 / F	I, L
	H2–D^b	1234N6789	M, I, L
	H2–K^b	1234Y678 / F	L, M, I, V
	H2–K^k	1E345678	I
	H2–L^d	1P3456789	L, M, F
	H2–D^d	1GP456789 / L / M	L, F, M, I
Human	HLA–A*0201	1L3456789	V, L
	HLA–A*0205	123456789	L
	HLA–A11	123456789, 10	K, R
	HLA–A31	123456789,10	R, K
	HLA–B*2705	1R3456789	R, Y, L, F, M
	HLA–B7	1P3456789	L
	HLA–B37	1D34V6789 / E M / I	L, F, I, M

[a] Non-bold numbers refer to residues where there is no sequence preference. Bold numbers indicate preference for hydrophobic residues and letters are the so-called anchor residues. The decamer sequences have been determined to contain the amino acids noted for C-terminal residues at either the 9th or 10th positions. (Table adapted from ref. 5.)

recently become possible with the ability to obtain single peptide–MHC complexes. Questions answered from these studies include the extent to which side chain and backbone atoms can contribute to peptide binding and stabilization of the liganded complex. The thermal stability of H2–K^b in complex with an alanine-substituted ovalbumin-derived octamer peptide (OVA-8; SIINFEKL) showed a strong correlation between alanine substitutions at 'anchor' positions and the extent of stabilization (see Fig. 2A of ref. 48). This same study also indicated a correlation between the calculated buried surface area of the substituted peptide ligand and its measured K_D (48). It remains to be seen how useful, or general, these relationships will be in helping to predict the relative affinities of related peptides.

4. Structure of MHC class I antigens

The class I proteins are heterodimers of a heavy (or α) chain (about 44 kDa) and a non MHC-encoded light chain, known as β_2-microglobulin (β_2m, about 12 kDa; see Fig. 3). The α chain is about 325 residues long, which includes a short membrane spanning region and cytoplasmic tail of approximately 55 amino acids. As might be expected from the conservation of amino acid sequence, the overall fold of the class I molecules, as well as the approximate orientation of β_2m subunit

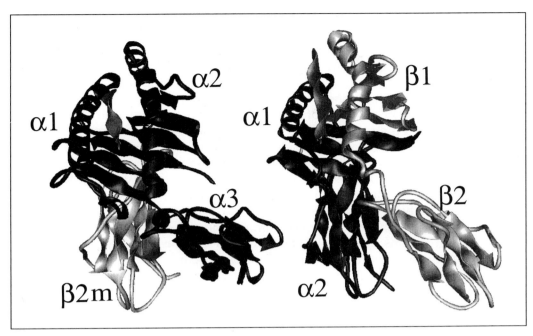

Fig. 3 A ribbon diagram showing the overall folds of MHC class I (left) and class II structures (right). The heavy or α chain is shown in black and the associated β₂m protein is shown in grey. The peptide-binding groove is viewed end on with the amino terminal binding region in the foreground. This orientation is viewed as if perpendicular to the extracellular surface of the APC.

relative to the α chain, is conserved between human and mouse molecules (9, 15, 20). The heavy chain consists of three domains, two similarly folded domains, α_1 and α_2, and an α_3 domain. The α_3 domain and the β_2m subunit have immunoglobulin-like folds that are thought to lie proximal to the cell membrane. A comparison of the overall fold of the HLA α_3 and β_3m domains show that they can be overlapped, with an average root–mean–square-difference (rmsd) of only 1.4 Å for backbone atoms, which is similar to the rmsd values calculated for similar comparisons made among (and with) the constant regions of immunoglobulins (0.6–1.1 Å, and 1.6 Å, respectively (9)). Also shared with immunoglobulin domains are conserved disulfide linkages made between Cys203 and Cys259 of the α_3 domain and between Cys25 and Cys80 of β_2m. Otherwise, the relative domain packing interactions between α_3 and β_2m is unlike the equivalent association in the constant domains of antibodies (10).

The α_1 and α_2 domains are quite novel and are not related to the immunoglobulin fold. They are closely associated and form a superdomain that sits on top of the α_3 and β_2m domains. The combined cross-sectional dimension of the α_1 and α_2 domains measures 50 by 40 Å and contains an approximately 25 Å long groove that corresponds to the peptide binding site and which is parallel to the longest dimension. The α_1 and α_2 domains each consist of four antiparallel β-strands straddled by a C-terminal α-helical region. The relative positions of the α_1 and α_2 domains about the peptide binding cleft can be described as pseudo two-fold symmetric with the rotation axis aligned perpendicular to the β-sheet and extending out between the helices. This α_1–α_2 association results in an extended eight-stranded antiparallel β-sheet traversed by two long α-helical segments that run antiparallel to each other and perpendicular with the strands of the β-sheet (Fig. 4).

The first 48 residues of the α chain form the first four strands of the large antiparallel β-sheet. This fold of the backbone exposes three loops (residues 13–20, 38–45, 86–93) on one side of the binding groove to solvent and possible interactions with accessory molecules. The helical region of α_1 is initiated with a 3_{10}-helix at residue 49 that terminates after a stretch of only four amino acids. Following an almost 90° bend, the chain then forms a long continuous stretch of regular α-helix (residues 58–87). The polypeptide chain remains highly exposed to solvent in forming a large loop (amino acids 86–92) before twisting under the two helical regions of the α_1–α_2 domains to form the fifth β-strand of the sheet. The resulting first-to-fifth β-strand interaction is part of the α_1–α_2 domain interface that forms the central floor of the peptide binding groove (see Fig. 4). The positioning of this domain interface in the ligand binding site is of some biological significance and may result from the association of these domains through gene duplication.

The heavy chain next winds around to form the final three strands of the

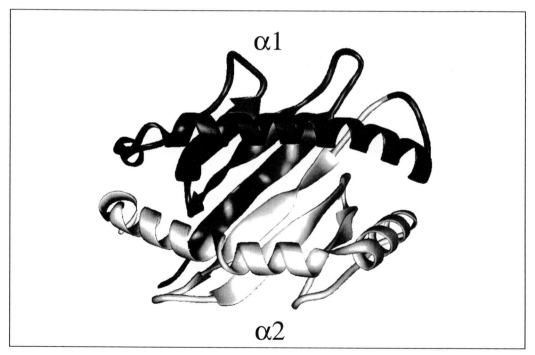

Fig. 4 A ribbon diagram showing the α_1 and α_2 domains of MHC class I in black and grey, respectively. The domain interface forms the base of the peptide binding groove.

antiparallel β-sheet and another two solvent exposed loops. This last set of β-strands in the α_2 domain contain key residues for binding the C-terminus and penultimate residue of the peptide ligand (residues 95, 97, 114, 116, 118, 123, 124). The three α_1 loops, extending out from under the α_1 domain helix, form a set of more concave surfaces than the indentations formed between the α_2 loops (residues 104–108 and 127–132). As a result, the depth and total surface area of the grooves and/or pockets formed on the side of the α_1 domain are greater than for those formed between the α_2 domain loops. Any potential binding sites for accessory molecules (or for other intermolecular interactions such as with superantigens for class II, refs 28, 49, 50) might, therefore, be expected to be greater and lesser on the sides of the α_1 and α_2 domains, respectively (Fig. 4).

An additional loop (residues 118–120) extends down from under the C-terminal region of the peptide-binding site. This loop helps to form a large channel that is filled with the side chains extending from a β-strand and a loop of the β_2m protein (amino acids 54–60; see Fig. 5). After completing the eight-stranded antiparallel β-sheet, the main chain rises to form a set of three contiguous α-helices of the α_2 domain. This helical stretch is broken into three distinct α-helices by two small bends created by two glycine residues at positions 151 and 162. The polypeptide chain, after leaving the α_2 domain, makes a large arch that is structurally analogous to the elbow region of Fabs before entering the Ig-like fold of the α_3 domain (9, 15).

The α_3 domain and the associated β_2m protein are thought to sit proximal to the cell membrane so as to hold the α_1 and α_2 domains away (about 60 Å) from the

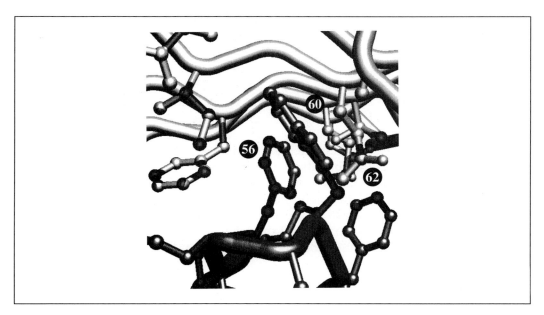

Fig. 5 Representation of the β_2m (black) interface with the α_1 and α_2 (grey) domains in a tube, ball, and stick representation. The region of contact shown sits directly under the first and fifth strands of the β-sheet designating the α_1 and α_2 domain interface.

cell membrane (see Fig. 3). This orientation is of functional significance since it allows for interaction of accessory factors (such as CD4 or CD8) with the α_3 domain (or the β_2m protein) while maximally exposing the antigenic peptide for TCR engagement (10, 11, 15, 30–32). The α_1 and α_2 superdomain, and the paired association of the α_3 domain with the β_2m, may then best be described as forming a 'head' and 'stalk' structure, respectively, off the extracellular face of the cell membrane, with the head group displaying the bound peptide for TCR recognition. Cell surface expression, as well as β_2m association, does not appear to require the conserved N-linked oligosaccharide attached to Asn 86 (9, 10) However, the symmetric nature of α_1 and α_2 domains is again emphasized by the apparent conservation of the structurally homologous glycosylation site at Asn 176. Also conserved between human and mouse is a third glycosylation site on Asn 256 in the α_3 domain. None of the covalently attached carbohydrate moieties have been seen clearly for the class I structures, presumably due to a combination of sequence and structural heterogeneity (9, 11, 15, 20, 51).

The β_2m makes several stabilizing contacts with the underside of the α_1 and α_2 domains. It does this by filling a hydrophobic groove, beneath the eight-stranded β-sheet of the two upper α_1–α_2 domains, with aliphatic side chains from positions 53, 54, 56, 60, and 62 (see Fig. 5). The direction of the β_2m backbone is nearly perpendicular to the direction of the α_1 and α_2 domain β-strands. Such a cross-stitching of side-chains from the crossing β-strand into the complementing groove might be expected to provide considerable stabilizing intermolecular interactions. The aromatic sidechains, Phe 56 and Phe 62, of β_2m lie at this α_1–α_2 domain interface, further implying the importance of the association of β_2m with the heavy chain for the integrity of the α_1–α_2 domain interface. Interestingly, although there is sequence conservation between human and mouse class I structures for the amino acids forming the interface between β_2m and α_1–α_2, the β_2m molecules of the mouse and human have relatively large shifts in the position and conformation of the backbone around residues 53, 54, 60, and 62. The structures do, however, maintain the side-chain interactions with the α_1–α_2 interface. The association of the β_2m with α_3 is also slightly variable when the human and mouse structures are compared (15). It is interesting to note that human β_2m is able to stabilize the murine H2–K^b better than its mouse counterpart, both in the absence and presence of bound peptide (12, 14) As part of β_2m's putative role in holding the α_1 and α_2 domains away from the membrane, β_2m forms several important contacts with the α_3 domain. In contrast to the contacts made with the underside of α_1 and α_2 domains, the interactions that β_2m makes with α_3 are a combination of main chain and side-chain hydrogen bonds as well as van der Waals contacts.

5. Peptide–MHC class I interactions

At present, nine structures of class I molecules with single peptides bound have been published or are in press. The class I molecules and peptide sequences bound in these complexes are compiled in Table 2.

Table 2 Single peptide–MHC class I structures

Class I	Peptide	Peptide source	Resolution (Å)	Ref.
HLA-Aw68	KRGGPIYKR	Influenza virus NP (91–99)	2.8	19
HLA-A2	ILKEPVHGV	HIV-1 reverse transcriptase (309–317)	2.5	23
	LLFGYPVYV	HTLV-1 Tax (11–19)	1.9	23
	GILGFVFTL	Influenza A matrix M1(58–66)	2.5	23
	TLTSCNTSV	HIV-1 gp120 (197–205)	2.6	23
	FLPSDFFPSV	Hepatitis B NP (18–27)	2.5	23
H2–K^b	RGYVYQGL	Vesicular stomatitis virus NP (52–59)	2.3	15, 20
	FAPGNYPAL	Sendai virus NP (324–332)	2.5	15
	SIINFEKL	Ovalbumin (257–264)	2.5	16
H2–D^b	ASNENMETM	Influenza virus NP (366–374)	2.4	21

When the crystal structures of the human and mouse class I antigens are overlapped, their bound peptides occupy the same relative positions in their respective binding sites. As illustrated by the peptide overlap shown at the bottom of Fig. 6, the ends of the peptides are highly restricted in conformation, while the structural organization in the middle appears much more sequence dependent (15–21, 23, 51, 52). For the mouse MHC class I H2–K^b antigen complexes (15, 16) (Table 2), the peptide backbone atoms have an rmsd of only 0.25 Å for the first three amino acid residues. Similarly, the five human HLA-A2 class I complexes show an rmsd of 0.22 Å for the first three residue positions. This overlap in main chain conformation diverges beyond the third, and over the next three to five amino acid residues, to give an average rmsd of 0.40 Å for the three peptides bound to H2–K^b and 1.22 Å for the five peptides bound to HLA-A2 structures. Towards the C-terminus, the backbone again becomes more constrained, giving an average rmsd of 0.11 Å and 0.22 Å for the last two residues of the mouse and human, respectively. In general, the amino acids bulge out from between what are the P3 and the P6 positions of the 9mer peptide in H2–K^b (15, 17), or between the P2 and the P8 positions for the HLA-B27, HLA-Aw68, and HLA-A2 complexes (18, 19, 23, 51, 52). This method for accommodating additional amino acids can be easily understood on the basis of the occurrence of the anchor residues (P2 in HLA-Aw68, B-27, and HLA-A2; P6 in H2–K^b (nonamer)) and the constraints imposed by the shape of the binding cleft. This is dramatically illustrated in the H2–D^b class I protein in complex with an influenza virus peptide (21). In this structure, the bulge

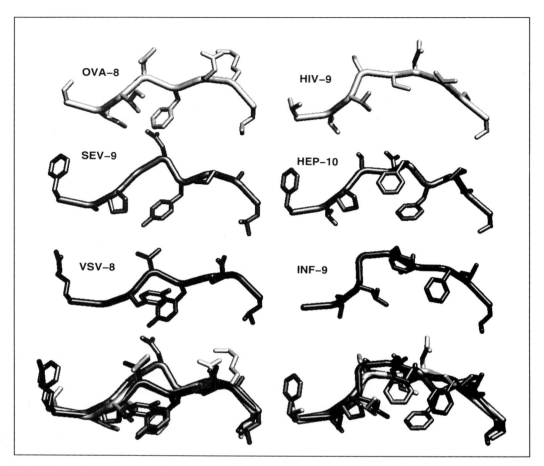

Fig. 6 Comparison of the peptide conformations for bound peptides of the class I structures of H2–K^b (left side) and HLA-A2 (right side) and an overlap of their structures (bottom). For the H2–K^b bound peptides, the 'anchor' residues that lie in pockets C and F are either Tyr (or Phe) and Leu, respectively. At the bottom is an overlap of all the peptides, illustrating that the conformation of ends of the peptide are sequence independent while the conformation of the peptides toward the centre is sequence dependent. The overlap of all three peptides at the bottom left of the figure also shows the bulge in the centre of the 9mer that is formed in the SEV-9 complex allowing for the accommodation of the additional amino acid. As in the H2–K^b structure, the ends of the peptides for the HLA-A2 complexes show a general conservation of 'anchor' residues. For both the human and mouse structures, the ends of the peptides are conformationally restricted. The backbones of the peptides show the greatest degree of overlap at the termini with greatest variability in the middle. The amino acid sequences for the peptides shown are as follows: OVA-8, SIINFEKL; SEV-9, FAPGNYPAL; VSV-8, RGYVYQGL; HIV-9, TLTSCNTSV; HEP-10, FLPSDFFPSV; INF-9, GILGFVFTL.

in the peptide is more prominent because of a high ridge in the C-terminal half of the peptide binding groove.

Due to their chemical nature, some residues will have a thermodynamic preference to become buried in the available pockets rather than interact with solvent. The pocket positions along the length of the cleft define the available shape complementarity that somewhat restricts the sequences of bound peptide. The chemical nature and size of the side chains making up the major (deep) pockets dictate which residues will serve as the principal contributors to ligand binding energy (i.e. anchors). For pockets that do not provide a high degree of shape complementarity for any one residue, the orientation of the side chain may be relatively flexible, or be somewhat restricted due to the interposal of water molecules as bridges between the peptide and the MHC molecule (15). As a result, peptides containing anchor type residues, but which are slightly out of register with the other pockets found within the groove, can still bind but allow deviations in the backbone geometry, to include bulges, so as to permit the anchor residues to occupy the pockets (15, 16, 17, 23, 51, 52; compare the SEV-9 peptide with the OVA-8 and VSV-8 peptides, Fig. 6). The effect is that the pockets, or any ridges in the cleft that prevent burying portions of the peptide, dictate the majority of the structural features of the peptide ligand that are left exposed for solvent or TCR interaction. In this somewhat indirect way then, the pockets found in the class I antigens define how much of the peptide will be accessible for recognition and specific binding of the TCR. Hence, these structural and chemical traits of the pockets distributed in the peptide binding cleft equip the class I with the binding determinants that act to restrict peptide presentation (17, 23).

It is also clear, from a qualitative analysis of all the class I antigen structures, that the imperfect complementarity of the binding groove for individual peptides is often compensated for by bound water molecules (15, 23, 51, 52). The buried waters are seen to add additional packing and hydrogen bond interactions between the peptide and receptor. By using water in this way, the class I antigens can bind a large number of peptide sequences in providing sufficient shape complementarity and binding energy. Water molecules also seem to play a more general and conserved role in maintaining relatively specific interactions with the ends of the bound peptides (15, 51, 52). In all the high-resolution structures reported, a conserved water molecule is found to participate in a hydrogen bonding network around the amino terminus of the peptide. A single water molecule helps to stabilize hydrogen bonds made between the amino terminus and several tyrosines (Tyr7, Tyr59, and Tyr171) by hydrogen bonding between the hydroxyl groups of two of the tyrosines (Tyr7, Tyr59) and the carboxyl group of Asp63 (see Fig. 7). It will be interesting to see how N-terminal modifications to peptides, such as the formylated bacterial and mitochondrial peptides, are accommodated in class I molecules. The amino acid change of tyrosine 171 to phenylalanine, found in the H2-M3a antigen, could allow for a different set of interactions with the N-formyl group; presumably, changes are needed to accommodate the additional atoms of the modified peptide (53–55).

In a similar way, the carboxyl group on the C-terminal end of the peptide makes conserved hydrogen bonds through a single water molecule and several conserved side chains of the class I molecule (see Fig. 7). Because this water molecule is more exposed to solvent than the water found near the amino terminus of the peptide, it is more likely to exchange with solvent and, therefore, would be expected to

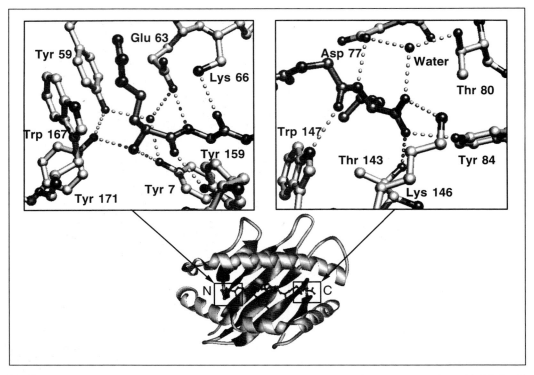

Fig. 7 Hydrogen bonding patterns for the amino and carboxyl terminus of the peptides of the H2-K^b(VSV-8) peptide complex. A conserved water molecule is found at both the N-terminus and the C-terminus of the class I peptide complexes. At the N-terminus (left panel) the water molecule indirectly participates in peptide binding through a pentacyclic hydrogen bonding system. At the C-terminus (right panel) the conserved water molecule is tetrahedrally co-ordinated by hydrogen bonds, while making direct hydrogen bonds with the peptide and the class I molecule.

contribute less to the overall binding of ligand. It will be interesting to see whether the surface exposure of this water molecule has any functional significance in TCR recognition of the class I antigen. There are two possible fates for this and other solvent-exposed water molecules found around the bound peptide: they may be displaced by intermolecular interactions made on binding TCR, or it might be that they are retained and become an integral part of the interface when the class I antigen is engaged by the TCR. In either scenario, the interactions made with the carboxyl group of the C-terminus might be expected to be a common feature in TCR binding.

A more significant consequence of the different peptide sequences that can be bound by the MHC molecule is the change of the molecular surface exposed for TCR interaction. The exposed side chains of each peptide will differ from one peptide sequence to another, as will the peptide interactions with the MHC molecule (see Fig. 6 and Fig. 10). Effective antigenic discrimination may be determined then by how well these surface features can be differentiated by individual T cell

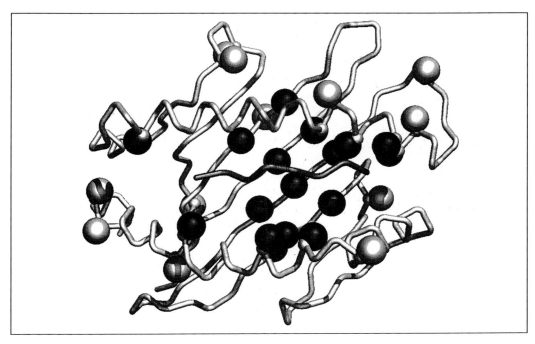

Fig. 8 The polymorphic positions found in the protein sequences of human and mouse class I. The positions that are found to play a role in peptide binding are represented by black spheres while the positions that have been implicated in either TCR, or TCR and peptide binding, are shown as grey spheres. It can be seen from the diagram that those mutations that give rise to the allele-specific ligand binding lie almost exclusively in the peptide binding groove, while those effecting TCR interactions are found on the surface of the complex.

receptors or, more subtly, in the dynamics of actually forming the MHC–TCR ternary complex.

A characteristic feature of the MHC class I proteins is that they show an extra-ordinarily high degree of amino acid diversity or polymorphism (56, 57). For example, there are 108 amino acid differences between H2-K^b and HLA-A2 (17). Of these, 52 residues differ in the α_1–α_2 domains (approximately 72% identity), 26 in the α_3 domain (approximately 72%), and 30 in the β_2-microglobulin (approximately 70%). Nevertheless, when the structures are overlapped, the two α_1–α_2 domains are nearly indistinguishable and the two structures have an average rmsd of only 0.66 Å.

Side chains for 13 of the 52 residues that differ in the α_1–α_2 domains point directly into the peptide binding site (refer to Fig. 8). Seven of these interact directly with the peptide, either through van der Waals or hydrogen bonds (17). These different residues determine the shape and chemical nature (but not the length) of the peptide binding groove. At the ends of the peptide binding groove are several conserved large, aromatic, and conserved amino acid side chains (Trp167 for the N-terminus of the groove and Tyr84, Tyr85, and Tyr123 at the C-terminus). These residues cap the ends of the binding groove and restrict the groove's length to about 25 Å. The polymorphic positions in the groove and the amino acids occupying them account for the differences in the allele-specific restriction of peptide sequences. Each allele then provides a unique set of binding constraints for ligands, which is observed in the consensus sequences derived from peptides eluted from individual MHC complexes (5, 41).

A comparison of the two H2–K^b single peptide structures and a later, but more extensive, analysis of five HLA-A2 single peptide structures showed that different peptides do not substantially change the shape of the pockets from one complex to another (15, 23). There are, however, slight alterations in the side chain conformations for residues lining or found at the edge of the peptide binding groove. Often accompanying these side chain movements are smaller conformational changes in the backbone of the class I molecule (15, 23). For H2–K^b, the side chains of residues 152 and 155 have different conformations in the VSV and SEV complexes, and the first α-helix of the α$_2$ domain appears to be shifted slightly (about 0.6 Å). In HLA-A2, similar changes arise for Arg97, Tyr116, and Trp167 and in the backbone between residues 144 and 151 in five different peptide complexes (23). These peptide-induced conformational changes in the MHC protein may allow for more structural information about the bound peptide to be imparted to the TCR (15). Structural differences in the MHC molecule have also been proposed to explain differences in antibody reactivity when different peptides are bound to the same MHC-molecule (17, 58, 59).

The amino acids that line the pockets of the peptide binding cleft determine the peptide sequence motifs selected for presentation on the cell surface. The finding that patterns of polymorphism at the *HLA-A, -B,* and *-C* loci are different suggest there is functional diversification between products of these loci that may have evolved to bind structurally distinct groups of peptides for specific T cell receptors (57).

6. Class I vs class II structures

The overall fold of the class I MHC antigens is quite well preserved in class II (Fig. 3 and Fig. 9). Unlike the class I structure, both chains that form the heterodimer, α and β, are encoded on the class II region of the MHC. The α and β chains are of equivalent size and share the same pseudosymmetric relationship and the 'head' and 'stalk' architecture described above for class I (26). The edge-to-edge relationship between the domains α$_1$ and α$_2$ of class I is part of the heterodimeric α–β interface for the class II molecule. Also, like the α$_1$–α$_2$ domain arrangement seen in class I molecules, the heterodimer α–β interface lies directly below the peptide binding site and may be of functional significance. Experimental evidence suggests that it would allow a structure-based mechanism for the modulation of the heterodimer formation which is dependent on peptide binding (60).

With respect to peptide binding, the most notable differences between class I and class II molecules seems to be at the ends of the peptide binding groove. Unlike the class I antigens, class II molecules have relatively open or extended

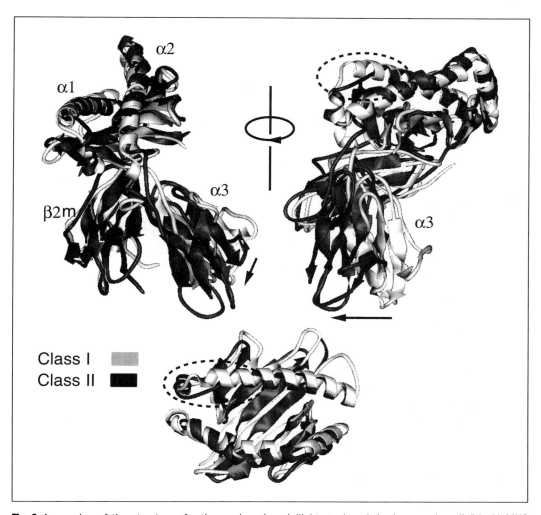

Fig. 9 An overlap of the structures for the murine class I (light grey) and the human class II (black) MHC molecules. The domains of class I are indicated; α_1, α_2, α_3, and β_2m. The ribbon diagram in the upper left corner shows the overlap to be remarkably close. The direction of the small displacement of the Ig-like domain shift for the β-strand of class II relative to the Ig-like domain position of the α_3 domain of class I is denoted by the straight arrows. A rotation about the vertical axis of these structures by 90° produces the structure in the upper right corner of the figure. Here the domain shift from the α_3 of class I in the Ig-like portion of class II is more easily seen. Also shown, designated by the dotted circle, is the location of an apparent deletion of the first helical portion of the α_1 domain in class II. This is more easily seen when the structures are viewed from the top of the peptide binding site (bottom structure).

ends to the binding cleft (26). As a result, a peptide having a length that would be expected to bulge out in the middle when bound to class I antigens is seen to form an extended backbone conformation that extends off the ends of the binding grooves of the class II antigen (26, 27). Also, unlike class I, many of the side chains seen for the HLA-DR1 influenza peptide complex occupy pockets of the class II

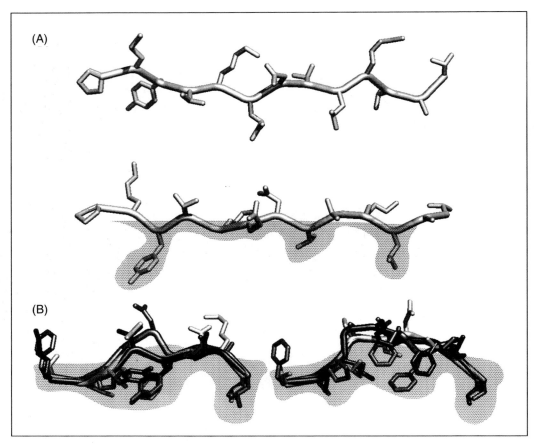

(A)

(B)

Fig. 10 (A) Two views of the 13mer peptide found in the HLA-DR1 class II complex. The top orientation is as if looking down into the peptide binding site. The bottom orientation is viewed from the side. The buried surface area is approximated and indicated by the shaded overlap. (B) Side views of peptides bound to H2-K^b (on the left) and HLA-A2 (on the right). The grey mask indicates those common portions of the peptides that are buried in the class I complexes.

peptide binding site and yet remain partially exposed for TCR interaction. This is a result of most of the pockets in the peptide binding groove extending on either side of the ligand's peptide backbone and not underneath, as in the class I peptide complexes. A likely consequence would be that the TCR might have more opportunity to read the peptide sequence and hence have more extensive peptide sequence preferences in class II than compared to class I restricted TCRs (see Fig. 10 and Fig. 11). It may be too early to generalize, but the class II peptide complex also appears unlike the class I structures in that, at least for the single structure of HLA-DR1 (27), it does not use water as extensively to bridge between the peptide and the MHC molecule and hence fill in the regions of poor complementarity. The HLA-DR1 does however bind peptide through various specific

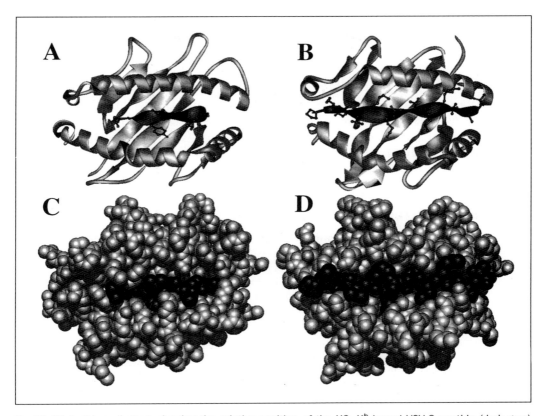

Fig. 11 (A) A ribbon diagram showing the relative position of the H2–K^b bound VSV-8 peptide (dark grey) between α_1 and α_2 domains (light grey). For the VSV-8 peptide, the peptide backbone conformation is found to twist up then back down. Unlike the OVA-8 peptide bound to class I H2–K^b and the influenza peptide bound in class II HLA-DR1, the twist in the backbone does not make a full revolution. (B) A ribbon representation of the class II HLA-DR1 showing the relative position of the peptide (dark grey) in the binding groove (light grey). The peptide binding site is made by both the α and β chains. The 13mer is shown covering the entire length of the top surface at the head of the class II molecule and extending out of the cleft. The backbone conformation of the bound peptide makes three complete twists. (C) A space-filling model of the H2-K^b-VSV-8 peptide complex with the MHC shown in grey and the bound peptide coloured dark grey. The peptide surface is shown to be highly buried (~80 per cent) and makes up only a small fraction of the total exposed surface on the top or distal end of the peptide–MHC class I complex. (D) A space-filling model of the HLA-DR1–influenza peptide complex with the peptide coloured dark grey and the MHC coloured grey. Compared to the class I structures, a large portion of the peptide is exposed for TCR interaction. The increase in peptide exposure comes from a longer peptide that is less buried in the binding site.

MHC Asn side-chain interactions with the peptide backbone as well as a short parallel β-structure and a conserved Trp interaction analogous to Trp147 of class I (27, 61).

In summary, it has been possible to propose that universal recognition codes can explain the interaction of peptides with both class I (15, 17, 20, 51, 52) and class II (26, 27, 61). The next stage is to understand the recognition of these MHC–peptide complexes by the TCR. Such studies are ongoing in several laboratories at present and will no doubt yield new insights and present new questions with respect to how their structures are related to their observed biological activities.

Acknowledgements

We are very grateful to Drs Lawrence Stern and Don Wiley for the co-ordinates of HLA-DR1 which were used to produce Figs 3, 9, 10, 11B, and 11D and Daved H. Fremont for the co-ordinates of the H2-K^b (OVA-8) complex used in Fig. 6. The authors have been supported by NIH grant CA-58896 (IAW) and NIH training grant AI-07244 (DAJ). DAJ is currently an American Cancer Society Fellow. This review is TSRI publication #8733-MB.

References

References marked * recommended for further reading.

1. Guillemot, F., Auffray, C., Orr, H. T., and Strominger, J. L. (1990). MHC antigen genes. In *Molecular immunology* (ed. B. D. Hames and D. M. Glover), p. 81. IRL Press, Oxford.
2. Gorer, P. A. (1937). The genetic and antigenic basis of tumour transplantation. *J. Pathol. Bacteriol.*, **44**, 691.
3. Snell, G. D., Daussett, J., and Nathenson, S. G. (1976). In *Histocompatibility*, (1st edn), p. 57. Academic Press, New York.
4. Van Bleek, G. M. and Nathenson, S. G. (1990). Isolation of an endogenously processed immunodominant viral peptide from class I H−2K^b molecule. *Nature*, **329**, 512.
*5. Falk, K. and Rotzschke, O. (1993). Consensus motifs and peptide ligands of MHC class I molecules. *Seminars in Immunology*, **5**, 81.
6. Killeen, N. and Littman, D. R. (1993). Helper T cell development in the absence of CD4-p56^{lck} association. *Nature*, **364**, 729.
7. Gabert, J., Langlet, C., Zamoyska, R., Parnes, J. R., Schmitt-Verhulst, A.-M., and Malissen, B. (1987). Reconsition of MHC class I specificity by transfer of the T cell receptor and Lyt-2 genes. *Cell*, **50**, 545.
8. Gay, D., Maddon, P., Sekaly, R., Talle, M. A., Godfrey, M., Long, E., Goldstein, G., Chess, L., Axel, R., Kappler, J., and Marrack, P. (1987). Functional interaction between human T cell protein CD4 and the major histocompatibility complex HLA-DR antigen. *Nature*, **328**, 62.
*9. Bjorkman, P. J., Saper, M. A., Samraoui, B., Bennett, W. S., Strominger, J. L., and Wiley, D. C. (1987*a*). Structure of the human class I histocompatibility antigen, HLA-A2. *Nature*, **329**, 506.
*10. Bjorkman, P. J., Saper, M. A., Samraoui, B., Bennett, W. S., Strominger, J. L., and Wiley, D. C. (1987*b*). The foreign antigen-binding site and T cell recognition regions of class I histocompatibility antigens. *Nature*, **329**, 512.
11. Saper, M. L., Bjorkman, P. J., and Wiley, D. C. (1991). Refined structure of the human histocompatibility antigen HLA-A2 at 2.6 Å resolution. *J. Mol. Biol.*, **219**, 277.
12. Jackson, M. R., Song, E. S., Yang, Y., and Peterson, P. A. (1992). Empty and peptide-containing conformers of class I major histocompatibility complex molecules expressed in *Drosophila melanogaster* cells. *Proc. Natl Acad. Sci. USA*, **89**, 12117.
13. Stura, E. A., Matsumura, M., Fremont, D. H., Saito, Y., Peterson, P. A., and Wilson,

I. A. (1992). Crystallization of multiple murine major histocompatibility complex-class-I H–2K^b with single peptides. *J. Mol. Biol.*, **228**, 975.

14. Matsumura, M., Saito, Y., Jackson, M. R., Song, E. S., and Peterson, P. A. (1992). *In vitro* peptide binding to soluble empty class I major histocompatibility complex molecules isolated from transfected *Drosophila melanogaster* cells. *J. Biol. Chem.*, **267**, 23589.

* 15. Fremont, D. H., Matsumura, M., Stura, E. A., Peterson, P. A., and Wilson, I. A. (1992). Crystal structure of two viral peptides in complex with murine MHC class I H–2K^b. *Science*, **257**, 919.

16. Fremont, D. H., Stura, E. A., Matsumura, M., Peterson, P. A., and Wilson, I. A. (1995). Crystal structure of an H2–k^b-ovalbumin peptide complex reveals the interplay of primary and secondary anchor positions in the major histocompatibility complex binding groove. *Proc. Natl Acad. Sci. USA*, **92**, 2479.

* 17. Matsumura, M., Fremont, D. H., Peterson, P. A., and Wilson, I. A. (1992). Emerging principles for the recognition of peptide antigens by MHC class I molecules. *Science*, **257**, 927.

18. Guo, H-C., Jardetzky, T. S., Garrett, T. P. J., Lane, W. S., Strominger, J. L., and Wiley, D. C. (1992). Different length peptides bind to HLA-Aw68 similarly at their ends but bulge out in the middle. *Nature*, **360**, 364.

19. Silver, M. L., Guo, H.-C., Strominger, J. L., and Wiley, D. C. (1992). Atomic structure of a human MHC molecule presenting an influenza virus peptide. *Nature*, **360**, 367.

20. Zhang, W., Young, A. C., Imarai, M., Nathenson, S. G., and Sacchettini, J. C. (1992). Crystal structure of the major histocompatibility complex class I H–2K^b molecule containing a single viral peptide: implications for peptide binding and T cell receptor recognition. *Proc. Natl Acad. Sci. USA*, **89**, 403.

* 21. Young, A. C., Zhang, W., Sacchettini, J. C., and Nathenson, S. G. (1994). The three-dimensional structure of H–2D^b at 2.4 Å resolution: Implication for antigen-determinant selection. *Cell*, **76**, 39.

22. Garboczi, D. N., Hung, D. T., and Wiley, D. C. (1992). HLA A2–peptide complexes: refolding and crystallization of molecules expressed in *Escherichia coli* and complexed with single antigenic peptides. *Proc. Natl Acad. Sci. USA*, **89**, 3429.

* 23. Madden, D. R., Garboczi, D. N., and Wiley, D. C. (1993). The antigenic identity of peptide–MHC complexes: A comparison of the conformations of five viral peptides presented by HLA-A2. *Cell*, **75**, 693.

* 24. Kozono, H., White, J., Clements, J., Marrack, P., and Kappler, J. (1994). Production of soluble MHC class II proteins with covalently bound single peptides. *Nature*, **369**, 151.

25. Wilson, I. A. (1994). Covalently linked ligand stabilizes expression of heterodimeric receptor. *Structure*, **2**, 561.

26. Brown, J. H., Jardetzky, T. S., Gorga, J. C., Stern, L. J., Urban, R. G., Strominger, J. L., and Wiley, D. C. (1993). Three-dimensional structure of the human class II histocompatibility antigen HLA-DR1. *Nature*, **364**, 33.

* 27. Stern, L. J., Brown, J. H., Jardetzky, T. S., Gorga, J. C., Urban R. G., Strominger, J. L., and Wiley, D. C. (1994). Crystal structure of the human class II MHC protein HLA-DR1 complexed with an influenza virus peptide. *Nature*, **368**, 215.

28. Jardetzky, T. S., Brown, J. H., Gorga, J. C., Stern, L. J., Urban, R. G., Chi, Y., Stauffacher, C., Strominger, J. L., and Wiley, D. C. (1994). Three-dimensional structure of a human class II histocompatibility molecule complexed with superantigen. *Nature*, **368**, 711.

29. Fields, B. A., Ysern, X., Poljak, R. J., Shao, X., Ward, E. S., and Mariuzza, R. A. (1994). Crystallization and preliminary x-ray diffraction study of a bacterially produced T cell antigen receptor Vα domain. *J. Mol. Biol.*, **239**, 339.

30. Wang, J., Yan, Y., Garret, T. P. J., Liu J., Rodgers, D. W., Garlick, R. L., Tarr, G. E., Husain, Y., Reinherz, E. L., and Harrison, S. C., (1990). Atomic structure of a fragment of human CD4 containing two immunoglobulin-like domains. *Nature*, **348**, 411.

31. Ryu, S.-E., Kwong, P. D., Truneh, A., Porter, T. G., Arthos, J., Rosenberg, M., Dai, X., Xuong, N., Axel, R., Sweet, R. W., and Hendrickson, W. A. (1990). Crystal structure of an HIV-binding recombinant fragment of human CD4. *Nature*, **348**, 419.

32. Leahy, D. J., Axel, R., and Hendrickson, W. A. (1992). Crystal structure of a soluble form of the human T cell coreceptor CD8 at 2.6 Å resolution. *Cell*, **68**, 1145.

33. Trowsdale, J. (1993). Genomic structure and function in the MHC. *Trends in Genetics*, **9**, 117.

*34. Monaco, J. J., Cho, S., and Attaya, M. (1990). Transport protein genes in the murine MHC: possible implications for antigen processing. *Science*, **250**, 1723.

*35. Trowsdale, J., Hanson, I., Mockridge, I., Beck, S., Townsend, A., and Kelly, A. (1990). Sequences encoded in the class II region of the MHC related to the 'ABC' superfamily of transporters. *Nature*, **348**, 741.

36. Martinez, C. K. and Monaco, J. J. (1991). Homology of the proteasome subunits to a major histocompatibility complex-linked LMP gene. *Nature*, **353**, 664.

37. Brown, M. G., Driscoll, J., and Monaco, J. J. (1991). Structural and serological similarity of MHC-linked LMP and proteasome (multicatalytic proteinase) complexes. *Nature*, **353**, 355.

38. Glynne, R., Powis, S. H., Beck, S., Kelly, A., Kerr, L.-A., and Trowsdale, J. (1991). A proteasome-related gene between the two ABC transporter loci in the class II region of the human MHC. *Nature*, **353**, 357

39. Kelly, A., Powis, S. H., Glynne, R., Radley, E., Beck, S., and Trowsdale, J. (1991). Second proteasome-related gene in the human MHC class II region. *Nature*, **353**, 667.

*40. Townsend, A. R. M., Gotch, F. M., and Davey, J. (1985). Cytotoxic T cells recognize fragments of the influenza nucleoprotein. *Cell*, **42**, 457.

*41. Falk, K., Rotzschke, O., Stevanovic, S., Jung, G., and Rammensee, H. G. (1991). Allele-specific motifs revealed by sequencing of self-peptides eluted from MHC molecules. *Nature*, **351**, 290.

42. Gammon, G., Sercarz, E., and Benichou, G. (1991). The specificity of the autoreactive T cell repertoire: the dominant self and the cryptic self. *Immunol. Today*, **12**, 193.

43. Williams, D. B., Ferguson, J., Gariepy, J., McKay, D., Teng, Y. T., Iwasaki, S., and Hozumi, N. (1993). Characterization of the insulin A-chain major immunogenic determinant presented by MHC class II I-A^d molecules. *J. Immunology*, **151**, 3627.

*44. Van Bleek, G. M. and Nathenson, S. G. (1991). The structure of the antigen-binding groove of major histocompatibility complex class I molecules determines specific selection of self-peptides. *Proc. Natl Acad. Sci. USA*, **88**, 11032.

45. Garrett, T. P. J., Saper, M. A., Bjorkman, P. J., Strominger, J. L., and Wiley, D. C. (1989). Specificity pockets for the side chains of peptide antigens in HLA-Aw68. *Nature*, **342**, 692.

46. Carbone, F. R. and Bevan, M. J. (1989). Induction of ovalbumin-specific cytotoxic T cells by *in vivo* peptide immunization. *J. Exp. Med.*, **169**, 603.

47. Pamer, E. G., Harty, J. T., and Bevan, M. J. (1991). Precise prediction of a dominant class I MHC-restricted epitope of *Listeria monocytogenes*. *Nature*, **353**, 852.

48. Saito, Y., Peterson, P. A., and Matsumura, M. (1993). Quantitation of peptide anchor residue contributions to class I major histocompatibility complex molecule binding. *J. Biol. Chem.*, **268**, 21309.

49. Dellabona, P., Peccoud, J., Kappler, J., Marrack, P., Benoist, C., and Mathis, D. (1990). Superantigens interact with MHC class II molecules outside of the antigen groove. *Cell*, **62**, 1115.

50. Soos, J. M., Russell, J. K., Jarpe, M. A., Pontzer, C. H., and Johnson, H. M. (1993). Identification of binding domains on the superantigen, toxic shock syndrome-1, for class II MHC molecules. *Biochem. Biophys. Res. Comm.*, **191**, 1211.

*51. Madden, D. R., Gorga, J. C., Strominger, J. L., and Wiley, D. C. (1991). The structure of HLA-B27 reveals nonamer self-peptides bound in an extended conformation. *Nature*, **353**, 321.

*52. Madden, D. R., Gorga, J. C., Strominger, J. L., and Wiley, D. C. (1992). The three-dimensional structure of HLA-B27 at 2.1 Å resolution suggests a general mechanism for tight peptide binding MHC. *Cell*, **70**, 1035.

53. Shawar, S. M., Cook, R. G., Rodgers, J. R., and Rich, R. R. (1990). Specialized functions of MHC class I molecules. I. An N-formyl peptide receptor is required for construction of the class I antigen Mta. *J. Exp. Med.*, **171**, 897.

54. Shawar, S. M., Vyas, J. M., Shen, E., Rodgers, J. R., and Rich, R. R. (1993). Differential amino-terminal anchors for peptide binding to H–2M3a or H–2K^b and H–2D^b. *J. Immun.*, **151**, 201.

55. Wang, C. R., Loveland, B. E., and Fischer–Lindahl, K. (1991). H-2M3 encodes the MHC class I molecule presenting the maternally transmitted antigen of the mouse. *Cell*, **66**, 335.

56. Trowsdale, J. and Campbell, R. D. (1992). Complexity in the major histocompatibility complex. *Eur. J. Immunogen.*, **19**, 45.

*57. Parham, P., Lomen, C. E., Lawlor, D. A., Ways, J. P., Holmes, N., Coppin, H. L., Salter, R. D., Wan, A. M., and P. D. Ennis. (1988). Nature of polymorphism in HLA-A, -B, and -C molecules. *Proc. Natl Acad. Sci. USA*, **85**, 4005.

58. Bluestone, J. A., Jameson, S., Miller, S., and Dick, R. (1992). Peptide-induced conformational changes in class I heavy chain alter major histocompatibility complex recognition. *J. Exp. Med.*, **176**, 1757.

59. Catipovic, B., Dal Porto, J., Mage, M., Johansen, T. E., and Schneck, J. P. (1992). Major histocompatibility complex conformational epitopes are peptide specific. *J. Exp. Med.*, **176**, 1611.

60. Jensen P. E. (1993). Acidification and disulfide reduction can be sufficient to allow intact proteins to bind class II MHC. *J. Immunology*, **150**, 3347.

61. Stern, L. J. and Wiley, D. C. (1994). Antigenic peptide binding by class I and class II histocompatibility proteins. *Structure*, **2**, 245.

5 | Molecular mechanisms of antigen processing

JOHN J. MONACO

1. Introduction

The mammalian immune system has developed methods for the surveillance of both extracellular and intracellular compartments for 'non-self'. This surveillance is effected by antigen-specific receptors on T lymphocytes, whose natural ligands are short peptides complexed with major histocompatibility complex (MHC)-encoded class I or class II molecules. MHC molecules are integral membrane proteins, specifically designed to bind a diverse set of peptides and to deliver them to the cell surface for recognition by T cells. The structural features of MHC molecules and their peptide binding specificities are discussed in Chapter 4 and in recent reviews (1–4). Here I will focus on the molecular mechanisms used to generate MHC-binding peptides and to deliver them to the appropriate MHC molecules.

Early experiments based on the ability to differentially induce or inhibit antigen presentation by MHC class I versus MHC class II molecules suggested that different mechanisms were utilized for the loading of these two classes of molecules with peptides (reviewed in refs 5–7). More recently, it has become possible to isolate and characterize the peptides found naturally associated with MHC molecules; the data derived from such studies are generally consistent with the notion of separate pathways for the derivation of class I- and class II-associated peptides. In general, MHC class I-associated peptides are derived from proteins that are either synthesized on the host cell's own ribosomes (3, 4), or which are artificially introduced into the cytoplasm of living cells (8, 9). In contrast, peptides found naturally associated with MHC class II molecules are derived predominantly from the extracellular portions of cell surface proteins, or from proteins endocytosed from the extracellular medium (10–14).

It is important to point out here that class I–peptide complexes are recognized by the CD8$^+$ (predominantly cytotoxic) subset of T cells, and that class II–peptide complexes are recognized by the CD4$^+$ (predominantly helper) subset. Thus, in terms of the biology of the immune system, the division of antigen processing into two pathways can be seen as one mechanism for ensuring that cytotoxic T cell responses are biased towards antigens from intracellular parasites (for example

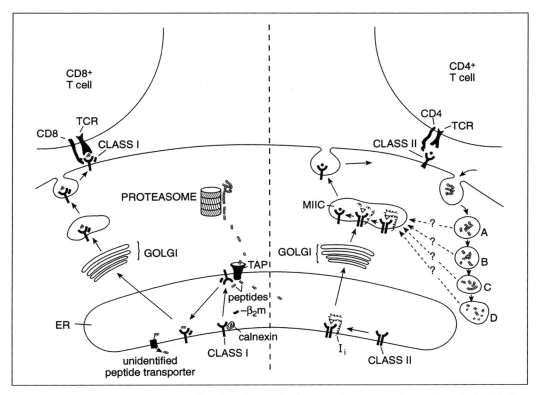

Fig. 1 Pathways of antigen processing. The class I or endogenous antigen processing pathway is depicted on the left, and the class II or exogenous antigen processing pathway is depicted on the right. In the class I pathway, cytoplasmic proteins are degraded by proteasomes, and the resulting peptides are transported into the lumen of the ER by the TAP transporter. Class I molecules synthesized in the ER originally associate with the chaperonin, calnexin, and then with the TAP transporter until they acquire an appropriate peptide. Peptide binding induces a conformational change in the class I molecule, which then travels to the cell surface via normal vesicular transport pathways, where they can be recognized by the antigen-specific receptors (TCR) of CD8$^+$ T cells. Transported peptides which are not bound by class I molecules are either degraded, or transported back into the cytoplasm by an as yet unidentified, ATP-dependent transport mechanism. For class II, newly synthesized α and β chains associate in the ER with the invariant chain (I$_i$), and localize to a peptide-binding compartment (MIIC) on their route to the cell surface, where I$_i$ is degraded and the CLIP peptide (triangles) is removed. Peptides derived from one or more of the vesicles along the endocytic/lysosomal pathway (A–D) enter the MIIC by an undefined mechanism, bind to class II, and are delivered to the surface by vesicular transport, where they can be recognized by the antigen-specific receptors (TCR) of CD4$^+$ T cells. See text for details.

viruses), and that helper T cell (antibody) responses are biased towards extra-cellular antigens (see Fig. 1). This arrangement makes intuitive sense, since:

1. Antibody is effective at recognizing and removing extracellular material from the circulation (by interacting with components of the reticuloendothelial system), but cannot interact with intracellular antigen.
2. It is desirable (from the host's point of view) for cytotoxic T lymphocytes (CTL) to recognize and destroy cells that contain replicating intracellular parasites, but not cells that have endocytosed non-replicating foreign proteins.

While the existence of largely separate class I and class II antigen processing pathways has been widely appreciated for almost 10 years, in the last three years the field has witnessed an extremely rapid pace of progress, due in large part to the application of molecular genetics to the dissection of these pathways. The detailed molecular mechanisms involved in each of the two antigen processing pathways are discussed separately below.

2. The class I (endogenous) antigen processing pathway

The evidence that degradation of native antigen into peptides in the class I pathway occurs predominantly in the cytoplasm, rather than within acidified intracellular compartments such as lysosomes, has been reviewed recently (5, 15). The first indications of cytoplasmic processing came from studies on the specificity of the CTL response to viral antigens; viral proteins produced during infection which do not enter the secretory pathway in their native form, such as influenza nucleoprotein (NP), are none the less presented at the cell surface for recognition by class I-restricted T cells (16–18). It was later shown that, in contrast to what is seen in the class II pathway, native antigens do not prime cells for lysis by CTL when added to the extracellular medium; they must be introduced directly into the cytoplasm of cells (8, 9). Furthermore, manipulations of cytoplasmic antigens which increase their rate of degradation also increase the efficiency with which they are presented to class I-restricted T cells (19, 20), and the synthesis of peptide fragments in the cytoplasm (from transfected minigenes) results in their presentation by class I (but usually not class II) molecules (21–23). Perhaps the most compelling evidence for cytoplasmic processing of antigen in this pathway, however, comes from the analysis of TAP transporter mutants (see below); cells that lack the ability to transport small peptides from the cytoplasm into the lumen of the endoplasmic reticulum (ER). These cells are severely deficient in the presentation of antigen to class I-restricted CTL.

2.1 MHC-linked transport protein (*Tap*) genes

Over the past 4 years, a number of cell lines with defects in the class I antigen processing pathway have been characterized. The mutation responsible for the defect in such cells resides in one (or both) of two MHC-linked genes named *Tap-1* and *Tap-2*.[1] As explained below, the products of these genes form a heterodimer that transports peptides from the cytoplasm into the ER lumen. The salient phenotypic characteristics (31–40) of cells with *Tap* gene defects are:

- deficient class I surface expression (generally about 5–10 per cent of wild-type levels);

---

[1] Tap stands for *T*ransporter *a*ssociated with *a*ntigen *p*rocessing, or *T*ransporter of *a*ntigenic *p*eptides (24). These genes were originally called *Ham-1* and *Ham-2* in the mouse (25); *mtp-1* and *mtp-2* in the rat (26); RING4 and RING11 (27), Y3 and Y1, or PSF-1 and PSF-2 (28–30) in humans.

- normal sequence and transcription of class I structural genes (including β_2-microglobulin, β_2m);
- failure to present native antigen to class I-restricted T cells;
- normal or enhanced presentation of short synthetic peptides to class I-restricted T cells;
- partial recovery of class I surface expression by incubation with appropriate class I-binding peptides (and a source of exogenous β_2m);
- partial recovery of class I surface expression by incubation at reduced temperatures (i.e. 26°C versus 37°C)

This phenotype results from an insufficient supply of appropriate class I-binding peptides in the ER. In the absence of such peptides, newly synthesized MHC class I molecules fail to obtain a stable conformation, and are detained for abnormally long periods of time in the ER (see Section 2.3). Those 'empty' class I heterodimers that do eventually reach the cell surface are unstable at 37°C (but not at 26°C), and rapidly dissociate into the component polypeptides. The released β_2m chains are free to diffuse into the medium, and the lone class I heavy chains that remain anchored in the cell membrane are conformationally incorrect and no longer react with anti-class I antibodies. If appropriate peptides are added, heavy chains, β_2m, and peptides can spontaneously assemble into stable class I molecules that can be recognized both by anti-class I antibodies and by CTL. At reduced temperature, release of empty class I heterodimers from the ER is enhanced, and these, being relatively stable at the lower temperature, accumulate at the cell surface.

As mentioned above, several cell lines with defective *Tap* genes have been isolated, and the phenotype of each is, with minor exceptions (see Section 2.1.2), identical. The best characterized of these antigen processing-defective cell lines is the mouse T cell line RMA-S (31, 32). The *Tap-2* gene in these cells harbours a point mutation that creates a premature termination codon (41), and transfection of mouse (42) or rat (43) *Tap-2* cDNA clones corrects the defect. In human B-lymphoblastoid cell line (B-LCL) 721.134, *TAP1* mRNA is absent (28), and transfection of a cloned copy of human *TAP1* cDNA under the control of an exogenous promotor corrects the antigen processing defect (29). The mouse *Tap-1* cDNA is also capable of correcting the defect in 721.134 (M. Attaya and J. J. M., unpublished observation). The human B-LCL 721.174 (ref. 33) and its derivative, T2 (ref. 44), contain a homozygous deletion of the MHC class II region, encompassing both *TAP* genes, and transfection of both genes is required to reverse the processing defect in these cells; single gene transfections of *TAP1* or *TAP2* fail to reverse the defect (29, 45, 46). *Tap-1*-deficient mice have also been produced by gene targeting, and have the expected phenotype, including a deficiency of CD8[+] T cells (47).

The two *Tap* genes are closely linked to one another in the class II region of the MHC (Fig. 2). Sequence analyses indicate that they are closely related to one another, as well as to a large family of transport protein genes derived from both prokaryotes and eukaryotes (see ref. 25 for other references), collectively called 'ABC transporters' (for *A*TP-*B*inding *C*assette). Well-known eukaryotic members

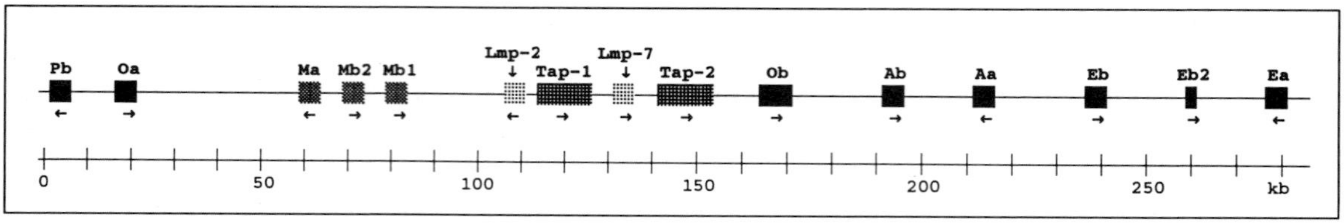

Fig. 2 Map of the class II region of the mouse MHC. The relative locations of all genes shown is identical for the human MHC, with the exception that there is only one copy of the class II β chain gene corresponding to *Mb1* and *Mb2* (*DMB* in humans). Transcriptional orientation is shown by arrows under each gene. Distances indicated are in kilobasepairs (kb); the map is approximately to scale. Other genes of unknown function that have been identified in this region are not shown.

of this family include the mammalian multidrug resistance (P-glycoprotein) genes, the cystic fibrosis (*CFTR*) gene, and the yeast mating factor transporter gene, *STE-6*. The distinguishing structural features of this family of molecules include four protein domains, residing on from one to four different polypeptide chains (48). Two domains are highly hydrophobic, and each contains either six or eight predicted transmembrane-spanning helixes. The remaining two domains each contain a pair of sequence motifs ('Walker boxes') believed to be diagnostic of nucleotide (ATP) binding sites. Consistent with the structural organization of other ABC family members, immunoprecipitation studies demonstrate that the TAP-1 and TAP-2 polypeptides (each of which contains a single transmembrane and single cytoplasmic domain), are physically associated with one another to form a heterodimer (49, 50).

Staining of cell sections with anti-TAP antibodies indicates that the TAP heterodimer is localized in the membranes of the ER and *cis*-Golgi (51). These data confirm the earlier indications, based on inhibition of antigen presentation with Brefeldin A (which blocks egress of class I molecules from the ER), that peptide enters the secretory pathway via a pre-Golgi compartment (52, 53).

2.1.1 Peptide transport *in vitro*

Based on the above data, it was predicted that the TAP protein mediates the transport of cytoplasmically-produced peptide fragments of antigen into the ER lumen. Hence, loss of function of this molecule would lead to the observed (class I-binding) peptide deficit in that compartment. Direct proof of TAP-dependent peptide transport has recently been obtained using whole cells permeabilized with streptolysin O (54, 55) as well as isolated microsomal vesicles (56–58). Efficient peptide transport in both systems is dependent on the presence of a functional TAP heterodimer, as demonstrated by the inability of TAP-negative cells or microsomes to perform this function, and by the ability to inhibit peptide transport with anti-TAP antibodies. The presence of ATP is also required, and this requirement is not met by non-hydrolysable ATP analogues. The rate of peptide transport is temperature-dependent, and very rapid at 37°C. The total peptide translocating capacity of TAP-transfected T2 cells (54) was estimated at approximately 2×10^4

peptides/min per cell. This must be considered a very rough estimate, since it is not known how the concentrations of peptide used in the above experiments compare to physiological concentrations, and whether or not delivery of peptides to the transporter *in vivo* might be mediated by accessory molecules rather than by diffusion (see Section 2.3).

For reasons not completely understood, peptide transport in these systems appears to cease after a few minutes, even in the presence of apparent excesses of ATP and peptide. Interestingly, peptide *accumulation* within the ER or microsomal vesicles is dependent on the availability of a lumenal mechanism for trapping the transported peptides inside the lumen once they have gained entry. Such mechanisms include the binding of peptide by endogenously synthesized class I molecules, or glycosylation of peptides containing *N*-linked glycosylation signal sequences. In the absence of such a trapping mechanism, transported peptide is lost over time. This loss may represent specific reabsorption into the cytoplasm of peptides that have failed to bind to MHC molecules, as it appears to also be temperature and ATP-dependent (57). The reabsorption of unbound peptide may partially explain the apparent cessation of transporter function after short incubations, if one assumes that the trapping mechanisms can be saturated, and peptide transport into and out of the ER then rapidly reaches equilibrium. Alternative explanations are also possible, such as exhaustion of supplies of a required co-factor (other than ATP) or feedback regulation on the transporter itself if too much peptide accumulates in the ER. The hypothesis that the transporter may be labile under the incubation conditions used is ruled out by the observation that pre-incubation of microsomes without peptide does not adversely affect transport when peptide is subsequently added.

Regardless of the reason for the transient nature of peptide transport in these systems, re-uptake of unbound peptide into the cytoplasm would be desirable for two reasons. First, it would prevent the (energetically costly) loss of large amounts of peptide to the extracellular medium, which might be expected to take place by bulk flow if intra-ER peptide concentrations were high. Second, maintaining relatively low concentrations of peptide in the compartment where class I molecules bind peptides would help ensure that high-affinity peptides preferentially occupy the binding site of class I molecules released to the cell surface (57).

It should also be mentioned that while rapid ATP-dependent peptide transport is dependent on functional TAP protein, TAP-negative cells or microsomes often allow a much slower, non-ATP-dependent entry of peptide. These latter observations provide a potential explanation for previous failures to observe TAP- and ATP-dependent peptide transport in similar systems (59, 60), where long incubations at physiological temperatures were employed. They also provide a potential explanation for some cases of non-TAP-dependent antigen presentation (see Section 2.1.2).

In summary, it is now reasonably certain that TAP functions as a specific peptide transporter. Preliminary analyses (57, 58) also suggest that the substrate specificity of this molecule correlates (roughly, at least) in terms of peptide length *and* sequence with the peptide-binding specificity of MHC class I molecules. Peptides shorter than about six to eight residues are not transported efficiently, but longer peptides (up to at least 16 residues, the upper limits have not been thoroughly evaluated) can be transported. Efficient transport also depends on the sequence of the substrate. Interestingly, sequence constraints may be dissociated from composition effects, in that two peptides of identical length and composition (but with reversed sequence) have been shown to differ dramatically in their ability to compete for transport of an antigenic peptide (56). The latter must be interpreted with some caution, however, since effects on competition for binding to lumenal class I versus competition for transport were not clearly separable in the above experiments.

The identity of the C-terminal residue of the peptide substrate appears to be a major determinant in transport efficiency (54, 57, 58), with a marked preference towards hydrophobic (and sometimes basic, see Section 2.4) C-terminal residues. As mentioned above, this correlates roughly with class I binding specificity. The mouse transporter prefers hydrophobic (aliphatic and aromatic) C-termini, while all mouse class I molecules whose sequence preferences are known prefer (only) aliphatic residues at this position. Other positions of the peptide must also play some role in transporter specificity, since peptides with identical C-termini can be transported with different efficiency. However, data on this point are still fragmentary.

2.1.2 Exceptions to the TAP transporter requirement

There are a number of indications that TAP-independent transport and/or presentation of peptides to class I restricted T cells can occur. One reasonably clearly defined alternative mechanism that has been revealed by analysis of TAP-defective cells involves the normal pathway of translocation of signal sequence-containing proteins into the ER. In fact, some of the first real evidence that the mutation in these cells resulted in a defect in peptide transport involved bypassing the defect by targeting peptides to the ER via the protein translocation pathway (61). In these studies, it was shown that a minigene encoding an influenza matrix protein epitope, introduced into cells by transfection, would sensitize wild-type (T1), but not mutant (T2) cells for lysis by specific CTL. In contrast, the same epitope, when fused with a signal peptide, could be presented by both cell types. Introduction of the signal sequence at the C-terminus of the peptide, rather than at the N-terminus, fails to rescue presentation by the mutant, indicating that the defect is not bypassed simply by increasing the overall hydrophobicity of the minigene-encoded peptide (62).

In (non-transfected) TAP-defective T2 cells, surface expression of HLA-B5 is drastically reduced, while surface expression of HLA-A2 class I molecules is approximately 30 per cent of wild type. All of the HLA-A2-bound peptides identified in such cells derive from cleaved signal sequences (63, 64). That this phenomenon has thus far only been observed for HLA-A2 probably reflects the peptide-binding specificity of the class I molecules themselves, that is to say, most class I molecules may not effectively bind highly hydrophobic signal sequence-derived peptides. Nevertheless, under certain circumstances, the signal sequence pathway can be physiologically relevant, and may even be responsible for the dominant response to certain antigens (65).

Some cells with *TAP* gene defects have been shown to be recognized efficiently by certain antigen-specific CD8$^+$ T cells which are not specific for signal sequence derived peptides, suggesting that some peptides may bypass the TAP requirement and gain entry to the ER lumen via other mechanisms (not involving the SRP pathway). For example, vesicular stomatitis virus (VSV)-infected RMA-S cells are efficiently killed by H2-K^b-restricted, VSV-specific CTL (66, 67). Endogenous expression of the VSV nucleocapsid protein gene is also sufficient to render these cells susceptible to CTL-mediated lysis. Similarly, Sendai virus-infected RMA-S is also recognized by specific CTL (68). In the former case, the same peptide is presented by wild-type RMA and mutant RMA-S cells, and in the latter case it was shown that presentation is sensitive to Brefeldin A, suggesting that the mechanisms involved at least overlap to some degree with the normal class I processing pathway. Interestingly, in both cases, prolonged duration of infection and increased quantities of antigen are required for adequate levels of recognition of RMA-S as compared to RMA cells. Thus, it is possible that the relevant peptides in these cases enter the ER via the slow, non-TAP-, non-ATP-dependent pathway observed in the *in vitro* peptide transport assays discussed above. However, VSV antigens are not presented to the same K^b-restricted CTL by VSV-infected T2 cells that have been transfected with the mouse K^b gene (although exogenous peptide sensitizes these cells, demonstrating the functional expression of K^b). The latter result raises the possibility that transport of the relevant peptides by RMA-S cells might occur via a TAP-1 homodimer, or via very small amounts of TAP-1/TAP-2 heterodimer produced by readthrough of the premature translation termination codon in the RMA-S *Tap-2* gene.

Finally, another TAP-independent pathway may exist only in certain cell types. Several studies indicate that, while soluble antigens taken into the endocytic pathway are not generally presented by class I molecules, the same antigens in particulate form taken up by phagocytosis *in vivo* are effectively presented by class I (69, and references therein). Peritoneal macrophages that have phagocytosed bacteria *in vitro* are capable of presentation of antigen to class I-restricted T cells (70). In addition, soluble antigen is presented to class I-restricted T cells after *in vitro* incubation with splenocytes, but only by a subset of cells that expresses both class I and class II molecules (71). The mechanisms involved (and the potential role of the TAP molecule in these cases) are currently unknown, but presentation in at least one of these systems (70) was not inhibitable by Brefeldin A. Based on the fact that mixtures of viable, MHC-mismatched macrophages with fixed MHC-matched cells present the relevant peptide to class I-restricted T cells (70), it was suggested that this phenomenon may be the result of phagocytosis and endosomal/lysosomal processing into peptides, followed by regurgitation of the peptides with subsequent binding to surface class I molecules. This potential pathway may also explain the Brefeldin A-insensitive presentation of Sendai virus antigen by K^b-transfected T2 cells infected with the virus (72).

2.1.3 Biological implications of transport mutants

The above points notwithstanding, the severity of the phenotype seen in the TAP mutants (generally less than about 10 per cent of wild-type class I surface expression, and the inability to present most antigens to $CD8^+$ T cells) indicates that the vast majority of class I bound peptides derive from the cytoplasm and gain entry to the ER via the TAP transporter. This implies that few peptides with the capacity to be bound by MHC class I molecules are generated within the ER (or indeed, within other compartments of the exocytic pathway), despite the known proteolytic activity in these compartments (73).[2] Several potential explanations exist for this observation (5), perhaps the most likely being either that TAP-mediated transport is directly coupled to MHC binding via a physical association of the two (TAP and MHC I) molecules, or that intra-ER protein degradation rarely produces nonameric peptides, which, in general, have considerably higher affinity (up to 1000-fold) for class I molecules *in vitro* than longer or shorter peptides. Thus, the latter would compete ineffectively with the more appropriately matched peptides supplied by the transporter. If indeed the latter explanation is correct, one further implication is that binding of longer peptides to class I molecules with subsequent 'trimming' does not commonly occur (despite the fact that the TAP transporter can transport such longer peptides). Analysis of peptides isolated from the HLA-A2 molecules of T2 cells, as discussed above (63, 64), supports this notion. Many of these peptides are longer than nine amino acids, indicating that longer peptides that *can* bind with high-affinity to class I molecules in the ER are not trimmed, and can be isolated from a significant proportion of class I molecules. The same peptides are found in greatly reduced proportions on normal cells, where they must compete for MHC I binding with peptides supplied by the transporter. Recently, evidence for the presence of an unusually long set of peptides in a subset of HLA-B27 molecules has appeared (75).

In summary, it would appear that the majority of class I bound peptides in normal cells, including those derived from cell surface and secreted molecules, are produced in the cytoplasm. This may occur if low levels of such proteins escape the cytoplasmic translational block normally imposed on them by the signal recognition particle (SRP), and are translated and released into the cytoplasm. Such molecules would presumably be targeted for rapid degradation due to the presence of hydrophobic transmembrane segments and/or leader sequences (which would lead to improper folding), and might thus be effectively presented even though produced at very low levels (compared to the correctly ER-targeted protein). The contribution of peptides that do not enter the ER via the TAP transporter to the

---

[2] However, (rare) examples of presentation of peptides produced within the ER will undoubtedly be found; one potential example being HIV-1 envelope protein (74).

pool of class I bound material is minor, but may none the less be important in the response to certain specific antigens.

2.2 MHC-linked proteasome subunit genes

If naturally occurring class I-bound peptides are derived from cytoplasmic sources, then that cellular compartment must necessarily contain the proteolytic machinery responsible for producing such peptides. As described below, this function is probably performed by large proteolytic particles called proteasomes. Proteasomes are large (20S, approx. 650 kDa), abundant (up to 1 per cent of soluble cellular protein) protein complexes found in the cytoplasm and nucleus of all eukaryotic cells (for reviews, see refs 76–78), and in the archaebacterium *Thermoplasma acidophilum* (79). Eukaryotic proteasomes are composed of more than 20 different protein subunits, arranged as four stacked rings of seven subunits each (see Fig. 3). The overall structure of the complex is evolutionarily well-conserved, as is the sequence of the individual subunits from yeast to man. The subunits can be grouped by sequence homology into two groups, α and β. Studies of *Thermoplasma* proteasomes, which unlike eukaryotic proteasomes are composed of only two different subunits, using immunoelectron microscopy suggest that the two outermost rings are composed of α subunits, and the two inner rings of β subunits (80).

Purified proteasomes have proteolytic activity *in vitro*, and the structure has multiple catalytic sites, with the capacity to hydrolyse peptide bonds involving a broad spectrum of amino acids. None of the subunits sequenced to date has striking homology to any known protease family, and the mechanism(s) of catalysis are still largely unknown. However, there is some data to suggest that the α subunits are structural or regulatory, and the β subunits determine the catalytic activities (81, 86). Digestion of protein substrates *in vitro* with purified proteasomes appears to yield predominantly short peptides, rather than free amino acids (82, 83).

Fig. 3 Diagrammatic representation of proteasome structure. Electron microscopic evidence suggests a cylindrical structure composed of four stacked rings, each containing seven subunits. The outermost rings are composed of α type subunits, and the innermost rings of β type subunits. Whether or not the apparently hollow 'core' of the cylinder is open to solvent is still controversial. Estimated dimensions are shown in angstroms (Å).

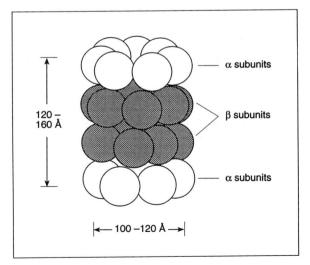

In the absence of further intervention, such peptides produced *in vivo* would then be expected to be rapidly degraded by cytoplasmic endo- and exo-peptidases.

Proteasomes probably represent the catalytic core of an even larger (26S, 1500 kDa) structure that is responsible for the ATP-dependent degradation of ubiquitinated proteins (84). However, only about 15 per cent of the total cellular pool of proteasomes is associated with the larger complex, and proteolysis by the 20S complex *in vitro* does not require either ubiquitin or ATP. Whether the 20S proteasome has a functional role *in vivo*, independent of the 26S particle, is still unknown. However, consistent with its role in the ubiquitin pathway (which is essential for cell viability), deletion of most proteasome subunit genes is lethal in yeast (85, 86).

The discovery that genes encoding two proteasome β subunits, *Lmp-2* and *Lmp-7*, map to the class II region of the MHC (87–94), immediately adjacent to the two *TAP* transporter genes, fuelled speculation that proteasomes are responsible for the production of peptides in the endogenous antigen processing pathway. Although earlier studies implicated the ubiquitin pathway in the generation of class I epitopes (20), there is still little direct evidence for a requirement for proteasomes in antigen processing. Perhaps the best evidence to date in favour of this hypothesis is that the presentation of at least one antigen (ovalbumin) to class I-restricted T cells requires the function of the ubiquitin pathway (95). However, it is formally possible that the requirement for ubiquitination in this system reflects not its role in targeting protein for degradation, but rather another unrelated function. It is also not clear how generalizable these results will be in relation to other antigens.

Interestingly, of all cloned mouse subunits tested thus far, only LMP-2 and LMP-7 are inducible with interferon-γ, and only these two subunits map to the MHC (E. Woodward and J. J. M., unpublished observations). Thus, it was first hypothesized that the LMP2 and LMP7 proteins were somehow specifically related to the immunological, as opposed to the housekeeping, functions of the complex. One plausible potential function was that these subunits served to couple the proteasome physically to the transporter. Immunohistochemical studies (96) were consistent with this hypothesis, in that a proportion of proteasomes (about 15 per cent) were found to be associated with the membranes of the ER. However, the hypothesis is complicated by several recent pieces of data. Analysis of MHC deletion mutants indicates that the two MHC-linked (LMP) proteasome subunits are not required for antigen processing (45, 46, 97). Thus, the two MHC-linked subunits are required for neither the (essential) degradative function (the deletion mutants are viable), nor the (presumed) antigen processing function of the complex. This is somewhat ironic in that the genetic linkage of these two subunits to the MHC was the original rationale for the proposal of a potential role of the complex in antigen processing.

So why are proteasome subunits linked to the MHC? The most viable remaining alternative is that addition[3] of these subunits somehow alters the specificity of the complex to produce peptides better suited for transport and/or binding to MHC

---

[3] It is not yet certain whether other subunits are displaced by LMP2 and LMP7, or whether these are simply added to the complex, but some (unpublished) evidence points to the former possibility.

class I molecules. It is interesting that in cells that express both LMP subunits, not all proteasome complexes contain these subunits (88, 98), but both LMP$^+$ and LMP$^-$ proteasomes have proteolytic activity. When tested on a panel of synthetic substrates, the relative activity of the two complexes for different substrates differs (99, 100). In general, LMP$^+$ proteasomes cleave after hydrophobic and basic amino acid residues more efficiently than LMP$^-$ proteasomes, while cleavage after acidic residues is similar. This is generally consistent with transporter and class I specificity as discussed above, but the observed effect is quantitative, not qualitative. Hence, effects of these subunits on presentation of antigen to T cells may be more subtle than can be measured by the techniques that have been used to date. The effect of the two MHC-linked subunits on the length of peptides produced by degradation of larger protein substrates has not yet been investigated. Furthermore, alternative post-translational processing of the MHC-linked LMP2 subunit may be involved in generating other structurally distinct proteasome forms (101, 102), whose activities have not yet been assessed.

In summary, the accumulated circumstantial evidence for the involvement of proteasomes in the class I antigen processing pathway is reasonably convincing:

1. Proteasomes are required in the ubiquitin-dependent pathway of protein degradation, and manipulations which target proteins into this pathway increase the efficiency with which they are presented to class I-restricted T cells.
2. For at least one antigen, an intact ubiquitination pathway is essential for presentation to class I-restricted T cells.
3. Two proteasome subunits are encoded in the MHC, immediately adjacent to the *TAP*-transporter genes, and these two subunits appear to modify the proteolytic action of the complex to produce peptides better suited for TAP-mediated transport (and class I binding), at least with respect to the identity of the C-terminal residue of the peptide product.
4. Although there is little direct information about the proteolytic specificity of the proteasome complex, the proteasome does indeed produce peptides (not free amino acids) from polypeptide substrates (82, 83). Kinetic data indicate that internal nonameric peptides (which require at least two cleavages to be produced from the intact substrate) can be produced from longer substrates without release of the substrate from the complex, suggesting that the (multiple) catalytic sites of proteasomes may be positioned such that peptides of this length could be preferentially produced.
5. Direct digestion of protein antigens *in vitro* with purified proteasomes can produce peptides identical to the naturally processed peptides of these antigens found associated with MHC class I molecules (M. Brown and J. J. M., unpublished observations, and C. Slaughter, personal communication).

2.3 Molecular chaperones in the class I pathway

Several remaining mechanistic questions in the class I pathway concern the handling of peptides once they have been produced. How do peptides get from the proteasome to the TAP transporter, and from the transporter to the class I molecule? The possibility of a direct physical association between proteasomes and the TAP transporter was discussed above. However, no direct evidence for physical association of these two structures is available, and they are not co-purified in standard immunoprecipitation experiments (50, 51, 87, 88, 90, 98). Thus, it is possible that simple diffusion mediates this transfer or that as yet uncharacterized peptide carrier proteins (chaperonins) are involved.

Peptides may be delivered from the transporter to newly synthesized class I molecules in the ER lumen, again either by direct physical interaction, diffusion, or carrier proteins. ER-resident chaperonins such as BiP and calnexin (p88, IP90) detain empty class I molecules in the ER (or cause them to recycle between the ER and *cis*-Golgi network, see ref. 103), presumably enhancing their ability to obtain peptide (104). These molecules also apparently protect unassembled class I heavy chains from rapid degradation within the ER (105), and/or may aid in the assembly of the peptide–β_2m-class I complex. Thus, in cells where class I heavy chains fail to bind peptides efficiently (resulting either from lack of a functional TAP transporter or from lack of β_2m), the normally transient association of class I with these chaperonins is prolonged, as is the length of time required for (and the proportion of) newly synthesized class I molecules to reach the cell surface.

2.4 Specificity and polymorphism in the endogenous antigen processing pathway

While it is clear that the peptide-binding specificity inherent in the structure of MHC class I molecules plays a significant role in determining which peptides derived from a given antigen will ultimately be presented to T cells (1–4), whether and to what extent this phenomenon of 'determinant selection' is constrained by the mechanisms of antigen processing is not yet known. Put simply, are MHC class I molecules allowed to choose from all possible peptides that can be derived from a given antigen, or is only a subset of such peptides available, determined by the ability of the antigen processing machinery to produce and transport them? That the latter is the case is suggested by several reports documenting the ability to generate CTL to 'cryptic' antigenic determinants by immunization with peptides. CTL made against exogenously added peptides derived from native antigen do not always recognize histocompatible cells synthesizing the native antigen (106, 107).

Since, as discussed above, the TAP transporter demonstrates selectivity for peptide length and sequence, one possible explanation of this effect is that those peptides corresponding to such cryptic determinants may fail to be transported. Hence, they would be unavailable for class I binding in cells synthesizing native antigen, but could bind to surface class I molecules when added exogenously. Another possible explanation is that these particular peptides are never generated from native antigen in the first place. The observation that certain antigenic peptides may or may not be presented, depending on their 'context' (that is to say, depending on the flanking amino acid residues) in their source proteins (108–110; and N. Shastri, unpublished data) is potentially consistent with either of these possibilities. For example, chimeric gene constructs were produced containing the coding sequence for a nonameric peptide derived from a murine cytomegalovirus protein (which is normally presented by the mouse L^d class I molecule), inserted at different sites within the hepatitis B virus core antigen (*HBcAg*) gene (108). When the peptide was flanked on either side by four or five residues derived from the authentic source protein, or by five alanine residues, it was presented equally well from either of two integration sites in the HBcAg. However, when the peptide was inserted without flanking residues, efficient presentation occurred only when it was integrated at one of the two sites. Thus, in this case, it may be that the nonameric peptide failed to be generated by the appropriate proteolytic cleavages when in an inappropriate context. However, it is also possible that under normal circumstances a longer peptide containing the nonamer epitope is produced and transported, and when the epitope is in a different context, a long peptide is still produced but fails to be transported because of its altered sequence (compared to the naturally occurring peptide).

Regardless of the exact mechanisms involved, if indeed selectivity in antigen processing constrains the repertoire of peptides available for class I binding, then one must consider the possibility that functional polymorphisms in the antigen processing machinery exist. Both of the transporter polypeptides (111–114), as well as both of the MHC-linked proteasome subunits (87, 89, 115) display structural polymorphism. Do different individuals have the potential for providing different repertoires of peptides from which their class I molecules may pick and choose? If so, this would have obvious implications for individual differences in immune responsiveness to particular antigens, and might play a role in predisposition to MHC associated diseases (116).

On the one hand, the apparent lack of species specificity in antigen processing would seem to argue against significant intraspecies variability in peptide transport. Cells of one species (for example human) transfected with class I genes derived from another species (for instance mouse) have generally been seen to present antigen normally to T cells restricted to the transfected gene product (117, 118), indicating that the peptides produced (and transported) in cells of both species are closely similar or identical. Furthermore, TAP-deficient cells transfected with mouse, rat, or human *TAP* genes apparently present the same peptides to class I-restricted T cells as the wild-type parents (119, 120).

Nevertheless, there is direct proof of functional transporter polymorphism in rats, and preliminary indications of antigen processing polymorphism in humans. The class I molecules of different strains of inbred rats, even though identical in sequence, can be distinguished by alloantibody, allospecific CTL, and antigen-specific class I-restricted CTL (121, 122). This difference was shown to be determined by a *trans*-acting, MHC-linked gene (distinct from the class I structural gene) originally called *cim* (for class I modifier). The *cim* gene now appears to be identical to *Tap*-2 (114, 123), and the *cim* phenotype results from transport of a different subset of peptides by the *cim*^{a} (*Tap-2*^{a}) as compared to the *cim*^{b} (*Tab-2*^{b}) allelic form of the transporter (58, 114). The net result is that a different set of peptides is presented to T cells by the identical class I molecules of the two strains.

An apparently similar phenomenon has been observed in human cell lines (124). HLA−B27$^+$ cell lines that differ in their ability to present a number of B27-restricted T cell epitopes to CTL have been identified. The B27 class I molecules in these lines are identical in sequence, and transfection experiments demonstrate that cloned B27 cDNA functions according to the genotype of the host cell. Thus, these experiments appear to define an antigen processing polymorphism due to *trans*-acting (non-class I) gene(s). A limited family study suggests that this polymorphism segregates with the MHC, consistent with the possibility that it results from polymorphism of the *TAP* or *LMP* genes. Further data will be required for formal proof of this hypothesis.

3. The class II (exogenous) pathway

As compared to the class I pathway described above, a less definitive (but more complicated) picture of the class II antigen processing pathway is currently available. Molecular dissection of this pathway has been difficult, probably owing both to its relative complexity, and to the fact that (as for class I) the immune system has made use of an evolutionarily pre-existing, and essential degradative system for the production of peptides for class II molecules. The latter implies an inability to generate specific (viable) mutants for essential components of the system, hampering the molecular genetic approach. Nevertheless, certain components of the system do appear to be non-essential for viability, and much has been learned from their analysis, as detailed below.

Class II molecules obtain their peptides from proteins exterior to the cell, via endocytosis (see Fig. 1). The exact subcellular compartment(s) in which these peptides are produced (early endosomes, late endosomes, lysosomes) is still a matter of some debate, as is the location in which class II molecules intersect this degradative pathway in order to bind peptides. There is also still some controversy regarding the ability of class II molecules which have previously bound peptides to recycle from the surface into endosomal compartments to exchange their bound peptides. Some of these issues will be discussed in more detail below.

3.1 Proteases involved in the class II pathway

That MHC class II (but not class I) molecules pass through an endosomal or early lysosomal compartment during their transit to the cell surface (125–129), and that the peptides found naturally associated with MHC class II molecules are derived predominantly from endocytosed material (10–14), suggest that such peptides are generated via proteolysis by the normal complement of endosomal and lysosomal degradative enzymes. Indeed, compounds which raise the pH of acidified intracellular compartments (such as chloroquine), and hence block endosomal and lysosomal proteolysis, have been shown to inhibit class II (but not class I) antigen processing. The class II pathway is also blocked by inhibitors of endosomal proteases (for example leupeptin). However, the relatively recent demonstration that normal trafficking of class II molecules is regulated by (leupeptin- and chloroquine-sensitive) processing of the invariant chain (see Section 3.2.3) emphasizes the need for caution in interpreting the results of studies using such inhibitors. Such studies are plagued by an inability to prove that the inhibitor is specific for the intended process, and has no effects on perhaps unanticipated elements and mechanisms that may be important.

Specific acid proteases, such as cathepsin D (130) can often produce, *in vitro*, fragments of antigen that do not require further processing for presentation to MHC class II-restricted T cells. However, other enzymes, such as trypsin, or even chemical agents like cyanogen bromide frequently can do the same. This is perhaps not surprising given the variability in length of the sequences flanking the 'core' epitope found in naturally occurring class II-bound peptides (10–14). Thus, attempts to reproduce the proteolytic events of the class II antigen processing pathway *in vitro* are of limited utility in defining the proteases involved *in vivo*.

In any event, it seems unlikely that a single specific protease, as is presumably the case in the class I pathway, will be identified. Rather, class II binding peptides are probably produced by the concerted action of a number of enzymes, the relative contributions of which may vary for different antigens. In fact, reports of differential inhibition of the presentation of different antigens by a panel of protease inhibitors (131) suggest that different epitopes may be produced by the action of different proteases.

3.2 The role of invariant chain (I_i)

Invariant chain is a multifunctional type II integral membrane protein that associates with MHC class II α and β chains shortly after their synthesis in the ER. I_i is not MHC-encoded, the corresponding gene being found on chromosome 18 in mice (132, 133) and chromosome 5 in humans (134). As its name implies, I_i is non-polymorphic in inbred mice (135), although the inbred variant can be distinguished from that found in other mouse subspecies (136). The association of I_i with class II occurs originally via the binding of three α and three β chains to an invariant chain trimer, thus forming a nine-chain complex (137). I_i has been reported to aid in the proper folding of class II molecules (138), to prevent the binding of peptides to class II molecules in the ER (139), and to direct the class II/I_i complex to endocytic compartments (140–142), where I_i is degraded and class II molecules bind exogenous peptides, subsequently trafficking to the cell surface.

Mice made deficient in I_i expression by gene targeting (143, 144) have severely reduced (4–10-fold) class II surface expression, drastically reduced numbers of CD4+ T cells, and are defective in presentation of exogenous native antigen (but not peptide). For recent reviews of I_i biology, see refs 145–147.

3.2.1 I_i as a molecular chaperone

Several reports suggest that I_i influences the conformation of class II αβ hetero-dimers. Improperly assembled or misfolded proteins are generally retained and degraded in the ER. Efficient egress from the ER of pairs of class II α and β chains expressed in COS cells is variably influenced by the presence of I_i (148). In general, surface expression of αβ pairs whose intrinsic association is relatively poor is greatly enhanced by the presence of I_i, whereas surface expression of other αβ pairs is not affected, suggesting that I_i aids in the attainment by the poorly matched pairs of a 'transportable' conformation. This is consistent with what is seen in I_i knockout mice (143, 144) in which the defect in surface class II expression results, at least in part, from inefficient transport of class II dimers from the ER/*cis*-Golgi. Interestingly, it has also been reported that I_i can impart stable conformational changes in class II heterodimers (independent of peptide binding) in early bio-synthetic compartments, as measured by reactivity with certain monoclonal anti-bodies (138). These conformational class II epitopes induced in early compartments may enhance the ability of class II molecules to bind peptides later in acidified endosomal compartments (149), and can be maintained in I_i-free class II molecules at the cell surface (150). Differences in glycosylation between the mature, cell surface αβ dimers were also noted in otherwise identical cells that do and do not express I_i (138, 151). Thus, the data suggest that I_i has a true chaperone function with respect to class II heterodimer folding, in addition to its effects on transport and peptide binding described below.

3.2.2 I_i as a targeting signal

The most compelling evidence that I_i contains an endosomal sorting signal in-dependent of class II α and β chains comes from transfection studies. In cells lacking class II molecules, the majority of I_i is retained within the ER (152, 153). However, a proportion (15–20 per cent) of the I_i molecules are found in endosomes (140), and these molecules acquire endo H-resistant carbohydrate, indicating that they have passed through the Golgi. No I_i is found on the cell surface, suggesting that it is specifically retained in endosomes. This conclusion is supported by the demonstration that I_i lacking the N-terminal (cytoplasmic) 15 amino acids fails to localize to endosomes, and is expressed on the cell surface (140). The latter results further imply that the endosomal localization signal is located at the amino-terminus of the I_i polypeptide (also see Section 3.2.3).

Although class II molecules can be expressed at the cell surface in the absence of I_i expression (154, 155), the co-expression of I_i with class II can enhance the efficiency of transport of class II to endosomes (141, 156, 157). However, as discussed above, mediation of this effect via a chaperone type function cannot be ruled out. More recent data suggest that I_i retains, rather than targets, the class II complex to endosomes. Blocking the release of I_i from newly synthesized class II heterodimers with protease inhibitors or lysosomotropic agents such as chloro-quine also blocks their appearance at the cell surface (142, 158, 159), and causes them to accumulate in endosomes. Such a 'retention' function might be analogous to the mechanism by which ER resident proteins are retained within the ER lumen (these actually cycle between the *cis*-Golgi and the ER), because it has been shown that cell surface class II/I$_i$ complexes are rapidly internalized into endosomes (160). Thus, 'retention' in this context may perhaps more precisely involve retrieval back into endosomes of I$_i$-containing complexes which have appeared on the cell surface. The proportion of class II/I$_i$ complexes that enter endosomes via the cell surface is currently unknown, but this pathway may in fact be the rule rather than the exception. Moreover, enhanced antigen presentation by I$_i$ does not depend on the endosomal localization signal contained in the cytoplasmic tail of I$_i$ (see Section 3.2.3), because this can be deleted from the gene prior to transfection without significant impairment of enhancing ability (161). However, in this case, class II/I$_i$ complexes which appear at the cell surface are still rapidly internalized into endosomes, suggesting that class II molecules delivered to endosomes via the cell surface are competent to acquire and present peptides to T cells.

Other arguments against a *unique* sorting function (i.e. not present in class II heterodimers) for I$_i$ derive from data demonstrating that at least some (162–164), but not all (164–167), exogenous antigens can be presented to T cells by cells lacking I$_i$ expression, and that in such cells, class II can still localize to endosomes (168).

3.2.3 Alternative splicing, translational initiation, and proteolytic processing of I$_i$

At least four forms of human invariant chain exist, that differ in the protein portion of the molecule. These have been called p31, p33, p41 and p43, based on their molecular sizes[4] (see Fig. 4). The p41 and p43 chains differ from p31 and p33 by the inclusion of an alternatively spliced exon in the C-terminal portion of the sequence, encoding a 64 amino acid residue domain with sequence homology to a repeating motif found in thyroglobulin (169, 170). Both p31 and p41 appear to enhance the accumulation of class II in endocytic vesicles (157) and can restore antigen presenting function in some cases where I$_i$ expression is required (166). However, in other cases, enhancement of antigen presentation is dependent on the p41 form (171). It is hypothesized that this functional difference may be due to the possibility that the thyroglobulin-like domain unique to p41 (and p43) might chaperone class II molecules deeper into the endocytic pathway, where peptides from antigens more resistant to degradation in early endosomal compartments might be obtained.

The p31 and p41 polypeptides differ from p33 and p43, respectively, in that the latter contain an additional 16 amino acid residues at the N-terminus of the protein resulting from the alternative usage of an upstream in-frame translational start site

---

[4] Unfortunately, p31 and p33 have sometimes been referred to as p33 and p35, respectively, in the literature, thereby creating some confusion as to whether what is referred to as p33 is the shorter or longer form of the molecule.

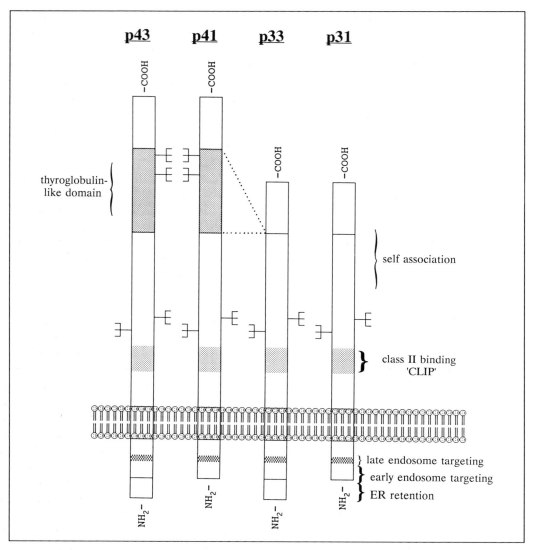

Fig. 4 The four forms of human invariant chain (I$_i$). Mice lack the amino-terminally extended forms (p43 and p33). The amino terminus (NH$_2$, bottom) extends into the cytoplasm, and the C-terminus (COOH, top) is lumenal. The locations of potential N-linked glycosylation sites (⌐⊥⌐) and of the CLIP peptide, implicated in binding to class II α and β chains, are indicated. The alternatively spliced thyroglobulin-like exon is indicated, as are portions of the molecules involved in ER retention and targeting to early and late endosomal compartments. The molecule also contains several O-linked glycosylation sites (not shown). See text for further details.

(169). This upstream AUG codon is present in human, but not in mouse, I_i cDNA (170), explaining the failure to observe p33 and p43 in mice. The p33 form is more strongly retained in the ER than the p31 form, suggesting that the N-terminal extension in the longer form contains an ER retention signal (141). Interestingly, an I_i (p31) variant lacking residues 2–11 of the cytoplasmic tail fails to accumulate in early endosomes, but appears in the late endosomal compartment (157). This implicates the N-terminal portion of p31 in regulating movement of the molecule from early to late endosomes. Removal of four more residues from the N-terminus results in the loss of endosomal targeting, and delivery of the molecule to the cell surface (140).

During its residency in endosomal compartments, the C-terminal portion of I_i is degraded in a number of steps, a process probably mediated at least in part by cathepsin B (172–174). This degradation results in the separation of the class II binding domain of I_i (see Section 3.2.4, and Fig. 4) from the endosomal retention signals, which results in the release of the class II dimer to the cell surface. As discussed above, leupeptin blocks the proteolytic processing of invariant chains, resulting in intracellular retention of class II molecules and inhibition of antigen presentation. The degradation of I_i also facilitates the exposure of the peptide binding site on $\alpha\beta$ dimers, albeit inefficiently (174).

In summary, the probable sequence of events in the transport of class II molecules to the cell surface are as follows. Trimers of invariant chains form in the ER, to which are added $\alpha\beta$ chain dimers to produce a nine subunit complex. Both invariant chains and individual class II α and β chains are detained in the ER, presumably by chaperonin-type molecules (perhaps identical to those involved in the class I pathway—calnexin and BiP), until the intact class II/I_i complex is generated. Signals in the N-terminal (cytoplasmic) portion of the I_i chain, and probably in the $\alpha\beta$ heterodimer, target the complex to early endosomes, either directly from the *trans*-Golgi network or via the cell surface, or both. These molecules are retained in early endosomes by virtue of the N-terminal portion (approximately 11 amino acids) of I_i. Masking or removal of this portion of I_i, perhaps related to partial degradation of the lumenal (C-terminal) portion of the molecule by endosomal hydrolases, allows the complex to then proceed into late endosomes and/lysosome-like compartments. Further degradation of I_i causes its dissociation from class II, at which time the class II molecule is released to the cell surface. While the molecule is traversing the endosomal pathway, the portion of I_i that blocks the peptide binding site of the class II molecule (see below) is also removed, and peptide binding occurs. The latter event transforms the class II dimer into a compact, SDS-resistant conformation, which may also be involved in releasing the molecule from the endosomal pathway to the cell surface.

3.2.4 I_i as an inhibitor of peptide binding

While free class II $\alpha\beta$ heterodimers will bind exogenously added peptides, class II/I_i complexes fail to do so, and exogenously added I_i competes with peptide for class II binding *in vitro* (139, 175, 176). Mutagenesis studies indicate that a region of I_i corresponding to residues 81–109 is required for binding of intact I_i to MHC class II heterodimers (177). Interestingly, a nested set of peptides derived from precisely this region of I_i is found associated with class II molecules purified from both normal (8–12) and mutant (see Section 3.3) cells. This set of peptides is collectively referred to as 'CLIP' (for class II-associated invariant chain peptides). CLIP binds with very high affinity to class II, and is an effective competitor for the binding of exogenous peptides to class II molecules *in vitro* (12, 178). Based on these data, it is tempting to speculate that the association of intact I_i with class II is mediated by binding of CLIP to the class II peptide binding site. The CLIP portion of I_i is located only 30 amino acid residues from the transmembrane region of I_i, but this is sufficient to reach the class II peptide binding site in either an extended (β sheet) or helical conformation. However, it seems unusual (but not impossible) that a single peptide sequence (CLIP) would bind with high affinity to the polymorphic peptide binding site of all class II molecules. Thus, it is not unreasonable to hypothesize that CLIP binds to a conserved region of the class II molecule distant from the peptide binding site, and that CLIP-mediated inhibition of peptide binding to class II is allosteric rather than steric. A definitive answer to this question should come from analysis of the crystal structure of class II/CLIP complexes.

3.3 Class II antigen processing mutants

Recently, several cell lines have been isolated that have a defect in the class II antigen processing pathway (179–182). The phenotype of these mutants is outwardly analogous to that discussed above (Section 2.2) for mutants in the class I pathway, in that they fail to present intact protein antigen, but not immunogenic peptides, to class II-restricted T cells. Presentation by all three (HLA-DP, -DQ and -DR) class II molecules is similarly affected. Although class II molecules are expressed at the surface of these cells, they are less stable than their counterparts found in the wild-type progenitor cell line, dissociating into individual α and β chains upon exposure to SDS. Furthermore, the class II molecules expressed by the mutants fail to express certain epitopes recognized by monoclonal antibodies. By analogy to the class I system, it was hypothesized that the mutants are impaired in the formation of complexes between class II molecules and peptides derived from endocytosed proteins. Subsequently, it was demonstrated that the peptide-binding groove of the majority of class II molecules of such mutants are occupied by CLIP, the nested set of peptides derived from residues 80–109 of I_i (178, 183). As discussed above, this class II/CLIP complex may represent a biosynthetic intermediate in normal cells. This intermediate apparently accumulates in the mutants because of a downstream block in the processing pathway. Interestingly, the genetic defect in the class II processing (cIIp) mutants was shown to map to the class II region of the MHC, near the *Tap* and *Lmp* genes (180). However, the cIIp mutants express normal amounts of class I molecules, suggesting that the defect in these cells is in a closely linked, but non-identical gene.

DMA and *DMB* are genes that map immediately centromeric to the *LMP-2* proteasome subunit gene in the human MHC (see Fig. 2). They were discovered by cross-hybridization to their mouse counterparts, *Ma* and *Mb*, respectively, and are related in sequence as well as in overall structure to MHC class II α and β chain genes (184, 185). Based on this structural homology, they were originally assumed to produce a heterodimeric class II molecule, whose function was thought to be to present antigenic peptides to T cells. Surprisingly, the defect in several cIIp mutants is reversed by transfection of the *DMB* (or *Mb*) genes (186). These data suggest that the *DM* molecule functions at an intracellular site, rather than at the cell surface, and is required for optimal loading of other class II molecules with exogenous peptides. Interestingly, a point mutation in the DR α chain gene results in a similar phenotype (but only with respect to HLA-DR), suggesting the possibility that physical interaction between class II molecules and DM molecules is required for DM function (187).

While the molecular basis of the DM requirement is currently unknown, several possibilities have been suggested (186). DM could be another class II chaperone, required for an essential trafficking or folding step in class II biosynthesis. Alternatively, despite the limited sequence similarity (14–25 per cent identity in the membrane distal domains) of DM and other class II molecules, it may be that HLA-DM assumes a class II-like structure, and is also a specialized peptide binding molecule. In this event, DM may act as a peptide carrier to rescue and/or deliver peptides from a degradative compartment to class II-containing endosomes. This would explain the accumulation of class II/CLIP complexes in the mutants, in that availability of other (non-CLIP) peptides would be limiting. Alternatively, the DM molecule may co-localize with class II and bind CLIP with very high-affinity, serving as a molecular 'sink' for such peptides, thereby favouring binding of immunogenic peptides to other class II molecules. The resolution of this issue will no doubt be determined in the very near future.

3.4 Presentation of endogenous antigen by MHC class II molecules

Although MHC class II molecules acquire peptides predominantly in the endocytic pathway, as described above, numerous reports of class II-restricted presentation of endogenous, cytoplasmic antigens to CD4$^+$ T cells have appeared (see ref. 2 for references therein). There are a number of potential ways in which this might come about:

1. It may be that competition for class II binding in the ER occurs between I_i and peptides either generated in that compartment, or transported into it via the TAP transporter. Some particularly high-affinity or abundant endogenously-derived peptides may attain occupancy of the class II peptide binding site in this manner. A perhaps less likely alternative is that such peptides could be carried to endosomes by bulk flow, or by specific carrier proteins, and bind to class II molecules in that compartment subsequent to the degradation of I_i.

2. Cytoplasmic proteins can be imported directly into lysosomes in an ATP-dependent process involving members of the hsp70 heat-shock proteins family (188).

3. Cytoplasmic material may enter the endosomal processing pathway via the processes of microautophagy (invagination of lysosomal membranes), or autophagy, in which cytoplasm and even intracellular organelles can be engulfed in double-membrane vesicles derived from smooth ER (189–191). These autophagic vacuoles are then acidified and can acquire lysosomal hydrolases either directly from the ER or by fusion with lysosomes.

4. Conclusions

The immune system has utilized, and in some cases, apparently adapted, components of the two major pre-existing pathways of intracellular protein degradation in order to supply MHC molecules with appropriate peptides. MHC class I molecules are most likely supplied with peptides produced by proteasomes, which are essential components of the ubiquitin-dependent protein degradation pathway and are believed to participate in most of the non-lysosomal protein turnover in eukaryotic cells. Evolutionarily, proteasomes clearly pre-date the mammalian immune system, and hence class I molecules have presumably evolved to bind the types of peptides produced by this structure, rather than vice versa. Two subunits of this complex are encoded by MHC genes. These two subunits share significant sequence homology with other non-MHC-linked proteasome subunits, and presumably arose by duplication and divergence of other proteasome subunit genes, but are unique in several respects. They are the only two subunits whose expression is known to be non-essential for normal cell growth, the only two known to be regulated by IFN-γ, and the only two that demonstrate electrophoretic polymorphism in inbred mice. Thus, it is hypothesized that their function is to regulate or alter the function of a subset of proteasomes in a manner beneficial to the immune system. However, expression of these two genes is not required for apparently normal operation of the class I antigen processing pathway. Although the MHC-encoded subunits have been demonstrated to produce subtle alterations in the proteolytic activity of the complex against model substrates, the biological significance of this observation is still uncertain. In any event, the peptide products resulting from degradation of intracellular protein by proteasomes are transported into the ER for binding to MHC class I molecules by the products of another pair of MHC-linked genes, the *Tap* genes, which are in close proximity to the two proteasome subunit genes.

In contrast, MHC class II molecules are supplied with peptides from the endosomal/lysosomal pathway of protein degradation. This pathway also pre-dates the immune system, and is responsible for the normal turnover of extracellular and cell surface proteins, as well as the catabolism of intracellular material and organelles under conditions of starvation. Class II molecules have adapted structures and subcellular trafficking patterns that allow them to salvage peptides from this pathway. Interestingly, a molecule closely related in structure to class II molecules (and encoded by MHC-linked genes) is required for the normal acquisi-tion of peptides by other ('classical') class II molecules, although the molecular mechanisms related to this requirement are still unknown.

Molecular chaperones are seen to play a role in both pathways, in retaining class I molecules within the ER until they have acquired peptides, and in preventing class II molecules from acquiring peptides in that location and directing them to endosomes for peptide loading. Thus, class I molecules have become specialized for presentation of intracellular antigen to T cells, whereas class II molecules specialize in presenting extracellular antigen. Though closely related in overall structure, these two classes of molecules have subtle differences that may be relevant to their specialized functions. For example, binding of peptides to class II molecules is enhanced at acid pH. Furthermore, peptide is required for the stable association of the two chains of the class I heterodimer, and both chains are required for stable peptide binding. This property, along with the relatively stringent requirements for peptide length, limits the ability of class I molecules to bind exogenous peptides at the cell surface, thus presumably protecting normal cells *in vivo* from being sensitized for CTL lysis by exogenous peptide.[5] Class II molecules, on the other hand, apparently have less stringent requirements for peptide length, and have (reduced, but sufficient) stability in the absence of peptide to be found on the cell surface under physiological conditions.

There are undoubtedly subtleties of MHC structure and antigen processing that remain to be discovered, not the least of which is the potential contribution of polymorphism in antigen processing to immune response defects and susceptibility to autoimmune disease. Furthermore, it seems likely that molecules and mech-anisms unique to cellular antigen processing pathways will be targets for potential subversion of the immune response by pathogenic microorganisms. The extra-ordinarily rapid recent progress in this field suggests that the remaining mysteries of antigen processing will soon yield to investigation.

Acknowledgements

The author is supported by grants from the National Institute of Allergy and Infectious Diseases.

References

References marked as * are recommended for further reading.

*1. Brown, J. H., Jardetzky, T. S., Gorga, J. C., Stern, L. J., Urban, R. G., Strominger, J. L., and Wiley, D. C. (1993). Three-dimensional structure of the human class II histocom-patibility antigen HLA-DR1. *Nature*, **364**, 33.

---

[5] However, exogenous peptide of the correct length added at high concentrations *in vitro* will render targets susceptible to CTL-mediated lysis.

*2. Germain, R. N. and Margulies, D. H. (1993). The biochemistry and cell biology of antigen processing and presentation. *Annu. Rev. Immunol.*, **11**, 403.

  3. Rammensee, H. G., Falk, K., and Rötzschke, O. (1993). Peptides naturally presented by MHC class I molecules. *Annu. Rev. Immunol.*, **11**, 213.

  4. Engelhard, V. (1994). Structure of peptides associated with class I MHC molecules. *Curr. Opin. Immunol.*, **6**, 13.

*5. Monaco, J. J. (1992). A molecular model of MHC class-I-restricted antigen processing. *Immunol. Today*, **13**, 173.

  6. Braciale, T. J., Morrison, L. A., Sweetser, M. T., Sambrook, J., Gething, M. J., and Braciale, V. L. (1987). Antigen presentation pathways to class I and class II MHC-restricted T lymphocytes. *Immunol. Rev.*, **98**, 95.

*7. Yewdell, J. W. and Bennink, J. R. (1990). The binary logic of antigen processing and presentation to T cells. *Cell*, **62**, 203.

  8. Moore, M. W., Carbone, F. R., and Bevan, M. J. (1988). Introduction of soluble protein into the class I pathway of antigen processing and presentation. *Cell*, **54**, 777.

  9. Yewdell, J. W., Bennink, J. R., and Hosaka, Y. (1988). Cells process exogenous proteins for recognition by cytotoxic T lymphocytes. *Science*, **239**, 637.

 10. Rudensky, A. Y., Preston-Hurlburt, P., Hong, S.-C., Barlow, A., and Janeway, C. A. (1991). Sequence analysis of peptides bound to MHC class II molecules. *Nature*, **353**, 622.

 11. Hunt, D. F., Michel, H., Dickinson, T. A., Shabanowitz, J., Cox, A. L., Sakaguchi, K., Appella, E., Grey, H. M., and Sette, A. (1992). Peptides presented to the immune system by the murine class II major histocompatibility complex molecule I-A^d. *Science*, **256**, 1817.

 12. Chicz, R. M., Urban, R. G., Lane, W. S., Gorga, J. C., Stern, L. J., Vignali, D. A. A., and Strominger, J. L. (1992). Predominant naturally processed peptides bound to HLA–DR1 are derived from MHC related molecules and are heterogenous in size. *Nature*, **358**, 764.

 13. Chicz, R. M., Urban, R. G., Gorga, J. C., Vignali, D. A. A., Lane, W. S., and Strominger, J. L. (1993). Specificity and promiscuity among naturally processed peptides bound to HLA–DR alleles. *J. Exp. Med.*, **178**, 27.

 14. Newcomb, J. R. and Cresswell, P. (1993). Characterization of endogenous peptides bound to purified HLA–DR molecules and their absence from invariant chain-associated αβ dimers. *J. Immunol.*, **150**, 499.

 15. Townsend, A. and Bodmer, H. (1989). Antigen recognition by class I-restricted T lymphocytes. *Annu. Rev. Immunol.*, **7**, 601.

 16. Braciale, T. J., Andrew, M. E., and Braciale, V. L. (1981). Heterogeneity and specificity of cloned lines of influenza-virus-specific cytotoxic T lymphocytes. *J. Exp. Med.*, **153**, 910.

 17. Bennink, J. R., Yewdell, J. W., and Gerhard, W. (1982). A viral polymerase involved in recognition of influenza-infected cells by a cytotoxic T cell clone. *Nature*, **296**, 75.

 18. Townsend, A. R. M. and Skehel, J. J. (1982). T cell clones that do not recognize viral glycoproteins. *Nature*, **300**, 655.

 19. Townsend, A. R. M., Bastin, J., Gould, K., and Brownlee, G. G. (1986). Cytotoxic T lymphocytes recognize influenza haemagglutinin that lacks a signal sequence. *Nature*, **324**, 575.

 20. Townsend, A., Bastin, J., Gould, K., Brownlee, G., Andrew, M., Coupar, B., Boyle, D., Chan, S., and Smith, G. (1988). Defective presentation to class I-restricted cytotoxic

T lymphocytes in vaccinia-infected cells is overcome by enhanced degradation of antigen. *J. Exp. Med.*, **168**, 1211.

21. Whitton, J. L. and Oldstone, M. B. A. (1989). Class I MHC can present an endogenous peptide to cytotoxic T lymphocytes. *J. Exp. Med.*, **170**, 1033.

22. Sweetser, M. T., Morrison, L. A., Braciale, V. L., and Braciale, T. J. (1989). Recognition of pre-processed endogenous antigen by class I but not class II MHC-restricted T cells. *Nature*, **342**, 180.

23. Anderson, K., Cresswell, P., Gammon, M., Hermes, J., Williamson, A., and Zweerink, H. (1991). Endogenously synthesized peptide with an endoplasmic reticulum signal sequence sensitizes antigen processing mutant cells to class I-restricted cell-mediated lysis. *J. Exp. Med.*, **174**, 489.

24. Bodmer, J. G., Marsh, S. G. E., Albert, E. D., Bodmer, W. F., Dupont, B., Erlich, H. A., Mach, B., Mayr, W. R., Parham, P., Sasazuki, T., Schreuder, G. M. Th., Strominger, J. L., Svejgaard, A., and Terasaki, P. I. (1992). Nomenclature for factors of the HLA system, 1991. *Tissue Ags.*, **39**, 161.

25. Monaco, J. J., Cho, S., and Attaya, M. (1990). Transport protein genes in the murine MHC: Possible implications for antigen processing. *Science*, **250**, 1723.

26. Deverson, E. V., Gow, I. R., Coadwell, W. J., Monaco, J. J., Butcher, G. W., and Howard, J. C. (1990). MHC class II region encoding proteins related to the multidrug resistance family of transmembrane transporters. *Nature*, **348**, 738.

27. Trowsdale, J., Hanson, I., Mockridge, I., Beck, S., Townsend, A., and Kelly, A. (1990). Sequences encoded in the class II region of the MHC related to the 'ABC' superfamily of transporters. *Nature*, **348**, 741.

28. Spies, T., Bresnahan, M., Bahram, S., Arnold, D., Blanck, G., Mellins, E., Pious, D., and DeMars, R. (1990). A gene in the human major histocompatibility complex class II region controlling the class I antigen processing pathway. *Nature*, **348**, 744.

29. Spies, T. and DeMars, R. (1991). Restored expression of major histocompatibility class I molecules by gene transfer of a putative peptide transporter. *Nature*, **351**, 323.

30. Bahram, S., Arnold, D., Bresnahan, M., Strominger, J. L., and Spies, T. (1991). Two putative subunits of a peptide pump encoded in the human major histocompatibility complex class II region. *Proc. Natl Acad. Sci. USA*, **88**, 10094.

31. Kärre, K., Ljunggren, H. G., Piontek, G., and Kiessling, R. (1986). Selective rejection of H–2-deficient lymphoma variants suggests alternative immune defense strategy. *Nature*, **319**, 675.

32. Townsend, A., Öhlén, C., Bastin, J., Ljunggren, H.-G., Foster, L., and Kärre, K. (1989). Association of class I major histocompatibility heavy and light chains induced by viral peptides. *Nature*, **340**, 443.

33. DeMars, R., Rudersdorf, R., Chang, C., Petersen, J., Strandtmann, J., Korn, N., Sidwell, B., and Orr, H. T. (1985). Mutations that impair a posttranscriptional step in expression of HLA-A and -B antigens. *Proc. Natl Acad. Sci. USA*, **82**, 8183.

34. Hosken, N. A. and Bevan, M. J. (1990). Defective presentation of endogenous antigen by a cell line expressing class I molecules. *Science*, **248**, 367.

35. Cerundolo, V., Alexander, J., Anderson, K., Lamb, C., Cresswell, P., McMichael, A., Gotch, F., and Townsend, A. (1990). Presentation of viral antigen controlled by a gene in the major histocompatibility complex. *Nature*, **345**, 449.

36. Townsend, A., Elliott, T., Cerundolo, V., Foster, L., Barber, B., and Tse, A. (1990). Assembly of MHC class I molecules analysed *in vitro*. *Cell*, **62**, 285.

37. Ljunggren, H. G., Stam, N. J., Öhlen, C., Neefjes, J. J., Höglund, P., Heemels, M.-T.,

Bastin, J., Schumacher, T. N. M., Townsend, A., Kärre, K., and Ploegh, H. (1990). Empty MHC class I molecules come out in the cold. *Nature*, **346**, 476.

38. Schumacher, T. N. M., Heemels, M. T., Neefjes, J. J., Kast, W. M., Melief, C. J. M., and Ploegh, H. (1990). Direct binding of peptide to empty MHC class I molecules on intact cells and *in vitro*. *Cell*, **62**, 563.

39. Elliott, T., Cerundolo, V., Elvin, J., and Townsend, A. (1991). Peptide-induced conformational change of the class I heavy chain. *Nature*, **351**, 402.

40. Rock, K. L., Gramm, C., and Benacerraf, B. (1991). Low temperature and peptides favor the formation of class I heterodimers on RMA-S cells at the cell surface. *Proc. Natl Acad. Sci. USA*, **88**, 4200.

41. Yang, Y., Früh, K., Chambers, J., Waters, J. B., Wu, L., Spies, T., and Peterson, P. A. (1992). Major histocompatibility complex (MHC)-encoded HAM2 is necessary for antigenic peptide loading on to class I MHC molecules. *J. Biol. Chem.*, **267**, 11669.

42. Attaya, M., Jameson, S., Martinez, C. K., Hermel, E., Aldrich, C., Forman, J., Fischer Lindahl, K., Bevan, M. J., and Monaco, J. J. (1992). *Ham-2* corrects the class I antigen processing defect in RMA-S cells. *Nature*, **355**, 647.

43. Powis, S. J., Townsend, A. R. M., Deverson, E. V., Bastin, J., Butcher, G. W., and Howard, J. C. (1991). Restoration of antigen presentation to the mutant cell line RMA-S by an MHC-linked transporter. *Nature*, **354**, 528.

44. Salter, R. D. and Cresswell, P. (1986). Impaired assembly and transport of HLA-A and -B antigens in a mutant T × B cell hybrid. *EMBO J.*, **5**, 943.

45. Arnold, D., Driscoll, J., Androlewicz, M., Hughes, E., Cresswell, P., and Spies, T. (1992). Proteasome subunits encoded in the MHC are not generally required for the processing of peptides bound by MHC class I molecules. *Nature*, **360**, 171.

46. Momburg, F., Ortiz-Navarrete, V., Neefjes, J., Goulmy, E., van de Wal, Y., Spits, H., Powis, S., Butcher, G. W., Howard, J. C., Walden, P., and Hämmerling, G. J. (1992). Proteasome subunits encoded by the major histocompatibility complex are not essential for antigen presentation. *Nature*, **360**, 174.

*47. Van Kaer, L., Ashton-Rickardt, P. G., Ploegh, H. L., and Tonegawa, S. (1992). *TAP1* mutant mice are deficient in antigen presentation, surface class I molecules, and CD4⁻8⁺ T cells. *Cell*, **71**, 1205.

*48. Hyde, S. C., Emsley, P., Hartshorn, M. J., Mimmack, M. M., Gileadi, U., Pearce, S. R., Gallagher, M. P., Gill, D. R., Hubbard, R. E., and Higgins, C. F. (1990). Structural model of ATP-binding proteins associated with cystic fibrosis, multidrug resistance and bacterial transport. *Nature*, **346**, 362.

49. Spies, T., Cerundolo, V., Colonna, M., Cresswell, P., Townsend, A., and DeMars, R. (1992). Presentation of viral antigen by MHC class I molecules is dependent on a putative peptide transporter heterodimer. *Nature*, **355**, 644.

50. Kelly, A., Powis, S. H., Kerr, L.-A., Mockridge, I., Elliott, T., Bastin, J., Uchanska-Ziegler, B., Ziegler, A., Trowsdale, J., and Townsend, A. (1992). Assembly and function of the two ABC transporter proteins encoded in the human major histocompatibility complex. *Nature*, **355**, 641.

51. Kleijmeer, M. J., Kelly, A., Geuze, H. J., Slot, J. W., Townsend, A., and Trowsdale, J. (1992). Location of MHC-encoded transporters in the endoplasmic reticulum and *cis*-Golgi. *Nature*, **357**, 342.

52. Nuchtern, J. G., Bonifacino, J. S., Biddison, W. E., and Klausner, R. D. (1989). Brefeldin A implicates egress from endoplasmic reticulum in class I restricted antigen presentation. *Nature*, **339**, 223.

53. Yewdell, J. W. and Bennink, J. R. (1989). Brefeldin A specifically inhibits presentation of protein antigens to cytotoxic T lymphocytes. *Science*, **244**, 1072.

54. Neefjes, J. J., Momburg, F., and Hämmerling, G. J. (1993). Selective and ATP-dependent translocation of peptides by the MHC-encoded transporter. *Science*, **261**, 769.

55. Shepherd, J. C., Schumacher, T. N. M., Ashton-Rickardt, P. G., Imaeda, S., Ploegh, H. L., Janeway, C. A., and Tonegawa, S. (1993). TAP1-dependent peptide transloca-tion *in vitro* is ATP dependent and peptide selective. *Cell*, **74**, 577.

56. Androlewicz, M. J., Anderson, K. S., and Cresswell, P. (1993). Evidence that trans-porters associated with antigen processing translocate a major histocompatibility com-plex class I-binding peptide into the endoplasmic reticulum in an ATP-dependent manner. *Proc. Natl Acad. Sci. USA*, **90**, 9130.

*57. Schumacher, T. N. M., Kantesaria, D. V., Heemels, M.-T., Ashton-Rickardt, P. G., Shepherd, J. C., Früh, K., Yang, Y., Peterson, P. A., Tonegawa, S., and Ploegh, H. L. (1994). Peptide length and sequence specificity of the mouse TAP1/TAP2 translocator. *J. Exp. Med.*, **179**, 533.

*58. Heemels, M.-T., Schumacher, T. N. M., Wonigeit, K., and Ploegh, H. L. (1993). Peptide translocation by variants of the transporter associated with antigen process-ing. *Science*, **262**, 2059.

59. Lévy, F., Gabathuler, R., Larsson, R., and Kvist, S. (1991). ATP is required for *in vitro* assembly of MHC class I antigens but not for transfer of peptides across the ER membrane. *Cell*, **67**, 265.

60. Koppelman, B., Zimmerman, D. L., Walter, P., and Brodsky, F. M. (1992). Evidence for peptide transport across microsomal membranes. *Proc. Natl Acad. Sci. USA*, **89**, 3908.

61. Anderson, K., Cresswell, P., Gammon, M., Hermes, J., Williamson, A., and Zweerink, H. (1991). Endogenously synthesized peptide with an endoplasmic reticu-lum signal sequence sensitizes antigen processing mutant cells to class I-restricted cell-mediated lysis. *J. Exp. Med.*, **174**, 489.

*62. Bacik, I., Cox, J. H., Anderson, R., Yewdell, J., and Bennink, J. R. (1994). TAP (transporter associated with antigen processing)-independent presentation of endo-genously synthesized peptides is enhanced by endoplasmic reticulum insertion sequences located at the amino- but not carboxyl-terminus of the peptide. *J. Immunol.*, **152**, 381.

*63. Henderson, R. A., Michel, H., Sakaguchi, K., Shabanowitz, J., Appella, E., Hunt, D. F., and Engelhard, V. H. (1992). HLA-A2.1-associated peptides from a mutant cell line: A second pathway of antigen presentation. *Science*, **255**, 1264.

64. Wei, M. L. and Cresswell, P. (1992). HLA-A2 molecules in an antigen-processing mutant cell contain signal sequence-derived peptides. *Nature*, **356**, 443.

65. Buchmeier, M. J. and Zinkernagel, R. M. (1992). Immunodominant T cell epitope from signal sequence. *Science*, **257**, 1142.

*66. Esquivel, F., Yewdell, J., and Bennink, J. (1992). RMA/S cells present endogenously synthesized cytosolic proteins to class I-restricted cytotoxic T lymphocytes. *J. Exp. Med.*, **175**, 163.

*67. Hosken, N. A. and Bevan, M. J. (1992). An endogenous antigenic peptide bypasses the class I antigen presentation defect in RMA-S. *J. Exp. Med.*, **175**, 719.

68. Zhou, X., Glas, R., Momburg, F., Hämmerling, G. J., Jondal, M., and Ljunggren, H.-G. (1993). Tap2-defective RMA-S cells present Sendai virus antigen to cytotoxic T lymphocytes. *Eur. J. Immunol.*, **23**, 1796.

*69. Carbone, F. R. and Bevan, M. J. (1990). Class I-restricted processing and presentation of exogenous cell-associated antigen *in vivo*. *J. Exp. Med.*, **171**, 377.

*70. Pfeifer, J. D., Wick, M. J., Roberts, R. L., Findlay, K., Normark, S. J., and Harding, C. V. (1993). Phagocytic processing of bacterial antigens for class I MHC presentation to T cells. *Nature*, **361**, 359.

71. Rock, K. L., Gamble, S., and Rothstein, L. (1990). Presentation of exogenous antigen with class I major histocompatibility complex molecules. *Science*, **249**, 918.

72. Zhou, X., Glas, R., Liu, T., Ljunggren, H.-G., and Jondal, M. (1993). Antigen processing mutant T2 cells present viral antigen restricted through H–2K^b. *Eur. J. Immunol.*, **23**, 1802.

73. Lippincott-Schwartz, J., Bonifacino, J. S., Yuan, L. C., and Klausner, R. D. (1988). Degradation from the endoplasmic reticulum: Disposing of newly synthesized proteins. *Cell*, **54**, 209.

74. Hammond, S. A., Bollinger, R. C., Tobery, T. W., and Siliciano, R. F. (1993). Transporter-independent processing of HIV-1 envelope protein for recognition by CD8$^+$ T cells. *Nature*, **364**, 158.

75. Urban, R. G., Chicz, R. M., Lane, W. S., Strominger, J. L., Rehm, A., Kenter, M. J. H., UytdeHaag, F. G. C. M., Ploegh, H., Uchanska-Ziegler, B., and Ziegler, A. (1994). A subset of HLA-B27 molecules contains peptides much longer than nonamers. *Proc. Natl Acad. Sci. USA*, **91**, 1534.

*76. Rivett, A. J. (1989). The multicatalytic proteinase of mammalian cells. *Arch. Biochem. Biophys.*, **268**, 1.

*77. Orlowski, M. (1990). The multicatalytic proteinase complex, a major extralysosomal proteolytic system. *Biochemistry*, **29**, 10289.

*78. Goldberg, A. L. and Rock, K. L. (1992). Proteolysis, proteasomes and antigen presentation. *Nature*, **357**, 375.

79. Zwickl, P., Lottspeich, F., Dahlmann, B., and Baumeister, W. (1991). Cloning and sequencing of the gene encoding the large (α-) subunit of the proteasome from *Thermoplasma acidophilum*. *FEBS Lett.*, **278**, 217.

80. Grziwa, A., Baumeister, W., Dahlmann, B., and Kopp, F. (1991). Localization of subunits in proteasomes from *Thermoplasma acidophilum* by immunoelectron microscopy. *FEBS Lett.*, **290**, 186.

81. Heinemeyer, W., Kleinschmidt, J. A., Saidowsky, J., Escher, C., and Wolf, D. H. (1991). Proteinase yscE, the yeast proteasome/multicatalytic-multifunctional proteinase: mutants unravel its function in stress induced proteolysis and uncover its necessity for cell survival. *EMBO J.*, **10**, 555.

82. Rivett, A. J. (1985). Purification of a liver alkaline protease which degrades oxidatively modified glutamine synthetase. *J. Biol. Chem.*, **260**, 12600.

83. Dick, L. R., Moomaw, C. R., DeMartino, G. N., and Slaughter, C. A. (1991). Degradation of oxidized insulin B chain by the multiproteinase complex macropain (proteasome). *Biochemistry*, **30**, 2725.

84. Driscoll, J. and Goldberg, A. L. (1990). The proteasome (multicatalytic protease) is a component of the 1500-kDa proteolytic complex which degrades ubiquitin-conjugated proteins. *J. Biol. Chem.*, **265**, 4789.

85. Fujiwara, T., Tanaka, K., Orino, E., Yoshimura, T., Kumatori, A., Tamura, T., Chung, C. H., Nakai, T., Yamaguchi, K., Shin, S., Kakizuka, A., Nakanishi, S., and Ichihara, A. (1990). Proteasomes are essential for yeast proliferation. *J. Biol. Chem.*, **265**, 16604.

86. Emori, Y., Tsukahara, T., Kawasaki, H., Ishiura, S., Sugita, H., and Suzuki, K. (1991).

Molecular cloning and functional analysis of three subunits of yeast proteasome. *Mol. Cell. Biol.*, **11**, 344.

87. Monaco, J. J. and McDevitt, H. O. (1982). Identification of a fourth class of proteins linked to the murine major histocompatibility complex. *Proc. Natl Acad. Sci. USA*, **79**, 3001.

88. Monaco, J. J. and McDevitt, H. O. (1984). H–2 linked low-molecular weight polypeptide antigens assemble into an unusual macromolecular complex. *Nature*, **309**, 797.

89. Monaco, J. J. and McDevitt, H. O. (1986). The LMP antigens: A stable MHC-controlled multisubunit protein complex. *Hum. Immunol.*, **15**, 416.

90. Brown, M. G., Driscoll, J., and Monaco, J. J. (1991). Structural and serological similarity of MHC-linked LMP and proteasome (multicatalytic proteinase) complexes. *Nature*, **353**, 355.

91. Glynne, R., Powis, S. H., Beck, S., Kelly, A., Kerr, L.-A., and Trowsdale, J. (1991). A proteasome-related gene between the two ABC transporter loci in the class II region of the human MHC. *Nature*, **353**, 357.

92. Ortiz-Navarrete, V., Seelig, A., Gernold, M., Frentzel, S., Kloetzel, P. M., and Hämmerling, G. J. (1991). Subunit of the '20S' proteasome (multicatalytic proteinase) encoded by the major histocompatibility complex. *Nature*, **353**, 662.

93. Martinez, C. K. and Monaco, J. J. (1991). Homology of proteasome subunits to a major histocompatibility complex-linked LMP gene. *Nature*, **353**, 664.

94. Kelly, A., Powis, S. H., Glynne, R., Radley, E., Beck, S., and Trowsdale, J. (1991). Second proteasome-related gene in the human MHC class II region. *Nature*, **353**, 667.

95. Michalek, M. T., Grant, E. P., Gramm, C., Goldberg, A. L., and Rock, K. L. (1993). A role for the ubiquitin-dependent proteolytic pathway in MHC class I-restricted antigen presentation. *Nature*, **363**, 552.

96. Rivett, A. J., Palmer, A., and Knecht, E. (1992). Electron microscopic localization of the multicatalytic proteinase complex in rat liver and in cultured cells. *J. Histochem. Cytochem.*, **40**, 1165.

97. Yewdell, J., Lapham, C., Bacik, I., Spies, T., and Bennink, J. (1994). MHC-encoded proteasome subunits LMP2 and LMP7 are not required for efficient antigen presentation. *J. Immunol.*, **152**, 1163.

*98. Brown, M. G., Driscoll, J., and Monaco, J. J. (1993). MHC-linked low molecular mass polypeptide subunits define distinct subsets of proteasomes: Implications for divergent function among distinct proteasome subsets. *J. Immunol.*, **151**, 1193.

99. Driscoll, J., Brown, M. G., Finley, D., and Monaco, J. J. (1993). MHC-linked *LMP* gene products specifically alter peptidase activities of the proteasome. *Nature*, **365**, 262.

100. Gaczynska, M., Rock, K. L., and Goldberg, A. L. (1993). γ-Interferon and expression of MHC genes regulate peptide hydrolysis by proteasomes. *Nature*, **365**, 264.

101. Martinez, C. K. and Monaco, J. J. (1993). Post-translational processing of a major histocompatibility complex-encoded proteasome subunit, LMP-2. *Mol. Immunol.*, **30**, 1177.

102. Patel, S. D., Monaco, J. J., and McDevitt, H. O. (1994). Delineation of the subunit composition of human proteasomes using antisera against the major histocompatibility complex-encoded LMP2 and LMP7 subunits. *Proc. Natl Acad. Sci. USA*, **91**, 296.

103. Hsu, V. W., Yuan, L. C., Nuchtern, J. G., Lippincott-Schwartz, J., Hämmerling, G. J., and Klausner, R. D. (1991). A recycling pathway between the endoplasmic reticulum and the Golgi apparatus for retention of unassembled MHC class I molecules. *Nature*, **352**, 441.

104. Degen, E., Cohen-Doyle, M. F., and Williams, D. B. (1992). Efficient dissociation of the p88 chaperone from major histocompatibility complex class I molecules requires both β₂-microglobulin and peptide. *J. Exp. Med.*, **175**, 1653.

105. Jackson, M. R., Cohen-Doyle, M. F., Peterson, P. A., and Williams, D. B. (1994). Regulation of MHC class I transport by the molecular chaperone, calnexin (p88, IP90). *Science*, **263**, 384.

106. Carbone, F. R., Moore, M. W., Sheil, J. M., and Bevan, M. J. (1988). Induction of cytotoxic T lymphocytes by primary *in vitro* stimulation with peptides. *J. Exp. Med.*, **167**, 1767.

107. Schild, H., Rötzschke, O., Kalbacher, H., and Rammensee, H.-G. (1990). Limit of T cell tolerance to self-proteins by peptide presentation. *Science*, **247**, 1587.

108. del Val, M., Schlicht, H.-J., Ruppert, T., Reddehase, M. J., and Koszinowski, U. H. (1991). Efficient processing of an antigenic sequence for presentation by MHC class I molecules depends on its neighboring residues in the protein. *Cell*, **66**, 1145.

109. Eisenlohr, L. C., Yewdell, J. W., and Bennink, J. R. (1992). Flanking sequences influence the presentation of an endogenously synthesized peptide to cytotoxic T lymphocytes. *J. Exp. Med.*, **175**, 481.

110. Cerundolo, V., Tse, A. G. D., Salter, R. D., Parham, P., and Townsend, A. (1991). CD8 independence and specificity of cytotoxic T lymphocytes restricted by HLA-Aw68.1. *Proc. R. Soc. Lond. (Biol.)*, **244**, 169.

111. Colonna, M., Bresnahan, M., Bahram, S., Strominger, J. L., and Spies, T. (1992). Allelic variants of the human putative peptide transporter involved in antigen processing. *Proc. Natl Acad. Sci. USA*, **89**, 3936.

112. Powis, S. H., Tonks, S., Mockridge, I., Kelly, A. P., Bodmer, J., and Trowsdale, J. (1993). Alleles and haplotypes of the MHC-encoded ABC transporters *TAP1* and *TAP2*. *Immunogenetics*, **37**, 373.

113. Marusina, K., Ginsburg, D. B., and Monaco, J. J. (1995). Allelic variation in the mouse *Tap-1* and *Tap-2* genes. (Submitted)

114. Powis, S. J., Deverson, E. V., Coadwell, W. J., Ciruela, A., Huskisson, N. S., Smith, H., Butcher, G. W., and Howard, J. C. (1992). Effect of polymorphism of an MHC-linked transporter on the peptides assembled in a class I molecule. *Nature*, **357**, 211.

115. Zhou, P., Cao, H., Smart, M., and David, C. (1993). Molecular basis of genetic polymorphism in major histocompatibility complex-linked proteasome gene (*Lmp-2*). *Proc. Natl Acad. Sci. USA*, **90**, 2681.

*116. Tiwari, J. L. and Terasaki, P. I. (1985). *HLA and disease associations*. Springer–Verlag, New York.

117. Falk, K., Rötzschke, O., and Rammensee, H.-G. (1992). Specificity of antigen processing for MHC class I restricted presentation is conserved between mouse and man. *Eur. J. Immunol.*, **22**, 1323.

118. Rojo, S., Lopez, D., Calvo, V., and Lopez de Castro, J. A. (1991). Conservation and alteration of HLA-B27-specific T cell epitopes on mouse cells. *J. Immunol.*, **146**, 634.

119. Lobigs, M. and Müllbacher, A. (1993). Recognition of vaccinia virus-encoded major histocompatibility complex class I antigens by virus immune cytotoxic T cells is independent of the polymorphism of the peptide transporters. *Proc. Natl Acad. Sci. USA*, **90**, 2676.

120. Yewdell, J. W., Esquivel, F., Arnold, D., Spies, T., Eisenlohr, L. C., and Bennink, J. R. (1990). Presentation of numerous viral peptides to mouse major histocompatibility complex (MHC) class I-restricted T lymphocytes is mediated by the human MHC-encoded transporter or by a hybrid mouse–human transporter. *J. Exp. Med.*, **177**, 1785.

121. Livingstone, A. M., Powis, S. J., Diamond, A. G., Butcher, G. W., and Howard, J. C. (1989). A *trans*-acting major histocompatibility complex-linked gene whose alleles determine gain and loss changes in the antigenic structure of a classical class I molecule. *J. Exp. Med.*, **170**, 777.

122. Powis, S. J., Howard, J. C., and Butcher, G. W. (1991). The major histocompatibility complex class II-linked *cim* locus controls the kinetics of intracellular transport of a classical class I molecule. *J. Exp. Med.*, **173**, 913.

*123. Monaco, J. J. (1992). Antigen presentation: Not so groovy after all. *Curr. Biol.*, **2**, 433.

124. Pazmany, L., Rowland-Jones, S., Huet, S., Hill, A., Sutton, J., Murray, R., Brooks, J., and McMichael, A. (1992). Genetic modulation of antigen presentation by HLA-B27 molecules. *J. Exp. Med.*, **175**, 361.

125. Guagliardi, L. D., Koppelman, B., Blum, J. S., Marks, M. S., Cresswell, P., and Brodsky, F. M. (1990). Co-localization of molecules involved in antigen processing and presentation in an early endocytic compartment. *Nature*, **343**, 133.

126. Neefjes, J. J., Stollorz, V., Peters, P. J., Geuze, H. J., and Ploegh, H. L. (1990). The biosynthetic pathway of MHC class II but not class I molecules intersects the endocytic route. *Cell*, **61**, 171.

127. Peters, P. J., Neefjes, J. J., Oorschot, V., Ploegh, H. L., and Geuze, H. J. (1991). Segregation of MHC class II molecules from MHC class I molecules in the Golgi complex for transport to lysosomal compartments. *Nature*, **349**, 669.

128. Harding, C. V. and Geuze, H. J. (1993). Immunogenic peptides bind to class II MHC molecules in an early lysosomal compartment. *J. Immunol.*, **151**, 3988.

*129. Neefjes, J. J. and Ploegh, H. L. (1992). Intracellular transport of MHC class II molecules. *Immunol. Today*, **13**, 179.

130. Diment, S. (1990). Different roles for thiol and aspartyl proteases in antigen presentation of ovalbumin. *J. Immunol.*, **145**, 417.

131. Puri, J. and Factorovich, Y. (1988). Selective inhibition of antigen presentation to cloned T cells by protease inhibitors. *J. Immunol.*, **141**, 3313.

132. Yamamoto, K., Floyd-Smith, G., Francke, U., Koch, N., Lauer, W., Dobberstein, B., Schäfer, R., and Hämmerling, G. J. (1985). The gene encoding the Ia-associated invariant chain is located on chromosome 18 in the mouse. *Immunogenetics*, **21**, 83.

133. Richards, J. E., Pravtcheva, D. D., Day, C., Ruddle, F. H., and Jones, P. P. (1985). Murine invariant chain gene: Chromosomal assignment and segregation in recombinant inbred strains. *Immunogenetics*, **22**, 193.

134. Claesson-Welsh, L., Barker, P. E., Larhammer, D., Rask, L., Ruddle, F. H., and Peterson, P. A. (1984). The gene encoding the human class II antigen-associated γ chain is located on chromosome 5. *Immunogenetics*, **20**, 89.

135. Jones, P. P., Murphy, D. B., Hewgill, D., and McDevitt, H. O. (1978). Detection of a common polypeptide chain in *I-A* and *I-E* sub-region immunoprecipitates. *Immunochem.*, **16**, 51.

136. Day, C. E. and Jones, P. P. (1983). The gene encoding the Ia antigen-associated invariant chain (I_i) is not linked to the *H–2* complex. *Nature*, **302**, 157.

137. Roche, P. A., Marks, M. S., and Cresswell, P. (1991). Formation of a nine subunit complex by HLA class II glycoproteins and the invariant chain. *Nature*, **354**, 392.

138. Anderson, M. S. and Miller, J. (1992). Invariant chain can function as a chaperone protein for class II major histocompatibility complex molecules. *Proc. Natl Acad. Sci. USA*, **89**, 2282.

139. Roche, P. A. and Cresswell, P. (1990). Invariant chain association with HLA-DR molecules inhibits immunogenic peptide binding. *Nature*, **345**, 615.

140. Bakke, O. and Dobberstein, B. (1990). MHC class II-associated invariant chain contains a sorting signal for endosomal compartments. *Cell*, **63**, 707.

141. Lotteau, V., Teyton, L., Peleraux, A., Nilsson, T., Karlsson, L., Schmid, S. L., Quaranta, V., and Peterson, P. A. (1990). Intracellular transport of class II MHC molecules directed by invariant chain. *Nature*, **348**, 600.

142. Loss, G. E. and Sant, A. J. (1993). Invariant chain retains MHC class II molecules in the endocytic pathway. *J. Immunol.*, **150**, 3187.

143. Viville, S., Neefjes, J., Lotteau, V., Dierich, A., Lemeur, M., Ploegh, H., Benoist, C., and Mathis, D. (1993). Mice lacking the MHC class II-associated invariant chain. *Cell*, **72**, 635.

144. Bikoff, E. K., Huang, L., Episkopou, V., van Meerwijk, J., Germain, R. N., and Robertson, E. J. (1993). Defective major histocompatibility complex class II assembly, transport, peptide acquisition, and CD4$^+$ T cell selection in mice lacking invariant chain expression. *J. Exp. Med.*, **177**, 1699.

*145. Sant, A. J. and Miller, J. (1994). MHC class II antigen processing: Biology of invariant chain. *Curr. Opin. Immunol.*, **6**, 57.

*146. Cresswell, P. (1992). Chemistry and functional role of the invariant chain. *Curr. Opin. Immunol.*, **4**, 87.

*147. Teyton, L. and Peterson, P. A. (1992). Invariant chain—a regulator of antigen presentation. *Trends in Cell Biol.*, **2**, 52.

148. Layet, C. and Germain, R. N. (1991). Invariant chain promotes egress of poorly expressed, haplotype-mismatched class II major histocompatibility complex AαAβ dimers from the endoplasmic reticulum/*cis*-Golgi compartment. *Proc. Natl Acad. Sci. USA*, **88**, 2346.

149. Humbert, M., Raposo, G., Cosson, P., Reggio, H., Davoust, J., and Salamero, J. (1993). The invariant chain induces compact forms of class II molecules localized in late endosomal compartments. *Eur. J. Immunol.*, **23**, 3158.

150. Rath, S., Lin, R.-H., Rudensky, A., and Janeway, C. A. (1992). T and B cell receptors discriminate major histocompatibility complex class II conformations influenced by the invariant chain. *Eur. J. Immunol.*, **22**, 2121.

151. Schaiff, W. T., Hruska, K. A., Bono, C., Shuman, S., and Schwartz, B. D. (1991). Invariant chain influences post-translational processing of HLA-DR molecules. *J. Immunol.*, **147**, 603.

152. Simonis, S., Miller, J., and Cullen, S. E. (1989). The role of the Ia-invariant chain complex in the posttranslational processing and transport of Ia and invariant chain glycoproteins. *J. Immunol.*, **143**, 3619.

153. Marks, M. S., Blum, J. S., and Cresswell, P. (1990). Invariant chain trimers are sequestered in the rough endoplasmic reticulum in the absence of association with HLA class II antigens. *J. Cell. Biol.*, **111**, 839.

154. Miller, J. and Germain, R. N. (1986). Efficient cell-surface expression of class II MHC molecules in the absence of associated invariant chain. *J. Exp. Med.*, **164**, 1478.

155. Sekaly, R. P., Tonnelle, C., Strubin, M., Mach, B., and Long, E. O. (1986). Cell-surface expression of class II histocompatibility antigens occurs in the absence of invariant chain. *J. Exp. Med.*, **164**, 1490.

156. Lamb, C. A., Yewdell, J. W., Bennink, J. R., and Cresswell, P. (1991). Invariant chain targets HLA class II molecules to acidic endosomes containing internalized influenza virus. *Proc. Natl Acad. Sci. USA*, **88**, 5998.

157. Romagnoli, P., Layet, C., Yewdell, J., Bakke, O., and Germain, R. (1993). Relationship between invariant chain expression and major histocompatibility complex class II transport into early and late endocytic compartments. *J. Exp. Med.*, **177**, 583.

158. Blum, J. S. and Cresswell, P. (1988). Role for intracellular proteases in the processing and transport of class II HLA antigens. *Proc. Natl Acad. Sci. USA*, **85**, 3975.

159. Neefjes, J. J. and Ploegh, H. L. (1992). Inhibition of endosomal proteolytic activity by leupeptin blocks surface expression of MHC class II molecules and their conversion to SDS resistant αβ heterodimers in endosomes. *EMBO J.*, **11**, 411.

160. Roche, P. A., Teletski, C. L., Stang, E., Bakke, O., and Long, E. O. (1993). Cell-surface HLA-DR-invariant chain complexes are targeted to endosomes by rapid internalization. *Proc. Natl Acad. Sci. USA*, **90**, 8581.

161. Anderson, M. S., Swier, K., Arneson, L., and Miller, J. (1993). Enhanced antigen presentation in the absence of the invariant chain endosomal localization signal. *J. Exp. Med.*, **178**, 1959.

162. Sekaly, R., Jacobson, S., Richert, J., Tonnelle, C., McFarland, H., and Long, E. (1988). Antigen presentation to HLA class II-restricted measles virus-specific T cell clones can occur in the absence of the invariant chain. *Proc. Natl Acad. Sci. USA*, **85**, 1209.

163. Peterson, M. and Miller, J. (1990). Invariant chain influences the immunological recognition of MHC class II molecules. *Nature*, **345**, 172.

164. Nadimi, F., Moreno, J., Momburg, F., Heuser, A., Fuchs, S., Adorini, L., and Hämmerling, G. (1991). Antigen presentation of hen egg lysozyme but not ribonuclease A is augmented by the MHC class II associated invariant chain. *Eur. J. Immunol.*, **21**, 1255.

165. Shastri, N., Malissen, B., and Hood, L. (1985). Ia-transfected L-cell fibroblasts present a lysozyme peptide but not the native protein to lysozyme-specific T cells. *Proc. Natl Acad. Sci. USA*, **82**, 5885.

166. Stockinger, B., Pessara, U., Lin, R. H., Habicht, J., Grez, M., and Koch, N. (1989). A role of Ia-associated invariant chains in antigen processing and presentation. *Cell*, **56**, 683.

167. Peterson, M. and Miller, J. (1992). Antigen presentation enhanced by the alternatively spliced invariant chain gene product p41. *Nature*, **357**, 596.

168. Simonsen, A., Momburg, F., Drexler, J., Hämmerling, G., and Bakke, O. (1993). Intracellular distribution of the MHC class II molecule and the associated invariant chain (I₁) in different cell lines. *Int. Immunol.*, **5**, 903.

169. O'Sullivan, D. M., Noonan, D., and Quaranta, V. (1987). Four Ia invariant chain forms derive from a single gene by alternate splicing and alternate initiation of transcription/translation. *J. Exp. Med.*, **166**, 444.

170. Koch, N., Lauer, W., Habicht, J., and Dobberstein, B. (1987). Primary structure of the gene for the murine Ia antigen-associated invariant chains (I₁). An alternatively spliced exon encodes a cysteine-rich domain highly homologous to a repetitive sequence of thyroglobulin. *EMBO J.*, **6**, 1677.

171. Peterson, M. and Miller, J. (1992). Antigen presentation enhanced by the alternatively spliced invariant chain gene product p41. *Nature*, **357**, 596.

172. Blum, J. S. and Cresswell, P. (1988). Role for intracellular proteases in the processing and transport of class II HLA antigens. *Proc. Natl Acad. Sci. USA*, **85**, 3975.

173. Pieters, J., Horstmann, H., Bakke, O., Griffiths, G., and Lipp, J. (1991). Intracellular transport and localization of major histocompatibility complex class II molecules and associated invariant chain. *J. Cell Biol.*, **115**, 1213.

174. Roche, P. A. and Cresswell, P. (1991). Proteolysis of the class II-associated invariant chain generates a peptide binding site in intracellular HLA-DR molecules. *Proc. Natl Acad. Sci. USA*, **88**, 3150.

175. Teyton, L., O'Sullivan, D., Dickson, P. W., Lotteau, V., Sette, A., Fink, P., and Peterson, P. A. (1990). Invariant chain distinguishes between the exogenous and endogenous antigen presentation pathways. *Nature*, **348**, 39.

176. Roche, P. A., Teletski, C. L., Karp, D. R., Pinet, V., Bakke, O., and Long, E. O. (1992). Stable surface expression of invariant chain prevents peptide presentation by HLA-DR. *EMBO J.*, **11**, 2841.

177. Freisewinkel, I., Schenck, K., and Koch, N. (1993). The segment of invariant chain that is critical for association with major histocompatibility complex class II molecules contains the sequence of a peptide eluted from class II polypeptides. *Proc. Natl Acad. Sci. USA*, **90**, 9703.

178. Riberdy, J. M., Newcomb, J. R., Surman, M. J., Barbosa, J. A., and Cresswell, P. (1992). HLA-DR molecules from an antigen-processing mutant cell line are associated with invariant chain peptides. *Nature*, **360**, 474.

179. Mellins, E., Smith, L., Arp, B., Cotner, T., Celis, E., and Pious, D. (1990). Defective processing and presentation of exogenous antigens in mutants with normal HLA class II genes. *Nature*, **343**, 71.

180. Mellins, E., Kempin, S., Smith, L., Monji, T., and Pious, D. (1991). A gene required for class II-restricted antigen presentation maps to the major histocompatibility complex. *J. Exp. Med.*, **174**, 1607.

181. Ceman, S., Rudersdorf, R., Long, E. O., and DeMars, R. (1992). MHC class II deletion mutant expresses normal levels of transgene encoded class II molecules that have abnormal conformation and impaired antigen presentation ability. *J. Immunol.*, **149**, 754.

182. Dang, L., Lien, L., Benacerraf, B., and Rock, K. (1993). A mutant antigen presenting cell defective in antigen presentation expresses class II MHC molecules with an altered conformation. *J. Immunol.*, **150**, 4206.

183. Sette, A., Ceman, S., Kubo, R. T., Sakaguchi, K., Appella, E., Hunt, D. F., Davis, T. A., Michel, H., Shabanowitz, J., Rudersdorf, R., Grey, H. M., and DeMars, R. (1992). Invariant chain peptides in most HLA-DR molecules of an antigen-processing mutant. *Science*, **258**, 1801.

184. Morris, P., Shaman, J., Attaya, M., Amaya, M., Goodman, S., Bergman, C., Monaco, J. J., and Mellins, E. (1994). An essential role for HLA-DM in antigen presentation by class II major histocompatibility molecules. *Nature*, **368**, 551.

185. Cho, S., Attaya, M., and Monaco, J. J. (1991). New class II-like genes in the murine MHC. *Nature*, **353**, 573.

186. Kelly, A. P., Monaco, J. J., Cho, S., and Trowsdale, J. (1991). A new human HLA class II-related locus, *DM*. *Nature*, **353**, 571.

187. Mellins, E., Cameron, P., Amaya, M., Goodman, S., Pious, D., Smith, L., and Arp, B. (1994). A mutant human histocompatibility leukocyte antigen DR molecule associated with invariant chain peptides. *J. Exp. Med.*, **179**, 541.

188. Chiang, H.-L., Terlecky, S. R., Plant, C. P., and Dice, J. F. (1989). A role for a 70-kilodalton heat shock protein in lysosomal degradation of intracellular proteins. *Science*, **246**, 382.

189. Gordon, P. B. and Seglen, P. O. (1988). Prelysosomal convergence of autophagic and endocytic pathways. *Biochem. Biophys. Res. Comm.*, **151**, 40.

190. Dunn, W. A. (1990). Studies on the mechanisms of autophagy: Formation of the autophagic vacuole. *J. Cell. Biol.*, **110**, 1923.
191. Dunn, W. A. (1990). Studies on the mechanisms of autophagy: Maturation of the autophagic vacuole. *J. Cell. Biol.*, **110**, 1935.

6 | B cell activation

GERRY G. B. KLAUS

1. Introduction

1.1 B cell responses to specific antigens

B lymphocytes are the precursors of antibody-forming cells (AFC), which express clonally distributed surface immunoglobulin (sIg) receptors for antigen (generally sIgM and sIgD). B cells are generated in the bone marrow throughout adult life: this process involves an ordered sequence of gene rearrangements which generates a primary B cell repertoire of some 10^6–10^7 specificities. Binding of certain polymeric, cross-linking antigens (so-called T-independent (TI) antigens) to sIg receptors is in itself sufficient to induce B cell activation and antibody formation. Generally, however, for T-dependent (TD) antigens (i.e. protein antigens) additional T cell-derived stimuli are required to elicit antibody formation. In either case, the clonal precursors bearing sIg receptors specific for epitopes on the antigen become activated, enter the cell cycle, proliferate, and differentiate into high-rate AFC, which, ultimately become plasma cells. For an overview of the biology of B cells see ref. 1

During responses to TD antigens, a certain proportion of the stimulated primary precursor population enters a distinct differentiation pathway, also involving extensive cellular proliferation, which results in the production of an expanded population of B memory (or secondary) B cells. This process involves the induction of high-rate somatic mutation in antibody V-regions and the selective expansion of variants with high affinity for the antigen in question: it occurs in specialized microenvironments in peripheral lymphoid tissues known as germinal centres, where B cells encounter antigen depots on specialized antigen-presenting cells (APC) called follicular dendritic cells (FDC). It is apparently a device evolved by the immune system for further refining the primary B cell V-region repertoire to generate the 'best fit' antibody response to a subsequent exposure to that antigen. The generation of memory is outside the scope of this review and will not be considered in detail further. It has been extensively reviewed elsewhere recently (2, 3).

1.2 Responses to polyclonal activators

Both B and T cells can also be activated in an antigen-independent fashion by polyclonal activators, which stimulate DNA synthesis and (in the case of B cells) sometimes Ig secretion as well. Such activators stimulate a much larger fraction of the lymphocyte population than specific antigens, so they have been widely used to study the biochemistry and molecular biology of lymphocyte activation. Some of these agents activate the cells via their antigen receptors (i.e. anti-T cell receptor (TCR), or anti-Ig antibodies), or via other cell surface receptors, whilst others invoke rather ill-defined pathways, such as bacterial lipopolysaccharide (LPS). In addition, many monoclonal antibodies (mAbs) to a diverse range of B cell surface molecules have been found either to induce B cell activation, or to modulate B cell activation induced by more classical activators. This approach has, therefore, yielded a great deal of information on the possible functions of different cell surface markers in the physiology of lymphocytes.

The primary aim of this chapter is to review the functions of the various cell surface molecules currently believed to play a role in B cell activation—of which there are many. Because of limitations of space, the review will be largely restricted to a survey of the control of B cell activation, that is to say, the stimuli which determine the entry of quiescent B cells into the cell cycle. It will not cover the complex interplay of cytokines involved in regulation of Ig secretion and isotype switching in any detail since this has recently been extensively reviewed elsewhere (4).

The chapter commences with a catalogue of the properties of the major cell surface molecules currently believed to be important in regulating the responses of mature B cells to antigens, and what is known about their mechanisms of signal transduction. It concludes with a discussion of the current understanding of the events which occur during T cell–B cell interaction. It will become apparent that some of the cell surface molecules described in the first part of the chapter play important roles in modulating B cell (and sometimes T cell) responses during T cell–B cell interaction.

1.3 Cell surface glycoproteins on B cells

B cells, like T cells, express a large number of cell surface glycoproteins through which they presumably communicate with their environment. Besides the antigen receptors, and receptors for a variety of cytokines, B cells express a variety of cell surface proteins defined by the CD (cluster of differentiation) nomenclature. These are better defined on human than on rodent B cells, because of the greater range of mAbs available. The major CD antigens expressed on human B cells (although not necessarily at all stages of their differentiation) include CD5, CD10, CD19, CD20, CD21, CD22, CD23, CD24, CD37, CD38, CD39, CD40, CD72, CD73, CD74, CD75, CD76, CD77, and CD78 (reviewed in refs 5, 6). In addition, all mature B cells constitutively express class I and class II MHC antigens, the leucocyte common antigen (CD45) and members of the integrin family of cell adhesion molecules (for example LFA-1 and ICAM-1). The definition of this plethora of cell surface molecules has inevitably led to an explosion of investigations on their possible functions in B cell physiology (also reviewed in ref. 5). In consequence, there now exists a large and confusing literature on this topic, which is difficult to summarize.

2. Surface immunoglobulin (sIg) receptors

During their differentiation from pre-B cells in the bone marrow, immature B cells first express sIgM receptors, and then subsequently (at 3–5 days of neonatal life in the mouse) gradually express sIgD. For reasons that are still not well-understood most mature peripheral B cells in both mouse and man express both sIgM and sIgD receptors (reviewed in ref. 7). Furthermore, as part of the memory pathway of differentiation individual B cells can potentially express sIg receptors of three distinct isotypes (for examples sIgM, sIgD, and sIgG1). In short, all classes of Ig can be expressed on the B cell surface and are capable of acting as antigen receptors. These receptors have at least three potential functions, whose relative importance may differ at different stages in the ontogeny of the cell:

1. The delivery of activating signals to the cell, which generally act in concert with other (for example T cell-derived) signals.
2. The delivery of inactivating (tolerizing) signals, to anergize or kill autoreactive clones (reviewed in ref. 8): this is probably as relevant in mature B cells as in immature ones, since potentially autoreactive cells could be generated through somatic mutations of antibody V-regions.
3. The concentration and uptake of antigens for processing and subsequent presentation via class II molecules to helper T cells (Section 10.1.3).

2.1 The effects of anti-Ig on mature B cells

The role of sIg receptors in B cell activation has been a matter of considerable debate. This is partly because most of the information about the consequences of engaging sIg receptors has emanated from experiments using various anti-Ig antibodies. In earlier studies, which employed polyclonal antibodies, conflicting results were obtained until the modulating effects of Fc receptor interactions (Section 4) were appreciated.

It is now clearly established that appropriate polyclonal anti-Ig antibodies (for example F(ab')$_2$ fragments of rabbit anti-mouse Ig) and many monoclonal anti-μ and anti-δ antibodies induce *abortive* activation of virtually all mature B cells (i.e their exit from G$_0$, but not full entry into the cell cycle as detected by increases in cell size and levels of class II MHC antigens). These effects are induced by relatively low concentrations of antibody (0.1–1 μg/ml). Substantially higher concentrations (10–100 μg/ml) of anti-Ig are required to induce significant levels of DNA synthesis and then only in a fraction of B cells (reviewed in refs 9, 10). Furthermore, not all monoclonal anti-Ig antibodies are mitogenic *per se*. The mitogenic potency of all forms of anti-Ig is markedly enhanced by the T cell-derived cytokine IL-4 (Section 10.3.1).

Two key properties of anti-Ig reagents which determine whether they stimulate DNA synthesis in B cells are:

- The capacity of the antibodies to effectively cross-link sIg receptors (which is particularly relevant in the case of mAb (see, for example, ref. 11)).

- Their capacity to interact with Fc receptors (FcRII) expressed on B cells, which inhibits activation, as discussed below.

The first point is strikingly illustrated by the effects of anti-Ig mAbs coupled to polymeric carriers, such as dextran; anti-Ig-dextran is mitogenic at some 1000-fold lower concentrations than soluble antibody (12). These findings strongly suggest that anti-Ig stimulation represents a polyclonal model of B cell activation by TI antigens, in particular, so-called TI-2 antigens. These are typically long chain polymers, such as bacterial capsular polysaccharides, with many repeating epitopes (for example levan and dextrans). Their structure is believed to facilitate extensive cross-linking of sIg receptors necessary for B cell activation via this pathway.

If this concept is correct, then anti-Ig (certainly when conjugated to a TI-2 carrier) should also induce B cells to differentiate into AFC. In fact, under conventional culture condition, soluble anti-Ig antibodies do not induce B cells to secrete immunoglobulins *in vitro*, even in the presence of cytokines which promote Ig secretion (for example IL-4 and IL-5 in the mouse). Instead, anti-Ig is a potent inhibitor of, for example, LPS-induced Ig secretion (13). It has been recently established that anti-Ig-dextran does indeed induce resting B cells to differentiate into AFC in the presence of T cell-derived lymphokines (14). In addition, solid-phase anti-Ig antibodies (15), and even soluble anti-μ mAbs (16), can prime mouse B cells to secrete Ig in response to IL-4 plus IL-5, for example. However, with soluble anti-μ it is important that the cells are washed free of the antibodies prior to lymphokine stimulation. Hence it is clear that, under appropriate conditions, the ligation of sIg receptors can prime B cells to secrete antibodies in response to additional stimuli.

2.2 Signal transduction via sIg receptors

There has been remarkable progress in our understanding of signal transduction via various receptors on both T and B cells. This is particularly true of the antigen receptors which have received a great deal of attention. However, it is worth stressing that a simple description of the early biochemical events which occur after ligation of any receptor is insufficient unless these can be placed within the context of the possible functions of that receptor in the physiology of the cell. This is especially true of sIg receptors, with their multiple functions already mentioned above. Thus, for example, there is very little information relating to how, in biochemical terms, one receptor can deliver either activating or tolerizing signals to the cell.

2.2.1 The B cell antigen receptor (BCR) complex

The cytoplasmic domains of the heavy chains of both sIgM and sIgD receptors consist of only three amino acids, indicating that additional polypeptide chains must be involved in coupling the receptors to the second messenger-generating system. Indeed it is now clear that both sIgM and sIgD in mouse and man are non-covalently associated with disulfide-linked heterodimers (consisting of two polypeptides called Ig-α and β), which are the products of the *mb-1* and *B29* genes, respectively (reviewed in refs 17–19). These are transmembrane proteins with extracellular Ig-like domains and with cytoplasmic tails of 61 and 48 amino acids, respectively. The cytoplasmic portions do not contain demonstrable protein tyrosine kinase (PTK) domains, but the overall structure of these proteins suggests that they are evolutionarily related to the proteins of the CD3 complex associated with the T cell antigen receptors. Most significantly, both proteins contain an ARH-1 (antigen receptor homology) motif. These tyrosine-containing motifs are found in components of several multichain receptors (for example on the γ, δ, ε, and ζ chains of the CD3 complex) and couple these receptors to downstream signalling molecules. In the B cell antigen receptor (BCR) complex there is evidence that the ARH-1 motifs of Ig-α, and -β are associated with different molecules. The former has been found to be associated with the src-family PTKs fyn and lyn, whilst β co-precipitates with two uncharacterized proteins of M_r 40 kDa and 42 kDa (reviewed in ref. 18). Hence, the αβ heterodimers, together with the sIg molecule comprise the BCR complex (Fig. 1). Initially it was assumed that the α-components associated with sIgM and sIgD are different, because IgD-α has a higher M_r than IgM-α (20). This appears to be due to differences in glycosylation, so that it is now

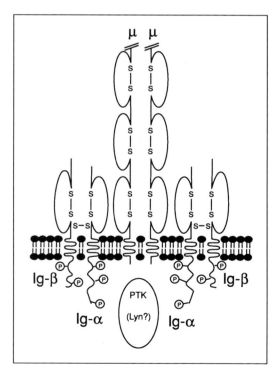

Fig. 1 The B cell antigen receptor complex. The figure shows part of the μ-chain of sIgM, with a C-terminal tail of only three amino acids, non-covalently associated with two identical hetero-dimers, consisting of the Ig-α and Ig-β chains. The latter have substantially longer cytoplasmic do-mains, the ARH-1 motifs of which become tyrosine phosphorylated after engaging the receptor. Seemingly identical α and β-chains are associated with all classes of sIg receptors. Also shown is a non-covalently associated PTK, which may be lyn, blk, or syk. The putative G-protein associated with the BCR complex is not shown, for the sake of simplicity. (Modified from ref. 17.)

believed that all classes of sIg receptors share common $\alpha\beta$ heterodimers (21). It still remains possible, however, that the BCR complex contains additional polypeptides which control signal transduction, for example the B cell equivalent of the ζ-chain associated with the TCR/CD3 complex (22).

2.2.2 Second messengers invoked by sIg receptors

The earliest biochemical events which occur after ligation of either sIgM or sIgD receptors on mature human or murine B cells (reviewed in refs 5, 10, 18) are:

- activation of (one or more) PTK, with resultant phosphorylation of a large number of (largely uncharacterized) proteins on tyrosine residues;
- an elevation in intracellular Ca^{2+} levels, $[Ca^{2+}]_i$;
- translocation (from cytosol to plasma membrane) and activation of the serine- and threonine-specific enzyme protein kinase C (PKC).

The last two of the above responses are the result of receptor-triggered hydrolysis of membrane phosphatidylinositol 4,5 bisphosphate (PIP$_2$), by a polyphosphoinositide-specific phosphodiesterase, or phospholipase C (PLC) (23, 24). The hydrolysis products (inositol 1,4,5 trisphosphate (IP$_3$) and 1,2 diacylglycerol) cause the release of Ca^{2+} from intracellular stores and the activation of PKC, respectively. The release of Ca^{2+} is short-lived and is followed by the prolonged influx of Ca^{2+} through the plasma membrane, caused by undefined mechanisms. The result is that $[Ca^{2+}]_i$ in anti-Ig stimulated B cells can remain elevated above resting levels for many minutes, and perhaps hours.

Activation of the (various) PTK seemingly associated with the BCR (see below) has been shown to result in the phosphorylation of a variety of potentially important substrates, many of them common to more conventional PTK-linked growth factor receptors. These include Ig-α and -β, CD19, p21^{ras} GAP, PLC-γ1 and 2, vav, phosphatidylinositol 3-kinase (PI-3K), and MAP kinase (25–31). Cross-linking of sIg also activates the low M_r G-protein p21^{ras} (32). The possibility that this might be associated with the BCR complex, first suggested by co-capping studies (33), has not been confirmed.

2.2.3 Biochemical coupling of the BCR complex to the PLC

The definition of the invariant chains of the BCR complex has, unfortunately, not shed any light on two key questions in B cell physiology. The first is, why do most B cells carry both sIgM and sIgD receptors? Secondly, the precise molecular details of the coupling of receptor complex to the PLC remain rather confused. There are currently two schools of thought (which are not mutually incompatible) with regard to the latter question:

- That (by analogy with Ca^{2+}-mobilizing receptors on other cell types, reviewed in ref. 34) ligation of the BCR complex induces the activation of an associated guanine nucleotide-regulatory (G) protein, which in turn induces activation of the PLC.

- Alternatively, that (in common with the TCR–CD3 complex and other growth factor receptors; ref. 22) sIg ligation activates one or more receptor-associated PTK, with resultant phosphorylation and activation of the $\gamma1$ or $\gamma2$-isoforms of PLC.

It has been clearly shown that permeabilizing B cells (which depletes the cells of endogenous GTP) uncouples sIg receptors from the PLC (35, 36). In short, the hydrolysis of PIP_2 in permeabilized B cells is absolutely dependent on the presence of GTP, hence providing formal evidence for G-protein involvement in sIg signalling. This is in distinct contrast with the situation in murine T cells, where signalling via the TCR complex is unaffected by permeabilization (37). There are, however, two problems concerning the involvement of a G protein in sIg signalling which have yet to be resolved. The first is that the structure of the BCR complex does not resemble those of classical G protein-coupled receptors, which generally have seven transmembrane domains. This issue is not a serious one, however, since there are now a number of examples of G-protein-linked receptors which do not conform to this rule (see, for example, refs 38, 39). The second problem is that the putative G-protein in B cells has not been identified. In some Ca^{2+}-mobilizing receptors the protein is a substrate for pertussis toxin, but this is not the case in B cells.

The alternative (currently more accepted) hypothesis is that stimulation of the PLC via sIg receptors is initiated via the activation of one or more, receptor-associated, src-related (or other) PTK, by analogy with the TCR/CD3 complex, for example. It is clear that ligation of the BCR complex induces very rapid activation of one or more PTK, leading to phosphorylation of the ARH-1 motifs of Ig-α and β chains (40). There is also evidence that lyn, which is one of the src family of PTK, is non-covalently associated with sIgM receptors and becomes activated after sIg stimulation (27). Two other src-related kinases (the B cell specific enzyme blk and $p60^{fyn}$) are also potential candidates in this signalling pathway, since cross-linking of either sIgM or sIgD activates the PTK activities of these enzymes (41). In addition, PTK72 (now known as syk), which is not a member of the src family, is also associated with the BCR complex (42). At the time of writing, the precise roles of these different kinases in BCR signal transduction had not been unravelled. A plausible hypothesis is that one of the src-family members phosphorylates the ARH-1 motifs of the associated proteins, thereby promoting the association of syk with the BCR complex and consequent activation of this PTK.

Treatment of human B cells or cell lines with PTK inhibitors (such as tyrphostin, genistein, or herbimycin A) abrogates both Ca^{2+} mobilization and inositol phosphate release induced by anti-Ig (30, 43). This suggests that PTK activation is the primary event leading to PLC activation. Unfortunately, such inhibitors are rarely absolutely specific. For example, we have found that treatment of mouse B cells with genistein for a few hours actually leads to reductions in the levels of inositol phospholipids (M. M. Harnett, and G. G. B. Klaus, unpublished data). In other words, the inhibition of Ca^{2+} mobilization seen with these inhibitors might reflect depletion of the substrate for the PLC, rather than implicating the prior activation of a src-like kinase in the initiation of phosphoinositide hydrolysis.

The resolution of the precise mechanisms which regulate signal transduction via BCR complex clearly requires further study. However, at present it remains possible that both src-related PTK (and syk) and (uncharacterized) G-proteins are involved in coupling the receptors to downstream events. How these two pathways relate to each other is unclear. In this context, it may be relevant that Pure and Tardelli (44) have recently shown that PTK inhibitors inhibit modulation and internalization of cross-linked sIg. This suggests the interesting possibility that activation of the PTK pathway may be critical in the capacity of B cells to take up antigens, preparatory to their processing and presentation to T cells.

2.2.4 The role of phosphoinositide-derived second messengers in B cell activation

There are several other B cell surface molecules which are linked to PIP_2 hydrolysis and Ca^{2+} mobilization (reviewed in ref. 5). These include MHC class II (Section 6), CD23 (Section 5), CD19/CD21 (Section 3), CD24, CD72 (Section 9.1) and Bgp95. It is, therefore, pertinent to ask what evidence there is that the second messengers generated following PLC activation play any role in driving B cells into the cell cycle. The problem has been approached experimentally by mimicking the effects of Ca^{2+} elevation and PKC activation by the use of Ca^{2+} ionophores and PKC-activating phorbol esters (such as phorbol myristic acetate, PMA), respectively. In mouse B cells, either of these agents alone induces abortive B cell activation (Section 2.1), and the two together induce cells to synthesize DNA (45). These results suggest that the two second messengers derived from PIP_2 breakdown synergize to induce B cell activation, and that this is one *potential* pathway of B cell activation. It has to be recognized, however, that this is an artificial system, since these agents cause massive biochemical responses that are unlikely to be seen following receptor stimulation. This is especially true of agents such as PMA, which causes irreversible activation and eventual down-regulation of PKC in many different cells. Morever, the generation of second messengers as a result of PIP_2 hydrolysis is clearly not the only pathway capable of inducing B cell activation, since LPS does not induce PIP_2 breakdown or Ca^{2+} mobilization in B cells (24).

An additional complication is that studies with anti-Ig coupled to dextran (Section 2.1) suggest that B cells can be activated via their sIg receptors in the absence of detectable PIP_2 hydrolysis or Ca^{2+} mobilization (46). These findings could be interpreted in one of two ways: either the available assays are not sensitive enough to detect subtle biochemical changes elicited by such a powerfully cross-linking ligand (which probably only needs to engage a fraction of the sIg receptors) or, perhaps another undefined (PTK-related?) second messenger system is of greater importance in regulating this mode of B cell activation. In fact, single-cell analyses show that persistent elevations of $[Ca^{2+}]_i$, rather than the initial large burst, correlate better with the capacity of anti-Ig preparations to induce commitment to DNA synthesis (47).

Certainly, a major problem in this field is understanding the long-term homeostasis of signal transduction. It is clear that both T and B lymphocytes need prolonged stimulation with mitogens to induce commitment to DNA synthesis. At present, it is impossible to say if this is maintained by products of PIP_2 breakdown (which seems improbable), or if additional second messenger systems are invoked to maintain the activation cascade. One possibility, which has thus far not been explored in lymphocytes, is that the initial burst of PIP_2 hydrolysis is followed by the breakdown of other phospholipids, most notably phosphatidylcholine. This is believed to provide an additional source of DAG to maintain PKC activation over prolonged periods in other cell types (48).

Another fundamental, related problem in understanding B cell activation (which has received scant attention) is whether the second messengers emanating from say, the BCR complex, are sufficient to induce B cell proliferation. By analogy with T cells, it remains possible that the proliferative signal(s) are in fact provided by cytokines, perhaps produced by B cells themselves and acting in an autocrine, or paracrine fashion. This possibility is in line with observations that anti-Ig-induced B cell proliferation is highly susceptible to the immunosuppressive drugs cyclosporine A and FK506 (49, 50). The mode of action of these drugs in T cells is to block cytokine (for example IL-2) gene transcription by preventing the nuclear translocation of the cytosolic component of the transcription factor NF-AT (reviewed in ref. 51). Recent studies from this laboratory and others have shown that anti-Ig (especially in conjunction with IL-4) does indeed induce NF-AT in mouse B cells (52, 53). Although there have been reports that B cells can make several cytokines (including IL-2), there is currently no clear indication that any of them play a role in directing B cell activation.

3. The CD19/CD21 complex

B cells express two receptors for split products of the third component of the complement system (C3); a C3b receptor (CR1 or CD35) and the C3dg receptor CR2 (CD21). There is a large body of evidence for the importance of the complement system in the induction of antibody responses. This has emerged both from patients who are complement deficient, and from experimental studies in rodents, for example with mice rendered C3-deficient by treatment with cobra venom factor. The latter have shown that both the production of primary antibody responses and the induction of B cell memory are C3-dependent (reviewed in ref. 54). It is, therefore, an attractive hypothesis that the binding of antigen–antibody–C3 complexes to both sIg and CR1/CR2 modulates B cell activation, and/or regulates the differentiation programme of the cells towards the memory pathway within germinal centres, for example.

The results of *in vitro* studies have provided ample support for this concept, and implicate CR2 as the most important C3 receptor on B cells. Both mAbs to CR2 and polymeric C3dg prime B cells to respond to anti-Ig, and co-crosslinking of sIgM and CR2 produces synergistic increases in $[Ca^{2+}]_i$ (reviewed in ref. 55).

However, CR2 (like sIg) has a rather short cytoplasmic tail, again suggesting the involvement of additional molecules in signal transduction via this receptor. Recent investigations by Fearon's group (56) have produced evidence that CR2 is indeed a member of another multimolecular, signal-transducing complex on B cells. This complex also contains the pan B cell marker CD19, TAPA-1 (the target for antiproliferative antibody) and possibly Leu-13 (Fig. 2; reviewed in ref. 57). While cross-linking of CD19, or CR2 alone, elicits a certain level of Ca^{2+} mobilization, co-crosslinking of either CD19 or CR2 with suboptimal concentrations of anti-Ig induced synergistic responses (58, 59). The ligand for CD19 is not known, but since this marker is expressed early in B cell ontogeny before the appearance of CR2, the complex may well play multiple roles in B cell biology.

Although both the BCR and CD19/CR2 complexes are Ca^{2+} mobilizing receptors, the biological consequence of engaging these two complexes differ substantially. Ligation of CD19 does not induce B cell activation, and indeed can inhibit the early activation steps induced by anti-Ig. A recent study by Carter et al. (60) has shown that although CD19 also activates a PLC and a PTK, the kinetics of PIP_2 hydrolysis provoked via this receptor was slower than via the BCR. Furthermore, CD19 and anti-μ generated different spectra of tyrosine phosphoproteins; most notably, CD19 ligation did not induce phosphorylation of PLC-γ1. These workers therefore concluded that the BCR and CD19/CR2 complex activate PLC via different mechanisms. This group has also recently shown that the cytoplasmic tail of CD19 contains a putative kinase insert domain, with characteristic YXXM motifs, involved in the binding of the p85 subunit of phosphatidylinositol 3-kinase (PI-3K). They showed that ligation of the BCR complex induces tyrosine phosphorylation of CD19 and its association with PI-3K (31). The current working hypothesis is, therefore, that the CD19/CD21 complex serves to amplify signalling via antigen receptors on B cells, and hence will play an important role in situations where the concentration of antigen is limiting. This idea is elegantly supported by the finding that co-ligation of CD19 and sIg (such as would occur after binding of Ag–Ab–C3 complexes) decreases the threshold for anti-Ig-induced B cell activation by two orders of magnitude (61).

4. FcγRII (CD32)

Receptors which bind the Fc region of IgG (FcγR) are widely distributed on different cell types. The major FcγR on B cells is FcRII, which binds aggregated IgG (i.e. antigen–antibody complexes) (reviewed in ref. 62). FcRII in the mouse is the product of a single gene; the extracellular domain is Ig-like and is identical on all cells which express the receptor. Alternative mRNA splicing yields two isoforms, FcRIIb1 (which is the form found on B cells) and FcRIIb2 (found on myeloid cells). The b1 isoform contains an insert of 47 amino acids not found in FcRIIb2, which is presumed to explain the differences in the functions of the two isoforms. Hence, the b2 form on macrophages mediates uptake of antigen–antibody complexes, while the b1 form on B cells does not.

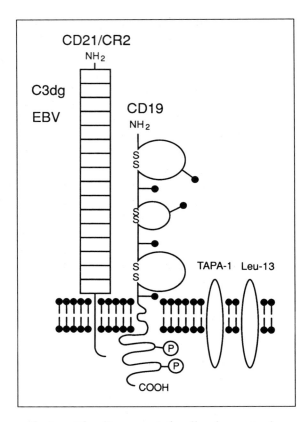

Fig. 2. The CD21/CD19 complex. CD19 and CD21 are non-covalently associated with each other on the surface of B cells. CD21 belongs to a family of complement-binding proteins and its extracellular domain consists of 15 short consensus repeats; the approximate binding sites of C3dg and Epstein–Barr virus (EBV) (in man) are shown. CD19 belongs to the Ig superfamily. The complex contains other proteins, including TAPA-1 (a member of the tetraspan family of proteins) and Leu-13. Phosphorylation sites within the kinase insert domain of CD19 are shown as P. (Modified from ref. 57.)

It has been known for many years that antibodies exert feedback control on antibody production *in vivo* (reviewed in refs 63, 64). The mechanisms involved are complex, but *inter alia* involve the interaction of Ag–Ab complexes with FcRII on B cells. The biochemical basis of this effect has been extensively studied using rabbit anti-Ig antibodies and mouse B cells, because rabbit Fc binds to the murine FcRII with high affinity under appropriate conditions. Unlike the $F(ab')_2$ fragments, intact rabbit anti-Ig is not mitogenic for B cells (65, 66). This is because *co-crosslinking* of FcRII with either sIgM or sIgD induces a dominant negative signal which aborts B cell activation and rapidly inhibits phosphoinositide hydrolysis (67). It is, therefore, probable that IgG rabbit anti-Ig mimics the effects of antigen–antibody complexes in feedback regulation of antibody responses. The biochemical mechanisms have been partly elucidated; studies in permeabilized B cells indicate that co-crosslinking of these two receptors functionally uncouples the BCR complex from its associated G-protein (Section 2.2.3), and hence aborts activation of the PLC (68). In addition, activation of B cells with $F(ab')_2$ (but not with IgG) anti-Ig induces phosphorylation of FcRII on serine residues, presumably via PKC (69), and also on a tyrosine residue (Y 309) within a 13 amino acid motif which is critically involved in the functions of FcRII (70). This motif probably interacts with additional protein(s) via their SH2 domains. The end-result of co-crosslinking the BCR and FcRII is selective inhibition of Ca^{2+} influx via the plasma membrane (and not IP_3-mediated Ca^{2+} release) (70, 71). Hence, these findings again highlight the importance of long-term Ca^{2+} homeostasis in B cells in regulating the egress of cells from G_0 and their entry into the cell cycle. It is also noteworthy that the levels of FcRII are regulated during B cell activation. Stimulation by conventional mitogens such as LPS increases their levels of FcRII and this is largely reversed by IL-4 (72). This may explain why IL-4 can substantially reverse the inhibitory effects of IgG rabbit anti-Ig in B cell activation (73).

In conclusion, it is now clear that sIg and FcRII form yet another example of a pair of interacting receptors on B cells. The biological consequence of the crosstalk between these two receptors is to dampen B cell activation, in contrast to the positive effects described for the CR2/CD19 complex above. Under physiological conditions, the effects of engaging FcRII are further modulated by additional influences, most notably from T cells, to ensure that the immune response generated in response to a particular antigen is appropriate, both in terms of its nature and its magnitude.

5. CD23

The low-affinity receptor for IgE (FcεRII, CD23) is a 45 kDa type II glycoprotein, belonging to the family of C-type lectins, which is widely distributed on many cell types in man, but largely restricted to B cells in the mouse (reviewed in ref. 74). The protein is susceptible to proteolysis, generating soluble fragments (sCD23), which are IgE-binding factors. On B cells, CD23 is a differentiation marker in that it is present only on mature sIgM/sIgD-positive cells and disappears as the cells mature towards AFC. The levels of CD23 are tightly regulated; T cell-derived stimuli, such as the CD40 ligand (see Section 8.2) and IL-4, markedly upregulate levels of CD23 on B cells and these effects are antagonized by IFN and TGF-β.

Most of the activities of CD23 are IgE-dependent. In B cells there is evidence that CD23 is involved in IgE-mediated antigen presentation to T cells. Ligation of CD23 by mAbs or by IgE-containing immune complexes can either promote or inhibit B cell activation. The importance of CD23 in controlling B cell activation is supported by the findings that ligation of the molecule induces phosphoinositide hydrolysis, which may also be G-protein regulated (75). The cleaved form of CD23 appears to act as a multifunctional cytokine (76). For example, in conjunction with IL-1α it has been shown to prevent apoptosis in a subpopulation of germinal centre B cells (77). Soluble CD23 also plays a role in the regulation of IgE synthesis; a recombinant 29 kDa fragment of CD23 augments IgE production *in vitro* and probably acts at a stage after the cells have switched to IgE (which is regulated by IL-4; Section 10.3.1). Recently, Yu *et al.* (78) have generated $CD23^{-/-}$ mice by gene targeting. These animals develop normal levels of T and B cells and respond to TD antigens, arguing that in the mouse CD23 is not involved in B cell growth. Interestingly, the mutant animals produce elevated and more sustained IgE antibody responses than normal mice. These results suggest that a major function of CD23 is to exert negative feedback control of IgE production perhaps via similar mechanisms to those utilized by FcγRII.

6. Class II MHC

B cells constitutively express class II antigens which, together with their capacity to focus antigens via their sIg receptors, make them highly efficient APC for CD4-positive T_H cells (Section 10.1.3). As discussed below, the initial phase of T cell–B cell interactions for the induction of antibody responses to most (protein) antigens is restricted to MHC class II, that is, it requires the recognition of peptide fragments of the antigen on class II molecules on B cells. The levels of class II on B cells are regulated via a variety of influences. Class II levels on murine B cells increase markedly after stimulation by virtually all polyclonal activators, including anti-Ig, LPS, and PKC-activating phorbol esters. One of the most powerful inducers of class II is IL-4 (79, 80). It is believed that the up-regulation of class II during B cell activation serves to maximize the capacity of these cells to present antigen and hence receive helper signals from T cells (81).

T cells interact with B cells via a variety of cell interaction molecules (Section 10.2.1), including CD4 molecules on T cells which bind to monomorphic determinants on class II. The possibility that this interaction results in the transduction of signals (to one or both cells) is, therefore, an attractive one. As usual, this question has been approached by the use of mAbs. The effects of anti-class II mAbs on B cells are complex; under some conditions they can inhibit polyclonal B cell activation (82). This may be explained by the findings of Cambier et al. (83) that in resting mouse B cells anti-Ia mAbs induce an elevation of cyclic AMP levels, accompanied by translocation of PKC to a nuclear location. cAMP is a well-known inhibitor of B cell activation (84). Conversely, when mouse B cells are pre-activated for 16–24 h with anti-Ig and IL-4, the cells can be induced to proliferate by exposure to immobilized anti-class II (85). Furthermore, the cross-linking of class II molecules on such pre-activated cells induced increases in $[Ca^{2+}]_i$ and phosphoinositide hydrolysis, apparently in a PTK-dependent manner (85). Anti-class II mAbs also induce similar responses in human B cells. However, in this case resting cells will respond to anti-class II with the same spectrum of second messengers induced by anti-Ig (86, 87).

It is, therefore, possible (but by no means proven) that the ligation of class II molecules during T cell–B cell interaction could provide a further source of second messengers to maintain B cell activation. However, this possibility has been somewhat diminished by the recent findings by Markowitz et al. (88) who found that class II-deficient mice could mount normal antibody responses to TI antigens, and their B cells could respond to B cell mitogens and CD40L-bearing, pre-activated T cells in vitro. It, therefore, seems that signalling via class II molecules does not play an absolutely essential role in B cell activation.

7. CD45 (leucocyte common antigen, murine Ly5)

CD45 is an abundant cell surface glycoprotein which, as its alternative name suggests, is present on all leucocytes. It consists of a complex family of 180–220 kDa, proteins which are heavily glycosylated. They are the products of a single gene, consisting of 33 exons. Alternative mRNA splicing of exons 4, 5, and 6 could generate eight potential isoforms which differ in their extracellular domains (reviewed in ref. 89). The predominant isoform on B cells is the largest form (containing the products of all three of these exons), which is generally called B220.

A large variety of mAbs to both human and rodent CD45 are available. Some of these react with all CD45 isoforms (pan CD45), whilst others are specific for restricted isoforms (for example CD45RA, reacting with the product of exon 4). However, an RA-specific mAb could potentially react with the RA, RAB, RABC, or RAC isoforms. The transmembrane and intracellular domains of all isoforms are identical; the latter consists of two tandem repeats of 300 amino acids with significant homology to protein tyrosine phosphatase PTPase-1B. Indeed, it is now well-established that CD45 is a membrane-bound PTPase (90). This observation has stimulated considerable interest in the functions of CD45 in various types of cells, because of the importance of PTK-phosphatase cascades in signal transduction (91). In T cells it has become clear that the presence of CD45 is essential for signal transduction via the TCR (92). It is believed that CD45 functions to regulate the activity of CD4/CD8 associated PTK p56^{lck}, and ultimately the activity of PLCγl. Less is known about the role of CD45 in the control of signal transduction in B cells. However, Justement et al. (93) have provided evidence recently that CD45 is also required for sIgM receptors to induce Ca^{2+} mobilization. They also observed that co-crosslinking of CD45 and sIgM on normal B cells led to dephosphorylation of the sIg-associated α and β chains. Justement et al. therefore propose that CD45 is part of the sIg signalling complex, being involved in dynamic regulation of signal transduction via these receptors. Certainly, a variety of mAbs against CD45 inhibit B cell activation by anti-Ig and via CD40 (94–96). In addition, cross-linking of CD45 with mAbs inhibits Ca^{2+} mobilization induced via sIg, class II, and CD19, but does not block anti-Ig induced tyrosine phosphorylation of major substrates (87, 96, 97). This suggests that, if CD45 acts by dephosphorylating proteins in B cells, one of its important substrates might be PLCγ1, rather than the PTK associated with sIg receptors. However, CD45 also regulates activation via CD40 (96), which may or may not not utilize the phosphoinositide pathway (see below). This indicates that CD45 may also regulate signal transduction via a variety of different receptors in subtly different ways. Certainly the availability of CD45$^{-/-}$ mice (98) should aid in our understanding of the functions of this important cell surface protein. Preliminary studies of B cells from such animals has indicated that they give very poor proliferative responses to anti-Ig, whilst their responses to LPS are normal. The existence of multiple isoforms of CD45, conceivably with particular isoforms interacting with different receptors, suggests that CD45 may act as a signal modulator for a wide range of receptors. Thus far, there is no clear evidence about the putative natural ligand(s) for CD45.

8. CD40

CD40 is expressed on all mature B cells, but not on plasma cells. It is also expressed on some tumour cells, on dendritic cells in T cell areas of lymphoid tissues and on follicular dendritic cells in B cell areas (reviewed in ref. 99). The monomer is a 48 kDa phosphoglycoprotein, the level of which on B cells is regulated by the same stimuli that regulate class II expression. For example, stimulation with anti-Ig, CD20, or IL-4 leads to up-regulation of CD40 levels on B cells. The molecule belongs to a novel cytokine receptor family, since it is homologous with the nerve growth factor and tumour necrosis factor-α receptors (reviewed in ref. 100).

8.1 Functional role of CD40

Monoclonal antibodies against CD40 augment the proliferation of human B cells induced by various agents including anti-Ig, PMA, or CD20 (99). CD40 ligation on human B cells does not induce Ca^{2+} mobilization (in contrast to murine B cells, ref. 101), but does induce the rapid phosphorylation of proteins on serine and threonine residues (96). However, various groups have recently found that anti-CD40 may also induce phosphoinositide hydrolysis, PTK activation, and PI-3K activation, at least in human B cell lines (102–104). Additionally, signals emanating via CD40 seem to interact with those emanating from IL-6 receptors. For example, IL-6 increases the phosphorylation of CD40 in CESS cells, and CD40 induces normal B cells to produce IL-6 (99).

In pre-activated (but not resting) human B cells, CD40 synergizes with IL-4 to promote B cell growth. This was one of the observations which led Banchereau *et al.* (105) to devise the first culture system which permits the development of long-term, factor-dependent B cell lines. In this system, human B cells are cultured with CD40 mAbs in the presence of mouse L cells transfected with FcRII (to present the mAb to B cells) and IL-4 or other cytokines.

Thus it is clear that engaging CD40 generates an important 'survival signal' to B cells and acts as a co-competence factor. This idea is in line with results from Liu *et al.* (106) who showed that B cells from germinal centres, which normally die rapidly *in vitro*, can be rescued from apoptotic death by stimulation with either anti-Ig or CD40. This suggests that, physiologically, one role for CD40 is in maintaining the persistence of B cells destined for the memory pathway. Similar data have been obtained from *in vitro* studies in the mouse where anti-CD40, especially in the presence of IL-4, prevents both spontaneous and anti-Ig-induced apoptosis of mature B cells (107, 108) and anti-μ induced apoptosis in the B cell lymphoma WEHI-231 (109).

8.2 The CD40 ligand

The receptor-like structure of CD40 suggested that its ligand is either a soluble cytokine, or a cell surface molecule present on cells that interact with B cells. It is now clear that the latter possibility is the correct one. The ligand for CD40 (CD40L) has been cloned in both mouse and man (110–113). It is a novel 33–39 kDa type II glycoprotein, related to tumour necrosis factor, which is transiently expressed on pre-activated, but not resting, $CD4^+$ T cells. Transfected cells expressing CD40L, or soluble constructs of CD40L, induce both human and murine B cells to synthesize DNA and, in the presence of appropriate cytokines, to secrete Ig and to switch to downstream isotypes such as IgE (110, 114, 115). Administration of anti-CD40L antibodies to experimental animals blocks antibody responses to TD but not TI antigens (116), thus highlighting the importance of this receptor–counterligand pair in the initiation of humoral immunity. Furthermore, mutations in the human CD40L are responsible for a severe immunodeficiency (X-linked hyper-IgM syndrome), again reinforcing the central importance of CD40L in the immune system (117, 118). There is, therefore little doubt that CD40L is the major (although perhaps not the only) cell surface molecule on T cells responsible for contact-mediated help (Section 10.2.3).

9. Other signalling molecules on B cells

There are a number of other B cell surface molecules, which may play a role in controlling B cell activation, whose properties will be briefly summarized here.

9.1 CD72 (murine Lyb2)

This is a 45 kDa type II glycoprotein, belonging to the same family as CD23 (119). Antibodies to mouse Lyb2 activate a fraction of resting B cells and, induce Ca^{2+} mobilization but are more mitogenic for pre-activated B cells (120). Anti-Lyb2 also blocks antibody responses to TD, but not TI antigens (121). In human B cells, CD72 mAbs induce the transition from G_0 to G_1 and act synergistically with certain stimuli such as immobilized anti-μ (122). CD72 may play a role during T cell–B cell interaction, since it has recently been demonstrated that its natural ligand is the CD5 antigen expressed predominantly on T cells (123). However, mice lacking CD5 as a result of gene targeting give normal antibody responses to both TI and TD antigens (124), arguing against an essential role for CD5/CD72 interactions during T cell–B cell collaboration.

9.2 CD20

CD20 is a pan B cell phosphoprotein, predicted to have four transmembrane domains (125). Monoclonal antibodies to this molecule can activate resting B cells or inhibit entry into cell cycle. Ligation of CD20 does not induce Ca^{2+} mobilization and induces only limited protein phosphorylation (5). However, the structure of CD20 suggests that it may function as an ion channel and Tedder *et al.* have suggested that CD20 may function to regulate calcium homeostasis.

9.3 CD22 (formerly Lyb8 in the mouse)

CD22 is a 135 kDa phosphoglycoprotein which is restricted to B cells. It is a member of the Ig superfamily, related to adhesion molecules such as CD33, MAG, and CEA (reviewed in ref. 126). In man, two alternative forms of CD22 (α and β) have been identified, whereas only the β form occurs in the mouse. Cross-linking of CD22 on human B cells augments anti-μ-induced B cell proliferation and Ca^{2+} mobilization. Only CD22$^+$ B cells give Ca^{2+} signals in response to anti-μ (127). This suggested an association between sIg and CD22, and *in vitro* kinase assays on mild detergent lysates of B cells have revealed that a small fraction of CD22 is indeed sIg-associated (128, 129). The cytoplasmic tail of CD22 also contains an ARH-1 motif, in line with the idea that CD22 is a signalling molecule. In addition, CD22 becomes tyrosine phosphorylated following BCR stimulation (129, 130).

A major function of CD22 is clearly as an adhesion molecule. Although a specific ligand for CD22 was originally believed to be the RO isoform of CD45 (131), more extensive studies have shown that in fact CD22 is a lectin, with specificity for N-linked oligosaccharides containing α2,6-linked sialic acids, present on a number of different proteins (132). Hence the precise functions of this interesting molecule in B cell biology remain unclear.

9.4 CD38

CD38 is an interesting protein, which is seemingly expressed in a quite different pattern in man and mouse. In the former, the protein is quite widely distributed on T and B cells at different stages of their differentiation (133). In the mouse, it seems to be present only on B cells and its levels increase with B cell activation (134). In this species, anti-CD38 is weakly mitogenic *per se* and synergizes with IL-4. The sequence of the protein shows significant homology with ADP-ribosyl cyclase from Aplysia, an enzyme which converts NAD$^+$ to cyclic ADP ribose. Indeed, work by several groups has now shown conclusively that the extracellular domain of CD38 has intrinsic ADP-ribosyl cyclase activity (135–137). The significance of this is that cADPR is a Ca^{2+}-mobilizing second messenger which acts on ryanodine receptors in the endoplasmic reticulum (which are distinct from IP$_3$ receptors) (reviewed in ref. 138). Hence, CD38 is an example of an ectoenzyme on the surface of lymphocytes. It is unclear what relevance its enzymatic activity has in regulating $[Ca^+]_i$ in these cells, given that NAD$^+$ is generally considered to be an intracellular metabolite.

10. T cell-dependent B cell activation

Unlike TI polysaccharide antigens, protein antigens and experimental antigens such as sheep erythrocytes (SRBC) in mice, do not have repeating epitopes. In consequence, these materials cannot cause sufficient cross-linking of sIg receptors to induce significant B cell activation. It has been known for many years that antibody responses to such antigens are absolutely dependent on T_H cells. Within recent years there has been considerable progress in our understanding of the nature of the helper signals delivered to B cells, which comprise both contact-mediated signals and soluble growth and differentiation factors (cytokines).

10.1 MHC-restricted and unrestricted B cell activation

10.1.1 Historical aspects

Classical experiments performed some 20 years ago indicated that so-called 'carrier determinants' of a protein antigen were recognized by T cells, whereas 'haptenic determinants' (for example those experimentally introduced by chemical modification of the protein) were recognized by B cells (reviewed in refs 139, 140). These pioneering investigations established the following:

- That hapten and carrier have to be on the same molecule for effective antibody formation (linked recognition). This was, retrospectively, the first real clue that T cells and B cells must interact physically, and it was believed to occur via a hapten–carrier bridge.

- That the activation of resting B cells by T_H cells is MHC-restricted. In other words, B cells and T cells must be from histocompatible individuals.

The MHC restriction in TD antibody responses maps to the class II genes of the MHC (the Ia antigens in the mouse, for example). Since macrophages are required for the responses to at least some TD antigens (such as SRBC), and macrophage–T cell interactions are also class II-restricted, for a time it was unclear whether MHC restriction in antibody formation operates at the level of macrophage–T cell and/or T cell–B cell interactions. B cells are known to bind native antigens (which can be readily demonstrated at the single cell level). However, it was then established that T cells recognize only processed (that is to say, degraded) proteins in the form of peptides bound to MHC class I or, in the case of $CD4^+$ T_H cells, class II molecules. It was not difficult to envisage that cells such as macrophages could carry out the necessary processing steps to generate peptides for presentation.

10.1.2 B cells as antigen-presenting cells

Ultimately the controversy about the level at which MHC restriction operates in antibody responses was resolved by the recognition that B cells are also very effective APC for T cells (141). This was demonstrated very elegantly by Lanzavecchia (142), who generated T cell clones and Epstein–Barr virus transformed B cell clones specific for tetanus toxoid. He found that B cells internalize and degrade tetanus toxoid and present peptides to T cells on their class II molecules. Furthermore, antigen-specific B cells are at least 1000-fold more effective at presenting this antigen to T cells than other APC. Hence, B cells capture and focus antigens in a very effective way via their sIg receptors. The hapten–carrier bridge hypothesis can, therefore, now be reinterpreted to state that B cells first recognize haptenic determinants on the native protein, process the protein, and present peptide fragments to an interacting T_H cell. However, resting B cells are relatively poor APC for activating resting T cells (143) and indeed may induce T cell tolerance, rather than activation (144). This is because resting B cells do not express the counter-receptors B7.1 and B7.2 which interact with important co-stimulatory receptors CD28 and/or CTLA-4 on T cells. B7 is markedly up-regulated following B cell activation, via CD40, for example (145–147). Activation also induces other changes in B cells, which increase their capacity to present antigen (such as up-regulation of class II levels, for example). In short, it is now abundantly clear that there is a complex two-way intercellular communication between T cells and B cells when the two cells meet (reviewed in refs 148, 149).

A plausible view of what may occur *in vivo* during the induction of a response to a TD antigen may therefore be summarized as follows:

1. T cells first encounter processed antigen on other APC, such as dendritic (interdigitating) cells in the T cell areas of spleen and lymph nodes, or macrophages.
2. Then, preactivated (CD40L$^+$ and cytokine-secreting) T_H cells interact with specific B cells, which have also processed the antigen and carry the relevant peptides on their class II molecules. This interaction generates the necessary contact-mediated and soluble signals to induce B cell proliferation and differentiation.

In model systems, it has been shown that only the activation of *resting* B cells requires antigen and contact-mediated, class II-restricted signals from T_H cells (reviewed in ref. 139). Once B cells are activated, subsequent phases of the response (i.e. B cell proliferation and differentiation to AFC) can be driven by non-specific, non-MHC-restricted soluble factors (cytokines). These are discussed below in Section 10.3. However, *pre-activated* T cells can also activate resting B cells via contact-mediated, non-MHC-restricted signals and render them responsive to lymphokines (Section 10.2.3). This is now believed to be largely due to fact that these cells express the CD40L and hence activate B cells via their CD40 receptors.

10.2 Events occurring during T cell–B cell interaction

For the sake of simplicity, the events occurring during T cell–B cell interaction will be divided into three phases. Clearly, however, the process is an extremely dynamic one, with the participation of a variety of cell surface molecules and soluble growth and differentiation factors (reviewed in ref. 140).

10.2.1 Low-affinity, reversible interactions

Both T and B cells express a number of classical cell adhesion molecules and their counterligands. These are summarized in Fig. 3a and include LFA-1/ICAM-1 (both expressed on both cell types), CD2/LFA-3, CD28/B7, and CD4/class II (reviewed in ref. 150). The latter involves the interaction of CD4 on T cells with monomorphic determinants on class II, which also occurs during the antigen-recognition phase of T–B cell interaction. It is believed that in the absence of antigen these molecules mediate low-affinity, reversible interactions of T cells with B cells, during which T cells survey the surface of the latter for relevant peptides.

10.2.2 High-affinity, MHC-restricted antigen recognition

Investigations of the antigen-dependent phase of T cell–B cell interaction have made use of model systems where T_H clones are mixed *in vitro* with either antigen-

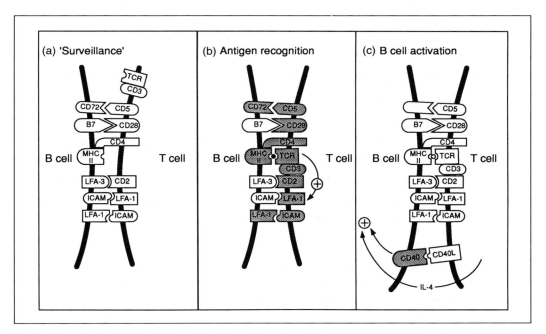

Fig. 3. Stages in T cell–B cell interaction. The process is arbitrarily divided into three stages. (a) Antigen-independent, T cell surveillance of B cells, resulting in low-affinity, reversible, cell–cell interactions, which may involve some, or all of the cell surface molecules shown. (b) Antigen recognition by the T cell. Once the B cell has processed the antigen and presented peptide to the T cell, antigen recognition leads to stable conjugation. Signal transduction through the TCR/CD3 complex rapidly enhances the affinity of LFA-1 for ICAM-1 (arrow). The shaded symbols depict potential signal-transducing molecules on both partners, which may play a role during this phase. (c) B cell activation; the activation of the T cell leads, after a few hours, to the appearance of a competence factor (which is probably CD40L) and to the secretion of cytokines (such as IL-4). The interaction of CD40L with CD40 on the B cell (especially in the presence of IL-4) leads to B cell activation and DNA synthesis, as well as up-regulation of co-stimulatory receptors such as B7.1 and B7.2. The programme for Ig secretion is controlled by the actions of additional cytokines (e.g. IL-4 and IL-5 in the mouse). (Modified from refs 148, 149.)

specific, or non-specific B cells (reviewed in ref. 139). These have shown that, once B cells have processed antigen, they form tight conjugates with MHC-compatible T cells, involving intimate areas of membrane contact. These interactions are monogamous (i.e. one T cell generally only binds one B cell). Two key events occur rapidly during this phase of cell interaction (Fig. 3b):

- The recognition of peptide by the TCR, on B cell class II, which provokes a similar cascade of second messengers in the T cell as in B cells stimulated with anti-Ig. This interaction is further stabilized by binding of CD4 to class II.

- Affinity conversion of LFA-1 on the T cell from a form with low affinity to one with a high-affinity form for ICAM-1. This effect can be induced by ligation of the TCR/CD3 complex, or by PMA, and hence is probably a consequence of PKC activation (151). It is likely that this is an important docking mechanism, which rapidly stabilizes the T cell–B cell conjugate.

A variety of cell surface molecules cluster to the site of cell interaction, including LFA-1, the TCR, and CD4. In addition, the microtubular organizing centre (MTOC) of the T cell becomes reorientated towards the area of cell contact (152). The MTOC is the site of nucleation of microtubules and hence is an indicator of cell polarity. It also co-localizes with the Golgi apparatus, where proteins are prepared for secretion. Kupfer *et al.* (153) have graphically demonstrated polarized concentration of cytokines (presumably within the Golgi) in T_H clones towards the area of B cell contact. It is, therefore, likely that the concentrations of cytokines within the area of T cell–B cell contact could reach substantial levels.

In experimental systems, T cell–B cell conjugates remain stable for more than 24 h. Since the LFA-1 affinity conversion is short-lived (151), it is likely that additional mechanisms serve to stabilize the conjugates over such long periods. These may include the *de novo* synthesis and up-regulation of molecules such as CD28 and CD2 on T cells following their activation. In addition, up-regulation of class II molecules on B cells during activation will increase the capacity of the B cell to present peptides and hence maximize antigen-dependent cell–cell contacts.

It will be apparent that several of the molecules involved in this early phase of T–B cell interaction are potential signal transducing molecules. Besides the antigen receptors of both partners, these include CD2, and CD28 on T cells and class II and CD72 on B cells (Fig. 3b). There is also some evidence that ligation of LFA-1 on B cells can modulate activation (154). The precise details of the biochemical events which occur during conjugation are, however, far from understood. Nevertheless, it is worth stressing that there must be a dynamic two-way dialogue between the cells, so that the traditional idea that T cells are the dominant partner in the interaction is no longer tenable.

10.2.3 Class II unrestricted, contact-dependent B cell activation

Chemically inactivated, *resting* T cells will bind to B cells, but do not cause B cell entry into the cell cycle (155). This indicates that simple engagement of the cell-interaction molecules discussed above is not sufficient to induce B cell activation.

Rather, T cells need to be activated for some 4–8 h before they activate B cells, and they will then do so in an MHC-unrestricted, antigen-independent manner. Several groups have shown that, once T cells are activated (by immobilized CD3 mAbs, for example), they will induce powerful polyclonal B cell activation, even when the T cells are fixed. Strikingly, even crude membrane preparations from such pre-activated cells are active (155–157). The appearance of this competence factor requires *de novo* RNA and protein synthesis, suggesting that it is due to the appearance of a novel protein (or proteins) in the T cell membrane with B cell activating properties. The factor has been called a competence factor because (at least in some studies) it induces only the G_0 to G_1 transition in B cells, and hence sensitizes the cells to progression factors (such as IL-4) which will drive DNA synthesis, and differentiation factors (such as IL-4 and IL-5 in the mouse) which will induce them to secrete Ig. However, in other studies, high concentrations of membranes from activated T_H cells that do not synthesize IL-4 (T_{H1} T cells, see below) drive DNA synthesis in the absence of added lymphokines (158).

The identity of the competence factor is now believed to be CD40L (Fig. 3c) which is transiently expressed on activated $CD4^+$ T cells. This protein, either in soluble form or when expressed on the membranes of irrelevant cells, causes B cell activation and, in the presence of appropriate cytokines, Ig secretion (refs 110, 159–162; Section 8.2).

To recap, it currently appears that MHC-restricted interactions between T and B cells are involved only in establishing stable, antigen-specific conjugates between the cells. Once the T cell is sufficiently activated to express the membrane-bound competence factor, and to secrete cytokines, subsequent phases of B cell activation can proceed in an MHC-unrestricted, antigen-independent fashion (Fig. 3). It is, therefore, conceivable that T cells which have become activated by antigen on other APC, such as dendritic cells, could activate B cells in a polyclonal fashion. *In vivo* this type of bystander help probably occurs rather infrequently, because T cells are more likely to form stable interactions with B cells which have also processed the relevant antigen. In addition, the expression of CD40L is clearly tightly regulated.

10.3 Cytokine-driven B cell proliferation and differentiation

As mentioned above, once B cells have become 'primed' by contact-mediated helper signals subsequent clonal expansion and differentiation to AFC are largely driven by cytokines. Most of the known interleukins have been found to have detectable effects (either on their own, or in combination with other agents) on B cells in *in vitro* assays (reviewed in ref. 163). In addition, γ-interferon (γ-IFN) and other factors, such as transforming growth factor-β (TGF-β) (164), may also play a role in modulating antibody responses. Given that cytokines are generally quite pleiotropic in their actions, the literature concerning this topic is, to say the least, confusing. Therefore, because of limitations of space, the following discussion will only briefly summarize the major activities of those cytokines whose role in B cell responses is well-established. The interested reader can obtain further in-depth information from other recent reviews (4).

10.3.1 Cytokines acting on resting B cells

The most important cytokine with clear actions on quiescent B cells is IL-4 (reviewed in ref. 165). On its own, IL-4 is an abortive B cell activator, causing slight cell enlargement and up-regulating class II, CD23, sIgM (in man and mouse, refs 166, 167) and sIgD (in mouse, ref. 167) levels. It synergizes with signals transmitted via sIg receptors, or CD40, and with PKC activators (167) to induce B cells to synthesize DNA. IL-4 also synergizes with pre-activated T cells (presumably via the CD40L) to drive B cell DNA synthesis. In mouse B cells, IL-4 does not induce PIP_2 hydrolysis, or its sequelae (73, 168), but it does synergize with anti-Ig to induce PKC translocation (169). It is therefore likely that IL-4 acts by potentiating and/or prolonging PKC-mediated signals in murine B cells. In contrast, in human B cells IL-4 has been reported to induce transient phosphoinositide breakdown, followed by prolonged elevation of cAMP (170).

IL-10 has some of the activities of IL-4 on resting B cells, causing up-regulation of class II but not CD23 levels. However, it does not synergize with anti-Ig to induce B cell activation (171). In man, it has been shown to also act as a growth and differentiation factor for pre-activated B cells (172). Recently, IL-13 (also a product of T_{H2} cells in the mouse) has been cloned and shown to have very similar effects to IL-4 on human B cells; it synergizes (weakly) with anti-μ and strongly with anti-CD40 to induce B cell proliferation, and drives both IgM and IgG secretion in the presence of CD40L (173, 174).

10.3.2 Cytokines acting as B cell growth and differentiation factors

The major lymphokines which act on pre-activated B cells, to induce either proliferation and/or differentiation to AFC, are IL-2, IL-4, IL-5, IL-6, and γ-IFN (reviewed in ref. 163). IL-4 and IL-5 act synergistically in the mouse to induce the production of all antibody isotypes, and IL-4 favours the production of IgG1 and IgE in LPS-activated B cells (175, 176). The production of IgE is absolutely dependent on IL-4 in this species (177), although other influences may play a role in man. The involvement of IL-5 in antibody responses in man is less apparent (178). IL-6 has clear activity as a B cell differentiation factor in man, although this is less striking in the mouse. γ-IFN favours the production of IgG2a in LPS-activated murine B cells, and antagonizes many of the effects of IL-4 (179).

10.3.3 Cytokine production by T_H subsets

Our understanding of how the plethora of cytokines found to act on B cells, many having apparently overlapping activities, regulates antibody responses under physiological conditions has been considerably advanced by the description of T_H subsets which produce differing spectra of lymphokines. This was first recognized in the mouse, by Mosmann and Coffman (reviewed in refs 180, 181). In summary, many long-term CD4-positive T cell clones can be grouped into either T_{H1} or T_{H2} cells; the major features of T_{H1} cells is that they produce IL-2 and γ-IFN, but not IL-4, IL-5, or IL-6. Conversely, whilst T_{H2} cells produce IL-4, IL-5, IL-6, and IL-10, they do not secrete γ-IFN or IL-2. Both types of clone have helper activity for B cells (they both express the CD40L after activation, for example, Section 8.3). However, as might be expected from their lymphokine secretion patterns, T_{H2} clones generally provide more effective help for antibody production, especially for IgG1 and IgE responses in the mouse. This is largely due to their production of IL-4 and IL-5, which are both required for the induction of all isotypes except IgE, where IL-4 is sufficient. On the other hand, the helper activity of T_{H1} clones is more variable. This is partly due to the fact that they produce γ-IFN which can be immunosuppressive. However, under appropriate conditions, T_{H1} clones do provide help, especially for IgG2a responses, which is regulated by γ-IFN (179). They also mediate delayed-type hypersensitivity (DTH) responses, which T_{H2} clones do not.

T_{H1} and T_{H2} clones also cross-regulate each others' helper functions. This is believed to be due to modulatory effects of different cytokines. Hence, as mentioned previously, γ-IFN antagonizes many of the effects of IL-4. In addition, IL-10 was first recognized by its capacity to inhibit cytokine production by T_{H1} clones (182), which it does by acting on antigen-presenting cells (183).

The picture that is emerging in this field is that *in vivo*, the type of T_H cell population which develops with time after immunization is determined by (*inter alia*) the nature and the persistence of the antigen. In this scenario, intracellular and viral pathogens which elicit strong cytotoxic and DTH responses,) would be expected to induce a predominant T_{H1} response. This would be accompanied by the production of complement-fixing antibodies, such as IgG2a in the mouse. Conversely, parasites which invoke strong antibody responses, and particularly IgE, and eosinophilia, would be expected to elicit responses dominated by T_{H2} cells. In other words, T_H subsets are believed to have evolved to provide help which will skew the immune responses towards one which is appropriate to the antigen in question.

11. Conclusions

Since the initial recognition of T cell–B cell collaboration some 20 years ago, there has been remarkable progress in our understanding of the events which occur during the induction of antibody responses. The aim of this chapter has been to provide the reader with an overview of the current state of the art, without, hopefully too much in the way of confusion! Clearly, although broad principles have emerged, there are still substantial gaps in our knowledge of how the activation of B cells is regulated in real life. This applies particularly to the precise functions of the surface antigens described in the early part of the chapter. It is satisfying, however, that the nature of the helper signals generated during T cell–B cell interactions are being elucidated—through the description of CD40L, for example. The ultimate goal of this field of endeavour must, of course, be to translate the enormous body of data gleaned from *in vitro* experiments to the microenvironments in which B cells encounter antigens and T$_H$ cells *in vivo*—a topic which has not been touched upon in this review.

References

References marked * are recommended for further reading.

1. Klaus, G. G. B. (1990). *B lymphocytes*, pp. 1–71. IRL Press, Oxford.
2. Maclennan, I. C. M., Liu, Y. J., and Johnson, G. D. (1992). Maturation and dispersal of B cell clones during T cell-dependent antibody responses. *Immunol. Rev.*, **126**, 143.
*3. Leanderson, T., Kallberg, E., and Gray, D. (1992). Expansion, selection and mutation of antigen-specific B cells in germinal centres. *Immunol. Rev.*, **126**, 47.
4. Callard, R. E. (ed.) (1990). *Cytokines and B lymphocytes*. Academic Press, London.
*5. Clark, E. A. and Lane, P. J. (1991). Regulation of human B-cell activation and adhesion. *Annu. Rev. Immunol.*, **9**, 97.
6. Moller, P., Eichelmann, A., and Moldenhauer, G. (1991). Surface molecules involved in B lymphocyte function. *Virchows Arch. A. Pathol. Anat. Histopathol.*, **419**, 365.
7. Calvert, J. E. and Cooper, M. D. (1988). Early stages of B cell differentiation. In *B lymphocytes in human disease* (ed. G. Bird and J. E. Calvert), pp. 77–101. Oxford University Press, Oxford.
8. Goodnow, C. C., Adelstein, S., and Basten, A. (1990). The need for central and peripheral tolerance in the B cell repertoire. *Science*, **248**, 1373.
9. DeFranco, A. L., Kung, J. T., and Paul, W. E. (1982). Regulation of growth and proliferation in B cell subpopulations. *Immunol. Rev.*, **64**, 161.
10. Klaus, G. G. B., Bijsterbosch, M. K., O'Garra, A., Harnett, M. M., and Rigley, K. P. (1987). Receptor signalling and crosstalk in B lymphocytes. *Immunol. Rev.*, **99**, 19.
11. Goroff, D. K., Stall, A., Mond, J. J., and Finkelman, F. D. (1986). *In vitro* and *in vivo* B lymphocyte-activating properties of monoclonal anti-delta antibodies. I Determinants of B lymphocyte activating properties. *J. Immunol.*, **135**, 2381.
12. Brunswick, M., Finkelman, F. D., Highet, P. F., Inman, J. K., and Dintzis, H. M. (1988). Picogram quantities of anti-Ig antibodies coupled to dextran induce B cell proliferation. *J. Immunol.*, **140**, 3364.
13. Andersson, J., Bullock, W. W., and Melchers, F. (1974). Inhibition of mitogenic stimulation of mouse lymphocytes by anti-mouse Ig antibodies. *Eur. J. Immunol.*, **4**, 715.
14. Peccanha, L. M., Snapper, C. M., Finkelman, F. D., and Mond, J. J. (1991). Dextran-conjugated anti-Ig antibodies as a model for T cell-independent type 2 antigen-mediated stimulation of Ig secretion *in vitro*. I Lymphokine dependence. *J. Immunol.*, **146**, 833.
15. Birkeland, M. L., Simpson, L., Isakson, P. C., and Pure, E. (1987). T-independent and T-dependent steps in the murine B cell response to anti-immunoglobulin. *J. Exp. Med.*, **506**, 519.
16. Phillips, C. and Klaus, G. G. B. (1992). Soluble anti-mu monoclonal antibodies prime resting B cells to secrete Ig in response to interleukins 4 and 5. *Eur. J. Immunol.*, **22**, 1541.
17. Reth, M., Hombach, J., Wienands, J., Campbell, K. S., Chien, N., Justement, L. B.,

and Cambier, J. C. (1991). The B-cell antigen receptor complex. *Immunol. Today*, **12**, 196.

*18. Cambier, J. C., Bedzyk, W., Campbell, K., Chien, N., Friedrich, J., Harwood, A., Jensen, W., Pleiman, C., and Clark, M. R. (1993). The B-cell antigen receptor: structure and function of primary, secondary, tertiary and quaternary components. *Immunol. Rev.*, **132**, 85.

*19. Borst, J., Brouns, G. S., De Vries, E., Verschuren, M. C., Mason, D. Y., and van Dongen, J. J. (1993). Antigen receptors on T and B lymphocytes: parallels in organization and function. *Immunol. Rev.*, **132**, 49.

20. Wienands, J., Hombach, J., Radbruch, A., Riesterer, C., and Reth, M. (1990). Molecular components of the B cell antigen receptor complex of class IgD differ partly from those of IgM. *EMBO J.*, **9**, 449.

21. Venkitaraman, A. R., Williams, G. T., Dariavach, P., and Neuberger, M. S. (1991). The B cell antigen receptor of the five immunoglobulin classes. *Nature*, **352**, 777.

22. Altman, A., Coggeshall, K. M., and Mustelin, T. (1990). Molecular events mediating T cell activation. *Adv. Immunol.*, **48**, 227.

23. Ransom, J. T., Harris, L. K., and Cambier, J. C. (1986). Anti-Ig induces release of inositol 1,4,5-trisphosphate, which mediates mobilization of intracellular Ca^{++} stores in B lymphocytes. *J. Immunol.*, **137**, 708.

24. Bijsterbosch, M. K., Meade, C. J., Turner, G. A., and Klaus, G. G. B. (1985). B lymphocyte receptors and polyphosphoinositide degradation. *Cell*, **41**, 999.

25. Bustelo, X. R. and Barbacid, M. (1992). Tyrosine phosphorylation of the vav proto-oncogene product in activated B cells. *Science*, **256**, 1196.

26. Gold, M. R., Chan, V. W., Turck, C. W., and DeFranco, A. L. (1992). Membrane Ig cross-linking regulates phosphatidylinositol 3-kinase in B lymphocytes. *J. Immunol.*, **148**, 2012.

27. Yamanishi, Y., Fukui, Y., Wongsasant, B., Kinoshita, Y., Ichimori, Y., Toyoshima, K., and Yamamoto, T. (1992). Activation of src-like protein tyrosine kinase lyn and its association with phosphatidylinositol 3-kinase upon antigen receptor-mediated signalling. *Proc. Natl Acad. Sci. USA*, **89**, 1118.

28. Gold, M. R., Sanghera, J. S., Stewart, J., and Pelech, S. L. (1992). Selective activation of p42 mitogen-activated protein (MAP) kinase in murine B lymphoma cell lines by membrane immunoglobulin cross-linking. Evidence for protein kinase C-independent and -dependent mechanisms of activation. *Biochem. J.*, **287**, 269.

29. Coggeshall, K. M., McHugh, J. C., and Altman, A. (1992). Predominant expression and activation-induced tyrosine phosphorylation of phospholipase C-gamma 2 in B lymphocytes. *Proc. Natl Acad. Sci. USA*, **89**, 5660.

30. Carter, R. H., Park, D. J., Rhee, S. G., and Fearon, D. T. (1991). Tyrosine phosphorylation of phospholipase C induced by membrane immunoglobulin in B lymphocytes. *Proc. Natl Acad. Sci. USA*, **88**, 2745.

31. Tuveson, D. A., Carter, R. H., Soltoff, S. P., and Fearon, D. T. (1993). CD19 of B cells as a surrogate kinase insert region to bind phosphatidylinositol 3-kinase. *Science*, **260**, 986.

32. Harwood, A. E. and Cambier, J. C. (1993). B cell antigen receptor cross-linking triggers rapid protein kinase C independent activation of p21ras. *J. Immunol.*, **151**, 4513.

33. Graziadei, L., Riabowol, K., and Barsagi, D. (1990). Co-capping of ras proteins with surface immunoglobulins in B lymphocytes. *Nature*, **347**, 396.

34. Cockcroft, S. (1987). Phosphoinositide phosphodiesterase: regulation by a novel guanine nucleotide binding protein, Gp. *TIPS*, **12**, 75.

35. Gold, M. R., Jakway, J. P., and DeFranco, A. L. (1987). Involvement of a G-protein in mIgM stimulated phosphoinositide breakdown. *J. Immunol.*, **139**, 3604.

36. Harnett, M. M. and Klaus, G. G. B. (1988). G protein coupling of antigen receptor-stimulated polyphosphoinositide hydrolysis in B cells. *J. Immunol.*, **140**, 3135.

37. Phillips, R. J., Harnett, M. M., and Klaus, G. G. B. (1991). Antigen receptor-mediated phosphoinositide hydrolysis in murine T cells is not initiated via G-protein activation. *Int. Immunol.*, **3**, 617.

38. Liang, M. N. and Garrison, J. C. (1991). The epidermal growth factor receptor is coupled to a pertussis toxin-sensitive guanine nucleotide regulatory protein in rat hepatocytes. *J. Biol. Chem.*, **266**, 13342.

39. Hide, M., Ali, H., Price, S. R., Moss, J., and Beaven, M. A. (1991). GTP-binding protein Gz: its downregulation by dexamethasone and its credentials as a mediator of antigen-induced responses in RBL-2H3 cells. *Mol. Pharmacol.*, **40**, 473.

40. Gold, M. R., Matsuuchi, L., Kelly, R. B., and DeFranco, A. L. (1991). Tyrosine phosphorylation of components of the B-cell antigen receptors following receptor crosslinking. *Proc. Natl Acad. Sci. USA*, **88**, 3436.

41. Burkhardt, A. L., Brunswick, M., Bolen, J. B., and Mond, J. J. (1991). Anti-immunoglobulin stimulation of B lymphocytes activates src-related protein-tyrosine kinases. *Proc. Natl Acad. Sci. USA*, **88**, 7410.

42. Hutchcroft, J. E., Harrison, M. L., and Geahlen, R. L. (1992). Association of the 72-kDa protein-tyrosine kinase PTK72 with the B cell antigen receptor. *J. Biol. Chem.*, **267**, 8613.

43. Lane, P. J. L., Ledbetter, J. A., McConnell, F. M., Draves, K., Deans, J., Schieven, G. L., and Clark, E. A. (1991). The role of tyrosine phosphorylation in signal transduction through surface Ig in human B cells. *J. Immunol.*, **146**, 715.

44. Pure, E. and Tardelli, L. (1992). Tyrosine phosphorylation is required for ligand-induced internalisation of the antigen receptor on B lymphocytes. *Proc. Natl Acad. Sci. USA*, **89**, 114.

45. Klaus, G. G. B., O'Garra, A., Bijsterbosch, M. K., and Holman, M. (1986). Activation and proliferation signals in mouse B cells. VIII. Induction of DNA synthesis in B cells by a combination of calcium ionophores and phorbol myristate acetate. *Eur. J. Immunol.*, **16**, 92.

46. Brunswick, M., Bonvini, E., Francis, M., Felder, C. C., Hoffman, T., and Mond, J. J. (1990). Absence of demonstrable phospholipid turnover in B cells stimulated with low mitogenic concentrations of dextran–anti-Ig conjugates. *Eur. J. Immunol.*, **20**, 855.

47. Yamada, H., June, C. H., Finkelman, F., Brunswick, M., Ring, M. S., Lees, A., and Mond, J. J. (1993). Persistent calcium elevation correlates with the induction of surface immunoglobulin-mediated B cell DNA synthesis. *J. Exp. Med.*, **177**, 1613.

48. Billah, M. M. and Anthes, J. C. (1991). The regulation and cellular functions of phosphatidylcholine hydrolysis. *Biochem. J.*, **269**, 281.

49. Wicker, L. S., Boltz R. C. Jr, Matt, V., Nichols, E. A., Peterson, L. B., and Sigal, N. H. (1990). Suppression of B cell activation by cyclosporin A, FK506 and rapamycin. *Eur. J. Immunol.*, **20**, 2277.

50. Klaus, G. G. B. (1988). Cyclosporine-sensitive and cyclosporine-insensitive modes of B cell stimulation. *Transplantation*, **46**, 11S.

*51. Rao, A. (1994). NF-AT$_p$: a transcription factor required for the co-ordinate induction of several cytokine genes. *Immunol. Today*, **15**, 274.

52. Choi, M. S. K., Brines, R. D., Holman, M. J., and Klaus, G. G. B. (1994). Induction of NF-AT (the nuclear factor of activated T lymphocytes in normal B lymphocytes by anti-immunoglobulin or CD40 ligand in conjunction with IL-4. *Immunity*, **1**, 179.

53. Venkataraman, L., Francis, D. A., Wang, Z., Liu, J., Rothstein, T. L., and Sen, R. (1994). Cyclosporin-A sensitive induction of NF-AT in murine B cells. *Immunity*, **1**, 189.

54. Klaus, G. G. B. and Humphrey, J. H. (1986). A re-evaluation of the role of C3 in B cell activation. *Immunol. Today*, **7**, 163.

55. Ahearn, J. M. and Fearon, D. T. (1989). Structure and function of the complement receptors, CR1 (CD35) and CR2 (CD21). *Adv. Immunol.*, **46**, 183.

56. Matsumoto, A. K., Kopickyburd, J., Carter, R. H., Tuveson, D. A., Tedder, T. F., and Fearon, D. T. (1991). Intersection of the complement and immune systems—a signal transduction complex of the lymphocyte-b containing complement receptor type-2 and CD19. *J. Exp. Med.*, **173**, 55.

*57. Fearon, D. T. (1993). The CD19–CR2–TAPA-1 complex, CD45 and signaling by the antigen receptor of B lymphocytes. *Curr. Opin. Immunol.*, **5**, 341.

58. Carter, R. H., Spycher, M. O., Ng, Y. C., Hoffman, R., and Fearon, D. T. (1988). Synergistic interaction between complement receptor type 2 and membrane IgM on B lymphocytes. *J. Immunol.*, **141**, 457.

59. Tsokos, G. C., Lambris, J. D., Finkelman, F. D., Anastassiou, E. D., and June, C. H. (1990). Monovalent ligands of complement receptor 2 inhibit whereas polyvalent ligands enhance anti-Ig-induced human B cell intracytoplasmic free calcium concentration. *J. Immunol.*, **144**, 1640.

60. Carter, R. H., Tuveson, D. A., Park, D. J., Rhee, S. G., and Fearon, D. T. (1991). The CD19 complex of B lymphocytes. Activation of phospholipase C by a protein tyrosine kinase-dependent pathway that can be enhanced by the membrane IgM complex. *J. Immunol.*, **147**, 3663.

*61. Carter, R. H. and Fearon, D. T. (1992). CD19: lowering the threshold for antigen receptor stimulation of B lymphocytes. *Science*, **256**, 105.

62. Fridman, W. H., Bonnerot, C., Daeron, M., Amigorena, S., Teillaud, J. L., and Sautes, C. (1992). Structural bases of Fc gamma receptor functions. *Immunol. Rev.*, **125**, 49.

63. Uhr, J. W. and Moller, G. (1968). Regulatory effect of antibody on the immune response. *Adv. Immunol.*, **8**, 81.

64. Sinclair, N. R. S. and Panoskaltsis, A. (1988). The immunoregulatory apparatus and autoimmunity. *Immunol. Today*, **9**, 260.

65. Phillips, N. E. and Parker, D. C. (1984). Cross-linking of B lymphocyte Fc gamma receptors and membrane immunoglobulin inhibits anti-immunoglobulin-induced blastogenesis. *J. Immunol.*, **132**, 627.

66. Klaus, G. G. B., Hawrylowicz, C. M., Holman, M., and Keeler, K. D. (1984). Activation and proliferation signals in mouse B cells. III. Intact (IgG) anti-immunoglobulin antibodies activate B cells but inhibit induction of DNA synthesis. *Immunology*, **53**, 693.

67. Bijsterbosch, M. K. and Klaus, G. G. B. (1985). Crosslinking of surface immunoglobulin and Fc receptors on B lymphocytes inhibits stimulation of inositol phospholipid breakdown via the antigen receptors. *J. Exp. Med.*, **162**, 1825.

68. Rigley, K. P., Harnett, M. M., and Klaus, G. G. B. (1989). Co-crosslinking of surface Ig and Fc gamma receptors on B lymphocytes uncouples the antigen receptors from their associated G-protein. *Eur. J. Immunol.*, **19**, 481.

69. Hunziker, W., Koch, T., Whitney, J. A., and Mellman, I. (1990). Fc receptor phosphorylation during receptor-mediated control of B cell activation. *Nature*, **345**, 628.

*70. Muta, T., Kurasaki, T., Misulovin, Z., Sanchez, M., Nussenzweig, M. C., and Ravetch, J. V. (1994). A 13 amino acid motif in the cytoplasmic domain of FcRIIB modulates B cell receptor signalling. *Nature*, **368**, 70.

71. Choquet, D., Partiseti, M., Amigorena, S., Bonnerot, C., Fridman, W. H., and Korn, H. (1993). Cross-linking of IgG receptors inhibits membrane immunoglobulin-stimulated calcium influx in B lymphocytes. *J. Cell Biol.*, **121**, 355.

72. Snapper, C. M., Hooley, J. J., Atasoy, U., Finkelman, F. D., and Paul, W. E. (1989). Differential regulation of murine B cell FcRII expression by CD4-positive T helper subsets. *J. Immunol.*, **143**, 2123.

73. O'Garra, A., Rigley, K. P., Holman, M., McLaughlin, J. B., and Klaus, G. G. B. (1987). B-cell-stimulatory factor 1 reverses Fc receptor-mediated inhibition of B-lymphocyte activation. *Proc. Natl Acad. Sci. USA*, **84**, 6254.

74. Delespesse, G., Sarfati, M., Wu, C. Y., Fournier, S., and Letelier, M. (1992). The low affinity receptor for IgE. *Immunol. Rev.*, **125**, 77.

75. Kolb, J. P., Renard, D., Dugas, B., Genot, E., Petit-Koskas, E., Sarfati, M., Delespesse, G., and Poggioli, J. (1990). Monoclonal anti-CD23 antibodies induce a rise in $[Ca^{2+}]_i$ and polyphosphoinositide hydrolysis in human activated B cells. Involvement of a Gp protein. *J. Immunol.*, **145**, 429.

76. Gordon, J., Liu, Y. J., Maclennan, I. C. M., Floresromo, L., Shields, J., and Bonnefoy, J. Y. (1991). CD23 and immune modulation. *Immunol. Today*, **12**, 206.

77. Liu, Y. J., Cairns, J. A., Holder, M. J., Abbot, S. D., Jansen, K. U., Bonnefoy, J. Y., Gordon, J., and Maclennan, I. C. M. (1991). Recombinant 25-kDa CD23 and interleukin-1 alpha promote the survival of germinal center b-cells—evidence for bifurcation in the development of centrocytes rescued from apoptosis. *Eur. J. Immunol.*, **21**, 1107.

*78. Yu, P., Kosco-Vilbois, M., Richards, M., Kohler, G., and Lamers, M. C. (1994). Negative feedback regulation of IgE synthesis by murine CD23. *Nature*, **369**, 753.

79. Roehm, N. W., Leibson, H. J., Zlotnik, A., Kappler, J., Marrack, P., and Cambier, J. C. (1984). Interleukin-induced increase in Ia expression by normal mouse B cells. *J. Exp. Med.*, **160**, 679.

80. Noelle, R., Krammer, P. H., Ohara, J., Uhr, J. W., and Vitetta, E. S. (1984). Increased expression of Ia antigens on resting B cells: an additional role for B-cell growth factor. *Proc. Natl Acad. Sci. USA*, **81**, 6149.

81. Bottomly, K., Jones, B., Kaye, J., and Jones, F. (1983). Subpopulations of B cells distinguished by cell surface expression of Ia antigens. Correlation of Ia and idiotype during activation by cloned Ia-restricted T cells. *J. Exp. Med.*, **158**, 265.

82. Forsgren, S., Pobor, G., Coutinho, A., and Pierres, M. (1984). The role of I-A/E molecules in B lymphocyte activation. I Inhibition of lipopolysaccharide-induced responses by monoclonal antibodies. *J. Immunol.*, **133**, 2104.

83. Cambier, J. C., Newell, M. K., Justement, L. B., McGuire, J. C., and Leach, K. L. (1987). Ia binding ligands and cAMP stimulate nuclear translocation of PKC in B lymphocytes. *Nature*, **327**, 629.

84. Cohen, D. P. and Rothstein, T. L. (1989). Adenosine 3′, 5′-cyclic monophosphate modulates the mitogenic responses of murine B lymphocytes. *Cell. Immunol.*, **121**, 113.

85. Cambier, J. C., Morrison, D. C., Chien, M. M., and Lehmann, K. R. (1991). Modeling of T cell contact-dependent B cell activation. IL-4 and antigen receptor ligation primes quiescent B cells to mobilize calcium in response to Ia cross-linking. *J. Immunol.*, **146**, 2075.

86. Mooney, N. A., Grillot-Courvalin, C., Hivroz, C., Ju, L. Y., and Charron, D. (1990). Early biochemical events after MHC Class II-mediated signalling on human B lymphocytes. *J. Immunol.*, **145**, 2070.

87. Lane, P. J., McConnell, F. M., Schieven, G. L., Clark, E. A., and Ledbetter, J. A. (1990). The role of class II molecules in human B cell activation. Association with phosphatidyl inositol turnover, protein tyrosine phosphorylation, and proliferation. *J. Immunol.*, **144**, 3684.

88. Markowitz, J. S., Rogers, P. R., Grusby, M. J., Parker, D. C., and Glimcher, L. H. (1993). B lymphocyte development and activation independent of MHC class II expression. *J. Immunol.*, **150**, 1223.

89. Trowbridge, I. S., Ostergaard, H. L., and Johnson, P. (1991). CD45: a leukocyte-specific member of the protein tyrosine phosphatase family. *Biochim. Biophys. Acta*, **1095**, 46.

90. Tonks, N. K., Charbonneau, H., Diltz, C. D., Fischer, E. H., and Walsh, K. A. (1988). Demonstration that the leukocyte common antigen CD45 is a protein tyrosine phosphatase. *Biochemistry*, **27**, 8695.

91. Hunter, T. (1989). Protein-tyrosine phosphatases: the other side of the coin. *Cell*, **58**, 1013.

92. Pingel, J. T. and Thomas, M. L. (1989). Evidence that the leukocyte-common antigen is required for antigen-induced T lymphocyte proliferation. *Cell*, **58**, 1055.

*93. Justement, L. B., Campbell, K. S., Chien, N. C., and Cambier, J. C. (1991). Regulation of B-cell antigen receptor signal transduction and phosphorylation by CD45. *Science*, **252**, 1839.

94. Hasegawa, K., Nishimura, H., Ogawa, S., Hirose, S., Sato, H., and Shirai, T. (1991). Monoclonal antibodies to epitope of CD45R (B220) inhibit IL-4 mediated B cell proliferation and differentiation. *Int. Immunol.*, **2**, 367.

95. Mittler, R. S., Greenfield, R. S., Schacter, B. Z., Richard, N. F., and Hoffmann, M. K. (1987). Antibodies to the common leukocyte antigen (T200) inhibit an early phase in the activation of human B cells. *J. Immunol.*, **138**, 3159.

96. Gruber, M. F., Bjorndahl, J. M., Nakamura, S., and Fu, S. M. (1989). Anti-CD45 inhibition of human B cell proliferation depends on the nature of the activation signals and the state of B cell activation. *J. Immunol.*, **142**, 4144.

97. Lane, P. J. L., Ledbetter, J. A., McConnell, F. M., Draves, K., Deans, J., Schieven, G. L., and Clark, E. A. (1991). The role of tyrosine phosphorylation in signal transduction through surface Ig in human B-cells—inhibition of tyrosine phosphorylation prevents intracellular calcium release. *J. Immunol.*, **146**, 715.

98. Kishihara, K., Penninger, J., Wallace, V. A., Kundig, T. M., Kawai, K., Wakeham, A., Timms, E., Pfeffer, K., Ohashi, P. S., Thomas, M. L., Furlonger, C., Paige, C. J., and Mak, T. W. (1993). Normal B lymphocyte development but impaired T cell maturation in CD45-exon6 protein tyrosine phosphatase-deficient mice. *Cell*, **74**, 143.

99. Clark, E. A. (1990). CD40—a cytokine receptor in search of a ligand. *Tissue Antigens*, **36**, 33.

*100. Smith, C. A., Farrah, T., and Goodwin, R. G. (1994). The TNF receptor superfamily of cellular and viral proteins: activation, co-stimulation and death. *Cell*, **76**, 959.

101. Klaus, G. G. B., Choi, M. S. K., and Holman, M. (1994). Properties of mouse CD40: ligation of CD40 activates B cells via a calcium-dependent, FK506-sensitive pathway. *Eur. J. Immunol.* **24**, 3229.

102. Faris, M., Gaskin, F., Parsons, J. T., and Man Fu, S. (1994). CD40 signaling pathway:

anti-CD40 monoclonal antibody induces rapid dephosphorylation and phosphorylation of tyrosine-phosphorylated proteins including protein tyrosine kinase lyn, fyn and syk and the appearance of a 28 kD tyrosine phosphorylated protein. *J. Exp. Med.*, **179**, 1923.

103. Ren, C. L., Morio, T., Fu, S. M., and Geha, R. S. (1994). Signal transduction via CD40 involves activation of ly kinase and phosphatidylinositol-3 kinase and phosphorylation of phospholipase Cγ2. *J. Exp. Med.*, **179**, 673.

104. Uckun, F. M., Schieven, G. L., Dibirdik, I., Chandan-Langlie, M., Tuel-Ahlgren, L., and Ledbetter, J. A. (1991). Stimulation of protein tyrosine phosphorylation, phosphoinositide turnover, and multiple previously unidentified serine/threonine-specific protein kinases by the Pan-B-cell receptor CD40/Bp50 at discrete developmental stages of human B-cell ontogeny. *J. Biol. Chem.*, **266**, 17478.

105. Banchereau, J., Depaoli, P., Valle, A., Garcia, E., and Rousset, F. (1991). Long-term human B-cell lines dependent on interleukin-4 and antibody to CD40. *Science*, **251**, 70.

*106. Liu, Y. J., Joshua, D. E., Williams, G. T., Smith, C. A., Gordon, J., and MacLennan, I. C. (1989). Mechanism of antigen-driven selection in germinal centres. *Nature*, **342**, 929.

107. Parry, S. L., Hasbold, J., Holman, M., and Klaus, G. G. B. (1994). Hypercrosslinking surface IgM or IgD receptors on mature B cells induces apoptosis that is reversed by costimulation with IL-4 and anti-CD40. *J. Immunol.*, **152**, 2821.

108. Parry, S. L., Holman, M. J., Hasbold, J., and Klaus, G. G. B. (1994). Plastic-immobilized anti-μ or anti-δ antibodies induce apoptosis in mature murine B lymphocytes. *Eur. J. Immunol.*, **24**, 974.

109. Tsubata, T., Wu, J., and Honjo, T. (1993). B cell apoptosis induced by antigen receptor crosslinking is blocked by a T cell signal through CD40. *Nature*, **364**, 645.

110. Armitage, R. J., Fanslow, W. C., Strockbine, L., Sato, T. A., Clifford, K. N., Macduff, B. M., Anderson, D. M., Gimpel, S. D., Davis-Smith, T., Maliszewski, C. R., Clark, E. A., Smith, C. A., Grabstein, K. H., Cosman, D., and Spriggs, M. K. (1992). Molecular and biological characterisation of a murine ligand for CD40. *Nature*, **357**, 80.

111. Graf, D., Korthauer, U., Mages, H. W., Senger, G., and Kroczek, R. A. (1992). Cloning of TRAP, a ligand for CD40 on human T cells. *Eur. J. Immunol.*, **22**, 3191.

112. Hollenbaugh, D., Grosmaire, L. S., Kullas, C. D., Chalupny, N. J., Braesch Andersen, S., Noelle, R. J., Stamenkovic, I., Ledbetter, J. A., and Aruffo, A. (1992). The human T cell antigen gp39, a member of the TNF gene family, is a ligand for the CD40 receptor: expression of a soluble form of gp39 with B cell co-stimulatory activity. *EMBO J.*, **11**, 4313.

113. Gauchat, J. F., Aubry, J. P., Mazzei, G., Life, P., Jomotte, T., Elson, G., and Bonnefoy, J. Y. (1993). Human CD40-ligand: molecular cloning, cellular distribution and regulation of expression by factors controlling IgE production. *FEBS Lett.*, **315**, 259.

114. Nonoyama, S., Hollenbaugh, D., Aruffo, A., Ledbetter, J. A., and Ochs, H. D. (1993). B cell activation via CD40 is required for specific antibody production by antigen-stimulated human B cells. *J. Exp. Med.*, **178**, 1097.

115. Maliszewski, C. R., Grabstein, K., Fanslow, W. C., Armitage, R., Spriggs, M. K., and Sato, T. A. (1993). Recombinant CD40 ligand stimulation of murine B cell growth and differentiation: cooperative effects of cytokines. *Eur. J. Immunol.*, **23**, 1044.

*116. Foy, T. M., Shepherd, D. M., Durie, F. H., Aruffo, A., Ledbetter, J. A., and Noelle, R. J. (1993). *In vivo* CD40–gp39 interactions are essential for thymus-dependent humoral immunity. II. Prolonged suppression of the humoral immune response by an antibody to the ligand for CD40, gp39. *J. Exp. Med.*, **178**, 1567.

117. DiSanto, J. P., Bonnefoy, J. Y., Gauchat, J. F., Fischer, A., and de Saint Basile, G. (1993). CD40 ligand mutations in X-linked immunodeficiency with hyper-IgM. *Nature*, **361**, 541.

*118. Aruffo, A., Farrington, M., Hollenbaugh, D., Li, X., Milatovich, A., Nonoyama, S., Bajorath, J., Grosmaire, L. S., Stenkamp, R., Neubauer, M., *et al.* (1993). The CD40 ligand, gp39, is defective in activated T cells from patients with X-linked hyper-IgM syndrome. *Cell*, **72**, 291.

119. Nakayama, E., Von Hoegen, I., and Parnes, J. R. (1989). Sequence of the Lyb-2B cell differentiation antigen defines a gene superfamily of receptors with inverted membrane orientation. *Proc. Natl Acad. Sci. USA*, **86**, 1352.

120. Grupp, S. A., Harmony, J. A., Baluyut, A. R., and Subbarao, B. (1987). Early events in B-cell activation: anti-Lyb2, but not BSF-1, induces a phosphatidylinositol response in murine B cells. *Cell. Immunol.*, **110**, 131.

121. Yakura, H., Shen, F. W., Kaemmer, M., and Boyse, E. A. (1981). Lyb2 system of mouse B cells: evidence for a role in the generation of antibody-forming cells. *J. Exp. Med.*, **153**, 129.

122. Kamal, M., Katira, A., and Gordon, J. (1991). Stimulation of B lymphocytes via CD72 (human Lyb-2). *Eur. J. Immunol.*, **21**, 1419.

123. Vandevelde, H., Vonhoegen, I., Luo, W., Parnes, J. R., and Thielemans, K. (1991). The B-cell surface protein CD72/Lyb-2 is the ligand for CD5. *Nature*, **351**, 662.

124. Tarakhovsky, A., Muller, W., and Rajewsky, K. (1994). Lymphocyte populations and immune responses in CD5-deficient mice. *Eur. J. Immunol.*, **24**, 1678.

125. Tedder, T. F., Streuli, M., and Schlossman, S. F. (1988). Isolation and structure of a cDNA encoding the B1 (CD20) cell surface antigen of human B lymphocytes. *Proc. Natl Acad. Sci. USA*, **85**, 208.

*126. Clark, E. A. (1993). CD22, a B cell-specific receptor, mediates adhesion and signal transduction. *J. Immunol.*, **150**, 4715.

127. Pezzutto, A., Rabinovitch, P. S., Dorken, B., Moldenhauer, G., and Clark, E. A. (1988). Role of the CD22 human B cell antigen in B cell triggering by anti-immunoglobulin. *J. Immunol.*, **140**, 1791.

128. Leprince, C., Draves, K. E., Geahlen, R. L., Ledbetter, J. A., and Clark, E. A. (1993). CD22 associates with the human surface IgM–B-cell antigen receptor complex. *Proc. Natl Acad. Sci. USA*, **90**, 3236.

129. Peaker, C. J. G. and Neuberger, M. S. (1993). Association of CD22 with the B cell antigen receptor. *Eur. J. Immunol.*, **23**, 1358.

130. Schulte, R. J., Campbell, M. A., Fischer, W. H., and Sefton, B. M. (1992). Tyrosine phosphorylation of CD22 during B cell activation. *Science*, **258**, 1001.

131. Stamenkovic, I., Sgroi, D., Aruffo, A., Sy, M. S., and Anderson, T. (1991). The Lymphocyte-B adhesion molecule CD22 interacts with leukocyte common antigen CD45RO on T-cells and alpha-2-6 sialyltransferase, CD75, on B-cells. *Cell*, **66**, 1133.

132. Sgroi, D., Varki, A., Braesch Andersen, S., and Stamenkovic, I. (1993). CD22, a B cell-specific immunoglobulin superfamily member, is a sialic acid-binding lectin. *J. Biol. Chem.*, **268**, 7011.

133. Jackson, D. G. and Bell, J. I. (1990). Isolation of a cDNA encoding the human CD38 (T10) molecule, a cell surface glycoprotein with an unusual discontinuous pattern of expression during lymphocyte differentiation. *J. Immunol.*, **144**, 2811.

134. Santos Argumedo, L., Teixeira, C., Preece, G., Kirkham, P. A., and Parkhouse, R. M. (1993). A B lymphocyte surface molecule mediating activation and protection from apoptosis via calcium channels. *J. Immunol.*, **151**, 3119.

135. Gelman, L., Deterre, P., Gouy, H., Boumsell, L., Debre, P., and Bismuth, G. (1993). The lymphocyte surface antigen CD38 acts as a nicotinamide adenine dinucleotide glycohydrolase in human T lymphocytes. *Eur. J. Immunol.*, **23**, 3361.

136. Summerhill, R. J., Jackson, D. G., and Galione, A. (1993). Human lymphocyte antigen CD38 catalyzes the production of cyclic ADP-ribose. *FEBS Lett.*, **335**, 231.

*137. Howard, M., Grimaldi, J. C., Bazan, J. F., Lund, F. E., Santos Argumedo, L., Parkhouse, R. M., Walseth, T. F., and Lee, H. C. (1993). Formation and hydrolysis of cyclic ADP-ribose catalyzed by lymphocyte antigen CD38. *Science*, **262**, 1056.

138. Galione, A. (1994). Cyclic ADP-ribose, the ADP-ribosyl cyclase pathway and calcium signalling. *Mol. Cell Endocrinol.*, **98**, 125.

139. Vitetta, E. S., Fernandez-Botran, R., Myers, C. D., and Sanders, V. M. (1989). Cellular interactions in the humoral immune response. *Adv. Immunol.*, **45**, 1.

*140. Parker, D. C. (1993). T cell-dependent B cell activation. *Annu. Rev. Immunol.*, **11**, 331.

141. Chesnut, R. W., Colon, S. M., and Grey, H. M. (1982). Requirements for the processing of antigens by antigen-presenting B cells. I. Functional comparison of B cell tumours and macrophages. *J. Immunol.*, **129**, 2382.

142. Lanzavecchia, A. (1985). Antigen-specific interaction between T and B cells. *Nature*, **314**, 537.

143. Frohman, M. and Cowing, C. (1985). Presentation of antigen by B cells: functional dependence on radiation dose, interleukins, cellular activation and differential glycosylation. *J. Immunol.*, **134**, 2269.

144. Parker, D. C. and Eynon, E. E. (1991). Antigen presentation in acquired immunological tolerance. *FASEB J.*, **5**, 2777.

145. Norton, S. D., Zuckerman, L., Urdahl, K. B., Shefner, R., Miller, J., and Jenkins, M. K. (1992). The CD28 ligand, B7, enhances IL-2 production by providing a costimulatory signal to T cells. *J. Immunol.*, **149**, 1556.

146. Damle, N. K., Linsley, P. S., and Ledbetter, J. A. (1991). Direct helper T cell-induced B cell differentiation involves interaction between T cell antigen CD28 and B cell activation antigen B7. *Eur. J. Immunol.*, **21**, 1277.

147. Ranheim, E. A. and Kipps, T. J. (1993). Activated T cells induce expression of B7/BB1 on normal or leukemic B cells through a CD40-dependent signal. *J. Exp. Med.*, **177**, 925.

*148. Clark, E. A. and Ledbetter, J. A. (1994). How B and T cells talk to each other. *Nature*, **367**, 425.

149. Noelle, R. J. and Snow, E. C. (1991). T-helper cell-dependent B-cell activation. *FASEB J.*, **5**, 2770.

150. Springer, T. A. (1990). Adhesion receptors of the immune system. *Nature*, **346**, 425.

151. Dustin, M. L. and Springer, T. A. (1989). T cell receptor crosslinking transiently stimulates adhesiveness through LFA-1. *Nature*, **341**, 619.

152. Kupfer, A., Swain, S. L., and Singer, S. J. (1987). The specific direct interaction of helper T cells and antigen-presenting B cells. II. Reorientation of the microtubule organizing center and reorganization of the membrane-associated cytoskeleton inside the bound helper T cells. *J. Exp. Med.*, **165**, 1565.

153. Kupfer, A., Mosmann, T. R., and Kupfer, H. (1991). Polarized expression of cytokines in cell conjugates of helper T cells and splenic B cells. *Proc. Natl Acad. Sci. USA*, **88**, 775.

154. Mishra, G. C., Berton, M. T., Oliver, K. G., Krammer, P. H., and Uhr, J. W. (1986). A monoclonal anti-mouse LFA-1 alpha antibody mimics the biological effects of B cell stimulatory factor-1 (BSF-1). *J. Immunol.*, **137**, 1590.

155. Noelle, R. J., McCann, J., Marshall, L., and Bartlett, W. C. (1989). Cognate interactions between helper T cells and B cells III. Contact dependent, lymphokine-independent induction of B cell cycle entry by activated helper T cells. *J. Immunol.*, **143**, 1807.

156. Brian, A. A. (1988). Stimulation of B cell proliferation by membrane-associated molecules from activated T cells. *Proc. Natl Acad. Sci. USA*, **85**, 564.

157. Hodgkin, P. D., Yamashita, L. C., Coffman, R. L., and Kehry, M. R. (1990). Separation of events mediating B-cell proliferation and Ig production by using T-cell membranes and lymphokines. *J. Immunol.*, **145**, 2025.

158. Hodgkin, P. D., Yamashita, L. C., Seymour, B., Coffman, R. L., and Kehry, M. R. (1991). Membranes from both Th1 and Th2 T cell clones stimulate B cell proliferation and prepare B cells for lymphokine-induced differentiation to secrete Ig. *J. Immunol.*, **47**, 3696.

159. Spriggs, M. K., Armitage, R. J., Strockbine, L., Clifford, K. N., Macduff, B. M., Sato, T. A., Maliszewski, C. R., and Fanslow, W. C. (1992). Recombinant human CD40 ligand stimulates B cell proliferation and immunoglobulin E secretion. *J. Exp. Med.*, **176**, 1543.

160. Lane, P. J. L., Brocker, T., Hubele, S., Padovan, E., Lanzavecchia, A., and McConnell, F. (1993). Soluble CD40 ligand can replace the normal T cell-derived CD40 signal to B cells in T cell-dependent activation. *J. Exp. Med.*, **177**, 1209.

161. Maliszewski, C. R., Grabstein, K., Fanslow, W. C., Armitage, R. J., Spriggs, M. K., and Sato, T. A. (1993). Recombinant CD40 ligand stimulation of B cell growth and differentiation: cooperative effects of cytokines. *Eur. J. Immunol.*, **23**, 1044.

162. Armitage, R. J., Macduff, B. M., Spriggs, M. K., and Fanslow, W. C. (1993). Human B cell proliferation and Ig secretion induced by recombinant CD40 ligand are modulated by soluble cytokines. *J. Immunol.*, **150**, 3671.

163. Armitage, R. J., Grabstein, K. H., and Alderson, M. R. (1990). Cytokine regulation of B cell differentiation. In *Cytokines and B lymphocytes* (ed. R. E. Callard), pp. 115–42. Academic Press, New York.

164. Defrance, T., Vanbervliet, B., Briere, F., Durand, I., Rousset, F., and Banchereau, J. (1992). Interleukin 10 and transforming growth factor beta cooperate to induce anti-CD40-activated naive human B cells to secrete immunoglobulin A. *J. Exp. Med.*, **175**, 671.

165. Paul, W. E. (1991). Interleukin-4: a prototypic immunoregulatory lymphokine. *Blood*, **77**, 1859.

166. Shields, J. G., Armitage, R. J., Jamieson, B. N., Beverley, P. C., and Callard, R. E. (1989). Increased expression of surface IgM but not IgD or IgG on human B cells in response to IL-4. *Immunology*, **66**, 224.

167. Klaus, G. G. B. and Harnett, M. M. (1990). Crosstalk between B cell surface immunoglobulin and interleukin-4 receptors: the role of protein kinase C and calcium-mediated signals. *Eur. J. Immunol.*, **20**, 2301.

168. Justement, L., Chen, Z., Harris, L., Ransom, J., Sandoval, V., Smith, C., Rennick, D., Roehm, N., and Cambier, J. (1986). BSF1 induces membrane protein phosphorylation but not phosphoinositide metabolism, Ca^{2+} mobilization, protein kinase C translocation, or membrane depolarization in resting murine B lymphocytes. *J. Immunol.*, **137**, 3664.

169. Harnett, M. M., Holman, M., and Klaus, G. G. B. (1991). IL-4 promotes anti-Ig-mediated protein kinase C translocation and reverses phorbol ester-mediated protein kinase C down-regulation in murine B cells. *J. Immunol.*, **147**, 3831.

170. Finney, M., Guy, G. R., Michell, R. H., Gordon, J., Dugas, B., Rigley, K. P., and Callard, R. E. (1990). Interleukin 4 activates human B lymphocytes via transient inositol lipid hydrolysis and delayed cAMP generation. *Eur. J. Immunol.*, **20**, 151.

171. Go, N. F., Castle, B. E., Barrett, R., Kastelein, R., Dang, W., Mosmann, T. R., Moore, K. W., and Howard, M. (1990). Interleukin-10, a novel B cell stimulatory factor— unresponsiveness of X-chromosome linked immunodeficiency-B cells. *J. Exp. Med.*, **172**, 1625.

172. Rousset, F., Garcia, E., Defrance, T., Peronne, C., Vezzio, N., Hsu, D. H., Kastelein, R., Moore, K. W., and Banchereau, J. (1992). Interleukin 10 is a potent growth and differentiation factor for activated human B lymphocytes. *Proc. Natl Acad. Sci. USA*, **89**, 1890.

173. McKenzie, A. N., Culpepper, J. A., de Waal Malefyt, R., Briere, F., Punnonen, J., Aversa, G., Sato, A., Dang, W., Cocks, B. G., Menon, S., *et al.* (1993). Interleukin 13, a T-cell-derived cytokine that regulates human monocyte and B-cell function. *Proc. Natl Acad. Sci. USA*, **90**, 3735.

174. Cocks, B. G., de Waal Malefyt, R., Galizzi, J. P., de Vries, J. E., and Aversa, G. (1993). IL-13 induces proliferation and differentiation of human B cells activated by the CD40 ligand. *Int. Immunol.*, **5**, 657.

175. Snapper, C. M., Finkelman, F. D., and Paul, W. E. (1988). Regulation of IgG1 and IgE production by interleukin 4. *Immunol. Rev.*, **102**, 51.

176. Coffman, R. L., Ohara, J., Bond, M. W., Carty, J., Zlotnik, A., and Paul, W. E. (1986). B cell stimulatory factor-1 enhances the IgE response of lipopolysaccharide-activated B cells. *J. Immunol.*, **136**, 4538.

177. Finkelman, F. D., Katona, I. M., Urban, J. F. J., Snapper, C. M., and Ohara, J. (1986). Suppression of *in vivo* polyclonal IgE responses by monoclonal antibody to the lympho-kine B-cell stimulatory factor 1. *Proc. Natl Acad. Sci. USA*, **83**, 9675.

178. Sanderson, C. J., Campbell, H. D., and Young, I. G. (1988). Molecular and cellular biology of eosinophil differentiation factor (IL-5) and its effects on human and mouse B cells. *Immunol. Rev.*, **102**, 29.

179. Snapper, C. M., Peschel, C., and Paul, W. E. (1988). Interferon gamma stimulates IgG2a secretion by murine B cells stimulated with bacterial lipopolysaccharide. *J. Immunol.*, **140**, 2121.

*180. Street, N. E. and Mosmann, T. R. (1991). Functional diversity of T lymphocytes due to secretion of different cytokine patterns. *FASEB J.*, **5**, 171.

181. Mosmann, T. R. and Coffman, R. L. (1989). Th1 and Th2 cells: different patterns of lymphokine secretion lead to different functional properties. *Annu. Rev. Immunol.*, **7**, 145.

182. Fiorentino, D. F., Bond, M. W., and Mosmann, T. R. (1989). Two types of mouse T helper cell. IV. Th2 clones secrete a factor that inhibits cytokine production by Th1 clones. *J. Exp. Med.*, **170**, 2081.

183. Fiorentino, D. F., Zlotnik, A., Vieira, P., Mosmann, T. R., Howard, M., Moore, K. W., and O'Garra, A. (1991). IL-10 acts on the antigen-presenting cell to inhibit cytokine production by Th1 cells. *J. Immunol.*, **146**, 3444.

7 | Advances in antibody engineering

MARTINE E. VERHOEYEN and JOHN H. C. WINDUST

1. Introduction

In 1894, Emil Behring reported the discovery of antibodies, then called antitoxins, and their therapeutic effects (1). About 80 years later, Kohler and Milstein established the techniques to reliably produce monoclonal antibodies (mAbs) by hybridoma cell lines (2). This event, together with the availability of increasingly sophisticated molecular biology and protein engineering techniques, opened up the exciting new science area of antibody engineering, and led to the creation of antibodies and antibody-derived molecules designed to suit possible applications not only in the medical but also industrial world.

Recombinant mAbs were first produced in animal cells, but more recently the emphasis has shifted towards antibody fragments produced by microorganisms, mainly *Escherichia coli*. Today, engineered antibodies are no longer derived exclusively from hybridoma genetic material, but can now be obtained by direct cloning from mixed cell populations such as spleen and peripheral blood lymphocytes (PBLs), creating large combinatorial libraries. Selection of the desired specificity from these large libraries has become a crucial step and has been developed into a fine art by means of phage display technology.

Emil Behring predicted that antitoxins might be produced one day without the aid of an animal body (1). Today, combinatorial libraries can be established from immunized or 'naive' sources and can be maintained indefinitely. They can be used over and over again to screen and select for new specificities without further need for animal experimentation or immunization. Antibody fragments with binding affinities in the nanomolar range have been isolated in this way and further improvement is feasible using mutagenesis and chain reshuffling. Thus, bypassing immunization is becoming more and more a reality.

2. Antibody structure

Immunoglobulins (Igs) typically are Y-shaped molecules whose basic unit consists of four polypeptide chains in two identical pairs, the heavy and light chains (see Fig. 1). At the protein level, each of these is folded into discrete domains. The N-

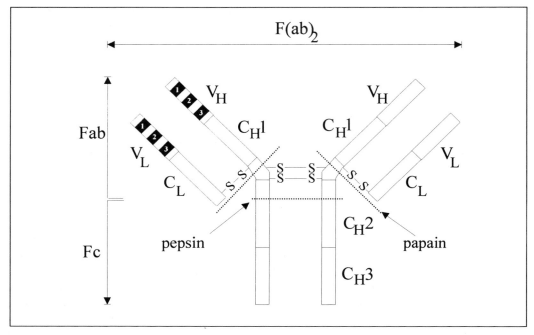

Fig. 1 Antibody molecule showing the CDRs in black and indicating the main cleavage sites for the proteolytic enzymes pepsin and papain.

terminal regions (V-domains) are variable in sequence, and within each V-domain are three hypervariable regions (complementarity determining regions or CDRs) responsible for the antibody specificity. Interestingly, despite their hypervariable amino acid sequences, structural variability of these CDR loops seems more restrained, since for all except CDR3 of the heavy chain there exists a limited number of main chain conformations or 'canonical structures' (3). The module formed by the V_H–V_L pair is referred to as the Fv of the antibody and is responsible for the binding of antigen. The C-terminal domains of both heavy and light chains are more conserved in sequence and are called the constant regions. Heavy chain constant regions consist of several domains, for example the heavy chains of the γ-isotype (which is the most abundant) possess three domains (C_H1, C_H2, and C_H3) and a hinge region which connects the C_H1 and C_H2 domains. The hinge is covalently linked to the hinge of the other heavy chain of the antibody molecule by disulfide bonds. In addition there is a carbohydrate chain attached to the C_H2 region (not shown). Light chains have one constant domain which packs against the C_H1 domain. The constant regions of the immunoglobulin are responsible for a variety of effector functions such as complement fixation and clearing by cytotoxic mechanisms. Digestion with papain causes proteolytic cleavage of the native antibody molecules in the hinge region and yields three fragments. One fragment is composed of the C_H2 and C_H3 domains and as it crystallizes easily is known as the Fc fragment. The other two fragments, designated the Fab (antigen binding) fragments, are identical and consist of the entire light chain combined with the Fd (the heavy chain variable region and C_H1 domain). When pepsin is used, the proteolytic cleavage is such that the two Fabs remain connected to form the $F(ab')_2$ fragment.

3. Antibody gene cloning

Many antibody genes have been successfully obtained from hybridoma cells via traditional cDNA cloning methods, but this route has its limitations. It is slow and, due to several factors, the majority of the mAbs and consequently the genes thus obtained, have been of murine origin. Furthermore, there has been no method to access the unrearranged *V*-gene repertoire of an unimmunized individual. The application of powerful cloning methods based on the polymerase chain reaction (PCR, ref. 4) has allowed direct access to a large number of heavy and light chain sequences and has had a profound effect on the speed of cloning. Whilst Larrick *et al.* (5) had demonstrated that the *V*-genes of a hybridoma could be amplified by PCR using a degenerate set of oligonucleotide primers, the key step involved the development of 'universal primers'. These primers, which hybridized to the sufficiently well conserved 5' and 3' regions of the *V*-genes, also incorporated restriction site sequences so that the *V*-genes could be force-cloned directly into an expression vector (6, 7). Subsequently, many families of PCR primers have been identified which have allowed the cloning of *V*-genes from human (7), mouse (8–10), or rat (11) hybridomas, and human (12) or rabbit (13) PBLs. It is now even possible to preserve the original pairing of V_H and V_L genes by carrying out 'in-cell' PCR of mRNA (14). The ability to clone *V*-genes from immunized or unimmunized sources, mRNA or genomic DNA, from IgG or IgM heavy chains, κ or λ light chains, memory or plasma cells and build designer antibodies is covered in a review by Hoogenboom *et al.* (15).

4. Recombinant antibody gene expression

4.1 Expression in animal cells

4.1.1 Myeloma cells

Myeloma cells are the most widely used cells for production of whole recombinant antibodies. Typically, vectors based on pSV_2^{gpt} and pSV_2^{neo} are used, with the antibody genes (usually in a genomic context) linked to the Ig heavy chain promoter and enhancer sequences (16). The heavy and light chain genes are usually on separate plasmids with different selection markers and are co-transfected into the myeloma cells by electroporation. Since co-transfection works very well, the use and selection for two different markers is not strictly necessary (17). Similar vectors were used by Riechmann *et al.* (18) for the production of Fv fragments in good yields (8 mg/litre), but using a human cytomegalovirus (CMV) promoter.

An alternative method to produce chimeric antibodies in myeloma cells has been described (19) that uses homologous recombination and gene targeting. Hybridoma cells with the desired specificity were used in an *in vivo* recombination experiment in which the mouse heavy chain constant region was replaced with a human heavy chain constant region. The same strategy should be applicable for replacement of the light chain constant region, and can be extended to variable domains as well, as was shown by Baker *et al.* (20). Thus very good producers could be used as hosts to express other specificities, perhaps coming from a poor producing background, or to switch to a more desirable isotype. Very high expression levels can be obtained using gene amplification systems.

Vector amplification using the dihydrofolate reductase (*DHFR*) marker gene has been used for the expression of antibodies in myeloma cells. However, since myeloma cells contain an endogenous *DHFR* gene, very high levels of the highly toxic methotrexate (MTX) were required to achieve good amplification (21).

A very elegant and highly successful system was described more recently by Bebbington *et al.* (22) which exploits the natural glutamine auxotrophy of most myeloma and hybridoma cell lines as the basis for an alternative gene amplification system. Glutamine synthetase (GS) is used as a selectable marker and its inhibitor, methionine sulfoximine (MSX), drives the amplification of GS-vector sequences (and incorporated Ig sequences). Control of Ig expression is from the hCMV–ME promoter–enhancer. Reported yields in excess of 500 mg/litre were obtained (in fed-batch air-lift fermenters) by application of this technique. This level of expression could be reached after only one round of selection with MSX and could be maintained for prolonged periods in the absence of MSX. The latter is in contrast to CHO/DHFR systems (see Section 4.1.2), where instability of production has been noticed both in the presence and absence of continued MTX selection (23).

4.1.2 Chinese hamster ovary (CHO) cells

The GS/MSX gene amplification system described above can also be used in CHO cells, and yields in excess of 200 mg recombinant antibody per litre of culture supernatant have been reported (24). An alternative system for high level expression of antibody genes makes use of DHFR⁻ CHO cells (25–27). The *DHFR* gene is used as a selectable marker and co-transfected heavy and light chain genes are amplified by subjecting transformants to increasing concentrations of MTX. Yields of up to 200 mg/litre were reported in shake flasks (26). Non-immunoglobulin promoters such as the human β-actin (26) and the adenovirus major late promoter (27) were used, the latter in combination with the SV40 enhancer. For expression in CHO cells, Ig cDNA rather than genomic DNA is used, although Fouser *et al.* (27) observed that retaining intron sequences had a beneficial influence on expression levels. In contrast to the myeloma/GS amplification system described above, continued selective pressure is required for the production of Ig to remain stable beyond six weeks of continuous culture (26).

4.1.3 COS monkey kidney cells

Stable expression of Ig genes in animal cells is time consuming. Small quantities of recombinant antibodies (or antibody fragments) can be generated relatively quickly for initial evaluation by using transient expression systems such as COS cells (28, 29). However, at least for antibody fragments, these systems have now been superseded by successful, high yielding microbial expression systems (see Sections 4.2 and 4.3).

4.1.4 Insect cells

In the last few years, Baculoviruses have joined the ever growing list of expression systems for high-level production of recombinant proteins, including immunoglobulins. An interesting feature of these systems is that the viral infection shuts down the host's own protein synthesis, thus favouring the production of viral protein and facilitating purification of recombinant protein. The first systems for antibody production involved the use of *Spodoptera frugiperda* (Sf9) cells infected with recombinant *Autographa californica* nuclear polyhedrosis virus (AcNPV) (30, 31). More recently, a group reported the use of an alternative baculovirus, *Bombyx mori* polyhedrosis virus (BmNPV), the host of which is *Bombyx mori*, the silkworm (32). Yields of 0.8 mg of antibody per silkworm larva were achieved. In contrast to mouse ascites, the material produced is free from contaminating immunoglobulins. However, a lot of other proteinaceous material is present in the haemolymph, which may render subsequent purification of antibody difficult.

4.2 Expression in *Escherichia coli*

The expression of antibodies in *E. coli* is a favourable alternative to animal cells for a number of reasons: *E. coli* has well-known genetics, the relevant transformation systems are efficient, good vectors are available, cell growth is fast, the growth medium is cheap, and there exists a wealth of expertise in the scale-up from laboratory-scale fermentations to production-scale. A disadvantage is that the molecules so produced are not glycosylated which may compromise the effector functions. However, this is relevant only in the case of whole antibodies for use *in vivo* and is of little importance for the production and use of antibody fragments.

Early attempts to produce whole antibodies in *E. coli* by direct secretion into the cytoplasm were disappointing (33, 34) since the resulting polypeptides were compartmentalized in an insoluble form in inclusion bodies. Whilst the expression level was high, functional antibodies could only be obtained in very low yield following an inefficient denaturing and refolding step. The key development in the area was the mimicking in *E. coli* of the normal folding and assembly pathway of antibodies within the eukaryotic cell. The cytoplasm is a harsh, reducing environment for the production of proteins with multiple disulfide bonds, and it was reasoned that directing them to the more favourable oxidizing conditions of the periplasm would be conducive to correct folding and assembly. This was achieved

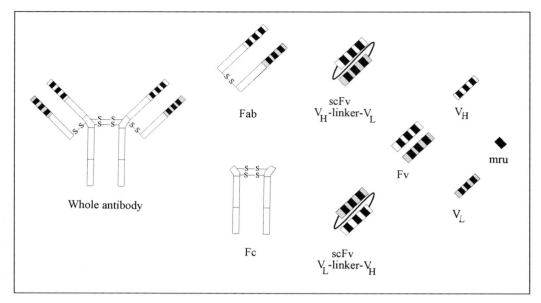

Fig. 2 Genetically engineered antibody fragments.

independently for both the Fab (35) and the Fv (36) fragments and represented the first demonstration of the successful production of foreign heterodimeric proteins in *E. coli*. Heavy and light chain genes, each preceded by a bacterial signal sequence were arranged in an artificial dicistronic operon under the control of a single inducible promoter. The translated proteins were produced simultaneously as precursors and exported to the periplasmic space, where they were precisely cleaved to yield the mature proteins which folded to give active fragment. This was then either retained within the periplasm (36) or gradually leaked out into the culture supernatant (35). This development led to the expression of an ever-expanding family of antibody fragments (see Fig. 2) the basic types of which are detailed below. For a comprehensive review of the expression of antibody fragments in *E. coli*, we recommend the reviews by Pluckthun (37, 38).

4.2.1 Fab

Recombinant Fab fragments produced in *E. coli* are stable species which have identical affinities to those produced by proteolysis (35). Interestingly, the yield of a functional Fab fragment is consistently smaller than that of the functional Fv fragment of the same antibody when expressed in *E. coli*. It has been suggested that this is due to the periplasmic folding process which is not proceeding as efficiently for the larger fragment as it does for the Fv (37). Nevertheless, Carter *et al.* (39) have reported the successful production of 1–2 g/litre of soluble, functional humanized Fab fragment largely as a result of the high cell densities used in a fermenter and very tight control of expression prior to induction. Surprisingly, despite the high concentration of Fab in the periplasmic space, dimerization to F(ab')$_2$ did not occur, but the latter could be made by *in vitro* chemical coupling of recovered Fab' with the single hinge cysteine in the free thiol state.

In contrast, Fd (1.28 g/litre) and light chain (2.88 g/litre) chimeric fragments specific for human carcinoembryonic antigen were obtained by Shibui *et al.* (40) as insoluble, but correctly processed, proteins in the periplasm using a jar fermenter system. They suggest that this is probably due to the efficiency of the ompF signal sequence causing secretion to the periplasm at a rate too great for proper folding of the Fab. Purification and refolding to give an active Fab was achieved with a yield of about 47 per cent.

4.2.2 Fv and single-chain Fv (scFv)

The Fv fragment is the smallest unit of an antibody which retains the complete binding site, and was previously only available by proteolysis in a select number of cases (41). The ability to produce it in a soluble, homogeneous and easily purifiable form by expression in *E. coli* (36) has resulted in it being a favourite material for structural studies of antibody binding sites by NMR (42, 43) and X-ray crystallography (44, 45). One advantage of the production of Fvs (and other fragments) in a soluble form is that it allows them to be purified, very often in a single step using affinity chromatography, from the culture supernatant or a periplasmic extract. In some cases, however, the antigen may be expensive or in short supply and, therefore, it would be advantageous to have a general method of purification. One of the most popular methods of achieving this is by engineer-ing a short peptide marker sequence or 'tail', usually on the C-terminus of the protein, which assists in the identification and purification of the recombinant protein (reviewed in ref. 46). The use of a polyhistidine tail (47) in combination with immobilized metal ion chromatography (IMAC) has proved popular (48, 49).

Sufficient numbers of recombinant Fvs now exist (36, 42, 43, 50), to enable crosswise comparisons of their properties to be undertaken. In particular, where the fragments are from related antibodies to the same antigen (50), protein en-gineering will allow the study of those properties which influence folding, binding, expression, etc. One such property of Fvs that has been highlighted is that their inherent sequence variability is reflected in different interaction energies between the V_H and V_L, resulting in a range of stabilities in the absence of antigen. This observation has led to the development of methods to stabilize them (51, 52), of which the scFv has proved the most popular as its construction employs the more straightforward approach. Single-chain Fvs consisting of the variable domains of an antibody connected by a genetically encoded peptide linker were first reported by Bird *et al.* (53) and Huston *et al.* (54). A linker of approximately 15 residues is required to span the 3.5 nm distance from the C-terminus of the V_H to the N-terminus of the V_L (or vice versa), depending on the exact start and end point chosen. The sequence (Gly$_4$Ser)$_3$ is commonly used but many others have been tried (reviewed in ref. 55). The effect of linker length on the stability, folding, and affinity of single-chain antibodies has revealed that in some cases a longer linker can have a small stabilizing effect on the antibody (56). Both V_H–linker–V_L and V_L–linker–V_H combinations have been expressed successfully (reviewed in refs 55, 57) sometimes for the same antibody (58). Only in one case was a dramatic difference in expression between the two orientations observed (59) where putting the V_H first reduced the expression level by 95 per cent compared with the opposite orientation. Comparison of the crystal structures of a scFv and its Fab counterpart showed that the association of V_H and V_L is indeed very similar in both molecules, with only slight changes noted (60).

Single-chain Fv fragments possess several attributes that have made them extremely popular and versatile reagents. They are small, easy to engineer, stable at low concentrations, penetrate rapidly into tissues, and are rapidly cleared from the body (61, 62). They have been shown to have binding affinities equivalent to, or within one order of magnitude of, those of the monoclonal antibodies from which they were derived (53, 54, 61, 63). Catalytic single-chain antibodies exhibit the same catalytic parameters as the parent monoclonal antibody (64). Single-chain Fv fragments have increasingly become the preferred partner in antibody-based fusion proteins (see Section 6.4). The expression of active scFv fusions with the minor coat protein of a filamentous bacteriophage (65) formed the basis of phage display technology (see Section 5.2).

4.2.3 Domain antibodies (dAbs)

Whilst an Fv is the minimum combining unit of an antibody, X-ray crystallography data have shown that the contributions to binding made by the V_H and V_L are not necessarily equal (66), therefore posing the question, is it possible to genetically engineer a smaller fragment? This was answered by Ward *et al.* (67) who discovered that isolated V_H domains of the anti-lysozyme antibody D1.3 expressed in *E. coli* bound lysozyme with an affinity only 10-fold lower than the whole antibody. However, the observed 'stickiness' (67) and lack of specificity (68) of V_Hs, due at least in part to the exposure of the hydrophobic residues that are normally buried in the V_H–V_L interface, is likely to limit their usefulness. The exposed hydrophobic face is also thought to be responsible for the poor solubility of these fragments and consequent low yield when expressed (43, 67). This hypothesis is supported by the fact that higher yields (10–30 mg/litre) have been reported only from periplasmic extracts that have undergone a solubilization and refolding treatment (69, 70). Engineering of the interface to increase solubility whilst maintaining affinity has so far proved elusive. However, it may be assisted by comparison with the V_H of camel IgG2 and IgG3 which have recently been shown to be completely devoid of a light chain yet show no difference in solubility as compared to normal IgGs (71, 72).

Brinkmann *et al.* (73) have reported the expression of individual V_H and V_L domains from the tumour specific mAb B3 as fusion proteins with *Pseudomonas* exotoxin. By assessing their capability to bind and kill target cells, these have been found useful for analysing the contributions each domain makes to the specific binding of antigen, the effects of mutations in the CDRs, and the folding properties of the individual variable domains and their dimerization to form Fv. Remarkably, in this case it would appear that the light chain is mainly responsible for the specificity of binding.

4.2.4 Factors influencing expression in *E. coli*

Now that the expression of antibody fragments in *E. coli* is considered routine, attention has increasingly been focused on the role of individual components or processes in expression in order to maximize the yield of functional protein. Different promoters have been tried (T7, tac, lac, lacUV5, araB, phoA, λ_{pRpL}) (35, 36, 43, 70, 74), but it appears that a stronger promoter does not necessarily lead to an increase in yield of functional product, rather it increases the amount of insoluble protein produced (74). It is likely that the limiting step is the folding and assembly of the fragment. The choice of signal sequence would seem less critical as a number have been tried with equal success (pelB, ompA, phoA, stII, ompF, bla) (35, 36, 39, 40) although in one case (40) use of the ompF sequence did give rise to an insoluble product.

Insolubility was a major problem in early work on antibody expression. Methods which had been developed for the production of soluble recombinant proteins in *E. coli* (75) were successfully applied to antibodies (76): one of the most general methods for increasing the level of soluble protein is lowering the temperature of the fermentation (77). The development of successful 'secretion' systems which can be scaled-up to produce grams of antibody (39) means that high-expression levels of recombinant protein no longer necessarily has to result in inclusion body formation (78). Studies on the effect of co-expressing folding catalysts on the yield of soluble fragment have so far shown only minor improvements (58). However, developments in our understanding of protein folding pathways and the involvement of molecular chaperones can now be directly applied to increasing the efficiency of the refolding process (79). Therefore, in particular cases where protein accumulation in inclusion bodies is preferred (for example when labelling fragments for NMR in expensive ^{13}C- or ^{14}N-containing media), methods are available to assist in obtaining a good yield of functional product.

Finally, the coding sequence itself can have a significant effect on the expression of antibody fragments. This has been illustrated by the dramatic effect that altering a domain (39), the composition of a peptide tag (48), or even a single codon (80) can have on expression. When the entire sequence is known, the option of creating a completely synthetic gene exists in which factors such as codon preference and avoidance of internal ribosome binding sites can be optimized (59).

4.3 Expression in other microorganisms

4.3.1 *Bacillus subtilis*

The ability of *Bacillus subtilis* to secrete proteins directly into the medium in high yield (81) is one of several reasons why it appears to be a very useful host for the expression of antibody fragments. The production of an anti-fluorescein scFv in

B. subtilis has been reported but the yield and stability of the secreted product were not mentioned (82). The major disadvantage of *B. subtilis* has been the high level of proteases secreted into the medium which degrade the product. This aspect was addressed by the construction of multiple protease deficient strains and resulted in the successful expression of an anti-digoxin scFv (83). Tests showed that the correctly processed, secreted scFv was stable in the medium for at least nine hours and could be recovered to high purity in a single step by affinity chromatography on oubain–Sepharose. Yields were up to 5 mg/litre in shake-flask culture and the functional yield of the antibody was approximately 98 per cent of the production yield. Affinity and ligand specificity studies demonstrated that the engineered scFv had almost identical properties as those of the parent monoclonal antibody. Intriguingly, the single-chain antidigoxin produced in *B. subtilis* has been found to differ in some of its physical characteristics (for example solubility at neutral pH, ease of elution from the affinity column) from that expressed in *E. coli* either intracellularly or via secretion (83). The reasons for this are unknown, but it is likely that such cross-comparisons of antibody fragments produced in different hosts will benefit our overall understanding of the intricacies of antibody fragment expression. Co-expression of V_H and V_L unexpectedly did not result in Fv production, presumably because, unlike *E. coli*, *B. subtilis* lacks a periplasmic space in which the highly concentration-dependent assembly reaction can take place (83). Thus whilst *B. subtilis* is suitable for the production of single-chain fragments, it is likely that its use for the expression of heterodimeric proteins will be limited.

4.3.2 *Staphylococcus*

The appeal of *Staphylococcus carnosus* is that, unlike other Gram-positive bacteria, it exhibits low exoproteolytic activity. This has been exploited by Pschorr *et al.* (84) who reported the production of a fusion protein consisting of a lipase pre-pro-protein gene fragment fused to a synthetic gene for the immunoglobulin REI variable domain in this organism. At the junction of the fusion partners was placed an oligohistidine sequence to aid purification by IMAC and a recognition sequence for *Neisseria gonorrheae* IgA protease for site-specific cleavage. Yields of 10 mg/litre of soluble, correctly folded protein secreted into the culture medium were reported.

4.3.3 *Streptomyces*

Streptomyces lividans is another example of a Gram-negative bacterium that has been used for secretory production of heterologous proteins. The genes encoding the Fv fragment of mAb HyHEL10 were successfully expressed by this organism under control of the streptomyces subtilisin inhibitor (ssi) promoter. Secretion was achieved by linking both the V_H and V_L genes to the ssi signal protein. Yields of functional Fv after purification were reported to be approx. 1 mg/litre of culture (85).

4.3.4 Yeast

Whole antibodies (86, 87) and Fab fragments (86, 88, 89) have been expressed and secreted by the yeast *Saccharomyces cerevisiae*, generally by co-expression of heavy and light chain immunoglobulin genes on either one or two plasmids within the same cell. Yields of functional, reassembled antibody or antibody fragments have remained persistently low however, (micrograms per litre) and only minor improvements have been made recently (90), in contrast to *E. coli* expression systems for antibody fragments which have become increasingly efficient. This factor, when coupled with others (for example likely plasmid instability, plus a tendency to hyperglycosylate expressed proteins), means that *S. cerevisiae* is probably not the best system for the expression of whole antibodies or antibody fragments. Whilst there has been a report of the successful expression of a scFv against fluorescein in *Schizosaccharomyces pombe* (91), other yeasts, in particular the methylotrophic yeasts (for instance *Pichia pastoris* and *Hansenula polymorpha*) have recently commanded attention (92). These have several features which make them potentially useful for the production of antibodies (strong, stringently controlled promoters, high levels of expression, a glycosylation pattern closer to the mammalian type), and it will be interesting to see whether these apparent benefits are realized in the near future.

4.3.5 Filamentous fungi

Among the wide variety of prokaryotic and eukaryotic hosts available for the production of heterologous gene products, certain species of filamentous fungi (for example *Aspergillus* spp, **Trichoderma** spp) possess features which make them exceptionally attractive. These include the ability to secrete large quantities (up to 25 g/litre) of homologous protein in culture, extensive fermentation experience for low-cost, large-scale production, and a long history of safe use for the manufacture of products for human consumption (93). Accurate processing of signal sequences, correct folding, and disulfide bond formation and, in some cases, glycosylation to a similar extent as a natural human protein has been observed (reviewed in ref. 94). Nyyssonen *et al.* (95) have reported the production of an antibody fragment in *Trichoderma reesei*: this represents the first expression of a heterologous dimeric protein in a filamentous fungus. A Fab was successfully secreted from *Trichoderma* cells that had been co-transfected with plasmids encoding the Fd and light chain sequences of a murine anti-2-phenyloxazolone. The yield was substantially increased when the Fd was fused to the *T. reesei* cellulase, cellobiohydrolase I (CBHI), being 1 mg/litre for the Fab and 150 mg/litre for the CBHI–Fab fusion protein. Immunologically active Fab, which could be released from the fusion protein by an extracellular *T. reesei* protease, had the same affinity as the parent immunoglobulin. By comparison with *E. coli* and *S. cerevisiae*, the genetics of filamentous fungi are in their infancy, but there is every incentive to press ahead rapidly in order to exploit fully the remarkable protein producing capacity of these organisms.

4.4 Expression in plants

Significant advances have been made in the techniques for plant transformation such that the introduction of foreign genes into model plant systems is now routine:

tobacco is most commonly used as it is easily transformed and regenerated. Hiatt *et al.* (96) were the first to express individual murine γ and κ antibody chains in transgenic tobacco plants. An interesting feature of this expression system is that when transgenic plants expressing γ or κ chains were crossed to produce F1 plants, these expressed the whole antibody. The yield was surprisingly high, being greater than 1 per cent of the total extractable protein (97), and the functional and catalytic activity of the antibody was the same or comparable with that of the ascites-derived one. The production of functional antibody by the transgenic plants was found to require the presence of native or heterologous signal sequences at the N-terminus of the nascent immunoglobulin chains (98). In addition to whole antibodies, both scFv (99–101) and V_H (102) antibody fragments have been expressed in plants. It was further shown that scFv can be produced in transgenic tobacco cell suspension cultures (100).

Antibody expression in plants is still at an early stage, but it offers a tantalizing glimpse of the future. It can be rapid; transient expression of whole antibodies or fragments in tobacco protoplasts can allow testing for activity to be undertaken only two days after obtaining the cDNA of interest (103). The antibodies produced appear to possess all the functional characteristics of those derived from hybridomas (although further study will be required to determine the effect of the difference in heavy chain glycosylation). Plant genetic material is readily stored in seeds which are extremely stable and require little or no maintenance (97). The ease with which genetic material can be exchanged by cross-fertilization may facilitate the construction of multimeric forms (eg. secretory IgA, ref. 104) and of bifunctional antibodies from the crossing of two transgenic plants. In theory, it also offers an opportunity to create any combination of heavy and light chain genes.

5. Combinatorial libraries and phage display

5.1 Combinatorial libraries

In the preceding sections we have seen how it is increasingly possible to clone the antibody repertoire by PCR, that the whole antibody molecule is not necessarily required, and the production of soluble antibody fragments in *E. coli* is now routine. Here we will see how these developments in concert with a powerful new screening method, promise to revolutionize the way in which antibodies are generated in the future.

The standard method of generating antibodies via immunization followed by hybridoma production has the disadvantage that, in some instances, it is very hard or even impossible to obtain mAbs against a certain specificity; this is especially true for human mAbs. For this reason, methods to access antibody genes directly from cells other than hybridomas were developed. The first step towards this goal was taken by cloning the repertoire of V_H genes from an immunized mouse spleen into bacteriophage lambda (105) or plasmid (67). This was then extended to the combinatorial approach initially developed for catalytic antibodies by Huse *et al.*

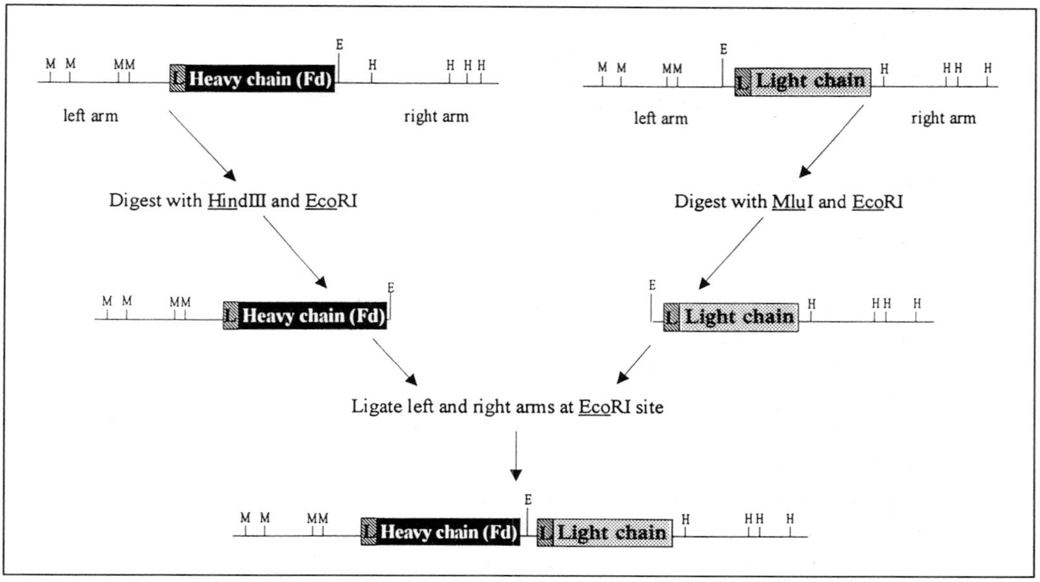

Fig. 3 Heavy (Fd) and light chain sequences generated by PCR are cloned into separate lambda vectors. The right arm of the heavy chain vector is removed by digestion with *Hind*III and the left arm of the light chain vector is removed by digestion with *Mlu*I. Subsequent digestion with *Eco*RI then allows ligation of the two arms via the central common restriction site. H, *Hind*III; M, *Mlu*I; E, *Eco*RI; L is the pelB leader sequence.

(106). Libraries of heavy (Fd) and light chain sequences generated by PCR of mouse spleen RNA were cloned into separate lambda vectors, in the left and right arms, respectively (see Fig. 3). For the light chain library, the left arm of lambda was cleaved with a restriction enzyme, leaving the right arm containing the light chain intact: for the heavy chain, the process was similar except it was the right arm which was cleaved. A random combinatorial library was generated by digestion at a common central restriction site followed by ligating the heavy and light chain containing left and right arms. Only those clones containing a left and right arm reconstituted viable phage. This approach allowed the production of libraries with 10^8–10^9 clones. However, screening of these libraries using a radioactive probe and nitrocellulose filter lifts allowed access only to approximately 10^6 clones per filter, making the screening extremely laborious. Combinatorial libraries in phage lambda have been used to isolate Fabs from immunized mouse spleen (106, 107) or from PBLs of immunized human donors (108, 109). Alternatively, random combinatorial libraries can be created by PCR assembly of pools of previously amplified V_H and V_L genes (for example from spleen mRNA) into scFv constructs, which can then be cloned and expressed in a suitable vector (110, Fig. 4). This method, combined with phage display to facilitate screening (see Section 5.2), has proved superior and has now become the method of choice. A disadvantage of both methods however, is that the original V_H and V_L pairing selected for high-affinity by immunization is lost. The chances of recovering this $V_H V_L$ combination or others of

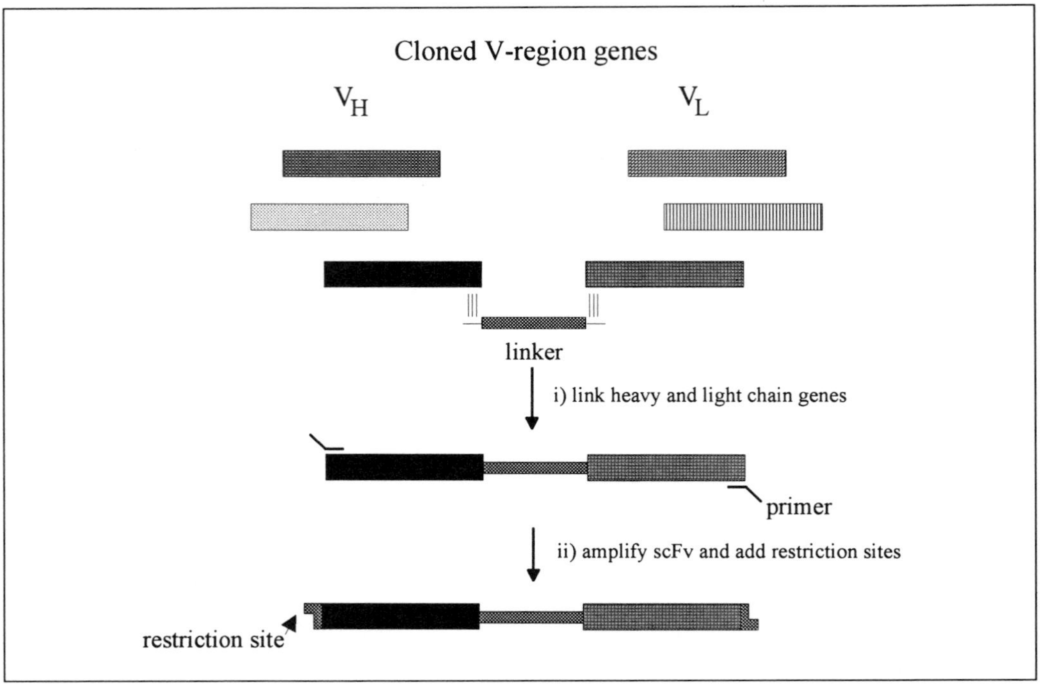

Fig. 4 PCR assembly (splicing by overlap extension). Previously amplified heavy and light chain genes are linked together by an oligonucleotide linker whose 5′ and 3′ ends are complementary with the 3′ and 5′ ends respectively of the heavy and light chains. The assembled scFv is then amplified using primers incorporating restriction enzyme sites to generate a product which after digestion can be ligated into an expression vector.

equally high affinity by this route and the size of the library required has been the subject of lively debate (111–114). Nevertheless, Portolano et al. (115) were able to isolate a high-affinity (about 10^{-9} M) human Fab fragment to thyroid peroxidase by screening a library of only 2×10^5 clones. In an alternative approach, Embleton et al. (14) used in-cell PCR assembly of V_H and V_L to generate an scFv, thus ensuring that the original *V*-gene combinations were preserved.

Combinatorial libraries can, in theory, be created in other hosts (for example plants), but ultimately the combinatorial approaches of selecting high-affinity antibodies from a large pool of non-specific antibodies will rely heavily on powerful screening methods. This has been achieved by the development of the phage display system in *E. coli* (see below) which aims to mimic the display of antibody on the surface of the B-cell.

5.2 Phage display of antibody fragments

McCafferty et al. (65) displayed scFv anti-lysozyme D1.3 on the surface of fd bacteriophage by creating a fusion with gene *III* (*gIII*), which encodes the minor

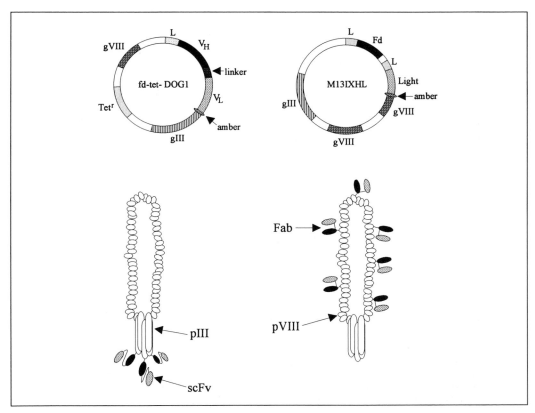

Fig. 5 Fusion of antibody fragment genes to either gIII or gVIII. Fusion to gIII results in 1–5 copies of antibody fragment per phage, whereas fusion to gVIII results in the display of 24 or more copies of antibody fragment per phage particle (108). Tet, tetracycline resistance; L, leader sequence; Light, light chain; amber, amber stop codon. Plasmid references: fd-tet-DOG1 (110), M13IXHL (122).

coat protein (pIII) of the phage and confirmed that the phage-bound antibody fragment retained the specificity of the parent D1.3 antibody. Antibody fragments have since been fused to the C-terminal domain of the pIII protein (116, 117) or to pVIII (118, 119) of phage (see Fig. 5). Fab fused to pVIII was the first heterodimeric fragment to be expressed on phage (118, 119) and took advantage of the natural anchoring of pVIII in the inner membrane of *E. coli* such that the heavy chain was able to fold correctly and associate with the light chain which was also secreted into the periplasm (Fig. 5). The decision whether to use *gIII* or *gVIII* as the fusion partner is to a certain degree dependent on the property for which the fragment library is being screened. As each phage only contains three to five copies of pIII, fusion to *gIII* enables discrimination to be made on the basis of affinity. The major coat protein is much more abundant with approximately 2700 copies per phage, resulting in multivalent display and selection on the basis of avidity (a combination of affinity and valency) rather than on affinity alone. This would hinder the discrimination between phage with different affinities but may be advantageous when screening for catalytic activity (119). It should be noted here that in practice there is a limit on how many antibody fragments can be displayed as pVIII fusions. Twenty-four or more are quoted (119), amounting to approx. one fusion protein molecule per 100 pVIII coat protein molecules. Bearing in mind that pVIII is the major coat protein, it can be envisaged that too many pVIII fusion events would interfere with the assembly of viable phage particles. Monovalent display of antibody fragments (110, 116, 117, 120) is preferable when searching for those of highest affinity, and is achieved with *gIII* fusions in phagemids. Phagemids are plasmids which contain the origin of replication for filamentous bacteriophage. They are double stranded, making cloning easier, and allow 100-fold higher efficiencies of transformation to be obtained compared with phage vectors, an important factor when trying to create large libraries. They lack the genetic instructions for packaging however, and, therefore, require the use of a helper phage whose pIII competes with the fusion protein for incorporation into the phage, resulting in an average of one fragment displayed per phage. Introduction of a peptide tag allows easy assaying whilst an amber stop codon (UAG) between the antibody sequence and *gIII* (ref. 121) or *gVIII* (ref. 122) allows the production of phage-displayed fragments when grown in supE strains of *E. coli*, or tagged, soluble fragments when grown in non-suppressor strains where the amber codon is read through as glutamine. In an alternative system, soluble Fab is obtained after removal of the *gIII* sequence by restriction digestion and religation of the vector (116).

The power of phage display lies in the ability to select those clones having the desired binding qualities from a vast excess of non-binders, quickly and simply. By adapting the process of biopanning (123), McCafferty *et al.* (65) showed that phages displaying antibody fragments on their surface can be greatly enriched over wild type by selection with antigen. The enrichment obtained in a single round could be as much as a thousand-fold (65) and a million-fold after two rounds (65, 117). By reinfecting *E. coli* with the purified phages and reselecting, rarer higher affinity fragments can be obtained (65, 110). Many rounds of screening will select for small differences in affinity.

Barbas *et al.* (116) demonstrated the effectiveness of the system via Fabs specific for antitetanus toxoid. Using the monovalent display approach, a 253-fold enrichment of a strong binding clone (10^{-9} M) over a weaker binding clone (10^{-7} M) could be obtained compared with a five-fold enrichment using the multivalent *gVIII* system. Garrard *et al.* (117) were able to discriminate between Fab phages having affinities in the low nanomolar range using monovalent display. They also demonstrated that the displayed Fab exhibits an affinity for its antigen that is almost identical to that of the free Fab.

Combined with combinatorial libraries, phage display offers an extremely elegant route to selecting good, high-affinity antibody fragments from a huge library. Two limiting factors exist however: first, the size of the antibody library (i.e. the number of each member) is limited by the phage titre; secondly, the diversity (i.e. the number of different members) of the library, is mainly limited by the frequency of bacterial transformation (124). Soderlind *et al.* (124) have used chaperonins to assist in the packaging of the phage vector: titres of packaged phagemid increased almost 200-fold when co-expressed with GroES or GroEL. The second factor was addressed by Hoogenboom *et al.* (121) who first proposed the use of combinatorial infection, whereby *E. coli* transformed with a repertoire of heavy (or light) chains encoded on phagemids is infected with a repertoire of light (or heavy) chains encoded by phage. These 'dual combinatorial' libraries, which are potentially more diverse than those encoded on a single vector and take advantage of the high efficiency of infection of *E. coli* cells, could potentially give rise to greater than 10^{11} phage displayed Fab per litre. Waterhouse *et al.* (125) have overcome the main disadvantage of this route, the fact that heavy and light chain genes would not be packaged within the same phage particle, by using the *lox*–Cre site-specific recombination system of bacteriophage P1 to lock together the heavy and light chain genes from two different replicons within an infected bacterium. This approach led to the construction of repertoires of human Fab fragments of approximately 6.5×10^{10} different members (126).

5.2.1 Bypassing immunization

Further progress towards the goal of constructing a phage library to mimic the immune repertoire was made by Marks *et al.* (127). A phage display library of single-chain antibodies derived from IgG and IgM heavy (V_H) and light (V_κ and V_λ) chain variable region mRNA from PBLs of unimmunized humans was created using PCR amplification. Antibody fragments specific for the hapten 2-phenyloxazol-5-one (phOx) and for the antigens bovine serum albumin and turkey-egg lysozyme were obtained from this library. Griffiths *et al.* (128) used the same phage library to isolate a further nine single-chain antibody fragments with high specificity against self-antigens. Sequencing of the fragments indicated that they were derived from a range of unmutated and somatically mutated *V*-genes. The specificity of these antibodies for cognate antigen was high in contrast with the poor specificities of mouse Fab fragments isolated from a naive library using *gVIII* fusion (129).

One of the limitations of repertoires based on *in vivo* rearranged *V* genes (as in splenocytes and PBLs) is that there will be a natural bias against certain specificities such as intracellular tumour- and other self-antigens (although antibody fragments against self-antigens have been isolated from such libraries (128). This potential problem can be circumvented by composition of 'synthetic' repertoires of *in vitro* rearranged *V* genes. Cloned V_H region sequences (unrearranged gene source) were linked to randomized CDR3 sequences with varying lengths, combined with a fixed light chain, and expressed from a phage display vector (130, 131). This resulted in 'single pot' libraries of approximately 10^8 different specificities from which antibody fragments were isolated that have proved difficult to obtain by other routes (131). However, the affinities of individual fragments were on the low side (micromolar range; refs 130, 131), although they were shown to be amenable to improvement by further mutagenesis and selection (see Section 5.2.2).

The principle of synthetic libraries was extended by Griffiths *et al.* (126) who assembled a huge library (6.5×10^{10} different specificities) using combinatorial infection (125) in which the V_H and V_L (V_κ and V_λ) genes of two separate synthetic libraries (both constructed as in refs 130, 131) were randomly combined on the same phage replicon (see also Section 5.2, Introduction). This yielded human Fab fragments with affinities up to 4 nM, thus giving weight to the hypothesis that there is a direct correlation between library size and (high) affinity.

5.2.2 Affinity maturation

There have been several different approaches to increasing the affinity of already selected fragments in a way which imitates the natural affinity maturation process. Methods of achieving this include growth of the phage in mutator strains (132) and PCR mutagenesis using an error prone polymerase, to create point mutations throughout the V-regions (129, 133). An alternative approach is to carry out chain shuffling, where a single heavy or light chain is combined with a library of partners in a PCR assembly reaction (see also Section 5.1), to generate new combinations in 'hierarchical libraries' (110). Marks *et al.* (134) used a phOx scFv (K_D 3.2×10^{-7} M) which they had previously isolated by screening a naive phage library, and, after shuffling the light chains, obtained an antibody with a 20-fold higher affinity. Shuffling the heavy chain whilst retaining HCDR3 resulted in a further 15-fold improvement, giving an affinity of 1.1×10^{-9} M, comparable to that of mouse hybridomas from the tertiary response to the same hapten.

5.2.3 Creating new specificities

It has been shown that the specificity of a human Fab against tetanus toxoid can be changed by randomizing CDR3 of the heavy chain (135). Selection using a monovalent display vector and a fluorescein affinity column resulted in the isolation of a new Fab which specifically bound this hapten with an affinity range of 10^{-7} M to 10^{-8} M, approaching that of the secondary response of mice immunized with free fluorescein. This forms part of a longer term approach to the generation of a completely naive combinatorial library which would be created by randomizing all six CDRs and placing them on varied framework regions.

5.2.4 Limitations to repertoire cloning

There are a few limitations placed on the success of repertoire cloning and phage display.

1. The first occurs in the choice of primers used in the amplification of the heavy and light chains. Where the sequence of the primer does not match the RNA or DNA, then that sequence or family of sequences will be absent from the final repertoire. New families of primers are being designed all the time so the chances of this happening are receding.
2. The choice of restriction sites included in the primers is also critical in order to minimize losses caused by cutting in the coding region when cloning into the

expression vector (136). This was also addressed by Clackson *et al.* (110) in the PCR assembly process for scFv (see Section 5.1) in which the primary PCR was carried out using primers without restriction sites and subsequently the whole construct was amplified using primers to the 5' V_H and 3' V_L regions which contained rare-cutting enzyme sequences.

3. Finally, the antibody fragment-fusion has to be successfully expressed on the phage surface and is subject to the constraints that were described in Section 4.2.4.

In conclusion, the extraction of high-affinity human antibodies to any specificity from a single huge pool of clones derived from an unimmunized source, whilst still difficult, is considerably closer than would have been predicted four years ago.

6. Applications of engineered antibodies

6.1 Structure–function studies

Protein engineering technology has revolutionized the approaches possible and the speed by which the relationship between structure and function of proteins can be studied and better understood. Site-directed mutagenesis (SDM) of predicted key residues, domain shuffling and swapping, hinge shuffling, swapping, shortening, lengthening, etc., have been successfully applied in systematic studies investigating various aspects of antibody function.

SDM studies and reshaping of variable regions have contributed to an increased understanding of antibody–antigen interactions, confirming the role of the CDRs as the primary contact points, but also establishing the importance of framework residues in supporting the variable loops in the correct conformation for antigen binding (137, 138; see also Section 6.2). In the case of a reshaped antilysozyme antibody it has been shown that the affinity of the original reshaped antibody could be increased 200-fold by the reintroduction of some of the original mouse framework residues, which were identified by molecular modelling as possible supports for the CDRs. The best reshaped antibody had an affinity which was only about four-fold lower compared with the original mouse antibody (138). Several groups have achieved modest increases (less than 10-fold) in the affinity for antigen by SDM of selected CDR residues (42, 139, 140), one exception being a study in which three CDR substitutions led to a 200-fold increase (141). Random point mutations mostly tend to yield antibodies which have lost their ability to bind antigen (142). Systematic SDM of. variable regions has also been used for mapping idiotypic determinants (143). The importance of certain variable region residues has been implied in functions other than antigen binding. For example, a report has been published in which a single framework region substitution in a lambda variable region was found to block Ig secretion (144). Other aspects of antibody function have been investigated by engineering of constant regions. This includes studies of complement activation, Fc receptor interactions, specificity of recognition by rheumatoid factors, and post-translational modification by glycosylation.

These latter areas have recently been reviewed (145, 146) and will, therefore, not be described here.

The capacity to produce genetically engineered antibodies has not only made it possible to study those antibodies which do not occur naturally (for example human–mouse chimeras, fusion proteins, fragments, etc.), but also to study antibodies which are normally only available in small quantities, for instance secretory IgD (also reviewed in 146).

6.2 Antibodies for *in vivo* use

6.2.1 'Humanized' antibodies

Although it was clear from the outset that mAbs had enormous potential for *in vivo* diagnosis and treatment of disease, this potential has failed to be realized. A major problem inherent to these antibodies, which are mainly from mouse origin, is that they provoke a human–antimouse antibody (HAMA) response upon administration to humans. This response does not only neutralize any therapeutic effect of the mAb but can also lead to conditions such as anaphylactic shock. Worst of all, once such a response has been primed, subsequent repetition of the treatment, which is often essential especially when targeting solid tumours, is severely compromised. It is further known that mouse antibodies are not very efficient in recruiting human effector functions. Human antibodies would be preferred, but unfortunately, unlike mouse mAbs, the production of human mAbs is by no means straightforward and is still fraught with many problems (147). Fortunately, 'human' antibodies can now be obtained by genetic manipulation of antibody genes.

The first attempts to 'humanize' mouse mAbs were by making 'chimeric' antibodies in which mouse variable regions were linked to human constant regions at the genetic level (148–150). Specificity and affinity for antigen are fully retained by this method and numerous mouse mAbs have been 'humanized' in this way to date (see ref. 145 for a recent review). Although not many chimeric antibodies have been fully tested in clinical trials, preliminary data show that they still can lead to a HAMA response, some more than others (151, 152). Indeed, it has been shown in mice that a considerable immune response is still directed to the heterologous variable regions in chimeric antibodies (153).

A more sophisticated approach came from Greg Winter's laboratory, in which the specificity of existing human mAbs was changed by replacing the residues involved in antigen binding (mainly the CDRs) with the corresponding mouse residues of a desired specificity. This technique was called 'CDR grafting' and the resulting antibodies were termed 'reshaped' human antibodies (16, 154, 155). As with chimeric antibodies, the whole procedure is carried out at the genetic level. The reshaped variable region genes are linked to appropriate human constant regions, and are usually expressed in myeloma cells. Of course alternative expression systems can be used (see Section 4). Although this technique is not as straightforward as simple 'chimerization', an ever-expanding range of non-human mAbs

Table 1 Reshaped human antibodies

Specificity	Reference
4-Hydroxy-3-nitrophenylacetyl (NP)	154
Hen-egg lysozyme (HEL)	155
CAMPATH-1	16
IL-2 receptor	156
IL-2 receptor	157
IL-6 receptor	158
Placental alkaline phosphatase (PLAP)	17
CD3	159, 160
CD3	161, 162
CD3/p185^{HER2}(bispecific)	163
CD4	164
CD5	165
CD18	166, 167
CD19	168
CD33	169
CD56	168
p185^{HER2}	170
Epidermal growth factor receptor (EGFR)	171
T-cell receptor α/β	172
Carcinoembryonic antigen (CEA)	173
Human milk fat globule (HMFG)	174
Herpes simplex virus glycoprotein B	175
Herpes simplex virus glycoprotein D	175
HTLV-IIIb, gp120	176
RSV glycoprotein	177
Human IgE	178
Human IgE	179
TNFα	180

have been reshaped successfully to date (see Table 1). Preliminary clinical data on some of these antibodies indicate their effectiveness in human patients and a lack of HAMA response (181, 182).

The rationale behind the reshaping of human variable regions lies in the observation that they all share a common structure, in which the CDRs sit on top of and are supported by a scaffold formed by the framework regions (FRs). Although it is mainly the CDRs which form contact with the antigen, the FRs are also sometimes involved in these interactions (66). Furthermore, there are numerous contacts between the FR and CDR residues which appear to support the CDRs in the conformation required for antigen binding, and it has been shown that loss of such contacts (for example because of a different FR residue) can lead to lower or even a total loss of antigen binding (16, 155). Various strategies have been adopted by different people to overcome this potential problem. Some continue to use the same set of human FRs and introduce potentially critical mouse residues into the human frameworks (these are usually identified using molecular modelling techniques). However, this approach often leads to diminished antibody affinity (17,

171, 176, 177). Although the binding affinity can be improved later (138), many researchers prefer the more reliable route of matching the mouse and human FRs, thus reducing the number of critical mouse residues to be retained to a minimum (156–158, 164, 165, 174, 175, 179, 180, 183). We have shown that if the framework regions match well, 'true' CDR grafting (that is to say, transplanting only the CDRs and making no other substitutions) can work and yield reshaped human antibodies with relative affinities indistinguishable from the parent mouse antibodies (174). How important it really is to limit the number of mouse residues in the FRs is not at all clear. This issue may be resolved only after clinical trials of the various reshaped human antibodies have been conducted. One way to circumvent this problem was suggested by Wu and Kabat who observed the existence of preserved 'ancient' FRs which are used by human, mouse, and rabbit Igs. They suggest these as ideal candidate FRs for CDR grafting (184).

An interesting alternative to humanizing by CDR grafting was described by Padlan (185) who proposed a procedure which 'humanizes' only the solvent accessible surface of the non-human antibody, retaining all the interior residues as in the original parent antibody. This concept was further developed and tested by others and has become known as 'resurfacing' or 'veneering' of antibodies (168). The resulting resurfaced antibodies had apparent affinities identical to their parent murine antibodies and are expected to have a dramatically reduced immunogenicity, rendering them suitable for diagnostic imaging. For prolonged therapeutic treatment, however, where larger doses are required, complete reshaping may still be preferable. Clinical evaluation will have to resolve this issue.

Further enhancement of the clinical potential of humanized antibodies can be achieved by engineering of their constant regions. For example, it was shown that a homodimeric form of a reshaped human antibody could be generated by the introduction of a cysteine in the CH3 region to allow interchain disulfide bonding. The avidity of the homodimer was the same as that of the monomer, but a dramatic improvement in the ability to internalize and retain radioisotope in target leukaemia cells was observed. Moreover, the homodimers were reported to exhibit a 100-fold increase in complement-mediated and antibody-dependent cell killing using human effector functions (186). In another study, a mutation was engineered in the constant region of a reshaped OKT3 antibody which led to a lower cell activation and thus diminished *in vivo* toxicity (161); a similar effect was achieved by abolishing the glycosylation site in the γ1 constant domain of a reshaped anti-CD3 human antibody (160).

6.2.2 Human antibodies

Despite the success in humanizing rodent mAbs and the encouraging results of preliminary clinical trials evaluating them, there remains a strong school of thought that antibodies encoded by naturally occurring human genes will be superior in performance *in vivo* compared to 'humanized' antibodies. A number of approaches to achieve this objective of obtaining 'genuine' human antibodies were developed by different groups.

As pointed out earlier, with the advent of combinatorial libraries and phage display cloning, rearranged and even unrearranged human antibody genes can now be cloned directly from human B lymphocytes and expressed as and assembled into functional human antibody fragments (108, 109, 114, 126–128, 130, 131). Selection is based on binding of the phage displayed antibody fragment to antigen. Alternatively, existing rodent antibody fragments of particular interest can be used as templates in a dual-step, chain-shuffling protocol, to guide the selection of human antibody fragments from phage display libraries (187). Cloned rodent Fd is combined with a repertoire of human light chains displayed on phage, and antigen binding phage is selected. The isolated human light chains are then paired with human Fd chains displayed on phage, again followed by selection. The resulting human antibody fragments are expected to have the same fine specificity and affinity as the original rodent antibody and, indeed, results obtained indicate that this is the case. This approach, therefore, provides a promising alternative to existing 'humanizing' processes, yet resulting in completely human antibody fragments.

There is some concern that human antibody fragments isolated from random libraries may contain harmful sequences (for example sequences associated with rheumatoid factors, superantigen, etc.). For *in vivo* use, it may, therefore, be advisable to check carefully the obtained sequences against known 'suspect' sequences and perhaps only to develop specificities derived from functional (in the sense of observed to be used in nature) V-region germline genes which, to the best of current knowledge, are unassociated with disease conditions.

A number of groups are pursuing the production of human antibodies by mice. Antibody phage display libraries from severe combined immune deficiency (SCID) mice injected with human PBLs and then challenged with appropriate antigens were successfully developed (188). Another group developed transgenic mice that produce antigen-specific human antibodies (189). The main drawback of the latter method, as it stands now, is the limited number of variable region genes (four V_H and four V_κ) incorporated in the minilocus transgenes. Even though 15 D segments, 6 J_H segments, and 5 J_κ segments are built in as well, this still provides a limited repertoire. Incorporation of all V-region germline genes in a minilocus currently seems a daunting prospect. Furthermore, development and breeding of these transgenic mice and hybridoma selection is a rather lengthy process compared to antibody phage display.

Overall, each of the approaches to obtain 'humanized' or genuine human antibodies has its pros and cons. CDR grafting is still the most commonly followed route and capitalizes on existing knowledge and resources with regard to the parental rodent antibodies. However, as technology evolves and matures, the scene may well be different in 5 or 10 years' time. Which approach is best with regard to efficacy and non-immunogenicity of the antibodies produced is unclear for the moment, and results of clinical trials on 'humanized' and human antibodies produced via different routes are keenly awaited.

6.3 Bispecific and bivalent antibodies

Traditionally, bispecific antibodies consisted of molecules containing two different Fv regions; for example, one that binds to target and another one that binds to a receptor molecule on an effector cell, thus promoting targeted cytolysis. Drug targeting could be achieved using the same principle, using anti-drug Fv as the second specificity. Production has been by a number of different methods, for example, chemical cross-linking, hybrid hybridomas (190) or transfectomas (191), or by hinge disulfide exchange (192). Unfortunately, none of these methods yields a product in sufficient purity and quantity. Genetic engineering technology has made the development of a new generation of bispecific antibodies possible, giving much greater purity, and antibodies which are not necessarily a mere combination of two different Fvs. For example, Traunecker *et al.* (193) developed double trans-fectomas secreting bispecific antibodies in a background mixture of only two other antibody species (rather than the usual mixture of ten different species obtained by hybrid hybridomas). A novel type of bispecific antibody was produced, consisting of the two N-terminal domains of CD4 as the first specificity and FvCD3 as the second, linked to the Fc region of IgG3. They were generated by a double transfection of J558L myeloma cells with two expression constructs, one encoding the two CD4 domains, the other encoding single-chain FvCD3 and both linked to the C_H2–C_H3 domains of IgG3. The C_H1 domain was omitted to encourage light chain-independent secretion of formed dimers; hence the outcome of only three species of antibody product. In the same paper (193) the authors describe the design and production of a single-chain bispecific molecule consisting of the two CD4 domains linked to single-chain FvCD3 which is, in turn, linked to a C_κ domain; they termed this molecule 'Janusin' after the Roman god Janus who had two faces. This type of construct is secreted from myeloma cells as a single species in good yields (about 10 mg/litre) and can be purified easily on anti-κ affinity columns. The material obtained was shown to be correctly folded with retention of functional CD4 and anti-CD3 binding sites. More recently, a method was described for the production of bispecific $F(ab')_2$ heterodimers using leucine zipper regions of the Jun and Fos proteins (194). The F(ab'-zipper)$_2$ homodimers are first expressed in Sp2/0 myeloma cells. The bispecific molecules are then assembled *in vitro* by reduction of the homodimers at the hinge to obtain monomers, followed by mixing with the homo-dimers of different specificity and reoxidizing to form heterodimers. This results in a mixture of products, the main component being the bispecific heterodimer, which can easily be separated from the other species by ABX chromatography and obtained in acceptable purity. Zipper peptides were also used to join *E. coli* produced single-chain Fvs with different specificities (195, 196), thus creating functional bispecific double binding sites. However, none of these methods results in the production of 100 per cent heterodimeric antibodies, homodimeric forms are always present. Fortunately, technology moves on, and novel ways of assembling variable-region domains were devised, by two independent groups, to get round this problem, leading to a new class of antibody-based molecules, termed 'diabodies'

(197) or multivalent Fvs (198). These molecules are essentially heterodimers consisting of 'mixed' scFvs (i.e. V_H and V_L of non-matching specificities: $V_H(A)–V_L(B)$ + $V_H(B)–V_L(A)$), preferentially co-expressed from one cell. The mixed scFvs spontaneously dimerize to form functional, bispecific molecules. Homodimers, if formed, do not bind antigen, and are, therefore, not a problem for purification. Dimerization is encouraged by short linkers (less than 10 amino acids), it was further shown that the linkers can be disposed of totally, still yielding functional bispecific molecules (197). Finally, it is also possible to produce genetically linked single-chain Fvs with different specificities, using a peptide linker to connect both units (199, 200).

Bivalent antibodies are sometimes required, often for avidity retention. Most methods described above can be used to produce bivalent antibodies. In addition, dimerization domains specifically promoting homodimer formation have been described. For example, helix–turn–helix motifs can be attached to scFvs. Formation of antiparallel four-helix bundles results in functional bivalent molecules, termed 'miniantibodies' (195, 196). Other examples of protein moieties able to promote homodimerization of antibody fragments when fused to them include subunits of protein A (201) and of alkaline phosphatase (202).

6.4 Bifunctional antibodies

Immunoglobulin domains can be genetically linked to other protein moieties such as oncogenic determinants (150), enzymes (150, 202, 203), metal chelating proteins (204, 205), growth factors (206), immunostimulatory proteins (207, 208), immunoglobulin-binding domains (201), signal transducing proteins (209), maltose-binding protein (210), and toxins (63). The resulting molecules have been shown to be biologically bifunctional and can be used in a diverse number of applications. The most common fusions are of the type F(ab')$_2$–protein and scFv–protein (or protein–scFv, ref. 211). It is also possible to replace the variable regions rather than the constant regions and still obtain biologically active, bifunctional molecules. The first such molecules described were CD4–Ig chimeric molecules consisting of all or the first two domains of CD4 genetically linked to IgG heavy chain constant domains (212–214). The resulting molecules were termed CD4 immunoadhesins, and were shown to possess the binding activity of CD4 for gp120 of HIV and to exhibit Ig effector functions.

It has also been shown that the variable domains in antibodies can be replaced by protein domains that, unlike CD4, are not members of the Ig superfamily and do not necessary fold in a way compatible with Ig constant region folding. An example of this was reported by Landolfi (215) who constructed a chimeric IL-2/Ig molecule which possesses the functional activities of both the IL-2 and Ig Fc region, they termed the resulting molecules 'immunoligands'. Such molecules could be useful in therapy if their binding specificity is unique to a disease state, for example a neoplasm.

6.5 Antibody domains as scaffolds for insertion of functional sites

6.5.1 'Antigenized' antibodies

'Antigenized' antibodies can be obtained by a process in which protein epitopes are inserted in one of the hypervariable loops, usually CDR3 of V_H. This can be achieved without adverse effect on V_H/V_L packing, folding, and secretion of intact Ig. It is a method which can provide a stable conformation for oligopeptides of biological interest and limit their flexibility. The CDR3 of V_H is seen as the most suitable surface loop for epitope insertion: it is the most flexible of the six CDRs (no canonical structures have been assigned to it) and is also the most variable in length and amino acid composition.

The principle was first demonstrated by Sollazo *et al.* (216) who inserted and expressed the epitope $(NANP)_3$ in the CDR3 of a V_H. This sequence is of particular interest as it is the dominant epitope expressed on the circumsporozoite surface (CS) antigen of the *Plasmodium falciparum* parasite, has been linked with protective immunity *in vitro* and *in vivo*, and is, therefore, considered as a candidate vaccine against malaria. Their results indicated that the inserted epitope folded with a configuration immunologically similar to that in the native protein and was more reactive than the corresponding synthetic peptide. In a further study this 'antigenized' antibody was used as an immunogen in rabbits, resulting in the production of antibodies which recognized $(NANP)_3$ peptide and which efficiently inhibited invasion of cultured liver cells by *P. falciparium in vitro* (217). Further information on this topic can be found in a review by Zanetti *et al.* (218).

6.5.2 Minibodies

Whereas in 'antigenized' antibodies whole antibody domains are utilized as scaffolds to insert functional groups, Pessi *et al.* (219) have gone beyond that and designed the so-called 'minibody': a metal-binding 61 amino acid protein based on only part of a variable-region structure. The molecule contains three β-strands from each of the two β-sheets of the heavy chain variable region of McPC603 and the CDR1 and CDR2 hypervariable loops. The metal-binding site was introduced by engineering one His in CDR1 and two His in CDR2. The protein was shown to be monomeric, it has a globular fold and is able to bind metal ions (Cu^{2+}, Zn^{2+}, Cd^{2+}, Co^{2+}). Initial problems with insolubility of the protein were overcome by mutagenesis and addition of a solubilizing motif (three Lys residues) at the N- or C-termini (220). Engineering of this metal-ion binding site into a catalytic site will be a major challenge and is probably not the optimal route to obtain catalytic antibodies (see Section 6.7). A more feasible application is to use phage displayed minibodies as scaffolds for the display of conformationally-constrained peptides which can be affinity-selected (221). This could form an alternative to existing peptide display technologies.

6.6 Molecular recognition units (MRUs)

In some cases, particularly anti-receptor antibodies, individual CDRs can display significant antigen binding and biological activity. The first two examples described were the CDR2 of the V_L of a mAb that binds reovirus type-3 receptor (222) and the CDR3 of the V_H of an antifibrinogen receptor mAb (223). It was soon recognized that the spacial arrangement of the active groups of amino acid side chains could be crucial for biological activity, a criterion difficult to fulfil with flexible linear peptides, hence the subsequent development of conformationally constrained loops. A crude way of constraining conformation can be achieved by dimerization of the peptide, for example by means of Cys residues at the end of the peptide. Indeed, the dimeric form (V_LSH) of the reovirus receptor MRU described above was shown not only to specifically interact with the receptor, but also to down-modulate the receptor and to inhibit DNA synthesis in the cells. The latter two effects were not seen with the monomer (222).

Further studies showed that constraint by cyclization can improve the biological characteristics of these peptides even more; a cyclic derivative of the anti-reovirus receptor CDR2 was shown to possess an affinity for the receptor which was at least 40-fold higher than the one observed for its linear analogue (224). However, linear and cyclic peptides are still proteinaceous compounds and because of that are likely to suffer from a number of disadvantages such as protease sensitivity, antigenicity, rapid clearance, restricted accessibility at blood–tissue barriers, etc., which could compromise their potential *in vivo* use. The next logical step, therefore, was the development of so-called non-peptide mimetics, which are small synthetic cyclic compounds containing only one or two peptide bonds and incorporating natural or unnatural amino acid side chain active groups, and which can assume only one configuration (225). A mimetic based on the V_L CDR2 of the antireovirus receptor was shown to possess the same binding properties and inhibitory effects on cells as the original mAb (225). It was also shown to be protease-resistant. Another report describes a human CD4 mimetic (based on the CDR2-like domain in CD4) that binds to gp120 of HIV and is also resistant to proteases (226). The particular interest of this report lies in the observation that peptides based on the same area of CD4 failed to show biological activity, emphasizing the importance of secondary structure.

Mimetics can be synthesized by modular component chemical synthesis. Because of their small size and synthetic nature it is thought unlikely that they will be immunogenic, and so they can be considered as potentially useful therapeutic agents. For more in-depth reading on this subject we recommend a review by Saragovi *et al.* (227).

6.7 Catalytic antibodies

Catalytic antibodies (abzymes) have commanded considerable interest since the first examples involving relatively simple acyl transfer reactions were reported in

1986 (228, 229). The list of transformations has grown impressively to include increasingly diverse and complex reactions (reviewed in ref. 230) and most recently includes the catalysis of a peptide-bond formation (231, 232) and ring closure reactions (233, 234). Baldwin and Schultz (235) used *in vitro* mutagenesis to replace a tyrosine residue with a histidine in the V_L region of the binding pocket of the monoclonal antibody MOPC315 and generated a hybrid Fv fragment consisting of an engineered V_L chain expressed in *E. coli* and a native V_H chain. This gave an approximately 10^5-fold increase in activity of the catalytic ability of the antibody. Iverson *et al.* (236) introduced the concept of metalloantibodies in which a co-ordination site for metals was created in the antigen binding pocket of a scFv antifluorescein antibody 4-4-20. Although the resulting protein was not reported to be catalytically active, it established the useful ability to put a metal ion binding site in an antibody, since metal ions can catalyse a variety of reactions. With the advent of combinatorial libraries, it should be possible to recombine this metal-binding light chain with any heavy chain of interest and screen for catalytic activity. Chen *et al.* (237) have recently reported that it is possible to obtain catalytic antibodies from a combinatorial library in λ phage. The increased use of combinatorial methods plus the development of new approaches to screening for catalytic activity (238, 239) will ensure the continuing advance in sophistication of catalytic antibodies and their wider application.

7. Conclusions

Progress in the area of antibody engineering over the last six years has been astonishing. The emergence of recombinant antibody fragments has meant that researchers are no longer constrained to whole antibodies or proteolytic fragments but have a choice of different sized fragments either on their own or coupled to a range of effector molecules. Their ready availability in a soluble, active form has led to a resurgence of interest in this area. An immediate impact has been on structural studies and this is now being reflected in an increased number of solved structures deposited in the Brookhaven Protein Database. Whilst *E. coli* will probably remain the expression system of choice for some time to come, certainly in the laboratory, there is an increasing number of alternatives which may prove to be more suitable for large-scale production. Efforts to bypass the immune system continue apace, largely through the intervention of phage display technology, and the goal of routine selection of high-affinity antibodies to any specificity from a single library is growing closer all the time. Recent developments in the expression of antibody fragments on the surface of bacteria (240) and the coat proteins of a retrovirus (241) are likely to further influence the speed and application of this technology. Engineered antibodies are no longer exclusively directed towards the medical area, although this remains dominant, but are proving useful reagents in a wider range of applications. For example, conjugates with alkaline phosphatase or biotin can now be made via recombinant routes rather than chemically, and abzymes are becoming increasingly sophisticated. In conclusion, these are exciting times for antibody engineers and prospects for this area are highly encouraging for the future.

References

References marked * are recommended for further reading.

1. Behring, E. A. (1894). In *Das neue Diphterieheilmittel*, p. 40. O. Hering, Berlin.
2. Kohler, G. and Milstein, C. (1975). Continuous cultures of fused cells secreting antibody of pre-defined specificity. *Nature*, **256**, 495.
*3. Chothia, C., Lesk, A. M., Tramontano, A., Levitt, M., Smith-Gill, S. J., Air, G., Sheriff, S., Padlan, E. A., Davies, D., and Tulip, W. R. (1989). The conformations of immunoglobulin hypervariable regions. *Nature*, **342**, 877.
4. Saiki, R. K., Scharf, S., Faloona, F., Mullis, K. B., Horn, G. T., Erlich, H. A., and Arnheim, N. (1985). Enzymatic amplification of beta-globulin genomic sequences and restriction site analysis for diagnosis of sickle-cell anemia. *Science*, **230**, 1350.
5. Larrick, J. W., Danielsson, L., Brenner, C. A., Abrahamson, M., Fry, K. E., and Borrebaeck, C. A. K. (1989). Rapid cloning of rearranged immunoglobulin genes from human hybridoma cells using mixed primers and the polymerase chain reaction. *Biochem. Biophys. Res. Commun.*, **160**, 1250.
6. Orlandi, R., Gussow, D. H., Jones, P. J., and Winter, G. (1989). Cloning immunoglobulin variable domains for expression by the polymerase chain reaction. *Proc. Natl Acad. Sci. USA*, **86**, 3833.
7. Larrick, J. W., Danielsson, L., Brenner, C. A., Wallace, E. F., Abrahamson, M., Fry, K., and Borrebaeck, C. A. K. (1989). Polymerase chain reaction using mixed primers: cloning of human monoclonal antibody variable region genes from single hybridoma cells. *Bio/tech.*, **7**, 934.
8. LeBoef, R. D., Galin, F. S., Hollinger, S. K., Peiper, S. C., and Blalock, J. E. (1989). Cloning and sequencing of immunoglobulin variable-region genes using degenerate oligodeoxyribonucleotides and polymerase chain reaction. *Gene*, **82**, 371.
9. Gavilondo-Cowley, J. V., Coloma, M. J., Vazquez, J., Ayala, M., Macias. A., Fry, K. E., and Larrick. J. W. (1990). Specific amplification of rearranged immunoglobulin variable region genes from mouse hybridoma cells. *Hybridoma*, **9**, 407.
*10. Leung, S., Dion, A. S., Pellegrini, M. C., Goldenberg, D. M., and Hansen, H. J. (1993). An extended primer set for amplification of murine kappa variable regions. *Biotechniques*, **15**, 286.
11. Kutemeier, G., Harloff, C., and Mocikat, R. (1992). Rapid isolation of immunoglobulin variable genes from cell lysates of rat hybridomas by polymerase chain reaction. *Hybridoma*, **11**, 23.
12. Marks, J. D., Tristem, M., Karpas, A., and Winter, G. (1991). Oligonucleotide primers for polymerase chain reaction amplification of human immunoglobulin variable genes and design of family-specific oligonucleotide probes. *Eur. J. Immunol.*, **21**, 985.
13. Suter, M., Blaser, K., Aeby, P., and Crameri, R. (1992). Rabbit single domain antibodies specific to protein C expressed in prokaryotes. *Immunol. Lett.*, **33**, 53.
*14. Embleton, M. J., Gorochov, G., Jones P. T., and Winter, G. (1992). In-cell PCR from mRNA: amplifying and linking the rearranged immunoglobulin heavy and light chain V-genes within single cells. *Nucleic Acids Res.*, **20**, 3831.

*15. Hoogenboom, H. R., Marks, J. D., Griffiths, A. D., and Winter, G. (1992). Building antibodies from their genes. *Immunol. Rev.*, **130**, 41.

*16. Riechmann, L., Clark, M., Waldmann, H., and Winter, G. (1988). Reshaping human antibodies for therapy. *Nature*, **332**, 323.

17. Verhoeyen, M., Broderick, L., Eida, S., and Badley, A. (1991). Reshaped human anti-PLAP antibodies. In *Monoclonal antibodies: applications in clinical oncology* (ed. A. Epenetos), p. 37. Chapman and Hall Medical, London.

18. Riechmann, L., Foote, J., and Winter, G. (1988). Expression of an antibody Fv fragment in myeloma cells. *J. Mol. Biol.*, **203**, 825.

19. Fell, H. P., Yarnold, S., Hellstrom, I., Hellstrom, K. E., and Folger, K. R. (1989). Homologous recombination in hybridoma cells: heavy chain chimeric antibody produced by gene targeting. *Proc. Natl Acad. Sci. USA*, **86**, 8507.

20. Baker, M. D., Pennell, N., Bosnoyan, L., and Shulman, M. J. (1988). Homologous recombination can restore normal immunoglobulin production in a mutant hybridoma cell line. *Proc. Natl Acad. Sci. USA*, **85**, 6432.

21. Dorai, H. and Moore, G. P. (1987). The effect of dihydrofolate reductase-mediated gene amplification on the expression of transfected immunoglobulin genes. *J. Immunol.*, **139**, 4232.

*22. Bebbington, C. R., Renner, G., Thomson, S., King, D., Abrams, D., and Yarranton, G. T. (1992). High level expression of a recombinant antibody from myeloma cells using a glutamine synthetase gene as an amplifiable selectable marker. *Bio/tech.*, **10**, 169.

23. Pallavacini, M. G., De Teresa, P. S., Rosette, C., Gray, J., and Wurm, F. M. (1990). Effects of methotrexate on transfected DNA stability in mammalian cells. *Mol. Cell. Biol.*, **10**, 401.

24. Bebbington, C. R. (1991). Expression of antibody genes in nonlymphoid mammalian cells. *Methods*, **2**, 136.

25. Wood, C. R., Dorner, A. J., Morris, G. E., Alderman, E. M., Wilson, D., O'Hara, R. M. Jr, and Kaufman, R. J. (1990). High level synthesis of immunoglobulins in Chinese hamster ovary cells. *J. Immunol.*, **145**, 3011.

26. Page, M. J. and Sydenham, M. A. (1991). High level expression of the humanized monoclonal antibody Campath-1H in Chinese hamster ovary cells. *Bio/tech.*, **9**, 64.

27. Fouser, L. A., Swanberg, S. L., Lin, B.-Y., Benedict, M., Kelleher, K., Cumming, D. A., and Riedel, G. E. (1992). High level expression of a chimeric anti-ganglioside GD2 antibody: genomic kappa sequences improve expression in COS and CHO cells. *Bio/tech.*, **10**, 1121.

28. Whittle, N., Adair, J., Lloyd, C., Jenkins, L., Devine, J., Schlom, J., Raubitschek, A., Colcher, D., and Bodmer, M. (1987) Expression in COS cells of a mouse–human chimaeric B72.3 antibody. *Protein Eng.*, **1**, 499.

29. De Sutter, K., Feys, V., Van De Voorde, A., and Fiers, W. (1992). Production of functionally active murine and murine:human chimeric F(ab')$_2$ fragments in COS-1 cells. *Gene*, **113**, 223.

*30. Hasemann, C. A. and Capra, J. D. (1990). High-level production of a functional immunoglobulin heterodimer in a baculovirus expression system. *Proc. Natl Acad. Sci. USA*, **87**, 3942.

31. Putlitz, J. Z., Kubasek, W. L., Duchene, M., Marget, M., von Specht, B.-U., and Domdey, H. (1990). Antibody production in baculovirus infected insect cells. *Bio/tech.*, **8**, 651.

*32. Reis, U., Blum, B., von Specht, B.-U., Domdey, H., and Collins, J. (1992). Antibody production in silkworm cells and silkworm larvae infected with a dual recombinant *Bombyx mori* nuclear polyhedrosis virus. *Bio/tech.*, **10**, 910.

33. Boss, M. A., Kenten, J. H., Wood, C. R., and Emtage, J. S. (1984). Assembly of functional antibodies from immunoglobulin heavy and light chains synthesised in *E. coli*. *Nucleic Acids Res.*, **12**, 3791.

34. Cabilly, S., Riggs, A. D., Pande, H., Shively, J. E., Holmes, W. E., Rey, M., Perry, L. J., Wetzel, R., and Heynekker, H. L. (1984). Generation of antibody activity from immunoglobulin polypeptide chains produced in *Escherichia coli*. *Proc. Natl Acad. Sci. USA*, **81**, 3273.

35. Better, M., Chang, C. P., Robinson, R. R., and Horwitz A. H. (1988). *Escherichia coli* secretion of an active chimeric antibody fragment. *Science*, **240**, 1041.

36. Skerra, A. and Pluckthun, A. (1988). Assembly of a functional immunoglobulin Fv fragment in *Escherichia coli*. *Science*, **240**, 1038.

*37. Pluckthun, A. (1991). Antibody engineering: advances from the use of *Escherichia coli* expression systems. *Bio/tech.*, **9**, 545.

*38. Pluckthun, A. (1992). Mono- and bivalent antibody fragments produced in *Escherichia coli*: engineering, folding and antibody binding. *Immunol. Rev.*, **130**, 151.

39. Carter, P., Kelley, R. F., Rodrigues, M. L., Snedecor, B., Covarrubias, M., Velligan, M. D., Wong, W. L. T., Rowland, A. M., Kotts, C. E., Carver, M. E., Yang, M., Bourell, J. H., Shepard, H. M., and Henner, D. (1992). High level *Escherichia coli* expression and production of a bivalent humanized antibody fragment. *Bio/tech.*, **10**, 163.

40. Shibui, T., Munakata, K., Matsumoto, R., Ohta, K., Matsushima, R., Morimoto, and Nagahari, K. (1993). High-level production and secretion of a mouse–human chimeric Fab fragment with specificity to human carcino embryonic antigen in *Escherichia coli*. *Appl. Microbiol. Biotechnol.*, **38**, 770.

41. Givol, D. (1991). The minimal antigen binding fragment of antibodies—Fv fragment. *Molec. Immunol.*, **28**, 1379.

42. Riechmann, L., Weill, M., and Cavanagh, J. (1992). Improving the antigen affinity of an antibody Fv-fragment by protein design. *J. Mol. Biol.*, **224**, 913.

43. Anthony, J., Near, R., Wong, S.-L., Iida, E., Ernst, E., Wittekind, M., Haber, E., and Ng, S.-C. (1992). Production of stable anti-digoxin Fv in *Escherichia coli*. *Molec. Immunol.*, **29**, 1237.

44. Bhat, T. N., Bentley G. A., Fischmann T. O., Boulot G., and Poljak, R. J. (1990). Small rearrangements in structures of Fv and Fab fragments of antibody D1.3 on antigen binding. *Nature*, **347**, 483.

45. Eigenbrot, C., Randal, M., Presta, L., Carter, P., and Kossiakoff, A. A. (1993). X-ray structures of the antigen-binding domains from 3 variants of humanized anti-p185[HER2] antibody 4D5 and comparison with molecular modeling. *J. Mol. Biol.*, **229**, 969.

46. Ford, C. F., Suominen, I. and Glatz, C. E. (1991). Fusion tails for the recovery and purification of recombinant proteins. *Prot. Exp. Pur.*, **2**, 95.

47. Hochuli, E., Bannwarth, W., Doebli, H., Gentz, R., and Stuber D. (1988). Genetic approach to facilitate purification of recombinant proteins with a novel metal chelate adsorbent. *Bio/tech.*, **6**, 1321.

48. Skerra, A., Pfitzinger, I., and Pluckthun, A. (1991). The functional expression of antibody Fv fragments in *Escherichia coli*: improved vectors and a generally applicable purification technique. *Bio/tech.*, **9**, 273.

49. Lindner, P., Guth, B., Wuelfing, C., Krebber, C., Steipe, B., Muller, F., and Pluck-thun, A. (1992). Purification of native proteins from the cytoplasm and periplasm of *Escherichia coli* using IMAC and histidine tails: a comparison of proteins and protocols. *Methods*, **4**, 41.

50. Pluckthun, A. and Pfitzinger, I. (1991). Comparison of Fv fragments of different phosphorylcholine binding antibodies expressed in *Escherichia coli*. *Ann. NY Acad. Sci.*, **646**, 115.

51. Glockshuber, R., Malia, M., Pfitzinger, I., and Pluckthun, A. (1990). A comparison of strategies to stabilise immunoglobulin Fv fragments. *Biochemistry*, **29**, 1362.

*52. Reiter, Y., Brinkmann, U., Webber, K. O., Jung, S.-H., Lee, B., and Pastan, I. (1994). Engineering interchain disulfide bonds into conserved framework regions of Fv frag-ments: improved biochemical characteristics of recombinant immunotoxins containing disulfide-stabilized Fv. *Protein Eng.*, **7**, 697.

*53. Bird, R. E., Hardman, K. D., Jacobson, J. W., Johnson, S., Kaufman, B. M., Lee, S.-M., Lee, T., Pope, S. H., Riordan, G. S., and Whitlow, M. (1988). Single-chain antigen-binding proteins. *Science*, **242**, 423.

54. Huston, J. S., Levinson, D., Mudgett-Hunter, M., Tai, M-S., Novotny, J., Margolies, M. N., Ridge, R. J., Bruccoleri, R. E., Haber, E., Crea, R., and Opperman, H. (1988). Protein engineering of antibody binding sites: recovery of specific activity in an anti-digoxin single-chain Fv analogue produced in *Escherichia coli*. *Proc. Natl Acad. Sci. USA*, **85**, 5879.

*55. Huston, J. S., Mudgett-Hunter, M., Tai, M.-S., McCartney, J., Warren, F., Haber, E., and Opperman, H. (1991). Protein engineering of single-chain Fv analogs and fusion proteins. In *Methods in enzymology* (ed. J. J. Langone), Vol. 203, p. 46. Academic Press, San Diego.

56. Pantoliano, M. W., Bird, R. E., Johnson, S., Asel, E. D., Dodd, S. W., Wood, J. F., and Hardman, K. D. (1991). Conformational stability, folding, and ligand-binding affinity of single-chain Fv immunoglobulin fragments expressed in *Escherichia coli*. *Biochemistry*, **30**, 10117.

57. Whitlow, M. and Filpula, D. (1991). Single-chain Fv proteins and their fusion proteins. *Methods*, **2**, 97.

*58. Knappik, A., Krebber, K., and Pluckthun, A. (1993). The effect of folding catalysts on the *in vivo* folding process of different antibody fragments expressed in *Escherichia coli*. *Bio/tech.*, **1**, 77.

59. Anand, N. N., Mandal, S., MacKenzie, C. R., Sandowska, J., Sigursjold, B., Young, N. M., Bundle, D. R., and Narang, S. A. (1991). Bacterial expression and secretion of various single-chain Fv genes encoding proteins specific for a *Salmonella* serotype B O-antigen. *J. Biol. Chem.*, **266**, 21874.

60. Zdanov, A., Li, Y., Bundle, D. R., Deng, S.-J., Mackenzie, C. R., Narang, S. A., Young, N. M., and Cygler, M. (1994). Structure of a single-chain antibody variable domain (Fv) fragment complexed with a carbohydrate antigen at 1.7 Å resolution. *Proc. Natl Acad. Sci. USA*, **91**, 6423.

61. Colcher, D., Bird, R., Roselli, M., Hardman, K. D., Johnson, S., Pope, S., Dodd, S. W., Pantoliano, M. W., Milenic, D. E., and Schlom, J. (1990). *In vivo* tumour targeting of a recombinant single-chain antigen-binding protein. *J. Natl Cancer Inst.*, **82**, 1191.

*62. Yokota, T., Milenic, D. E., Whitlow, M., and Schlom, J. (1992). Rapid tumor penetra-tion of a single-chain Fv and comparison with other immunoglobulin forms. *Cancer Res.*, **52**, 3402.

*63. Chaudhary, V. K., Queen, C., Junghans, R. P., Waldmann, T. A., FitzGerald, D. J., and Pastan, I. (1989). A recombinant immunotoxin consisting of two antibody variable domains fused to *Pseudomonas* exotoxin. *Nature*, **339**, 394.

64. Gibbs, R. A., Posner, B. A., Filpula, D. R., Dodd, S. W., Finkelman, M. A. J., Lee, T. K., Wroble, M., Whitlow, M., and Benkovic, S. J. (1991). Construction and characterisation of a single-chain catalytic antibody. *Proc. Natl Acad. Sci. USA*, **88**, 4001.

*65. McCafferty, J., Griffiths, A. D., Winter, G., and Chiswell, D. J. (1990). Phage antibodies: filamentous phage displaying antibody variable domains. *Nature*, **348**, 552.

66. Amit, A. G., Mariuzza, R. A., Phillips, S. E. V., and Poljak, R. J. (1986). Three-dimensional structure of an antigen–antibody complex at 2.8 Å resolution. *Science*, **233**, 747.

*67. Ward, E. S., Gussow, D., Griffiths, A. D., Jones, P. T., and Winter, G. (1989). Binding activities of a repertoire of single immunoglobulin variable domains secreted from *Escherichia coli*. *Nature*, **341**, 544.

68. Berry, M. J. and Davies, J. (1992). Use of antibody fragments in immunoaffinity chromatography: comparison of Fv fragments, V_H fragments and paralog peptides. *J. Chromatography*, **597**, 239.

69. Udaka, K., Chua, M.-M., Tong, L.-H., Karush, F., and Goodgal, S. H. (1990). Bacterial expression of immunoglobulin V_H proteins. *Mol. Immunol.*, **27**, 25.

70. Power, B. E., Ivancic, N., Harley, V. R., Webster, R. G., Kortt, A. A., Irving, R. A., and Hudson, P. J. (1992). High-level temperature-induced synthesis of an antibody V_H domain in *Escherichia coli* using the pelB secretion signal. *Gene*, **113**, 95.

*71. Hamers-Casterman, C., Atarhouch, T., Muyldermans, S., Robinson, G., Hamers, C., Bajyana Songa, E., Bendahman, N., and Hamers, R. (1993). Naturally occurring antibodies devoid of light chains. *Nature*, **363**, 446.

*72. Muyldermans, S., Atarhouch, T., Saldanha, J., Barbosa, J. A. R. G., and Hamers, R. (1994). Sequence and structure of V_H domain from naturally occurring camel heavy chain immunoglobulin lacking light chains. *Protein Eng.*, **7**, 1129.

73. Brinkmann, U., Gallo, M., Brinkmann, E., Kunwar, S., and Pastan, I. (1993). A recombinant immunotoxin that is active on prostate cancer cells and that is composed of the Fv region of monoclonal antibody PR1 and a truncated form of *Pseudomonas* exotoxin. *Proc. Natl Acad. Sci. USA*, **90**, 547.

74. Skerra, A. and Pluckthun, A. (1991). Secretion and *in vivo* folding of the Fab fragment of the antibody McPC603 in *Escherichia coli*: influence of disulphides and *cis*-prolines. *Protein Eng.*, **4**, 971.

75. Schein, C. H. (1989). Production of soluble recombinant proteins in bacteria. *Bio/tech.*, **7**, 1141.

76. Buchner, J. and Rudolph, R. (1991). Renaturation, purification and characterization of recombinant Fab-fragments produced in *Escherichia coli*. *Bio/tech.*, **9**, 157.

77. Takagi, H., Morinaga, Y., Tsuchiya, M., Ikemura, H., and Inouye, M. (1988). Control of folding of proteins secreted by a high expression secretion vector, pIN-III-ompA: 16-fold increase in production of active subtilisin E in *Escherichia coli*. *Bio/tech.*, **6**, 948.

*78. Schein, C. H. (1993). Must we live with inclusion bodies? In *Miami short reports* (ed. K. Brew, G. A. Petsko, F. Ahmad, H. Bialy, S. Black, A. Fernandez, R. E. Fenna, E. Y. C. Lee, and W. J. Whelan), p. 29. IRL Press, Oxford.

79. Buchner, J., Brinkmann, U., and Pastan, I. (1992). Renaturation of a single-chain immunotoxin facilitated by chaperones and protein disulfide isomerase. *Bio/tech.*, **10**, 682.

80. Duenas, M., Ayala, M., Vazquez, J., Ohlin, M., Borrebaeck, C. A. K., and Gavilondo, J. V. (1993). Immunoglobulin V-region sequence can determine expression of bacterial Fab antibody fragments. In *Miami short reports* (ed. K. Brew, G. A. Petsko, F. Ahmad, H. Bialy, S. Black, A. Fernandez, R. E. Fenna, E. Y. C. Lee, and W. J. Whelan), p. 87. IRL Press, Oxford.

81. Doi, R. H., Wong, S.-L., and Kawamura, F. (1986). Potential use of *Bacillus subtilis* for secretion and production of foreign proteins. *Trends Biotechnol.*, **4**, 232.

82. Pantoliano, M. W., Alexander, P., Dodd, S. W., Bryan, P., Rollence, M., Wood, J. F., and Fahnestock, S. (1989). The characterisation of single-chain antibodies synthesised in *Bacillus subtilis*. *J. Cell Biochem.*, **13A**, 91.

83. Wu, X.-C., Ng, S.-C., Near, R., and Wong, S.-L. (1993). Efficient production of a functional single-chain antidigoxin antibody via an engineered *Bacillus subtilis* expression-secretion system. *Bio/tech.*, **11**, 71.

84. Pschorr, J., Bieseler, B., and Fritz, H.-J. (1994). Production of the immunoglobulin variable domain REI$_v$ via a fusion protein synthesized and secreted by *Staphylococcus carnosus*. *Biol. Chem. Hoppe–Seyler*, **375**, 271.

85. Ueda, Y., Tsumoto, K., Watanabe, K., and Kumagai, I. (1993). Synthesis and expression of a DNA encoding the Fv domain of an anti-lysozyme monoclonal antibody, HyHEL10, in *Streptomyces lividans*. *Gene*, **129**, 129.

86. Wood, C. R., Boss, M. A., Kenten, J. H., Calvert, J. E., Roberts, N. A., and Emtage, J. S. (1985). The synthesis and *in vivo* assembly of functional antibodies in yeast. *Nature*, **314**, 446.

87. Horwitz, A. H., Chang, C. P., Better, M., Hellstrom, K. E., and Robinson, R. R. (1988). Secretion of functional antibody and Fab fragment from yeast cells. *Proc. Natl Acad. Sci. USA*, **85**, 8678.

88. Edqvist, J., Keranen, S., Penttila, M., Straby, K. B., and Knowles, J. K. C. (1991). Production of functional IgM Fab fragments by *Saccharomyces cerevisiae*. *J. Biotechnol.*, **20**, 291.

*89. Better, M. and Horwitz, A. H. (1993). *In vivo* expression of correctly folded antibody fragments from microorganisms. In *Protein folding: in vivo and in vitro* (ed. J. L. Cleland), pp. 201–217. ACS Symposium series.

90. Kotula, L. and Curtis, P. J. (1991). Evaluation of foreign gene codon optimization in yeast: expression of a mouse Ig kappa chain. *Bio/tech.*, **9**, 1386.

91. Davis, G. T., Bedzyk, W. D., Voss, E. W., and Jacobs, T. W. (1991). Single-chain antibody (SCA) encoding genes: one step construction and expression in eukaryotic cells. *Bio/tech.*, **9**, 165.

*92. Buckholz, R. G. and Gleeson, M. A. G. (1991). Yeast systems for the commercial production of heterologous proteins. *Bio/tech.*, **9**, 1067.

93. Dunn-Coleman, N. S., Bodie, E. A., Carter, G. L., and Armstrong, G. L. (1992). Stability of recombinant strains under fermentation conditions. In *Applied molecular genetics of filamentous fungi* (ed. J. R. Kinghorn and G. Turner), p. 152. Blackie Academic and Professional Press, London.

94. Gwynne, D. I. (1992). Foreign proteins. In *Applied molecular genetics of filamentous fungi* (ed. J. R. Kinghorn and G. Turner), p. 132. Blackie Academic and Professional Press, London.

95. Nyyssonen, E., Penttila, M., Harkki, A., Saloheimo, A., Knowles, J. K. C., and Keranen, S. (1993). Efficient production of antibody fragments by the filamentous fungus *Trichoderma reesei*. *Bio/tech.*, **11**, 591.

*96. Hiatt, A., Cafferkey, R., and Bowdish, K. (1989). Production of antibodies in transgenic plants. *Nature*, **342**, 76.

97. Hiatt, A. and Ma, J. K.-C. (1992). Monoclonal antibody engineering in plants. *FEBS Lett.*, **307**, 71.

98. Hein, M. B., Tang, Y., McLeod, D. A., Janda, K. D., and Hiatt, A. (1991). Evaluation of immunoglobulins from plant cells. *Biotechnol. Progr.*, **7**, 455.

99. Owen, M., Gandecha, A., Cockburn, B., and Whitelam, G. (1992). Synthesis of a functional anti-phytochrome single-chain Fv protein in transgenic tobacco. *Bio/tech.*, **10**, 790.

100. Firek, S., Draper, J., Owen, M. R. L., Gandecha, A., Cockburn, B., and Whitelam, G. C. (1993). Secretion of a functional single-chain Fv protein in transgenic tobacco plants and cell suspension cultures. *Plant Mol. Biol.*, **23**, 861.

101. Tavladoraki, P., Benvenuto, E., Trinca, S., Demartinis, D., Cattaneo, A., and Galeffi, P. (1993). Transgenic plants expressing a functional single-chain Fv-antibody are specifically protected from virus attack. *Nature*, **366**, 469.

102. Benvenuto, E., Ordas, R. J., Tavazza, R., Ancora, G., Biocca, S., Cattaneo, A., and Galeffi, P. (1991). 'Phytoantibodies': a general vector for the expression of immunoglobulin domains in transgenic plants. *Plant Mol. Biol.*, **17**, 865.

103. Hiatt, A. and Pinney, R. (1992). 'Plantibodies': expression of monoclonal antibodies in plants. In *Antibody engineering: a practical guide* (ed. C. A. K. Borrebaeck), pp. 159. W. H. Freeman and Co.

*104. Ma, J. K.-C., Lehner, T., Stabila, P., Fux, C. I., and Hiatt, A. (1994). Assembly of monoclonal antibodies with IgG1 and IgA heavy chain domains in transgenic tobacco plants. *Eur. J. Immunol.*, **24**, 131.

105. Sastry, L., Alting-Mees, M., Huse, W. D., Short, J. M., Sorge, J. A., Hay, B. N., Janda, K. D., Benkovic, S. J., and Lerner, R. A. (1989). Cloning of the immunological repertoire in *Escherichia coli* for generation of monoclonal catalytic antibodies: construction of a heavy chain variable region-specific cDNA library. *Proc. Natl Acad. Sci. USA*, **86**, 5728.

*106. Huse, W. D., Sastry, L., Iverson, S. A., Kang, A. S., Alting-Mees, M., Burton, D. R., Benkovic, S. J., and Lerner, R. A. (1989). Generation of a large combinatorial library of the immunoglobulin repertoire in phage lambda. *Science*, **246**, 1275.

107. Caton, A. J. and Koprowski, H. (1990). Influenza virus hemagglutinin-specific antibodies isolated from a combinatorial library are closely related to the immune response of the donor. *Proc. Natl Acad. Sci. USA*, **87**, 6450.

108. Mullinax, R. L., Gross, E. A., Amberg, J. R., Hay, B. N., Hogrefe, H. H., Kubitz, M. M., Greener, A., Alting-Mees, M., Ardourel, D., Short, J. M., Sorge, J. A., and Shopes, B. (1990). Identification of human antibody fragment clones specific for tetanus toxoid in a bacteriophage λ immunoexpression library. *Proc. Natl Acad. Sci. USA*, **87**, 8095.

109. Persson, M. A. A., Caothien, R. H., and Burton, D. R. (1991). Generation of diverse high-affinity human monoclonal antibodies by repertoire cloning. *Proc. Natl Acad. Sci. USA*, **88**, 2432.

*110. Clackson, T., Hoogenboom, H. R., Griffiths, A. D., and Winter, G. (1991). Making antibody fragments using phage display libraries. *Nature*, **352**, 624.

*111. Winter, G. and Milstein, C. (1991). Man-made antibodies. *Nature*, **349**, 293.

*112. Burton, D. R., Barbas III, C. F., Persson, M. A. A., Koenig, S., Chanock, R. M., and Lerner, R. A. (1991). A large array of human monoclonal antibodies to type 1 human immunodeficiency virus from combinatorial libraries of asymptomatic seropositive individuals. *Proc. Natl Acad. Sci. USA*, **88**, 10134.

*113. Gherardi, E. and Milstein, C. (1992). Original and artificial antibodies. *Nature*, **357**, 201.

*114. Burton, D. R. and Barbas, C. F., III (1992). Antibodies from libraries. *Nature*, **359**, 782.

115. Portolano, S., Seto, P., Chazenbalk, G. D., Nagayama, Y., McLachlan, S. M., and Rapoport, B. (1991). A human Fab fragment specific for thyroid peroxidase generated by cloning thyroid lymphocyte-derived immunoglobulin genes in a bacteriophage lambda library. *Biochem. Biophys. Res. Comm.*, **179**, 372.

*116. Barbas, C. F., III, Kang, A. S., Lerner, R. A., and Benkovic, S. J. (1991). Assembly of combinatorial antibody libraries on phage surfaces: the gene III site. *Proc. Natl Acad. Sci. USA*, **88**, 7978.

117. Garrard, C. J., Yang, M., O'Connell, M. P., Kelley, R. F., and Henner, D. J. (1991). Fab assembly and enrichment in a monovalent phage display system. *Bio/tech.*, **9**, 1373.

118. Chang, C. N., Landolfi, N. F., and Queen, C. (1991). Expression of antibody Fab domains on bacteriophage surfaces. Potential use for antibody selection. *J. Immunol.*, **147**, 3610.

119. Kang, A. S., Barbas, C. F., III, Janda, K. D., Benkovic, S. J., and Lerner, R. A. (1991). Linkage of recognition and replication functions by assembling combinatorial antibody Fab libraries along phage surfaces. *Proc. Natl Acad. Sci. USA*, **88**, 4363.

120. Breitling, F., Dubel, S., Seehaus, T., Klewinghaus, I., and Little, M. (1991). A surface expression vector for antibody screening. *Gene*, **104**, 147.

*121. Hoogenboom, H. R., Griffiths, A. D., Johnson, K. S., Chiswell, D. J., Hudson, P., and Winter, G. (1991). Multi-subunit proteins on the surface of filamentous phage: methodologies for displaying antibody (Fab) heavy and light chains. *Nucleic Acids Res.*, **19**, 4133.

122. Huse, W. D., Stinchcombe, T. J., Glaser, S. M., Starr, L., MacLean, M., Hellstrom, K. E., Hellstrom, I., and Yelton, D. E. (1992). Application of a filamentous phage pVIII fusion protein system suitable for efficient production, screening and mutagenesis of Fab antibody fragments. *J. Immunol.*, **149**, 3914.

*123. Parmley, S. F. and Smith, G. P. (1988). Antibody-selectable filamentous fd phage vectors: affinity purification of target genes. *Gene*, **73**, 305.

124. Soderlind, E., Lagerqvist, A. C. S., Duenas, M., Malmborg, A.-C., Ayala, M., Danielsson, L., and Borrebaeck, C. A. K. (1993). Chaperonin assisted phage display of antibody fragments on filamentous bacteriophages. *Bio/tech.*, **11**, 503.

*125. Waterhouse, P., Griffiths, A. D., Johnson, K. S., and Winter, G. (1993). Combinatorial infection and *in vivo* recombination: a strategy for making large phage antibody repertoires. *Nucleic Acids Res.*, **21**, 2265.

*126. Griffiths, A. D., Williams, S. C., Hartley, O., Tomlinson, I. M., Waterhouse, P., Crosby, W. L., Kontermann, E., Jones, P. T., Low, N. M., Allison, T. J., Prospero, T. D., Hoogenboom, H. R., Nissim, A., Cox, J. P. L., Harrison, J. L., Zaccolo, M., Gherardi, E., and Winter, G. (1994). Isolation of high affinity human antibodies directly from large synthetic repertoires. *EMBO J.*, **13**, 3245.

*127. Marks, J. D., Hoogenboom, H. R., Bonnert, T. P., McCafferty, J., Griffiths, A. D., and Winter, G. (1991). By-passing immunization: human antibodies from V-gene libraries displayed on phage. *J. Mol. Biol.*, **222**, 581.

*128. Griffiths, A. D., Malmqvist, M., Marks, J. D., Bye, J. M., Embleton, M. J., McCafferty,

J., Baier, M., Holliger, K. P., Gorrick, B. D., Hughes-Jones, N. C., Hoogenboom, H. R., and Winter, G. (1993). Human anti-self antibodies with high specificity from phage display libraries. *EMBO J.*, **12**, 725.

129. Gram, H., Marconi, L.-A., Barbas C. F. III, Collett, T. A., Lerner, R. A., and Kang, A. S. (1992). *In vitro* selection and affinity maturation of antibodies from a naive combinatorial immunoglobulin library. *Proc. Natl Acad. Sci. USA*, **89**, 3576.

*130. Hoogenboom, H. R. and Winter, G. (1992). By-passing immunization—Human antibodies from synthetic repertoires of germline VH gene segments rearranged *in vitro*. *J. Mol. Biol.*, **227**, 381.

131. Nissim, A., Hoogenboom, H. R., Tomlinson, I. M., Flynn, G., Midgley, C., Lane, D., and Winter, G. (1994). Antibody fragments from a 'single pot' phage display library as immunochemical reagents. *EMBO J.*, **13**, 692.

132. Schaaper, R. M. (1988). Mechanisms of mutagenesis in the *Escherichia coli* mutator mutD5: role of DNA mismatch repair. *Proc. Natl Acad. Sci. USA*, **85**, 8126.

*133. Hawkins, R. E., Russell, S. J., and Winter, G. (1992). Selection of phage antibodies by binding affinity. Mimicking affinity maturation. *J. Mol. Biol.*, **226**, 889.

134. Marks, J. D., Griffiths, A. D., Malmqvist, M., Clackson, T. P., Bye, J. M., and Winter, G. (1992). Bypassing immunization: building high-affinity human antibodies by chain shuffling. *Bio/tech.*, **10**, 779.

*135. Barbas, C. F., III, Bain, J. D., Hoekstra, D. M., and Lerner, R. A. (1992). Semisynthetic combinatorial antibody libraries: a chemical solution to the diversity problem. *Proc. Natl Acad. Sci. USA*, **89**, 4457.

136. Chaudhary, V. K., Batra, J. K., Gallo, M. G., Willingham, M. C., FitzGerald, D. J., and Pastan, I. (1990). A rapid method of cloning functional variable region antibody genes in *Escherichia coli* as single-chain immunotoxins. *Proc. Natl Acad. Sci. USA*, **87**, 1066.

137. Hasemann, C. A. and Capra, J. D. (1991). Mutational analysis of arsonate binding by a CRI_{A+} antibody. *J. Biol. Chem.*, **266**, 7626.

*138. Foote, J. and Winter, G. (1992). Antibody framework residues affecting the conformation of the hypervariable loops. *J. Mol. Biol.*, **224**, 487.

139. Roberts, S., Cheetham, J. C., and Rees, A. R. (1987). Generation of an antibody with enhanced affinity and specificity for its antigen by protein engineering. *Nature*, **328**, 731.

140. Xiang, J. and Chen, Z. (1992). Genetic engineering of high affinity anti-human colorectal tumour mouse/human chimeric antibody. *Immunology*, **75**, 209.

141. Sharon, J. (1990). Structural correlates of high antibody affinity: three engineered amino acid substitutions can increase the affinity of an anti-*p*-azophenylarsonate antibody 200-fold. *Proc. Natl Acad. Sci. USA*, **87**, 4814.

142. Chen, C., Roberts, V. A., and Rittenberg, M. B. (1992). Generation and analysis of random point mutations in an antibody CDR2 sequence: many mutated antibodies lose their ability to bind antigen. *J. Exp. Med.*, **176**, 855.

143. Sollazzo, M., Castiglia, D., Billetta, R., Tramontano, A., and Zanetti, M. (1990). Structural definition by antibody engineering of an idiotypic determinant. *Protein Eng.*, **3**, 531.

144. Dul, J. L. and Argon, Y. (1990). A single amino acid substitution in the variable region of the light chain specifically blocks immunoglobulin secretion. *Proc. Natl Acad. Sci. USA*, **87**, 8135.

*145. Morrison, S. L. (1992). *In vitro* antibodies: Strategies for production and application. *Annu. Rev. Immunol.*, **10**, 239.

*146. Shin, S.-U., Wright, A., Bonagura, V., and Morrison, S. L. (1992). Genetically-engineered antibodies: tools for the study of diverse properties of the antibody molecule. *Immunol. Rev.*, **130**, 87.

147. Kalsi, J. K. and Isenberg, D. A. (1992). Immortalisation of human antibody producing cells. *Autoimmunity*, **13**, 249.

148. Morrison, S. L., Johnson, M. J., Herzenberg, L. A., and Oi, V. T. (1984). Chimaeric human antibody molecules: mouse antigen-binding domains with human constant region domains. *Proc. Natl Acad. Sci. USA*, **81**, 6851.

149. Boulianne, G. L., Hozumi, N., and Shulman, M. J. (1984). Production of functional chimaeric mouse/human antibody. *Nature*, **312**, 643.

150. Neuberger, M. S., Williams, G. T., and Fox, R. O. (1984). Recombinant antibodies possessing novel effector functions. *Nature*, **312**, 604.

151. Begent, R. J., Lederman, J. A., Bagshawe, K. D., Green, A. J., Kelley, A. M. B., Lane, D., Secher, D. S., Dewji, M. R., and Baker, T. S. (1990). Chimeric B72.3 antibody for repeated radioimmunotherapy of colorectal carcinoma. *Antibody Immunoconjug. Radiopharm.*, **3**, 86.

152. Meredith, R. F., Khazaeli, M. B., Plott, W. E., Brezovick, I. A., Russell, C. D., Wheeler, R. H., Spencer, S. A., and Lobuglio, A. F. (1992). Comparison of two mouse/human chimeric antibodies in patients with metastatic colon cancer. *Antibody Immunoconjug. Radiopharm.*, **5**, 75.

*153. Bruggeman, M., Winter, G., Waldmann, H., and Neuberger, M. S. (1989). The immunogenicity of chimeric antibodies. *J. Exp. Med.*, **170**, 2153.

154. Jones, P. T., Dear, P. H., Foote, J., Neuberger, M. S., and Winter, G. (1986). Replacing the complementarity-determining regions in a human antibody with those from a mouse. *Nature*, **321**, 522.

*155. Verhoeyen, M., Milstein, C., and Winter, G. (1988). Reshaping human antibodies: grafting an antilysozyme activity. *Science*, **239**, 1534.

156. Queen, C., Schneider, W. P., Selick, H. E., Payne, P. W., Landolfi, N. F., Duncan, J. F., Avdalovic, N. M., Levitt, M., Junghans, R. P., and Waldmann, T. A. (1989). A humanized antibody that binds to the interleukin 2 receptor. *Proc. Natl Acad. Sci. USA*, **86**, 10029.

157. Nakatani, T., Lone, Y.-C., Yamakawa, J., Kanaoka, M., Gomi, H., Wijdenes, J., and Noguchi, H. (1994). Humanization of mouse anti-human IL-2 receptor antibody B-B10. *Protein Eng.*, **7**, 435.

158. Sato, K., Tsuchiya, M., Saldanha, J., Koishihara, Y., Ohsugi, Y., Kishimoto, T., and Bendig, M. M. (1994). Humanization of a mouse anti-human interleukin-6 receptor antibody comparing 2 methods for selecting human framework regions. *Mol. Immunol.*, **31**, 371.

159. Routledge, E. G., Lloyd, I., Gorman, S. D., Clark, M., and Waldmann, H. (1991). A humanized monovalent CD3 antibody which can activate homologous complement. *Eur. J. Immunol.*, **21**, 2717.

160. Bolt, S., Routledge, E., Lloyd, I., Chatenoud, L., Pope, H., Gorman, S. D., Clark, M., and Waldmann, H. (1993). The generation of a humanized, non-mitogenic CD3 monoclonal-antibody which retains *in vitro* immunosuppressive properties. *Eur. J. Immunol.*, **23**, 403.

161. Alegre, M.-L., Collins, A. M., Pulito, V. L., Brosius, R. A., Olson, W. C., Zivin, R. A., Knowles, R., Thistlethwaite, J. R., Jolliffe, L. K., and Bluestone, J. A. (1992).

Effect of a single amino acid mutation on the activating and immunosuppressive properties of a 'humanized' OKT3 monoclonal antibody. *J. Immunol.*, **148**, 3461.

162. Woodle, E. S., Thistlethwaite, J. R., Jolliffe, L. K., Zivin, R. A., Collins, A., Adair, J. R., Bodmer, M., Athwal, D., Alegre, M.-L., and Bluestone, J. A. (1992). Humanized OKT3 antibodies: successful transfer of immune modulating properties and idiotype expression. *J. Immunol.*, **148**, 2756.

163. Shalaby, M. R., Shepard, H. M., Presta, L., Rodrigues, M. L., Beverley, P. C. L., Feldman, M., and Carter, P. (1992). Development of humanized bispecific antibodies reactive with cytotoxic lymphocytes and tumor cells overexpressing the HER2 proto-oncogene. *J. Exp. Med.*, **175**, 217.

164. Gorman, S. D., Clark, M. R., Routledge, E. G., Cobbold, S. P., and Waldmann, H. (1991). Reshaping a therapeutic CD4 antibody. *Proc. Natl Acad. Sci. USA*, **88**, 4181.

165. Studnicka, G. M., Soares, S., Better, M., Williams, R. E., Nadell, R., and Horwitz, A. H. (1994). Human-engineered monoclonal antibodies retain full specific binding activity by preserving non-CDR complementarity-modulating residues. *Protein Eng.*, **7**, 805.

166. Daugherty, B. L., DeMartino, J. A., Law, M.-F., Kawka, D. W., Singer, I. I., and Mark, G. E. (1991). Polymerase chain reaction facilitates the cloning, CDR-grafting and rapid expression of a murine monoclonal antibody directed against the CD18 component of leukocyte integrins. *Nucleic Acids Res.*, **19**, 2471.

167. Singer, I. I., Kawka, D. W., DeMartino, J. A., Daugherty, B. L., Elliston, K. O., Alves, K., Bush, B. L., Cameron, P. M., Cuca, G. C., Davies, P., Forrest, M. J., Kazazis, D. M., Law, M. F., Lenny, A. B., MacIntyre, D. E., Meurer, R., Padlan, E. A., Pandya, S., Schmidt, J. A., Seamans, T. C., Scott, S., Silberklang, M., Williamson, A. R., and Mark, G. E. (1993). Optimal humanization of 1B4, an anti-CD18 murine monoclonal antibody, is achieved by correct choice of human V-region framework sequences. *J. Immunol.*, **150**, 2844.

*168. Roguska, M. A., Pedersen, J. T., Keddy, C. A., Henry, A. H., Searle, S. J., Lambert, J. M., Goldmacher, V. S., Blattler, W. A., Rees, A. R., and Guild, B. C. (1994). Humanization of murine monoclonal antibodies through variable domain resurfacing. *Proc. Natl Acad. Sci. USA*, **91**, 969.

169. Co, M. S., Avdalovic, N. M., Caron, P. C., Avdalovic, M. V., Scheinberg, D. A., and Queen, C. (1992). Chimeric and humanized antibodies with specificity for the CD33 antigen. *J. Immunol.*, **148**, 1149.

170. Carter, P., Presta, L., Gorman, C. M., Ridgway, J. B. B., Henner, D., Wong, W. L. T., Rowland, A. M., Kotts, C., Carver, M. E., and Shepard, H. M. (1992). Humanization of an anti-P185^{HER2} antibody for human cancer-therapy. *Proc. Natl Acad. Sci. USA*, **89**, 4285.

171. Kettleborough, C. A., Saldanha, J., Heath, V. J., Morrison, C. J., and Bendig, M. M. (1991). Humanization of a mouse monoclonal antibody by CDR-grafting: the importance of framework residues on loop conformation. *Protein Eng.*, **4**, 773.

172. Shearman, C. W., Pollock, D., White, G., Hehir, K., Moore, G. P., Kanzy, E. J., and Kurrle, R. (1991). Construction, expression and characterization of humanized antibodies directed against the human α/β T-cell receptor. *J. Immunol.*, **147**, 4366.

*173. Gussow, D. and Seemann, G. (1991). Humanization of monoclonal antibodies. In *Methods in enzymology* (ed. J. J. Langone), Vol. 203, p. 99. Academic Press, San Diego.

174. Verhoeyen, M. E., Saunders, J. A., Price, M. R., Marugg, J. D., Briggs, S., Broderick,

E. L., Eida, S. J., Mooren, A. T. A., and Badley, R. A. (1993). Construction of a reshaped HMFG1 antibody and comparison of its fine specificity with that of the parent mouse antibody. *Immunology*, **78**, 364.

175. Co, M. S., Deschamps, M., Whitley, R. J., and Queen, C. (1991). Humanized antibodies for antiviral therapy. *Proc. Natl Acad. Sci. USA*, **88**, 2869.

176. Maeda, H., Matsushita, S., Eda, Y., Kimachi, K., Tokiyoshi, S., and Bendig, M. M. (1991). Construction of reshaped human antibodies with HIV-neutralizing activity. *Hum. Antibod. Hybridomas*, **2**, 124.

177. Tempest, P. R., Bremner, P., Lambert, M., Taylor, G., Furze, J. M., Carr, F. J., and Harris, W. J. (1991). Reshaping a human monoclonal antibody to inhibit human respiratory syncytial virus infection *in vivo*. *Bio/tech.*, **9**, 266.

178. Presta, L. G., Lahr, S. J., Shields, R. L., Porter, J. P., Gorman, C. M., Fendly, B. M., and Jardieu, P. M. (1993). Humanization of an antibody directed against IgE. *J. Immunol.*, **151**, 2623.

179. Kolbinger, F., Saldanha, J., Hardman, N., and Bendig, M. M. (1993). Humanization of a mouse anti-human IgE antibody: a potential therapeutic for IgE-mediated allergies. *Protein Eng.*, **6**, 971.

180. Corti, A., Barbanti, E., Tempest, P. R., Carr, F. J., and Marcucci, F. (1994). Idiotope determining regions of a mouse monoclonal antibody and its humanized versions — Identification of framework residues that affect idiotype expression. *J. Mol. Biol.*, **235**, 53.

181. Hale, G., Dyer, M. J. S., Clark, M. R, Phillips, J. M., Marcus, R., Riechmann, L., Winter, G., and Waldmann, H. (1988). Remission induction in non-Hodgkin lymphoma with reshaped human monoclonal antibody CAMPATH-1H. *The Lancet*, **2**, 1394.

182. Hird, V., Verhoeyen M., Badley, R. A., Price, D., Snook, D., Kosmas, C., Gooden, C., Bamias, A., Meares, C., Lavender, J. P., and Epenetos, A. A. (1991). Tumour localisation with a radioactively labelled reshaped human monoclonal antibody. *Br. J. Cancer*, **64**, 911.

183. Hsiao, K. C., Bajorath, J., and Harris, L. J. (1994). Humanization of 60.3, an anti-CD18 antibody—importance of the L2 loop. *Protein Eng.*, **7**, 815.

184. Wu, T. T. and Kabat, E. A. (1992). Possible use of similar framework region amino acid sequences between human and mouse immunoglobulins for humanising mouse antibodies. *Mol. Immunol.*, **29**, 1141.

185. Padlan, E. A. (1991). A possible procedure for reducing the immunogenicity of antibody variable domains while preserving their ligand binding properties. *Mol. Immunol.*, **28**, 489.

186. Caron, P. C., Laird, W., Sung Co, M., Avdalovic, N. M., Queen, C., and Scheinberg, D. A. (1992). Engineered humanized dimeric forms of IgG are more effective antibodies. *J. Exp. Med.*, **176**, 1191.

*187. Jespers, L. S., Roberts, A., Mahler, S. M., Winter, G., and Hoogenboom, H. R. (1994). Guiding the selection of human antibodies from phage display repertories to a single epitope of an antigen. *Bio/Tech.*, **12**, 899.

*188. Duchosal, M. A., Eming, S. A., Fischer, P., Leturcq, D., Barbas, C. F., III, McConahey, P. J., Caothien, R. H., Thornton, G. B., Dixon, F. J., and Burton, D. R. (1992). Immunization of hu–PBL–SCID mice and the rescue of human monoclonal Fab fragments through combinatorial libraries. *Nature*, **355**, 258.

*189. Lonberg, N., Taylor, L. D., Harding, F. A., Trounstine, M., Higgins, K. M., Schramm, S. R., Kuo, C.-C., Mashayekh, R., Wymore, K., McCabe, J. G., Munoz-O'Regan, D.,

O'Donnell, S. L., Lapachet, E. S. G., Bengoechea, T., Fishwild, D. M., Carmack, C. E., Kay, R. M., and Huszar, D. (1994). Antigen-specific human antibodies from mice comprising four distinct genetic modifications. *Nature*, **368**, 856.

190. Milstein, C. and Cuello, A. C. (1983). Hybrid hybridomas and their use in immuno-histochemistry. *Nature*, **305**, 537.

191. Lenz, H. and Weidle, U. H. (1990). Expression of heterobispecific antibodies by genes transfected into producer hybridoma cells. *Gene*, **87**, 213.

192. Nisonoff, A. and Mandy, W. J. (1962). Quantitative estimation of the hybridization of the rabbit antibodies. *Nature*, **194**, 355.

*193. Traunecker, A., Lanzavecchia, A., and Karjalainen, K. (1991). Bispecific single-chain molecules (Janusins) target cytotoxic lymphocytes on HIV infected cells. *EMBO J.*, **10**, 3655.

194. Kostelny, S. A., Cole, M. S., and Tso, J. Y. (1992). Formation of a bispecific antibody by the use of leucine zippers. *J. Immunol.*, **148**, 1547.

*195. Pack, P. and Pluckthun, A. (1992). Miniantibodies: use of amphipathic helices to produce functional, flexibly linked dimeric Fv fragments with high avidity in *Escherichia coli*. *Biochem.*, **31**, 1579.

*196. Pack, P., Kujau, M., Schroeckh, V., Knüpfer, U., Wenderoth, R., Riesenberg, D., and Plückthun, A. (1993). Improved bivalent miniantibodies, with identical avidity as whole antibodies, produced by high cell density fermentation of *Escherichia coli*. *Bio/Tech.*, **11**, 1271.

*197. Holliger, P., Prospero, T., and Winter, G. (1993). 'Diabodies': Small bivalent and bispecific antibody fragments. *Proc. Natl Acad. Sci. USA*, **90**, 6444.

*198. Whitlow, M., Filpula, D., Rollence, M. L., Feng, S.-L., and Wood, J. F. (1994). Multivalent Fvs: characterization of single-chain Fv oligomers and preparation of a bispecific Fv. *Protein Eng.*, **7**, 1017.

199. Mallender, W. D. and Voss, E. W., Jr. (1994). Construction, expression, and activity of a bivalent bispecific single-chain antibody. *J. Biol. Chem.*, **269**, 199.

200. Gruber, M., Schodin, B. A., Wilson, E. R., and Kranz, D. M. (1994). Efficient tumor-cell lysis mediated by a bispecific single-chain antibody expressed in *E. coli*. *J. Immunol.*, **152**, 5368.

201. Ito, W. and Kurosawa, Y. (1993). Development of an artificial antibody system with multiple valency using an Fv fragment fused to a fragment of protein A. *J. Biol. Chem.*, **268**, 20668.

202. Ducancel, F., Gillet, D., Carrier, A., Lajeunesse, E., Ménez, A., and Boulain, J.-C. (1993). Recombinant colorimetric antibodies construction and characterization of a bifunctional F(ab)$_2$/alkaline phosphatase conjugate produced in *Escherichia coli*. *Bio/Tech.*, **11**, 601.

203. Williams, G. T. and Neuberger, M. S. (1986). Production of antibody-tagged enzymes by myeloma cells: application to DNA polymerase I Klenow fragment. *Gene*, **43**, 319.

204. Das, C., Kulkarni, P. V., Constantinescu, A., Antich, P., Blattner, F. R., and Tucker, P. W. (1992). Recombinant antibody-metallothionein: design and evaluation for radioimmunoimaging. *Proc. Natl Acad. Sci. USA*, **89**, 9749.

205. Sawyer, J. R., Tucker, P. W., and Blattner, F. R. (1992). Metal-binding chimeric antibodies expressed in *Escherichia coli*. *Proc. Natl Acad. Sci. USA*, **89**, 9754.

206. Shin, S.-U. and Morrison, S. L. (1990). Expression and characterization of an antibody binding specificity joined to insulin-like growth factor 1: potential applications for cellular targeting. *Proc. Natl Acad. Sci. USA*, **87**, 5322.

207. Fell, H. P., Gayle, M. A., Grosmaire, L., and Ledbetter, J. A. (1991). Genetic construction and characterization of a fusion protein consisting of a chimeric F(ab') with specificity for carcinomas and human IL-2. *J. Immunol.*, **146**, 2446.

208. Savage, P., So, A., Spooner, R. A., and Epenetos, A. A. (1993). A recombinant single chain antibody interleukin-2 fusion protein. *Br. J. Cancer*, **67**, 304.

209. Eshhar, Z., Waks, T., Gross, G., and Schindler, D. G. (1993). Specific activation and targeting of cytotoxic lymphocytes through chimeric single chains consisting of antibody-binding domains and the γ or ζ subunits of the immunoglobulin and T-cell receptors. *Proc. Natl Acad. Sci. USA*, **90**, 720.

210. Bregegere, F., Schwartz, J., and Bedouelle, H. (1994). Bifunctional hybrids between the variable domains of an immunoglobulin and the maltose-binding protein of *Escherichia coli*. Production, purification and antigen-binding. *Protein Eng.*, **7**, 271.

211. Tai, M.-S., Mudgett-Hunter, M., Levinson, D., Wu, G.-M., Haber, E., Opperman, H., and Huston, J. S. (1990). A bifunctional fusion protein containing Fc-binding fragment B of Staphylococcal protein A amino terminal to antidigoxin single-chain Fv. *Biochemistry*, **29**, 8024.

*212. Capon, D. J., Chamow, S. M., Mordenti, S. A., Marsters, S. A., Gregory, T., Mitsuya, H., Byrn, R. A., Lucas, F., Wurm, F. M., Groopman, J. E., Broder, S., and Smith, D. H. (1989). Designing CD4 immunoadhesins for AIDS therapy. *Nature*, **337**, 525.

213. Traunecker, A., Schneider, J., Kiefer, K., and Karjalainen, K. (1989). Highly efficient neutralization of HIV with recombinant CD4-immunoglobulin molecules. *Nature*, **339**, 68.

214. Byrn, R. A., Mordenti, J., Lucas, C., Smith, D., Marsters, S., Johnson, J. S., Cossum, P., Chamow, S. M., Wurm, F. M., Gregory, T., Groopman, J. E., and Capon, D. J. (1990). Biological properties of a CD4 immunoadhesin. *Nature*, **344**, 667.

215. Landolfi, N. S. (1991). A chimeric IL-2/Ig molecule possesses the functional activity of both proteins. *J. Immunol.*, **146**, 915.

216. Sollazzo, M., Billetta, R., and Zanetti, M. (1990). Expression of an exogenous peptide epitope genetically engineered in the variable domain of an immunoglobulin: implications for antibody and peptide folding. *Protein Eng.*, **4**, 215.

217. Billetta, R., Hollingdale, R. M., and Zanetti, M. (1991). Immunogenicity of an engineered internal image antibody. *Proc. Natl Acad. Sci. USA*, **88**, 4713.

*218. Zanetti, M., Rossi, F., Lanza, P., Filaci, G., Lee, R. H., and Billetta, R. (1992). Theoretical and practical aspects of antigenized antibodies. *Immunol. Rev.*, **130**, 125.

219. Pessi, A., Bianchi, E., Crameri, A., Venturini, S., Tramontano, A., and Sollazzo, M. (1993). A designed metal-binding protein with a novel fold. *Nature*, **362**, 367.

*220. Bianchi, E., Venturini, S., Pessi, A., Tramontano, A., and Sollazzo, M. (1994). High level expression and rational mutagenesis of a designed protein, the Minibody. From an insoluble to a soluble molecule. *J. Mol. Biol.*, **236**, 649.

221. Sollazzo, M. (1993). Genetic strategies for creating and refining molecular diversity in polypeptides. *Chem. Today*, **11**, 34.

222. Williams, W. V., Moss, D. A., Kieber-Emmons, T., Cohen, J. A., Myers, J. N., Weiner, D. B., and Greene, M. I. (1989). Development of biologically active peptides based on antibody structure. *Proc. Natl Acad. Sci. USA*, **86**, 5537.

223. Taub, R., Gould, R. J., Ciccarone, T. M., Hoxie, J., Friedman, P. A., Shattil, S. J., and Garsky, V. M. (1989). A monoclonal antibody against the platelet fibrinogen receptor contains a sequence that mimics a receptor recognition domain in fibrinogen. *J. Biol. Chem.*, **264**, 259.

224. Williams, W. V., Kieber-Emmons, T., Von Feldt, J., Greene, M. I., and Weiner, D. B. (1991). Design of bioactive peptides based on antibody hypervariable region structures. *J. Biol. Chem.*, **266**, 5182.

225. Saragovi, H. U., Fitzpatrick, D., Raktabuhr, A., Nakanishi, H., Kahn, M., and Greene, M. (1991). Design and synthesis of a mimetic of an antibody complementarity region. *Science*, **253**, 792.

226. Chen, S., Chrusciel, R. A., Nakanishi, H., Raktabutr, A., Sato, A., Weiner, D. B., Hoxie, J., Saragovi, H. U., Greene, M. I., and Kahn, M. (1992). Design and synthesis of a β-turn mimetic that inhibits HIV gp120 binding and infection of human lymphocytes. *Proc. Natl Acad. Sci. USA*, **89**, 5872.

*227. Saragovi, H. U., Greene, M. I., Chrusciel, R. A., and Kahn, M. (1992). Loops and secondary structure mimetics: development and applications in basic science and rational drug design. *Bio/tech.*, **10**, 773.

228. Pollack, S. J., Jacobs, J. W., and Schultz, P. G. (1986). Selective chemical catalysis by an antibody. *Science*, **234**, 1570.

229. Tramontano, A., Janda, K. D., and Lerner, R. A. (1986). Catalytic antibodies. *Science*, **234**, 1566.

*230. Benkovic, S. J. (1992). Catalytic antibodies. *Annu. Rev. Biochem.*, **61**, 29.

231. Jacobsen, J. R. and Schultz, P. G. (1994). Antibody catalysis of peptide bond formation. *Proc. Natl Acad. Sci. USA*, **91**, 5888.

232. Hirschmann, R., Smith, A. B., Taylor, C. M., Benkovic, P. A., Taylor, S. D., Yager, K. M., Sprengeler, P. A., and Benkovic, S. J. (1994). Peptide-synthesis catalyzed by an antibody containing a binding site for variable amino-acids. *Science*, **265**, 234.

233. Janda, K. D., Shevlin, C. G., and Lerner, R. A. (1993). Antibody catalysis of a disfavoured chemical transformation. *Science*, **259**, 490.

234. Li, T. Y., Janda, K. D., Ashley, J. A., and Lerner, R. A. (1994). Antibody-catalyzed cationic cyclization. *Science*, **264**, 1289.

235. Baldwin, E. P. and Schultz, P. G. (1989). Generation of a catalytic antibody by site-directed mutagenesis. *Science*, **245**, 1104.

236. Iverson, B. L., Iverson, S. A., Roberts, V. A., Getzoff, E. D., Tainer, J. A., Benkovic, S. J., and Lerner, R. A. (1990). Metalloantibodies. *Science*, **249**, 659.

237. Chen, Y.-C. J., Danon, T., Sastry, L., Mubaraki, M., Janda, K. D., and Lerner, R. A. (1993). Catalytic antibodies from combinatorial libraries. *J. Am. Chem. Soc.*, **115**, 357.

238. Tawfik, D. S., Green, B. S., Chap, R., Sela, M., and Eshhar, Z. (1993). catELISA: a facile general route to catalytic antibodies. *Proc. Natl Acad. Sci. USA*, **90**, 373.

239. Lesley, S. A., Patten, P. A., and Schultz, P. G. (1993). A genetic approach to the generation of antibodies with enhanced catalytic activities. *Proc. Natl Acad. Sci. USA*, **90**, 1160.

*240. Fuchs, P., Brietling, F., Dubel, S., Seehaus, T., and Little, M. (1991). Targeting recombinant antibodies to the surface of *Escherichia coli*: fusion to peptidoglycan associated lipoprotein. *Bio/tech.*, **9**, 1369.

241. Russell, S. J., Hawkins, R. E., and Winter, G. (1993). Retroviral vectors displaying functional antibody fragments. *Nucleic Acids Res.*, **21**, 1081.

8 | The complement system

KENNETH B. M. REID

1. Introduction

Complement can be activated by two main routes, the classical and alternative pathways (Fig. 1). The major complement component C3 (present at 1.3 mg/ml plasma) is common to both pathways. C3 plus the other 12 glycoproteins shown in Fig. 1 and Table 1 constitute the 13 components of the pathways. A recently described third route, the lectin pathway, bypasses the C1 stage of the classical pathway but utilizes all the other components of the classical pathway (Fig. 1). Control of the activated components is mediated partly via at least eight control proteins present in plasma (Table 1) and partly by a variety of membrane-bound control proteins and receptors (Table 2). These membrane proteins bind activated components or fragments of the activated components generated by further limited proteolysis. Many of the activation and control steps in the system involve the splitting of only one or two specific peptide bonds, and therefore it is useful to identify, at an early stage, which components and control proteins are enzymes. Four of the components of the classical pathway and lectin pathway are pro-enzymes: subcomponents C1r and C1s of the C1 complex, the MASP enzyme of the MBP–MASP complex, and component C2. Activation of these four serine proteases leads to the formation of the enzyme complexes which split C3 and C5 (Fig. 1). Two of the components of the alternative pathway are also serine proteases; factor B which circulates in its pro-enzyme form and factor D which has, to date, been found only in its activated form in plasma. When factor B is complexed to the C3b portion of activated C3, or $C3(H_2O)$ (see Sections 3.1 and 5.1), it can be split and activated by factor D to yield eventually the enzyme complexes of the alternative pathway which split C3 and C5 in exactly the same fashion as the enzyme complexes generated by the classical pathway. After the splitting of C5, by the complexes from either pathway, no other proteolytic splitting is considered to take place and the lytic C5b–9 complex is generated by a self-assembly mechanism. Two other proteases, factor I and anaphylatoxin inactivator, play crucial roles in the control of activated products of C3, C4, and C5, as outlined in Sections 5 and 6. With the exception of the anaphylatoxin inactivator, which has carboxypeptidase-B-like activity, all the enzymes associated with the system are serine proteases.

The classical pathway of human complement can be activated by the interaction of the C1 complex with immune complexes or aggregates containing IgG1, IgG2,

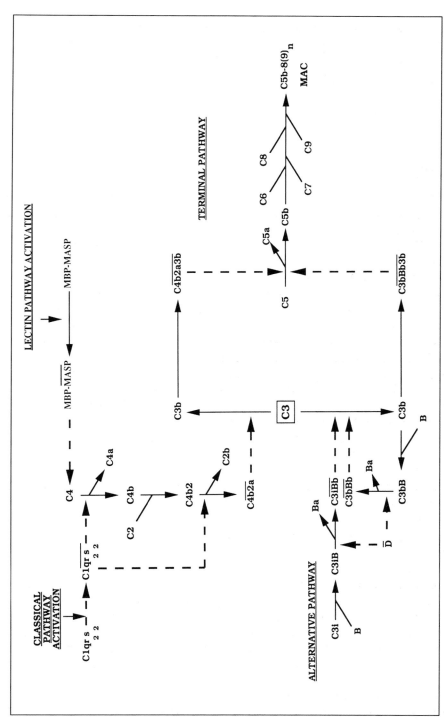

Fig. 1 The major activation steps in the classical and alternative pathways of the serum complement system are shown. The classical and lectin pathways are shown in the upper part of the diagram and the amplification loop of the alternative pathway is shown in the lower part. Complexes containing enzymatic activity are denoted by overbars. Dashed lines represent enzymatic cleavages. C3i is a 'C3b-like' form of C3. The system is down-regulated, in solution, by: C1-Inh (for C1r and C1s); C4bp and factor I (for C4b); factors H and I (for C3b); anaphylatoxin inactivator (for C3a, C4a, C5a); S-protein (for C5b—9 complex). The system is also down-regulated at cell-surfaces by: CR1 and factor I (for C3b and C4b); DAF (for C3b and C4b); MCP and factor I (for C3b); CD59 (for C5b—8). Unlike other regulators of the system, properdin, which stabilizes C3bBb, has an enhancing effect on activation. It is not yet known how the activity of the MBP—MASP complex of the lectin pathway is regulated, but it is likely that MASP is regulated by C1-inh.

Table 1 Plasma proteins involved in activation and control of the complement system

	Mol. wt	Approximate serum concentration (μg/ml)	Number of chains in plasma form prior to activation	+denotes enzymic site in activated form (and natural substrate split)	Chromosomal localization
Classical pathway components					
C1q	460 000	80	18 (six A + six B + six C)	−	1p34.1–1p36.3
C1r	166 000	50	2 (identical)	+(C1s)	12p13
C1s	166 000	50	2 (identical)	+ (C4, C2)	12p13
C4	200 000	600	3 (α + β + γ)	−	6p21.3
C2	102 000	20	1	+ (C3)	6p21.3
C3	185 000	1300	2 (α + β)	−	19
Alternative pathway components					
Factor D	24 000	1	1	+ (B)	unknown
Factor B	92 000	210	1	+ (C3)	6p21.3
C3	185 000	1300	2 (α + β)	−	19
Terminal components					
C5	191 000	70	2 (α + β)	−	9q32–9q34
C6	120 000	64	1	−	5h
C7	110 000	56	1	−	5h
C8	151 000	55	3 (α + β + γ)	−	1p34(α and β) 9q(γ)
C9	71 000	59	1	−	5p13
Control proteins in plasma[a]					
C1-inhibitor	110 000	200	1	−	11p11.2–11q13
C4-binding protein	557 000	250	8 (7α + 1β)	−	1q32
Factor H	150 000	480	1	−	1q
Factor I	88 000	35	2 (α + β)	+ (C4b, C3b)	4q24–4q26
Anaphylatoxin inactivator	310 000	35	6 (2α + 2β + 2γ)	+ (C3a, C4a, C5a)	unknown
Properdin	106–112 000	20	2, 3 or 4 (identical)	−	Xp11.23
S-protein	83 000	505	1	−	17q11
SP-40,40	80 000		1	−	8p21

[a] Membrane associated receptors and regulatory proteins (detailed in Table 2) also play a role in control of the complement system as outlined in the text.

IgG3, or IgM, and this is probably the primary mode of activation *in vivo*. However, it is clear that non-immunological activation of C1 can take place via polyanions (such as bacterial lipopolysaccharides, DNA, and RNA), small polysaccharides, viral membranes, etc. Recently it has been shown that the serum lectin, mannan-binding protein (MBP), can also cause activation of the classical pathway upon binding to suitable carbohydrate ligands and utilizing a newly described serine protease in the lectin pathway which bypasses C1 (Fig. 1). This provides an antibody-independent route of activating the classical pathway and may be

Table 2 Membrane-associated molecules which act as receptors/regulators for fragments[a] of activated complement components

Membrane molecule	Mol. wt[b]	Fragment specificity	Cell distribution and chromosomal localization (where known)
Complement receptor type 1 (CR1) CD35	250 000 222 000 190 000 160 000 (four structural allotypes)	C3b, C4b, iC3b	Erythrocytes, monocytes, macrophages, neutrophils, B lymphocytes, some T lymphocytes, kidney podocytes, dendritic cells. 1q32
Complement receptor type 2 (CR2) CD21	145 000	C3d, C3dg, iC3b, C3u	B lymphocytes, spleen dendritic cells. 1q32
Membrane co-factor protein (MCP) CD46	45–70 000	C3b, C3a	Platelets, monocytes, B and T lymphocytes. 1q32
Decay acceleration factor (DAF[c]) CD55	70 000	C4b2a or C3bBb	Platelets, erythrocytes, leucocytes. 1q32
CD59 (protectin, HRF20)	20 000	C5b–8	Erythrocytes, renal cells, glomerular epithelial cells. 11p13
Complement receptor type 3 (CR3) CD11b/CD18	165 000 (α) 95 000 (β)	iC3b	Monocytes, neutrophils, and natural killer (NK) cells 16p11–16p13.1 (CD11) 21q22.1 (CD18)
p150,95	150 000 (α) 95 000 (β)	iC3b	Tissue macrophages, monocytes, and neutrophils
C3a/C4a receptor	—[d]	C3a, C4a	Mast cells, granulocytes
C5a receptor	50 000	C5a, C5a–des–arg	Mast cells, granulocytes, monocytes, macrophages, platelets
C3e-receptor	—	C3e	Neutrophils, monocytes

[a] Receptors for the collagen region of intact C1q and receptors for intact factor H have been found on monocytes, B lymphocytes, and neutrophils.
[b] All appear to be single chain molecules except CR3 and p150,95 which have non-covalently linked α and β chains.
[c] A low level of DAF is also found in body fluids.
[d] Not known.

important in the innate immune defence. Activation of the alternative pathway, on the other hand, is considered to be primarily mediated by a wide range of non-immunoglobulin activators such as certain strains of Gram-positive and Gram-negative bacteria, lipopolysaccharides, fungi, yeast, tumour cells, etc. Complexes of IgG, IgA, and IgE have been implicated in alternative pathway activation, but are considered to play perhaps an enhancing role rather than to be a major cause of activation. Once the C3 convertases from either pathway have been formed, the major event in the activation process takes place, the conversion of C3 to C3b plus C3a. Many C3 molecules are split at a single activation site due to the several amplification factors inherent in the system. The large fragment C3b (176 000 mol. wt) is utilized in a variety of reactions:

- in the covalent binding to the C4b2a complex in such a way that it becomes capable of splitting C5 (Fig. 1);
- in the covalent binding to the C3bBb complex to allow it to split C5 (Fig. 1);

- to amplify the system by 'feeding-back' into the alternative pathway by interacting with pro-enzyme factor B to generate more of the C3bBb complex;
- in the attachment to the initial activating particle and aiding opsonization and phagocytosis of the bound particle;
- it can also become covalently bound to a variety of surfaces as outlined in Section 5.1.

It is considered that C3b itself, or more likely C3i or C3(H_2O) (a 'C3b-like' form of C3), may act as the recognition molecule during the initial steps of activation of the alternative pathway (Fig. 1). The smaller fragment of C3a (9000 mol. wt) is one of the three anaphylotoxins (C3a, C4a, C5a) generated during activation of the pathway and thus has an important role as a mediator of inflammation.

The salient features concerned with the structure and activation of each of the components, control proteins, and receptors associated with the complement system are covered briefly in the following sections, but attention will be focused on the data obtained from the recently described cDNA and genomic clones. In most cases, the initial cDNA clones were obtained from liver cDNA libraries except for the properdin clone (spleen cDNA library), the factor D clone (adipocyte library), the complement receptor 1 clone (tonsil cDNA library), and the complement receptor 3 clone (U937 cell-line cDNA library).

2. Activation of the classical pathway

2.1 Structure, activation, and control of C1

The C1 complex is composed of subcomponents C1q, C1r, C1s in the molar ratio 1:2:2. The C1q molecule, which has no enzymic activity, is involved in the recognition and binding of activators of the classical pathway which allows subsequent activation of the C1r and C1s pro-enzymes (1). C1q is one of the largest complement components (460 000 mol. wt.) and it has an unusual shape. When viewed in the electron microscope it appears to be composed of six peripheral globular 'heads', each of which is joined by a collagen-like connecting strand to a fibril-like central portion (Fig. 2). The molecule is composed of 18 polypeptide chains (six A, six B, and six C), each chain being approximately 25 000 mol. wt and containing 225 amino acid residues. Each of the three types of chain contains a region of 81 amino acids of collagen-like (Gly–Xaa–Yaa–)$_n$ repeating sequence starting close to the N-terminal end followed by a C-terminal portion of about 136 amino acids that is non-collagen-like in nature. Amino acid sequence data, electron microscopy measurements, and other physical evidence has led to the proposal that the chains interact in sets of three (one A, one B, and one C) to form a triple helix involving their collagen-like N-terminal regions with non-collagen-like C-terminal portions forming one of the globular 'heads'. The (Gly–Xaa–Yaa–)$_n$ repeating nature of the N-terminal sequences of the A and C chains is interrupted by one residue approximately half-way along each collagen region. In view of the

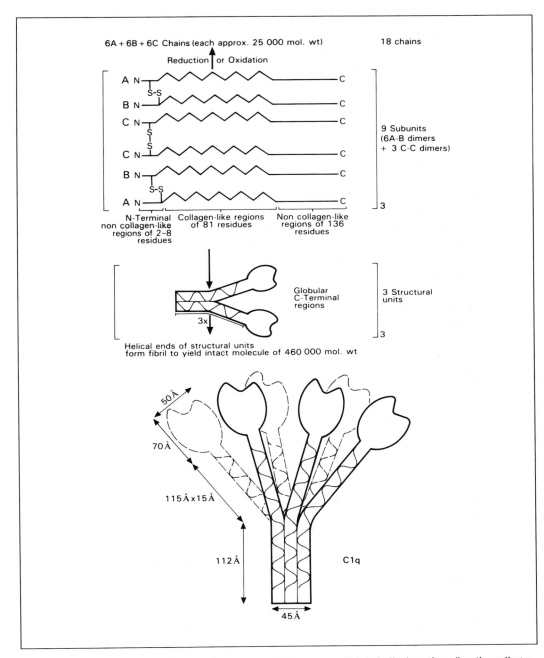

Fig. 2 Structure of human C1q. The wavy line represents the proposed triple-helical sections (i.e. the collagen-like connecting strands and fibril-like central portion) having lengths of 115 + 112 Å = 227 Å as estimated by electron microscopy [cf. a value of 232 Å predicted from the sequence studies (3)]. In the intact molecule shown at the bottom of the figure, ——— indicates portions of the molecules pointing towards the reader and − − − − indicates portions pointing away from the reader.

requirement for glycine at every third position for true triple-helix formation, this probably accounts for the manner in which the collagen-like regions of the molecule appear to be distributed and 'bend' at the point where the connecting strands merge to form the fibril-like central portion. Thus, one C1q molecule appears to be composed of six triple helices that are aligned in parallel throughout half their length and then diverge for the remainder of their length of triple-helical structure to form the connecting strands, each of which extends into one globular 'head' region composed of the 136 C-terminal amino acids of one A, one B, and one C chain (2, 3). Neutron diffraction studies indicate that the model shown (Fig. 2) is close to the conformation the molecule adopts in solution (4) and thus is informative as to how C1q may interact with C1r and C1s.

The Ca^{2+}-dependent $C1r_2$–$C1s_2$ complex is composed of two 83 000 mol. wt chains of pro-enzyme C1r and two 83 000 mol. wt chains of pro-enzyme C1s. Both pro-enzymes have a similar appearance when viewed in the electron microscope,

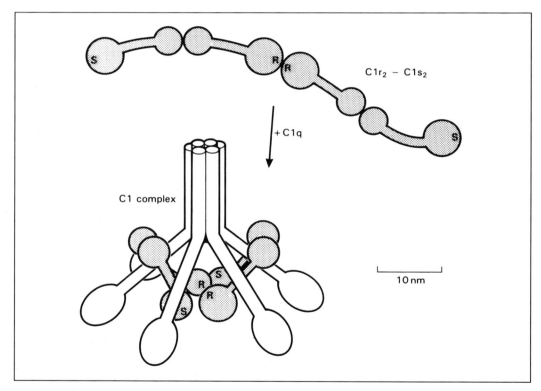

Fig. 3 Models proposed for $C1r_2$–$C1s_2$ and the C1 complex. R and S denote the larger, catalytic domains of C1r and C1s, respectively. The slightly inverted S-shaped $C1r_2$–$C1s_2$ complex consists of a central $C1r_2$ part (composed of two monomers of C1r in mutual contact through their globular catalytic domains) with a molecule of C1s attached to each end of $C1r_2$. The C1 complex model is formed by placing the rod-like $C1r_2$–$C1s_2$ complex across the C1q arms, then bending each end of the $C1r_2$–$C1s_2$ complex around two opposite C1q arms so that both C1s catalytic domains come into contact with the centrally located catalytic domains of C1r. (Figure adapted from ref. 5.)

both having two globular domains connected by an elongated structure. In each case the larger of the globular domains (Fig. 3) is considered to contain the catalytic site in activated C1r (denoted as C1r) or activated C1s (C1s) while smaller domains are classed as interaction domains (5). It is likely that the almost rod-like $C1r_2-C1s_2$ complex interacts with C1q to yield a C1 complex in which the pro-enzymes C1r and C1s are primarily associated with the collagen-like connecting strands. Various models for the C1q complex have been proposed (1, 6). A typical model suggests that the $C1r_2-C1s_2$ complex adopts a distorted 'figure of 8' structure with the interaction domains being located outside the C1q collagen-like strands while the catalytic domains are closely associated within the collagen-like strands (see ref. 5 and Fig. 3).

Activation of the C1 complex is under the control of the C1-inhibitor (C1-Inh) which is a single-chain glycoprotein of 104 000 mol. wt with an unusually high (49 per cent) carbohydrate content (7). On the interaction of probably two or more of the 'heads' of C1q with a suitable activator of the classical pathway, a conformational change may be induced within the C1 complex which allows the autoactivation of pro-enzyme C1r to take place to yield activated C1r. This is followed by the activation of pro-enzyme C1s by the C1r. In both pro-enzyme C1r and pro-enzyme C1s, a single peptide bond is split to yield disulfide-bonded chains of 56 000 mol. wt (the N-terminal non-catalytic domain) and 27 000 mol. wt (the C-terminal catalytic domain) found in C1r and C1s. The C1-Inh rapidly forms a 1:1 complex with both C1r and C1s, probably via the active site in the catalytic chains. This prevents over-activation of the system and removes C1r and C1s from the activator–C1 complex, leaving the collagen regions of C1q free to interact with the C1q receptor found on a wide variety of cells.

2.2 Activation of the classical pathway via mannan binding protein (MBP)

Mannan binding protein is a serum lectin which shows strong structural similarity to the complement protein C1q in terms of being composed of multiple globular 'heads' linked by collagen-like 'tails'. Therefore the overall shape of the molecule is very similar to that of C1q (8, 9). The globular heads of MBP are capable of binding to carbohydrate structures (with peripheral N-acetylated glucosamine or mannosamine residues) on the surfaces of a range of microorganisms such as *Escherichia coli, Salmonella montevideo, S. typhimurium*, and also to the HIV virus. It was first noted that MBP could utilize complement when Ikeda *et al.* (1987) (10) showed that MBP activated complement in the lysis of mannan-coated red blood cells by serum and that the lysis required the presence of component C4. Upon binding to carbohydrate ligands, MBP can activate the pro-enzyme $C1r_2-C1s_2$ complex without the involvement of C1q (9, 11). However, it was later shown that MBP is associated in serum with a newly described 100 000 mol. wt serine protease and that the complex of MBP with this protease could activate C4 and C2 after the

MBP had bound to carbohydrate (12, 13). This new serine protease has been termed MASP (MBP-associated serine protease) and has been found to have a similar overall structure to that of C1r and C1s (14, 15). It is probable that the MBP–MASP complex plays a more significant physiological role in complement activation than MBP–C1r$_2$/C1s$_2$ interaction since most of the C1r$_2$–C1s$_2$ complex present in blood would be expected to be associated with C1q. The importance of this lectin-mediated activation of the classical pathway of complement has still to be fully assessed, but it may be important in immunodeficient individuals, or the very young, in providing antibody-independent protection against a wide variety of pathogens (16). Although this route of complement activation is defined, in general terms, as 'lectin-mediated', it should be noted that, to date, MBP is the only lectin that has been shown to be involved in complement activation.

2.3 Molecular cloning of C1q

The first cDNA clone for one of the chains of C1q (the B chain) was isolated from a liver cDNA library (17), but use of the cloned cDNA in a Northern blotting study of mRNA isolated from various tissues and cell lines indicated that very high levels of C1q B-chain mRNA are synthesized by cultured monocytes, whereas relatively weak signals are obtained from liver mRNA. The finding of many positive clones for a second chain of the human C1q (the A chain) in a monocyte cDNA library along with the demonstration of the synthesis of a C1q molecule apparently identical to that found in serum by cultured monocytes and macrophages (18), indicated that macrophages, whether circulating or resident, may be a major source of serum C1q. Overlapping cosmid clones containing the genes for the A-, B-, and C-chains of human C1q have been characterized and have established that the genes are aligned 5' to 3', in the same transcriptional orientation, in the order A–C–B on a 24 kb stretch of DNA on the 1p34.1–1p36.3 region of chromosome 1 (19, 20). The A-, B-, and C-chain genes have very similar structures. Each is approximately 3 kb in length and contains one intron which is located within a codon for a glycine residue positioned half-way along the collagen-like region in each chain. These glycine residues are positioned just prior to the point where the triple-helical portions of the C1q molecule are interrupted and appear to 'bend' (as viewed in the electron microscope; see Fig. 2). The positioning of these introns suggests that they are involved in the conservation of an important structural feature in the C1q molecule. The C1q chains have a remarkable overall similarity in size, collagen-like content, and gene organization to the pulmonary surfactant protein, SP-A (21). However, the lack of any strong amino acid sequence homology within the globular regions of C1q with that of the surfactant protein indicates that the C1q chains are not closely related to those of SP-A. The SP-A is a member of a subgroup of the C-type lectin-proteins containing collagen-like sequences which includes conglutinin, mannose-binding protein, lung surfactant protein D, and collectin CL-43 (22).

Two types of C1q deficiency have been described. One type is characterized by a non-functional form of C1q which has antigenic activity, whereas in the other type no functional or antigenic activity is seen. Both types of deficiency lead to the development of immune-complex-related disease. The use of the cDNA from the normal B chain as a probe to isolate genomic clones from a patient suffering from the second type of deficiency indicates that in one family this deficiency may be a consequence of the inability to transcribe the B chain gene (23). An intriguing observation has been made showing that patients suffering from systemic lupus erythematosus (SLE) have high levels of a non-functional, low molecular weight form of C1q which appears similar to the defective form of C1q seen in the first type of C1q deficiency (24). It is not clear if this abnormal low molecular weight form of C1q plays a role in the hindrance of the processing of immune complexes (for instance by blocking the binding of C1q-coated complexes to the C1q receptor found on a wide variety of lymphoid cells) or if it is merely a passive by-product generated by over-production of one or more of the C1q chains. Further studies using the cDNA probes for the C1q chains and the development of a probe for the C1q-receptor may help clarify this area.

2.4 Molecular cloning of C1r and C1s

The availability of cDNA clones for pro-enzymes C1r and C1s has provided the complete derived amino acid sequences for these serine proteases, thus rapidly clarifying some of the unusual structural features of these molecules (25, 26). In both C1r and C1s, the A (non-catalytic) heavy chain is of approximately 55 000 mol. wt and originates from the N-terminal end of the 83 000 mol. wt pro-enzyme. The A chain, in each case, contains a potential growth factor domain and two different pairs of internal repeating sequences. The C-terminal B chains of about 27 000 mol. wt, contain the active site characteristic of other serine proteases. The first set of repeating homology units (occupying approximately positions 10–80 and 185–255) in the A chains of both C1r and C1s do not appear to have counterparts in any other known serum proteins (apart from the serine protease associated with MBP; see Section 2.6) and may be involved in the activation and control of specificity of C1r and C1s. The second set of repeating homology units, each of about 60 amino acids occupying the C-terminal end of the A chain), are homologous to the 60-amino acid repeat found in the C3b/C4b binding proteins as discussed in Section 6. The $C1r_2$–$C1s_2$ complex would contain eight of the latter type repeating units presumably located in a symmetrical fashion on or near the outside of the 'cage' of the collagen-like C1q 'arms' (Fig. 3). It is possible that they may fulfil a role in the interaction with C4 (the substrate for activated C1s). It has been shown that the *C1r* and *C1s* genes are orientated divergently (i.e. 3' to 3'), with the distance between their 3' ends being only 9.5 kb (27). Both genes are approximately 10–12 kb long and have been localized to the region 12p13 on chromosome 12 (28). Initial studies on the intron/exon boundaries of the *C1r* gene indicates that domains I to V of the non-catalytic A-chain are each encoded by separate exons, while the entire serine proteinase domain, which represents the catalytic B-chain, is encoded by a single exon. The B-chain of C1s also appears to be encoded by a single exon. Homozygous deficiencies of *C1r* (several cases) and of *C1s* (one case) have been reported and are generally associated with susceptibility to immune-complex-related, SLE-like, disease.

2.5 Molecular cloning of mannan binding protein (MBP)

MBP is found in liver and serum as a range of oligomeric forms which contain between 6 and 18 identical chains of 32 kDa. Subunits of 96 kDa are formed from three polypeptide chains and between two to six of the subunits yield oligomers of 200–576 kDa. The liver form of MBP is composed mainly of two to three subunits, while a range of oligomers, containing three to six subunits, are seen in serum MBP preparations (9, 29, 30). Only the higher oligomers, containing four to six of the 96 kDa subunits, can efficiently initiate complement activation (9, 11). Cloning studies, at the cDNA and genomic levels, indicate that two forms of MBP (A and C) are present in both the rat and mouse (29, 31, 32), while only type of sequence has been described in humans (33, 34). The human MBP sequence shows approximately 62 per cent identity with both the A and C forms seen in rodents. The MBP polypeptide chain shows an overall structural similarity to those of C1q in that they have:

- An N-terminal region containing cysteine residues involved in the formation of disulfide bonds.
- A region of Gly–Xaa–Yaa collagen-like sequence which is involved in the formation of triple-helical structure
- A C-terminal 'globular' region.

Unlike C1q, the C-terminal globular region is involved in binding carbohydrates and it contains 18 invariant residues characteristic of the C-type lectin carbohydrate recognition domain (CRD).

The gene encoding human MBP is located within a cluster of C-type lectins containing collagen-regions on the long arm of chromosome 10, at 10q22–23 (35). The gene is approximately 7 kb long and contains four exons which encode the N-terminal cysteine-rich region (exon 1), the collagen-like region (exons 1 and 2), an α-helical 'neck' region (exon 3), and the C-type lectin domain (exon 4) (33, 34).

2.6 Molecular cloning of mannan binding protein-associated serine protease (MASP)

MBP is considered to associate in plasma with a serine protease which has been designated 'MBP-associated serine protease' (MASP) (12), or 'P100' (36), and which has been found to have structural and functional similarities to C1r and C1s. Activation of the pro-enzyme form of MASP, by the interaction of the MBP–MASP complex with carbohydrate structures on the surface of microorganisms, allows the activated MASP to split C4 and C2, in the same manner as C1s, and thus provides a route for complement activation which is independent of antibody and the C1 complex. Human, and mouse, MASP have been cloned at the cDNA level and shown to have about 38 per cent sequence identity to C1r and C1s and to show a similar overall organization of domains (14, 15, 37).

2.7 Molecular cloning of C1-inhibitor (C1-Inh)

The molecular cloning of C1-Inh along with extensive studies at the protein level have shown that it circulates as a single chain glycoprotein of 478 residues, the polypeptide portion of the molecule accounting for only 51 per cent of its 104 000 mol. wt (7). The 49 per cent carbohydrate is probably located at about 17 positions involving asparagine, serine, and threonine residues within the N-terminal 120 residues of the molecule. The threonine-linked carbohydrate is particularly located within a region where a tetrapeptide sequence of the type –Gly–Pro–Thr–Thr– is repeated seven times. C1-Inh has extensive homology over its C-terminal residues 120–478 with other members of the serpin (serine proteinase inhibitor) gene family of inhibitors of serine esterases, which includes antithrombin III and α1-antitrypsin. Like these other serpins, C1-Inh forms a proteolytically-inactive stoichiometric complex with target proteases. It is the only plasma inhibitor considered to interact with C1r and C1s under physiological conditions. The role of the exceptionally high carbohydrate content of C1-Inh is not entirely clear since studies involving deglycosylation of the molecule indicate that this procedure does not markedly interfere with the role of C1-Inh in preventing C1r autoactivation within the C1 complex or its interaction with activated C1s. This implies a possible role for the carbohydrate in clearance mechanisms after the inhibitor has bound C1r and C1s.

The *C1-Inh* gene has been mapped to chromosome 11, p11.2–q.13 (7). The gene is 17 kb long and is organized into eight exons and seven introns (38, 39), but, surprisingly, it shows no obvious similarity in its intron–exon structure to other members of the serpin superfamily. The gene contains several unusual features, such as a high density of Alu elements and the presence of regions which are considered to be capable of forming Z-DNA and H-DNA structures (39). There are 17 Alu elements within introns 3–7 and it has been suggested that they could play a negative regulatory role causing repression of the *C1-Inh* gene as well as being involved in the generation of deletions and duplications known to occur within the *C1-Inh* genes of patients with hereditary angioneurotic oedema (HANE). Approximately 15 per cent of HANE patients show rather large partial deletions or duplications within the *C1-Inh* gene, probably as a result of the presence of Alu repeats (39, 40). In HANE, C1-Inh deficiency is inherited as an autosomal dominant trait which is associated with attacks of localized, increased vascular permeability (41). There are two forms of the deficiency, one in which the plasma concentration of the apparently normal inhibitor is markedly reduced (type I HAE), the other in which a dysfunctional protein is found in normal or raised concentrations (type II HAE). The molecular defects involved in the disease are heterogeneous, and the most successful treatment involves treatment with androgens which raises the levels of normal gene product. The availability of cDNA and genomic probes, and determination of the structure of the normal *C1-Inh* gene, has led to the finding that the point mutations, as well as the deletion or insertions mentioned above, shown by most HANE patients, can be located anywhere within the 17 kb-long *C1-Inh* gene, although there is some clustering of the mutations (42).

3. Activation of the alternative pathway

The non-antibody-dependent activation of the alternative pathway is considered to be dependent upon the spontaneous hydrolysis, at a low level, of the intra-molecular thiol–ester bond present in C3. The hydrolysis of this bond by water induces a conformational change in the hydrolysed C3, designated $C3(H_2O)$ or C3i, with the result that it adopts a C3b-like conformation and function (43). On interaction of C3i with pro-enzyme factor B, a Mg^{2+}-dependent C3iB complex is formed. The enzyme factor D, which circulates in the blood in its activated form, then splits one bond in the pro-enzyme B portion of the C3iB complex to yield the activated complex C3iBb which is the initial convertase of the alternative pathway (Fig. 1). Activation of C3 by this initial convertase allows freshly hydrolysed C3b to be deposited on to suitable sites on targets such as microorganisms. If the freshly activated C3b binds to host cells, then the down-regulatory proteins, as outlined in Section 6, prevent the formation of, or inactivate, the C3bBb complex. The C3b which is bound to invading microorganisms may evade down-regulatory mechanisms and be allowed to form C3bBb which, in turn, amplifies the activation of C3. The target microorganism can, therefore, become coated with a large number of C3b molecules, which are recognized by phagocytic cells, and the continuing complement activation can lead to membrane-lysis via the terminal components.

3.1 Factor D

Factor D present in the blood is an enzymically active, single-chain serine protease of 24 000 mol. wt. It does not appear to circulate in a pro-enzyme form. Thus the regulation of factor Ds enzymic activity depends upon its high specificity, that is to say, it will only act efficiently on a single arginyl–lysyl bond in pro-enzyme factor B after factor B has formed a Mg^{2+}-dependent complex with C3b, or $C3(H_2O)$. The cDNA cloning of human factor D (44) has demonstrated that factor D is probably identical to adipsin. Adipsin was first described in the mouse as a serine protease which is constitutively secreted from cultured rodent adipocyte cell lines. In both rodents and man, adipose tissues appear to be a major site of synthesis of factor D/adipsin. However, it is clear that human factor D is also synthesized by U937 cells, HepG2 cells, and blood macrophages (45). Decreases in adipsin expression are seen in obese rodent models of a genetic or neuroendocrine origin (46, 47), and it will be of interest to see if the level of factor D/adipsin varies in any aspects of human obesity. The mouse adipsin/factor D gene has been characterized (48), but no information is available yet concerning the structure of the human gene or its chromosomal localization. Elements present in a region of approximately 950 bp of DNA, located 5' to the transcription start site, are responsible for the control of the expression of adipsin/factor D in adipose tissue and also for the response to a gene product, or chemical, that induces obesity (48). Homozygous deficiency of factor D/adipsin in humans is rare and usually results in recurrent *Neisseria* infections (49).

3.2 Properdin—a positive regulator of the alternative pathway

Properdin is the only known normal positive regulator of the alternative pathway by virtue of its ability to bind and stabilize the inherently labile enzyme complexes, C3bBb and C3bBb3b, and also to protect these complexes from factors I and H (50). Properdin is present in the blood as a mixture of cyclic polymers (mainly dimers, trimers, and tetramers) formed by the interactions of a single 53 000 mol. wt. chain. The monomer form has not been observed *in vivo*. Human properdin has been cloned at the cDNA (51) and genomic (52) levels. The derived amino acid sequence shows that the 441-amino-acid long chain of properdin is composed of six repeating motifs (each 60 amino acids long), of a type first described in the cell–cell adhesion molecule thrombospondin. These repeats are flanked at the N- and C-terminal ends by discrete domains which are considered to interact with each other to form the properdin oligomers. The human properdin gene spans approximately 6.0 kb of DNA and is organized into ten exons with the first five repeating motifs, of the thrombospondin type (TSR modules), being precisely encoded by exons 3 to 8. These TSR modules are also found in the complement proteins C6, C78, C8 (α and β) and C9, and also in the circumsporozoite protein of a number of species of malaria parasites as well as in properdin and thrombospondin. The fact that these proteins are involved in interactions at the cell-surface is suggestive that cell binding may take place via a Val–Thr–Cys–Gly sequence motif found in certain TSRs and considered to be important in the interaction with the membrane protein CD36 (reviewed by Frazier; ref. 53). Individuals deficient in properdin (either by having zero or low levels of normal properdin, or showing the presence of a dysfunctional molecule) are at risk to fatal fulminant meningococcal infections, which illustrates the importance of efficient and continued activation of the alternative pathway in combating such infections. These forms of properdin deficiency are X-linked recessive. This is consistent with the properdin gene being mapped to the short arm of the X chromosome at Xp11.3–Xp11.23 (54, 55)

4. Complement class III products of the MHC

Components of C4, C2, and factor B are designated the class III complement genes in the MHC and, along with the 21-hydroxylase genes, form a tight cluster in the HLA system on chromosome 6 in man and 17 in mouse. The class I and II antigens encoded on chromosome 6 are highly polymorphic and are involved in the regulation of the immune response (see other chapters in this volume). These cell-surface molecules are related to other proteins in the immunoglobulin superfamily (56). The complement class III proteins are also polymorphic, but show no structural similarity to the class I and II gene products. The genes for C2 and factor B code for serine–esterase-type proteases. Component C4 is quite unlike C2 and factor B, but shows structural homology with C3, C5, and the protease inhibitor α_2-macroglobulin. However, none of these latter proteins are encoded by genes linked to the MHC.

4.1 Component C4: structure and activation

Component C4 is synthesized as a single chain precursor molecule of approximately 200 000 mol. wt. The primary site of synthesis is probably hepatocytes, although synthesis by macrophages has also been reported. Before secretion, the pro-C4 is glycosylated (57), sulfated, an internal thiol–ester bond is formed, and the molecule is processed into three disulfide-linked chains ordered in the pro-molecule as β (70 000 mol. wt), α (95 000 mol. wt), and γ (33 000 mol. wt). The predicted amino acid sequence of pro-C4 (derived from the cDNA sequence) indicates that the processing events involve splitting and trimming of short sections containing basic amino acids at the β–α and α–γ junctions (58–61) (Fig. 4). On activation of C4 by C1s in the C1 complex, a single bond is split in the α chain to yield the anaphylatoxin C4a (9000 mol. wt) and the large fragment C4b. The freshly-activated C4b has the capacity to bind covalently to hydroxyl or amino groups through a reactive acyl group of a glutamyl residue in its α' chain (62), thus yielding ester or amide bonds (as later shown for C3b in Fig. 9). In this manner, C4b can become covalently bound to either amino or hydroxyl groups of small molecules or cell-surface bound protein or carbohydrate. The predominant type of bond formed in either case is dependent upon which of the two forms of C4 (C4A or C4B) is involved in the reaction, the C4A form reacting more rapidly with amino groups than the C4B form and vice-versa for hydroxyl groups (63, 64). As discussed later, sequencing of the cDNAs for the common and rare allotypes of C4 has allowed the probable identification of the four residues in each of the two forms of C4 which determine the specificity of the covalent binding via the thiol–ester bond. As well as C4, the C3 component, α_2-macroglobulin and the pregnancy zone protein are all members of a family of proteins containing a thiol–ester bond (65). The essential features of the activation and binding via this bond are discussed more fully in Sections 4.3 and 5.1.

4.2 Component C2 and factor B: activation and role in formation of the C3 convertases

In the classical pathway, a single-chain pro-enzyme form of component C2 (120 000 mol. wt) associates, probably via its N-terminal C2b domain, with C4b in

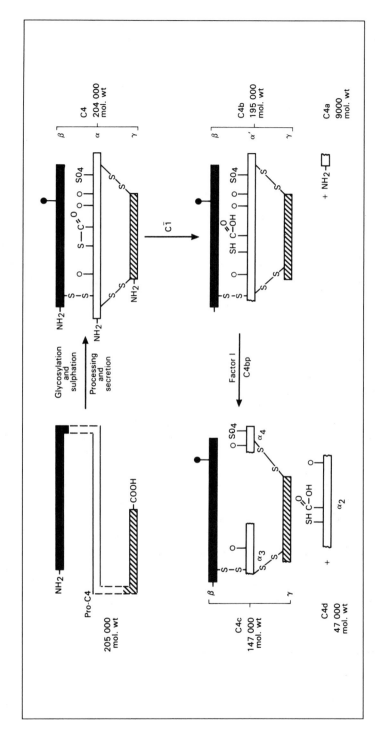

Fig. 4 The biosynthesis, processing, and activation of C4. Small areas of the pro-molecule split out by processing enzymes are denoted by a broken line. C$\overline{1}$s in the C1 complex (composed of C1q plus the activated C1r$_2$—C1s$_2$ complex) activates C4 by hydrolysis of a single bond in the α chain which yields the C4a fragment and simultaneously causes formation of a reactive acyl group and a free sulfhydryl group from the interchain thiol—ester bond. The acyl group can undergo covalent binding reactions as outlined in Fig. 9. Inactivation of C4b by factor I (in the presence of co-factor C4bp) is caused by the splitting of two bonds in the α' chain to yield C4c and C4d. ●, high mannose type oligosaccharide; ♀, complex fucosylated biantennary type oligosaccharide (57).

an Mg^{2+}-dependent interaction. The C2 is then split at one point by C1s in the C1 complex to yield the non-catalytic C2b chain of 30 000 mol. wt and the C-terminal catalytic chain C2a of 70 000 mol. wt. The C2b is not required for the C3 convertase activity mediated via the catalytic site in C2a in the C4b2a complex (Fig. 1). In the alternative pathway, the pro-enzyme factor B (90 000 mol. wt) associates via its N-terminal Ba domain (66), with C3b (or a 'C3b-like' form of C3 in the initial events of the alternative pathway activation) in an Mg^{2+}-dependent interaction. Factor B is split by factor D to yield the non-catalytic Ba of 30 000 mol. wt and the C-terminal catalytic chain of Bb of 60 000 mol. wt. Factor D is the smallest of the complement components (24 000 mol. wt) and the only component considered to lack carbohydrate. It is synthesized by monocytes and the cell lines U937 and HepG2 in a form which is indistinguishable from that found in plasma (45). It is present in low concentrations (1 µg/ml) in this active form in the plasma and is a highly specific protease with a tryptic-like specificity for a single Arg–Lys bond in factor B when the latter is bound to C3b. Thus factor D is homologous in function to C1s of the classical pathway and factor B and C3b are homologous in function to C2 and C4b of the classical pathway (Fig. 1). Component C2 and factor B are unusual serine proteases (67) in that their catalytic chains of 60 000–70 000 mol. wt are much larger than those of the classical serine proteases (about 24 000 mol. wt). At the protein level they are similar in length, Ba and C2b being 234 and 223 residues long and Bb and C2a being 505 and 509 residues long, respectively (68–70).

The enzyme complexes from either pathway which split C3 can 'decay' by release of C2a or Bb from C4b or C3b, respectively. This 'decay' is accelerated by the interaction of the complexes with the control proteins, C4b-binding protein (C4bp) and factor H, which can also act as co-factors in the splitting and inactivation of C4b, or C3b, by the control enzyme factor I. These control mechanisms, along with control by various membrane-associated molecules, are described in Section 6.

While the roles of the C1 complex in recognition and activation of the initial steps of the classical pathway leading to the formation of the C4b2a complex are quite well defined, identification of the precise mechanisms of activation of the alternative pathway to the C3bBb stage has been elusive. Part of the problem arose from the fact that a large number of diverse alternative pathway activators have been identified such as lipopolysaccharides, viruses, bacteria, fungi, tumour cells, red blood cells, and immune aggregates (71), thus providing no obvious common feature responsible for activation. It has been suggested that what is common to these varied activators is the presence of 'protected sites' (71) for the deposition of C3b and the subsequent formation of the C3bBb complex which would then be protected from the rapid decay and inactivation brought about by the control protein factor H and protease factor I. Thus a high turnover of C3 and deposition of the C3b can be achieved at an activation site by overcoming the inhibitory effects of the control proteins. Stabilization of the C3/C5 convertases composed of C3b and Bb is mediated by properdin.

In both pathways, as well as being involved in the formation of the C5 convertases, the freshly-activated C3b and C4b play an important role in opsonization by coating microorganisms or particles with large numbers of C3b or C4b molecules during complement activation. The opsonized microorganisms are phagocytosed and cleared from the blood more efficiently by macrophages. The C3b binding to antibody–antigen complexes results in their solubilization, making them less likely to be deposited in blood vessels and to cause tissue injury.

4.3 Molecular genetics of C4

C4 is highly polymorphic being coded for by at least two separate loci, *C4A* and *C4B*, with at least 35 allotypes being assigned to date (72). In most individuals, the genes coding for the C4A and C4B types are found 10 kb apart on each chromosome (about 70 per cent frequency), but expression of only a *C4A* or a *C4B* gene by one or both chromosomes is also observed (about 10–15 per cent frequency in each case), as is duplication of the *C4A* or *C4B* loci on either chromosome (about 1–2 per cent frequency). Total deficiency of C4 is rarely seen. The relatively frequent occurrence of null alleles, defined as *Q0* ('quantity zero') at both loci results in partial C4 deficiency states being quite common. It is important to elucidate any functional differences between the two types of C4 and the many alleles if conclusions are to be drawn regarding the possible association of C4 type with certain disease states (73). In the human system, after activation by C1s, the C4A and C4B types are both functionally active, but the C4B protein shows a several-fold higher activity in the haemolytic assay system. This difference is probably attributable to a difference in the covalent binding of activated C4 to the many carbohydrate groups on the cell-surface of the antibody-coated red blood cells used in the assay (74, 75); activated C4 binds more efficiently to proteins rather than carbohydrates while the reverse is true for activated C4B. This difference in binding efficiencies is markedly illustrated when the reaction rates of binding to small molecules is measured; C4A allotypes in general react about 300 times more rapidly with the amino group of glycine than C4B allotypes, while C4B allotypes react 10 times more rapidly with the hydroxyl group of glycerol than C4A allotypes (64). As well as differing in chemical reactivity, the C4A and C4B proteins differ in antigenicity; the *C4A* alleles normally carry the blood group Rodgers while *C4B* alleles usually carry the so far Chido antigen. Although 35 different allelic forms of *C4A* and *C4B* have been identified by electrophoresis, this number seems likely to increase by the subdivision of these identified forms by serological studies, restriction enzyme digestion studies (72) and nucleotide sequencing of the gene (76). The *C4A* gene is approximately 22 kb long and is split into 41 exons (76) and it is likely that the *C4B* gene is also split into 41 exons. Differences in the size of the *C4B* gene occur between individuals due to the length of the intron separating exons 9 and 10.

Protein and DNA sequencing of *C4A* and *C4B* alleles has allowed identification of those residues probably involved in determining the differences observed between the C4A and C4B forms in terms of their covalent binding and antigenicity. It has been clearly established (58, 59) that differences between several common alleles

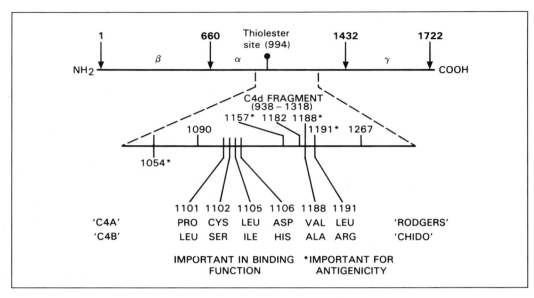

Fig. 5 Isotypic residues in C4A and C4B considered to be important in determining covalent binding specificity are located at positions 1101, 1102, 1105, and 1106 (numbering based on the single chain pro-C4 molecule) which are approximately 106 amino acids C-terminal to the thiol–ester site at position 994. Residues marked with an asterisk (*) at positions 1054, 1157, 1188, and 1191 are important in determining the two Rodgers and six Chido determinants with some contribution to Chido antigenicity also being made from the residues within the 1101–1106 region. (Adapted from ref. 77.)

are restricted to only 15 of the 1722 amino acid residues of the single chain pro-C4 molecule, and that 12 of these differences lie within a section of 230 residues in the α' chain, C-terminal to the thiol–ester bond (refs 58, 59, 77 and Fig. 5). Six of the twelve residues appear to be type-specific and, therefore, may account for the difference in properties seen between C4A and C4B allotypes. Studies involving the isolation of the different allelic forms of *C4*, using a monoclonal antibody column which bound the Rodgers positive (Rg⁺) and Chido positive (Ch⁺) proteins with different affinities, have helped to assign the precise residues involved in the functional and antigenic activities (63). It was found that all C4B proteins bound glycine and glycerol with the same efficiency and that nearly all C4A proteins bound glycine 100 times more rapidly than C4B but did not bind glycerol. However, the results obtained for two rare C4 allotypes, C4A1 and C4B5, allowed the final segregation of the residues involved in the differing chemical reactivity and antigenicity of the *C4* alleles (63, 64). These rare allotypes behave functionally as typical C4A and C4B proteins, but display a reversal of the Chido and the Rodgers antigenicity, the C4A1 being Rg⁻Ch⁺ and the C4B5 being Rg⁺Ch⁻ (76, 77). Comparison of derived amino acid sequences from the polymorphic sections in the α' chain from both the common and rare allotypes has allowed the conclusion that only four amino acids, within positions 1101–1106 of the pro-C4 molecule, are responsible for the different chemical reactivities of C4A and C4B (Fig. 5). The rare allotype C4A1 has Pro, Cys, Leu, and Asp at positions 1101, 1102, 1105, and 1106, typical of C4 allotypes, but Ala and Arg at positions 1188 and 1191, typical of most C4B allotypes. The C4B5 has Leu, Ser, Ile and HIs at 1101–1106, typical of C4B allotypes, but has Val and Leu at 1188 and 1191, typical of most C4 allotypes. Thus residues 1101, 1102, 1105, and 1106 (Fig. 5) are involved in covalent binding and the probable locations of the two Rg and six Ch antigenic determinants can also be deduced (76, 77). It has been shown (78) that substitution of the histidine residue at 1106, by aspartic acid, converts the functional activity of C4B to that of C4A. This observation has been extended by Sepp et al. (1993) (79) who expressed variants of C4B with a variety of different residues in the positions 1101–1106. Variants with a histidine at position 1106 showed C4B-like properties, while those with aspartic acid, alanine, or asparagine, at this position, were C4A-like. It was concluded that histidine 1106 is important in catalysing the reaction of the thiol–ester bond with hydroxy-group-containing compounds. When other amino acids are in this position, the thiol–ester bond becomes apparently more reactive with amino groups.

The isotypic residues considered to be involved in determining the covalent binding function are separated by 106 amino acids from the putative thiol–ester site. Thus, it seems likely that a conformational change takes place in C4 after its activation by C1 which would bring the important isotypic residues close to the thiol–ester site (considered to be in the middle of a hydrophobic pocket) and thus control the type of binding mediated via the active acyl group released from the thiol–ester. In view of the central role played by C4 in the complement system, the differential reactivity of the C4A and C4B allotypes could be of considerable importance in determining the effectiveness of an individual's complement system in disease states (80).

As discussed above, deficiency of either C4A or C4B is relatively common (10–15 per cent gene frequency), is linked to the MHC, and is defined as a null allele (Q0). It is of some clinical importance to obtain an understanding of the molecular basis for null alleles since, for example, C4AQ0 alleles are found at an approximately 3-fold greater frequency in SLE patients than in normal individuals (73). An association with C4AQ0 alleles has also been noted in insulin-dependent diabetes and myasthenia gravis. In all three types of disease the association is predominantly with the haplotype HLA A1, B8, DR3, BfS, C2C, C4AQ0 B1 (73). The precise role that the presence of a C4 null allele might play in these disease states is not clear, but in SLE it could be related to an individual's ability to solubilize immune complexes. In this context it has been shown that C4A is much more efficient than C4B in its capacity to inhibit immune precipitation in human serum (81). Therefore the absence of C4A, which reacts preferentially with immune complexes, may result in a predisposition to the elimination of defective immune complexes. Further evidence is provided by the finding that drugs which induce an SLE-like condition as a side effect also inhibit C4 in vitro and that this inhibition is greater for C4A than for C4B (82).

Two C4-like genes also occur in the mouse. One codes for mouse complement component C4 and the other codes for Slp (sex-limited protein), a homologue of C4 which is not split by C1s and, therefore, is haemolytically inactive. The molecular cloning of mouse C4 (60, 61, 83) and Slp (60, 83, 84) has shown that they have 90 per cent homology at the derived amino acid sequence level, they both appear to undergo essentially identical intracellular processing and both contain a thiol–ester bond. However, only C4 is cleaved and activated by C1s. The C4 and Slp genes of the C4 mouse H2 region provide a useful model for expression studies since the C4 levels between strains of mice can vary up to 20-fold and in most strains of mice the synthesis of Slp is regulated by testosterone. However, in some wild-derived strains of mice, Slp is expressed constitutively in the same manner as C4. In these strains there are C4–Slp recombinant genes composed of a 5′ region derived from the C4 gene and a 3′ region derived from the Slp gene (83), suggesting that the C4 derived 5′ region in these recombinant genes is responsible for the testosterone-independent expression.

4.4 Molecular genetics of component C2 and factor B

Despite the overall general similarity of component C2 and factor B at the protein level and the similar sizes of their mRNAs (C2 mRNA is 2.9 kb and factor B mRNA is 2.6 kb in liver) the genes differ considerably in size. The factor B gene (Bf) is 6 kb in length (69) while the C2 gene, due to the presence of more intron sequence, is approximately 18 kb long (67, 70). The 6 kb of the Bf gene is split into 18 exons, the N-terminal Ba fragment of 30 000 mol.wt being encoded by four exons. The Ba fragment is composed of three internal homology regions, each of about 60 amino acids, which are encoded by the three exons found after the exon encoding the leader sequence. A similar situation is true for C2b, the N-terminal portion of C2. The precise function of these repeating homology units is not known, but they are found in most of the complement proteins which interact with C3b or C4b (as discussed in Section 6). It seems likely that the Ba and C2b portions of factor B and C2 are directly involved in the initial binding of C3b and C2b, respectively, prior to the activation of factor B and C2 to form the C3 convertases (66).

The factor B and C2 enzymes are unusual types of serine proteases in that their catalytic chains are over twice the size of the catalytic chains of other serine proteases. The catalytic peptide Bb of 60 000 mol. wt is encoded by 13 exons. The sequences encoded by the first five of these exons shows significant homology with the A domains of von Willebrand factor (vWF), but no homology with other serine proteases, except C2, which may imply a role for the N-terminal portions of Bb and C2a in the binding to C3b in the formation of the C3 and C5 convertases. The C-terminal half of the Bb catalytic chain is homologous to the catalytic chains of other serine proteases, each of the functionally important parts of the active site being contained in separate exons. This region of the Bf gene shows a close correlation with the exon organization of other serine proteases of the trypsin/chymotrypsin family except for the presence of an extra exon in factor B which has no homologous counterpart except perhaps in the C2 gene. It is possible that the section of protein coded for by this exon may function to limit the specificity of factor B to cleavage of C3 and C5. Alternatively, it may play a role in the manner that factor B is activated.

The mapping of the class III complement genes on human chromosome 6 established that the *C2* and *Bf* genes are extremely closely linked (67). Further studies showed that the initiation site of the *Bf* gene lies only 421 bp from the poly(A) addition site of the *C2* gene (85). However, although the two genes lie very close to each other there is a 10-fold difference between the amounts of their gene products found in plasma. The major site of synthesis of plasma factor B is the liver, but synthesis by mononuclear phagocytes may serve an important role in local defence mechanisms. Factor B synthesis (reviewed in ref. 86) has been shown to be increased in monocytic cells, or hepatocyte-derived cells, by a variety of factors which might be found in inflammatory response, such as interleukin 1 (87), γ-interferon (88), and lipopolysaccharides. Stimulation of factor B synthesis by these agents in a system involving the transfection of mouse fibroblasts with the *Bf* gene has also been reported (87). These studies indicate that expression of the *Bf* gene is mainly controlled at a pre-translational level. Some understanding of the mechanisms underlying the regulation of the expression of *Bf* and identification of the DNA sequences essential for its constitutive and inducible expression has been achieved by deletion analysis of the 5' flanking region of the *Bf* gene (85). This has indicated the presence of *cis*-acting DNA elements which are essential for the cell specific and inducible expression of this gene (85) and these are outlined in Fig. 6.

Deficiency of C2 is the most common (1 in 10 000) genetic deficiency of the complement system in western Europeans while factor B deficiency is almost unknown. In type I C2-deficiency, no C2 protein is found whereas in type II deficiency, C2 protein is synthesized but there is a block of C2 secretion. Unlike complete C4 deficiency, which is very rare and occurs on a variable HLA background, the relatively common C2 deficiency is inherited as an autosomal dominant trait and over 90 per cent of C2-deficient individuals belong to the common haplotype *HLA-A25, B18, C2Q0, BfS, C4A4, C4B2, Dr2*. The molecular basis for type I C2 deficiency has been established in an individual with this common haplotype (89). It is caused by a 28 bp genomic deletion which results in skipping of exon 6 during RNA splicing and the generation of a premature translation termination codon. Almost half of the homozygous C2-deficient individuals present with symptoms resembling SLE or other autoimmune disorders (and even heterozygous deficient individuals) have a predisposition towards diseases related to an inability to deal efficiently with immune complexes.

Both the *BF* and *C2* genes are polymorphic, allelic forms having been described on the basis of differences in electrophoretic mobility of the gene products. These are *C2C* 97 per cent, *C2A* < 1 per cent, *C2B* 2 per cent; *BF* 28 per cent (F^a 11 per cent plus F^b 17 per cent), *BS* 70 per cent, plus seven rare alleles; in each case the percentages refer to the proportions of each allele found in the general population. The use of DNA probes has allowed further subdivision of the electrophoretic

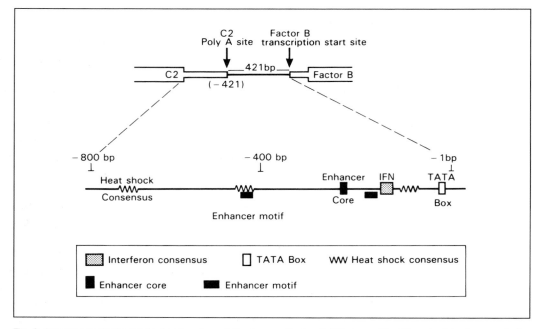

Fig. 6 Sequences in the 5′ flanking region of the human factor B (*Bf*) gene. The 5′ end of the human factor B gene lies only 421 bp from the polyadenylation site in the *C2* gene. The sequence of 1.6 kb of DNA encompassing the 3′ end of the *C2* gene and the 5′ end of the factor B gene indicates the presence of a number of regions which may play a role in the regulation of gene expression; the TATA box (nucleotides -28 to -25); a possible enhancer core element (-234 to -227); enhancer motifs, i.e. regions which show homology to sequences in cytomegalovirus enhancer (-472 to -455, -171 to -153); a 29 bp sequence (-154 to -127) homologous to the consensus sequence in the promoters of human genes stimulated by α-interferon; three regions each of 14 bp (-757 to -744, -467 to -454, and -98 to -85) which are similar to the consensus sequences found in the promoters of genes coding for heat-shock proteins. (Taken from ref. 85.)

types and also provided markers for the assessment of susceptibility to disease association with HLA. To date, four useful restriction fragment-length polymorphisms (RFLPs) have been detected in the C2/factor B region of genomic DNA (reviewed in ref. 68):

● an *Sst*I multiallelic polymorphism in the 5′ end of the *C2* gene allowing the subdivision of haplotypes carrying the *C2C* and factor *BF* alleles;

● a *Bam*HI polymorphism in the middle of the *C2* gene;

● a *Taq*I polymorphism in the 3′ region of the *C2* gene which gives a marker for the *BF* subtypes

● an *Msp*I polymorphism at the 5′ end of the *BF* gene which distinguishes between *F* and *S* alleles. This polymorphism brings about an arginine/glutamine change at the protein level which accounts for the charge differences seen in electrophoretic typing.

4.5 Molecular map of the MHC class III region

The MHC occupies approximately 4000 kb of DNA and maps to the short arm of chromosome 6 in man and chromosome 17 in mice. Many of the MHC genes are highly polymorphic and play key roles in regulating the immune response. The gene products have been divided into three classes, primarily on the basis of their structure and function. The polymorphic class I and class II molecules are involved in antigen presentation during an immune response (see Chapter 5). The class III molecules are a rather heterogeneous collection and in many cases their functions are unknown. Over 36 genes have, to date, been characterized within a 680 kb stretch of the 1.1 Mb region which separates the class I and class II regions and is generally termed the class III region. The class III genes include the complement components C2, *BF*, *C4A*, and *C4B*. These are clustered in the order cen–*C4B*–*C4A*–*B*–*C2*–tel on a 120 kb stretch of DNA, with the *C4B* gene lying approximately 400 kb from (the HLA region *DRα* while the *C2* gene lies about 600 kb from *HLA-B* (90). The *CYP21* gene (encoding the microsomal 21-hydroxylase P-450 enzyme) and CYP21 pseudogenes lie immediately 3' to the *C4B* and *C4A* genes, respectively. Recently three new genes designated *RD*, *G11*, and *OSG*, have also been located within the complement cluster (see Fig. 7 and ref. 91). Although several of the class III molecules such as the complement components (C2, C4, and factor B), the heat-shock protein (HSP70), and tumour necrosis factor (TNFα and β) are involved in the immune response, it is not known if many of the newly described class III genes, such as *RD*, *G11*, and *OSG*, also have the potential to affect the immune response. Nevertheless, it is of some importance to define polymorphic markers within these newly identified coding sequences because of the considerable clinical interest in the marked association between particular alleles in the MHC region with increased susceptibility to certain diseases, especially those related to auto-immunity (73, 80). However, in view of the linkage disequilibrium displayed by particular alleles in this region, it is not clear if this association is due to the products of the established and/or novel class III genes or of particular *HLA-B* or *HLA-D* loci or even of unidentified loci within this region. Since certain MHC haplotypes (e.g. *B8DR3*) appear with increased frequency in many of the disease-associated haplo-types, it is important to characterize these disease-associated haplotypes in detail, especially taking into account any of the new genes found in regions between the known class III genes and class I and II genes.

5. Components C3 and C5

5.1 Component C3

Component C3 is the most abundant of the complement proteins in plasma (1.3 mg/ml) and plays a central role in the activation of the complement system since it participates in both pathways (Fig. 1). Many of the biological functions associated with the complement system, such as opsonization and anaphylatoxin

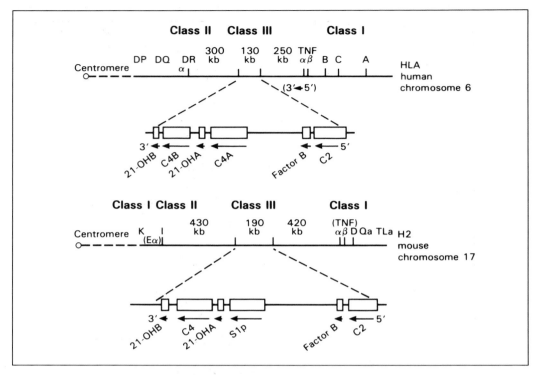

Fig. 7 Map of the MHC in man (*HLA*) on the short arm of chromosome 6 and of the MHC in mouse (*H2*) on chromosome 17. The class II regions are expanded to show alignment of the complement genes (see text).

activities, can be mediated by fragments of C3 generated during activation and regulation of the C3 component by specific complement enzymes and other tryptic-like proteases found in plasma (see ref. 92 and Fig. 9).

Pro-C3 is synthesized as a single chain of 185 000 mol. wt (1.7 per cent carbo-hydrate). It is then processed to yield the α chain (115 000 mol. wt) and β chain (70 000 mol. wt) disulfide-linked dimer found in plasma. The cDNA studies (93, 94) on human C3 show that the two chains are synthesized in the order β–α and that they are separated by a highly basic –Arg–Arg–Arg–Arg– sequence which is not present in mature C3 and, therefore, must be removed during proteolytic processing. Activation of C3 by the C3 convertase from either pathway (Fig. 8) proceeds in an identical fashion with the arginyl residue at position 78 in the N-terminal region of the α chain being split to yield C3a (9000 mol. wt) and C3b (176 000 mol. wt). The C3b fragment contains the large C-terminal portion of the α chain (the α′ chain) disulfide-linked to the β chain and, like C4b, contains a thiol–ester bond within its α′ chain.

In a similar manner to C4b, the freshly activated, metastable C3b can bind covalently, via an acyl group derived from the thiol–ester, to suitable acceptor molecules located near the site of activation and which contain hydroxyl- or amino-

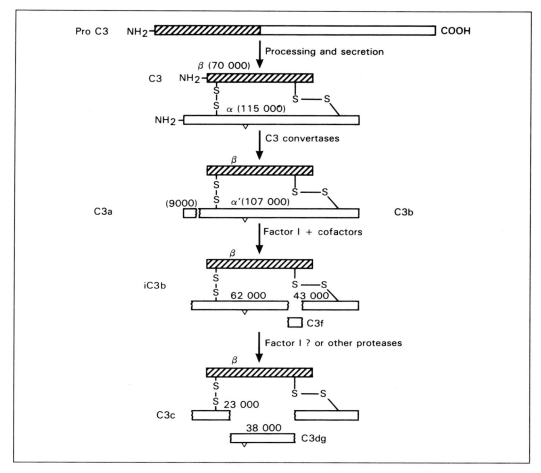

Fig. 8 Synthesis, activation, and degradation of human C3. Pro-C3 (185 000 mol. wt) is processed to yield a disulfide cross-linked dimer (C3) consisting of an α chain (115 000 mol. wt) and a β chain (70 000 mol. wt). After activation by the C3 convertases, as discussed in the text, C3b may bind covalently to suitable surfaces via the thiol—ester site (denoted). Once C3b is formed it is rapidly split two positions in the α' chain by factor I (in the presence of a co-factor) to form iC3b and C3f (2000 mol. wt) is liberated. iC3b is then slowly cleaved at a single site possibly by factor I to form C3c (143 000 mol. wt) and C3dg (38 000 mol. wt). However, the precise involvement of factor I in this step is not absolutely clear. iC3b can also be split by other enzymes, e.g. leucocyte elastase and trypsin, to form C3d (which is smaller than C3dg) plus C3c. Due to the covalent binding, via the thiol—ester site, degradation of surface-bound C3b results in retention of iC3b and C3dg or C3d on the surface while C3c dissociates.

groups (Fig. 9). This type of binding reaction was first described by Müller-Eberhard *et al.* in 1966 (95), but a clear understanding of the covalent binding of metastable C3b, and C4b, via the thiol—ester bond only emerged from much later studies during the period 1977–1980 (reviewed in refs 96, 97). These studies showed that on treatment of C3, or C4, with small radioactive nucleophiles such as CH_3NH_2, the radiolabel was incorporated into the penultimate glutamic acid in

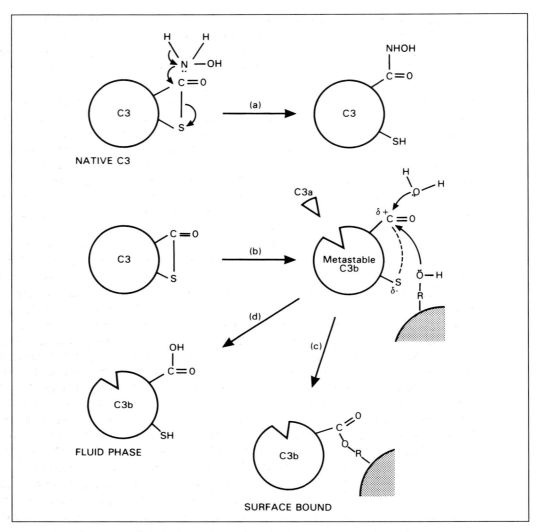

Fig. 9 The covalent binding reaction of activated C3 (or C4). The thiol–ester bond in the α chain of C3 (or C4) can undergo a variety of reactions: (a) it can be rapidly activated by small amines such as NH₂OH [or slow hydrolysis by H₂O can take place giving C3(H₂O)]; (b) proteolytic activation by the C3 convertases of either pathway gives metastable C3b; (c) the reactive acyl group can react with hydroxyl groups (or amino groups) on cell or particle surfaces; (d) the reactive acyl group can react with water to form fluid phase C3b.

the sequence–Gly–*Cys*–Gly–Glu–*Glu*–Asn– present in the α chain. The residues in italics are considered to form an activatable internal β-cysteinyl-γ-glutamyl thiol–ester in C3 and C4. Further supportive evidence was that the binding of one mole of nucleophile exposed one mole of free-SH and this could be identified by the use of radiolabelled iodoacetamide. In addition, it was known that native C3 contains no free sulfhydryl groups while the α' of C3b contains one. In native C3 the thiol–ester is quite stable, having a half-life of approximately 230 h in isotonic neutral buffer at 37 °C (43), while the half-life of the thiol–ester in freshly activated metastable C3b is about 60 μsec. Thus, a remarkable degree of reactivity is imparted to the thiol–ester bond in C3b and C4b, presumably by a conformational change, by loss of the C3a or C4a polypeptide. In solution, the freshly activated C3b reacts primarily with water as the most abundantly available small nucleophile. This type of mechanism also helps to explain why native C3 decays spontaneously in aqueous solutions and at an accelerated rate in the presence of chaotropic agents. The spontaneously decayed C3 appears to be uncleaved C3 in which the thiol–ester has been hydrolysed, and is usually designated C3i or C3(H_2O). The generation of C3(H_2O), which has a 'C3b-like' function, can then allow the formation of the initial C3 convertase of the alternative pathway [i.e. of the form C3 (H_2O)], Bb, which can produce many molecules of C3b (refs 43, 96, and see Fig. 9) to be used in the amplification of pathway activation.

The complete amino acid sequences of both human and mouse pro-C3 have been derived from the cDNA and genomic sequencing data (93, 94). There is 79 per cent identity at the protein level and 77 per cent identity at the nucleotide level. As might be expected in terms of retention of function, the 27 half-cysteine residues are identically placed in the proteins from both species and the sequence around the reactive thiol–ester bond is highly conserved. The human C3 gene, on chromosome 19, is 42 kb long and contains 41 exons (98–100). A number of regions within 1 kb upstream of the transcription start site of the C3 gene show homology to known transcriptional regulatory sequences—such as the responsive elements for interferon γ, interleukin-6, nuclear factor kB, oestrogen, glucocorticoids, and thyroid hormone. The presence of these regions is consistent with C3 being an acute phase protein which shows a marked rise in its plasma concentration during inflammatory conditions and the fact that C3 synthesis and mRNA levels appear to be raised by most of the agents mentioned above. The C3 β-chain is encoded by exons 1–16 over a region of 13 kb while the α chain is encoded by exons 16–41 over a region of 28 kb. The view that C3 and C4 arose from a common ancestor is supported by the finding that, on comparison of analogous exons present in C3 and C4, 39 out of the 41 exons were found to be of a similar size. C3 is synthesized in a wide variety of tissues, but the liver is probably the major site of synthesis with cells of the monocyte/macrophage type perhaps being of importance at sites of inflammation. Common electrophoretic variants of C3 (C3S and C3F) have been found but at different frequencies in all the major racial groups. Over 20 allelic variants have also been described but these are rare. At least 15 cases of homozygous C3 deficiency have been reported and most of the affected individuals showed recurrent bacterial infections and evidence of immune complex disease. In two patients the molecular basis for the deficiency has been established. A splice-site mutation caused the deficiency in one case and a deletion of 800 bp from the region of the C3 gene encoding the α chain in the other case (101). The deletion appears to have arisen from homologous recombination between two flanking Alu repeats.

5.2 Component C5

Like C3 and C4, C5 is synthesized as a single-chain precursor which is subsequently processed by cleavage and removal of a short sequence of basic residues prior to forming a disulfide-linked dimer composed of an α chain (115 000 mol. wt) and a β chain (75 000 mol. wt). The complete cDNA sequence of human pro-C5 has been determined (102), and the derived protein sequence shows strong homology to the sequences of C3, C5, and α_2-macroglobulin indicating that they have a common evolutionary origin (65). However, unlike the other proteins, C5 does not contain an internal thiol–ester bond and thus does not show covalent binding activity upon activation. Instead, the freshly-activated C5b portion of C5 is considered to have a metastable binding site with a specificity for C6 due to the transient expression of hydrophobic residues. Thus it acts as the nucleus for the self-assembly of the membrane-attack complex without further proteolytic cleavages (103). The C5 convertases, of the classical and alternative pathways, are formed from the C3 convertases by the addition of C3b. In the classical pathway, the action of the C4b2a enzyme on C3 generates C3b which binds covalently to the serine residue at position 1217 of C4b to form the C4b3b2a complex (104). The bound C3b, along with C4b, acts as a co-factor which allows the efficient binding of C5 prior to its splitting by the C2a enzyme within the C4b3b2a complex. Similarly, in the alternative pathway, the covalent binding of freshly activated C3b to the C3b protein of the C3bBb complex yields C3b$_2$Bb which is the C5 convertase of the alternative pathway (105). The C5 is activated in an identical manner by the C5 convertases of either pathway which results in the C5a peptide (11 200 mol. wt) being split off from the α chain by the cleavage of a single arginyl bond, leaving the large (180 000 mol. wt) C5b fragment free to interact with C6. The C5a, like C3a and C4a, is an anaphylatoxin but differs from them in the susceptibility of its activity to plasma carboxypeptidase B (anaphylatoxin inactivator); the C3a and C4a are rapidly inactivated by the carboxypeptidases while the des-Arg C5a (the C5a lacking the C-terminal arginine) retains most of its activity *in vivo*. The C5 gene is located on chromosome 9q32–34 and is at least 50 kb long (106).

6. Regulation of complement activation by control of the C3 and C5 convertases. Factor I and the members of the RCA (regulator of complement activation) cluster

The complement system is controlled by a variety of inhibitors, enzymes, and regulatory proteins (Table 1) to prevent damage to self-tissue and to prevent all the components being consumed in a single event. The regulation of both the C3 and C5 convertases is mediated by plasma and membrane proteins which can interact with the C3b, or C4b, to cause dissociation of the appropriate convertase (i.e. to accelerate decay) or to act as co-factor in the limited proteolysis of C3b, or C4b, by the enzyme factor I. The genes for factor H, C4BPα and C4BPβ, CR1, CR2, MCP, and DAF have all been mapped to the long arm of human chromosome 1 in the q32 region. The products of these genes are all known to interact with C3b and/or C4b and, therefore, this linkage group is known as the 'regulator of complement activation' (RCA) cluster (107). These complement regulatory proteins (C4bp, H, CR1, CR2, MCP, DAF, along with the factor B, C2, C1r, and C1s enzymes) all contain three or more repeating homology units, each of approximately 61 amino acids, known as short consensus repeats (SCRs) or complement control protein (CCP) modules. These modules conform to a consensus sequence having a framework of highly conserved residues consisting of one tryptophan, two prolines, and four half-cysteine residues. Two other positions show conserved glycine residues, and at other positions a bulky hydrophobic amino acid (tyrosine or phenylalanine) is often found. The presence of the SCRs in the enzymes C2 and factor B and in the regulatory proteins appears to correlate with the binding of C3b or C4b. The enzymes of the C1 complex, C1r and C1s, also each contain two of the repeating units, indicating that eight of the repeating units could be aligned along the collagen-like region of the C1q molecule in the $C1q–C1r_2–C1s_2$ complex. This could possibly provide a site for interaction with C4 during activation or C4b and C3b during regulation (6). With respect to C4bp, factor H, CR1, CR2, MCP, and DAF, the argument that the SCRs in these proteins interact with C3b and/or C4b is very strong since these proteins are composed primarily of SCRs and there is good evidence from protein engineering studies which show that particular SCRs, or particular residues within a given SCR, are involved in the binding of C3b or C4b, as is discussed below. However, a number of non-complement proteins, with no known interaction with C3b or C4b, have also been found to contain the 60 amino acid homology repeat; human $β_2$-glycoprotein contains five repeats, human and mouse interleukin-2 (Il-2) receptor each contain two repeats, the β-subunit of factor XIII of the blood clotting system contains 10 repeats, and the selectins, LAM-1, ELAM-1, and GMP-140, contain 2, 6, and 9 repeats, respectively (reviewed in refs 108, 109). The functions of the non-complement proteins containing the repeats are quite varied. For example, the IL-2 receptor is a 55 000 mol. wt glycoprotein found on the membranes of antigen- or mitogen-stimulated T cells and in the presence of Il-2 is involved in the expansion of T cell populations, while the $β_2$ glycoprotein readily associates with heparin, lipoproteins, and platelets and the β-subunit of factor XIII also plays a binding role.

Electron microscopy studies of factor B, C2, H, and C4bp have provided measurements concerning the overall dimension of these molecules, and the three-dimensional structure, in solution, has been determined for one of the SCRs of factor H by two-dimensional NMR and simulated annealing (110). The NMR-study shows that the framework likely to be shown by all SCRs is a compact structure (3.8 × 2.0 × 10 nm) with a β-sandwich arrangement, with one face made up of three β-strands hydrogen-bonded to form a triple-stranded region at its centre and the other face formed from two separate β-strands. Molecular models of the proteins containing large numbers of repeats (eight or more) can be proposed if one assumes a similar disulfide-bonding (1 to 3, 2 to 4) in these repeats. Perhaps the C4bp molecule (with eight contiguous SCR repeats running from the N-terminal end of each of its seven identical disulfide-bonded α-chains) gives the clearest picture of the type of structure likely to be imparted by the repeats. The major form of C4bp has a mol. wt. of 570 000 and contains seven α-chains (each 70 kDa) and one β-chain (45 kDa). However, minor forms with compositions of $\alpha_7\beta_0$ and $\alpha_6\beta_1$ are also found. In the electron microscope, C4bp appears as a spider-like structure with seven flexible 'tentacles' (each 3 nm × 33 nm) joined to a central 'core'. Each tentacle and a portion of the 'core' is considered to represent an α chain of C4bp. The β chain is composed primarily of three SCRs and is located at the 'core' of the molecule. The C-terminal 58 residues of each of the seven chains are considered to form part of the stable α-helical structure disulfide-bonded core (111) along with the α-chain, while the N-terminal 491 amino acids, containing the eight repeating units, are considered to provide an array of compact β-stranded structures like 'beads on a string'. It seems likely that the eight SCRs in the C4bp α-chain would form eight similar structural domains each of approximately 4 nm arranged in a linear and tandem fashion (8 × 4 = 32 nm) along the 'tentacle' structure (estimated to be 33 nm long from the electron microscopy studies), the N-terminal SCR being at the extremity. Factor H and β_2 glycoprotein I, which contain twenty and five SCRs, respectively, also show elongated structures, in-dicating that this may be a general feature of proteins containing large numbers of these repeats—especially in the case of the largest allotype of CR1, which has thirty SCRs. It is probable that the role of the repeat units is to confer a general structural framework that can be utilized in a variety of binding reactions in much the same way as the different immunoglobulin domains can fulfil a varied number of binding and functional roles.

Gene mapping studies from several laboratories have established the order of the RCA genes (Fig. 10). They lie within an approximately 900 kb of DNA, in the order MCP, CR1, CR2, DAF, C4BPα, C4BPβ (reviewed in ref. 112), with all the genes being aligned in the same 5' to 3' transcriptional orientation which is suggestive that evolution of the RCA cluster evolved by gene duplication. Recently a C4BP-like gene has been mapped adjacent to the 3' end of the C4BPβ gene, a CR1-like gene mapped to the 3' end of the CR1 gene, and an MCP-like sequence mapped between the CR1-like and CR1 genes, but the products of these recently identified genes have not yet been characterized. It appears that the factor H gene is at least 500 kb distant from the other members of the RCA cluster and, as discussed in Section 6.2, there also appears to be a cluster of factor H related genes, along with the factor XIIb gene.

6.1 Factor I

Factor I is involved in the regulation of the C3/C5 convertases of either the classical or alternative pathway. It is an enzyme which specifically cleaves at one or two

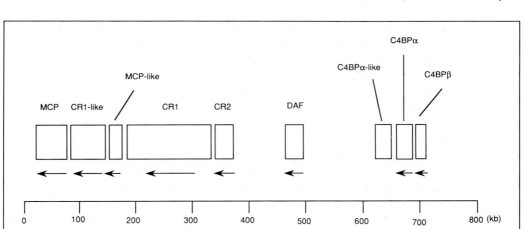

Fig. 10 Alignment of the genes within the RCA cluster on human chromosome 1. (112).

positions the α' chains of the activation products (C4b and C3b) of components C4 and C3 (i.e. after association of the C4b or C3b with one of the appropriate co-factor proteins, (Table 3). These specific cleavage reactions cause rapid loss of the biological activities associated with C3b and C4b, which includes their role in the C3/C5 convertases. Biosynthetic studies utilizing human hepatoma cell lines showed that factor I is synthesized as a single chain precursor of 88 000 mol. wt (after about 15 per cent glycosylation) which is then processed to yield the active form composed of two disulfide chains of 50 000 and 38 000 mol. wt found circulating in plasma. The isolation of cDNA clones (113) covering the entire coding region (1750 bp) of the pre-pro-factor I molecule and the derived amino acid sequence shows that there is a putative leader peptide of 18 residues followed by a cysteine rich (29 Cys residues) heavy chain of 317 residues and a catalytic light chain of 240 residues, of the serine proteinase type. Six potential N-linked glycosylation sites (three in each chain) are also seen. Activation of the precursor I probably involves the limited proteolysis and trimming of a highly basic peptide sequence (–Arg–Arg–Lys–Arg–) located at the C terminus of the heavy chain. The amino acid sequence of the heavy chain is of a 'mosaic' nature similar to that found in other serum serine proteases. One striking feature is the presence of both class A and class B low density lipoprotein (LDL) receptor repeats with conserved cysteine residues similar to those found in other complement proteins such as C9.

Only a few cases of complete deficiency of factor I have been reported. As would be expected from the regulatory role of factor I, these individuals have little or no native C3 in their plasma due to the uncontrolled formation of the C3bBb complex which splits C3. These patients showed impaired resistance to bacterial infection, clearly illustrating one of the important biological roles played by the complement system. The gene has been mapped to the 4q25 region of human chromosome 4 (114).

Table 3 Proteins involved in the control and regulation of C3b and C4b[a]

Types of control reaction

A C3b + co-factor–C3b–co-factor complex $\xrightarrow[\text{by factor I}]{\text{2 cleavages}}$ iC3b + C3f + co-factor

B iC3b–C3c + C3dg (possible involvement of factor I plus CR1)

C C4b + co-factor–C4b–co-factor $\xrightarrow[\text{by factor I}]{\text{1st cleavage (iC4b–co-factor complex)}}$ 2nd cleavage by factor I
   $\downarrow$
   C4c + C4d + co-factor

D Acceleration of decay of C4b2a or C3bBb

Reaction	Substrate or pathway	Plasma control proteins		Membrane associated control proteins		
		C4bp	**H**	**CR1**	**MCP**	**DAF**
A	Soluble C3b	(+)[b]	+	+	(+)[d]	−
	Surface-bound C3b	−	+	+	(+)[d]	−
B	Soluble iC3b	−	−	−	−	−
	Surface-bound iC3b	−	(+)[b]	(+)[c]	−	−
C	Soluble C4b	+	−	+	(+)[d]	−
	Surface-bound C4b	+	−	+	−	−
D	Accelerated decay of C4b2a	+	−	+	−	+
	Accelerated decay of C3bBb	−	+	+	−	+

[a] Table adapted from refs 139 and 143.
[b] Not in physiological conditions.
[c] Due to the susceptibility of iC3b to random proteolysis, the role of CR1 and factor I in breakdown of iC3b requires further assessment.
[d] MCP (gp45-70) supports only the first of the two cleavages in reactions A and C.

6.2 Plasma control proteins: C4b-binding protein and factor H

C4b-binding protein (C4bp) regulates the C3 convertase (C4b2a) of the classical pathway via its high-affinity for both fluid phase C4b and surface-bound C4b. By virtue of this binding capacity, C4bp can inhibit the formation of the convertase and it can also accelerate the decay of the convertase by displacing C2a from C4b such that it is readily cleaved in two positions by factor I to yield C4c and C4d (Fig. 4 and Table 3).

C4bp is the largest of the complement proteins being approximately 570 000 mol. wt. Human C4bp is composed of seven identical α-chains of 71 000 mol. wt and a single β-chain of 45 000 mol. wt. The N-terminal 491 residues of each α-chain (111) are composed of eight repeating, but not identical, homology regions of 60 amino acids of the SCR type (i.e. similar to the SCRs in the N-terminal regions of factor B and C2). This is considered to be a general feature of most proteins involved in binding to C4b or C3b (108, 109). The C-terminal region of 55 amino acids of each chain is not related to the repetitive homology regions and appears to be involved in the formation of a closely disulfide-linked 'core' holding the seven chains together. The β-chain is composed of three SCRs as judged by protein sequence and cDNA cloning studies (115). The repetitive regions appear to impart an unusual elongated structure to the α-chains of the C4bp molecule since, in the electron microscope, C4bp is seen as a 'spider-like' shape composed of seven thin, elongated and flexible arms linked to the small ring-like 'core' (116, 117). The binding sites for C4b appear to be located near the peripheral end of each arm (i.e. the NH_2-end of each chain). Ultracentrifuge studies indicate that, at physiological ionic strength, C4bp has four binding sites for C4b, each having an association constant of $1.2 \times 10^7 \ M^{-1}$, while at low ionic strength up to six binding sites can be detected (118). A cDNA clone for the α-chain of mouse C4bp has been isolated from a liver cDNA library (119). The predicted amino acid sequence for the α-chain of mouse C4bp indicates that each of its chains is composed of six internal SCRs of 60 amino acids (unlike the human C4bp which has eight) and a C-terminal region of 55 amino acids which shares no homology with the repeating units. There is extensive homology between the mouse and human C4bp at the amino acid level (61 per cent) as well as at the nucleotide level for both the coding and 3′ untranslated regions. There are two major differences found between the mouse and human C4bp α-chains. Firstly, the area corresponding to residues 232–289 and 329–392 (repeat regions 5 and 6) in human C4bp are absent in mouse C4bp, thus accounting for the size difference between the polypeptide chains. Second, there is an absence of four cysteines in mouse C4bp, two of which are involved in the interchain disulfide linkages in human C4bp, thus accounting for the lack of covalent association between the mouse C4bp α-chains.

The human *C4bp* gene has been mapped to chromosome 1 while the mouse *C4bp* is on either chromosome 1 or 3. The human *C4bp* α-chain gene is approximately 30 kb long and is split into twelve exons (120) with each of the SCRs, defined at the protein level, being encoded by separate exons. The *C4bp* β-chain gene spans 13 kb of DNA and is split into seven exons with exons 3 and 4 each encoding a complete SCR (SCRs I and II) while the third SCR is split over exons 5 and 6. No deficiencies of C4bp have yet been described.

Factor H is a single chain glycoprotein of 150 000 mol. wt which binds to C3b and thereby regulates many of the functions associated with C3b. Its binding greatly accelerates the decay of the C3bBb and C3bBbP complexes in plasma (Table 3). A cell-surface bound form of factor H (also having a molecular weight of 150 000) may function as C3b receptor on certain monocytic and lymphoblastoid cell lines and monocytes and tonsil lymphocytes (92). It is likely that factor H also regulates the C5 convertase since it competes with the binding of C5 to C3b. In an analogous manner to that of C4bp–C4b interaction in the classical pathway, factor H acts as a co-factor in the cleavage of C3b, and also of $C3(H_2O)$, by factor I, allowing factor I to split the α′ chain of C3b in two positions to yield iC3b. Further proteolysis of iC3b by trypsin-like enzymes yields C3c and C3d. As well as acting as the main soluble plasma regulatory protein involved in the control of C3b, factor H is also found in a membrane-bound form, of approximately 150 000 mol. wt, on monocytes, lymphocytes, Raji, and U937 cell lines.

The plasma forms of mouse and human factor H are each composed entirely of

20 SCRs (121, 122) thus displaying the characteristics found for other C3b/C4b binding proteins (108, 109). Again, the presence of a large number of these repeating units appears to generate an unusually shaped molecule since physical and electron microscopy studies indicate that factor H is very elongated. Functional studies on the human factor H (123) indicate that the binding activity for C3b probably is located near the N-terminal end of the molecule (i.e. within repeats 1–6). Northern blot analysis indicates the presence in human liver of a full-length mRNA of 4.3 kb for factor H. Two smaller mRNAs (1.8 kb and 1.4 kb) are also detected by factor H cDNA probes (124). The 1.8 kb transcript codes for a 43 kDa polypeptide which corresponds to the N-terminal 431 residues of intact factor H and is probably derived by alternative splicing (125). The 1.4 kb transcript appears to code for two glycoproteins which, due to different usage of glycosylation sites, have molecular weights of 42 kDa and 37 kDa (126). These proteins contain five SCRs, two of which are very similar to SCRs present in factor H. Therefore, these proteins seem likely to be derived from a gene related to the factor H gene. Another glycoprotein of 37.5 kDa, which contains five SCRs that are clearly related to use in factor H, has also been described (127), but its function is unknown. Two other factor H-related plasma proteins, of 29 kDa and 24 kDa, which each contain four SCRs have been characterized at the cDNA level (126). There, therefore, appears to be a gene family composed of the factor H gene plus at least three factor H-related genes. However, the functional roles of the products of the factor H-related genes have not yet been determined.

There is only one report of factor H deficiency and this is for an incomplete deficiency since factor H levels at about 5 per cent of the normal level were observed in the patient examined. Three genetic variants (*FH1*, *FH2*, and *FH3*) which differ in isolectric point have been described for human factor H, while in mice two variants have been found. The genetic variants in both species are encoded by co-dominant alleles at a single autosomal locus on chromosome 1 (q-region) in both species.

6.3 Membrane-associated regulatory proteins

6.3.1 Complement receptor type 1

Complement receptor type I (CR1) is an integral membrane, single chain glycoprotein found on a wide range of cells which include erythrocytes, monocytes/macrophages, neutrophils, B lymphocytes, and some T lymphocytes and podocytes (92). CR1 shows binding and co-factor activities for the regulation of both C4b and C3b (Table 3), but its principal role is considered to lie in the elimination of immune complexes. After the complexes become coated with C3b during complement activation, they can bind to CR1 on erythrocytes and thus be removed efficiently from the circulation via the liver and spleen. As C3b is degraded by the action of the control proteins CR1, or H, and enzyme 1, fragments of C3b are generated which are recognized by other complement factors involved in clearance mechanisms (Table 3).

Human CR1 from erythrocytes was the first complement receptor to be purified to homogeneity. This allowed a limited amount of protein sequence to be obtained for the generation of oligonucleotide probes in order to screen a tonsillar cDNA library and isolated cDNA clones for one of the structural allotypes (128). The protein and cDNA data indicate that the extracellular portion of the membrane-bound molecule is composed principally of four tandem long homologous repeating structures (LHRs) each of approximately 450 amino acid residues (LHRs A–D). Alignment of the repeats shows that LHR-B is almost identical to LHR-C over the first 259 residues while LHR-D differs over that region. From residues 260 to 450 the LHRs C and D are almost identical while LHR-B is different. Within each LHR there are seven repeating homology units of approximately 60 amino acids similar to the SCRs found in other C3b/C4b binding proteins. The whole CR1-A chain (the longest form of CR1) contains 30 SCRs, and the only non-repeating structures found in the molecule are 25 residues which may form a membrane-spanning segment followed by an anchor of four positively-charged residues and a 43 residue C-terminal cytoplasmic tail (128). The site in CR1 which binds C4b has been localized to SCR1 and/or SCR2 and the site which binds C3b has been localized to SCR8 and/or SCR9 (129). Protein engineering studies have identified three residues within SCR2 and a stretch of five residues within SCR-9 which appear important for the binding of C4b, and iC3, respectively (130), which emphasizes the probable requirement of contributions from two contiguous SCRs to allow CR1 to interact with its ligands.

Four genetic variants of human CR1 are found which differ in molecular weight and gene frequency: CR1-A, 190 000 mol. wt (0.83 gene frequency); CR1-B, 220 000 (0.16); CR1-C, 160 000 (0.01); CR1-D, 250 000 (< 0.01). No antigenic or functional differences have been demonstrated despite the differences in molecular weights between the four variants (reviewed in ref. 131). It has also been noted that the variation in molecular weights does not appear to be due to a variation in carbohydrate content. In view of the demonstration in CR1 of the presence of long homologous repeats of 450 amino acids, it is likely that the size polymorphism (30 000 mol. wt between allotypes) is determined at the genomic level and may have been generated by unequal crossing over. Genomic studies on the two most common allotypes appear to confirm this view since the size polymorphism (about 40 kDa) between the allotype is determined by the presence, or absence, of regions encoding an entire LHR (100, 132). The CR1-B allotype (designated as the S allele) is encoded by 158 kb of DNA which includes the coding information for SLHRs. However, the coding sequences for the second LHR of the CR1-B allotype are not present in the CR1-A allotype (F allele), which thus accounts for the difference in molecular weight seen between the two most common allotypes. All the variants have been shown to be encoded by a co-dominant allele at a single locus on chromosome 1 (131).

No complete deficiency states for CR1 have been reported but the number of CR1 molecules found on red cells appears to be genetically influenced, being an autosomal co-dominantly inherited trait which has been linked to an RFLP of the

CR1 gene (131). Individuals with low numbers of CR1 might be expected to have an impaired ability to clear circulating immune complexes and thus be at risk of developing an immune complex-mediated disease. It has been reported that patients with SLE have a relative deficiency of CR1 on their erythrocytes (133). However, this situation is considered to be due to a combination of factors; an inherited deficiency which correlates with the absence of the RFLP seen in individuals with normal numbers of CR1 on their erythrocytes (133), and a second (acquired) deficiency which is associated with complement activation at the cell-surface. This acquired loss of CR1 has been seen in several other immune-type diseases.

6.3.2 Complement receptor type 2

Complement receptor type 2 (CR2) is a membrane glycoprotein of 145 000 mol. wt (about 20 per cent carbohydrate in the form of 8–11 *N*-linked oligosaccharides). It is located in B lymphocytes and spleen follicular dendritic cells. It binds to the C3d region of C3 and the solubilized receptor will bind to C3u, C3b, iC3b, C3d, and C3dg (92, 131). The main binding site appears to involve a Leu–Tyr–Asn–Val–Glu–Ala hexapeptide sequence in the C3d region. In addition to recognizing C3d, the CR2 also serves as a receptor for the human Epstein–Barr virus (EBV), but the site on CR2 which recognizes EBV is probably different from the site which recognizes the C3 fragments (for a review see ref. 134). CR2 may play a role in B-cell activation (134) since it is phosphorylated when B cells are stimulated with phorbol esters or antibody to surface Ig (135), a property it shares in common with CR1. It has been established, by cDNA studies, that the extracellular portion of CR2 can be composed of 15 or 16 SCRs since alternative splicing events can eliminate the use of the region encoding SCR11 (136). SCRs 1 and 2 of CR2 appear to contain the binding sites for C3dg and the viral protein gp350/200. Protein engineering studies have identified specific residues within these two SCRs which are important for efficient binding (137). Both *CR1* and *CR2* have been mapped to chromosome 1 at band 1q32 (138).

6.3.3 Membrane co-factor protein

Membrane co-factor protein (MCP) is present on a wide variety of cells, which includes lymphocytes, monocytes, platelets, epithelial cells, endothelial cells, and fibroblasts. MCP possesses co-factor activity for the factor I cleavage of C3b, or C4b, but it has a preferential affinity for protein-bound C3b (139). Unlike the decay acceleration factor (DAF), it does not display decay-accelerating activity. However, it has a similar overall structure to DAF, being composed of four SCRs, followed by a 25-residue-long, *O*-glycosylated, region (rich in Thr and Ser) plus a 23-residue-long membrane spanning region and a C-terminal cytoplasmic domain of 33 residues (140). Nevertheless, MCP differs from membrane-bound DAF in that it is bound to the membrane by a hydrophobic transmembrane sequence rather than a GPI anchor. Four different cDNA isoforms of MCP have been characterized (141, 142), three of which appear to be expressed and, along with other possible variants, may correlate with the different forms of MCP found in tissues, tumours, and serum. The *MCP* gene is 43 kb long, maps to within 100 kb of the 3' end of the *CR1* gene, and the intron–exon boundaries of the gene have been established (143).

6.3.4 Decay acceleration factor

Decay acceleration factor (DAF) is mainly found as a single-chain integral membrane glycoprotein of 70 000 mol. wt (about 35 per cent carbohydrate present as *O*-linked sugars) which has a very wide tissue distribution, the major sites being platelets, erythrocytes, and leucocytes. Low levels (about 60 ng/ml) of a soluble form of DAF have also been found in plasma, saliva, and urine. DAF accelerates the decay of the C3 convertase of both pathways but does not exhibit any co-factor activity in the presence of factor I (Table 3). It appears to bind only to C3b or C4b molecules on the membrane of the same cell as the DAF molecule is itself located. It is possible that its role is to prevent assembly of the convertases rather than to dissociate complexes that have already been formed (124). An unusual structural feature of the membrane form of DAF is that it is anchored to the cell-surface by a glycophospholipid anchor. Two classes of cDNA clones encoding DAF have been characterized and found to differ by an apparent internal deletion of 118 bp within the C-terminal portion of the coding region (144). It appears that this deleted segment represents an unspliced intron composed of Alu consensus sequence (144). Analysis of the cDNA data suggested that mRNAs for two different DAF proteins could be generated by means of inclusion or exclusion of the apparent 'intron' sequence by splicing. The unspliced DAF mRNA would encode a protein of 440 amino acids (including a signal peptide of 34 amino acids) in which residues 328–366 would be encoded by the intron (145). The spliced DAF mRNA would yield a shorter protein of 381 amino acids with a different C terminus produced as a result of a frameshift brought about by the splicing. The predicted C-terminal domains of the two forms of DAF differ markedly in that the long form of DAF (unspliced mRNA) would be hydrophilic in nature while that of the short form (spliced mRNA) would be hydrophobic in nature. It was proposed that the predicted short form of DAF with its hydrophobic C-terminal region represents the major (90 per cent) membrane form of DAF (the hydrophobic portion being cleaved off and replaced by the glycophospholipid anchor), while the predicted long form, which lacks a hydrophobic tail, would represent the minor (10 per cent) soluble form of DAF found in body fluids. The cDNA studies also showed that the N-terminal portion of the mature protein is composed of four SCRs, followed by a 70 residue-long, *O*-glycosyated, region (rich in Thr and Ser) and then the C-terminal hydrophilic or hydrophobic tail. The gene encoding human DAF is approximately 40 kb long and contains 11 exons, with SCRs I, II, and IV each being precisely encoded by single exons (145). SCR III is encoded by two exons due to a splice junction after the second nucleotide in the Gly codon at position 34 in the SCR. This is also the case for the SCRs encoded by two exons in human *CR1*, *CR2*, *MCP*, and mouse factor H. The Ser/Thr rich region is encoded within three separate exons (6–9). The finding of an Alu type sequence within exon 10 is consistent with the generation of two classes of DAF cDNA clones (as discussed above), one of which contains a 118 bp insert near the 3' end of the cDNA, thus confirming the view that the longer soluble minor form of DAF arises by alternative splicing.

7. The C3a, C4a, and C5a anaphylatoxins.

The C3a, C4a, and C5a anaphylatoxins are peptides, 77 amino acids long, which are released from the N-terminal ends of the α-chains of C3, C4, and C5, respectively, by (in each case) the splitting of a single Arg–Xaa band during complement activation. These peptides mediate inflammatory responses by binding and triggering receptors which cause vascular permeability changes, induction of smooth muscle contraction, histamine release, ATP release, chemotaxis, and superoxide production. The spasmogenic activities are in the order C5a>C3a>C4a, and the receptors are found on quite a wide variety of cell types such as monocytes, macrophages, granulocytes, endothelial cells, and platelets. The C5a receptor is distinct from the C3a/C4a receptor, although they are often expressed in the same cell type. The activities of the anaphylatoxins are controlled in blood by the removal of the arginine, found at the C-terminal end of each anaphylatoxin, by serum carboxypeptidase N. The des-Arg forms are inactive except in the case of C5a-des-Arg which appears to retain its neutrophil stimulatory activities and thus indicates that C5a is probably the most important of the anaphylatoxins in terms of stimulation of host defence mechanisms. It is known that C5a stimulates chemotaxis and superoxide production in neutrophils by binding to the C5a receptor which is coupled to a GTP-binding protein complex. The cDNA encoding the C5a receptor has been cloned from U937 and HL-60 cells (146, 147). Its analysis has shown that C5a is a member of the rhodopsin superfamily, having the characteristic seven transmembrane helices plus four short extracellular segments. This is consistent with functional data indicating that the C5a receptor acts via a GTP-binding protein with which it forms a strong complex (148). The C5a receptor shows 35 per cent identity in amino acid sequence to the receptor for the bacterial chemotactic peptide f–Met–Lau–Phe which agrees with the fact that the two receptors mediate similar but not identical responses in neutrophils and monocytes.

8. Complement receptor type 3 and p150,95: members of the LFA-1 family of cell adhesion molecules designated the leucocyte-specific β_2 integrins

Complement receptor type 3 (CR3) (previously defined as Mac 1) p150,95, and LFA-1 are members of the integrin family of cell-surface antigens (for review see ref. 149). They each have a common β chain (95 000 mol. wt) linked covalently to one of three distinct types of α chain. The molecular weights of the α chains of CR3, p150,95, and LFA-1 are 165 000, 150 000, and 180 000, respectively. The common β_2 chain of the leucocyte specific integrins thus gives the molecules: $\alpha_M\beta_2$ for CR3; $\alpha_X \beta_2$ for p150,95 and $\alpha_L \beta_2$ for LFA-1. Using the CD nomenclature, $\beta_2 =$ CD18 and α_L, α_M, $\alpha_X =$ CD 11a, 11b, 11c. 'CD' denotes 'cluster of differentiation' antigen and is based on the designation of human leucocyte cell-surface antigens identified with monoclonal antibodies, each antigen being given a CD number (for review see ref. 150) Each of these three different types of $\alpha\beta$ complexes participates in some form of cell adhesion activity:

1. LFA-1 is involved in the interaction between T cells and their targets and has ICAM-1 and ICAM-2 as ligands.
2. CR3 has affinity for iC3b, produced by degradation of C3b by factors I and H, this interaction possibly promoting phagocytic activity (151). Fibrinogen, factor X, ICAM-1 and lipopolysaccharides are also considered to be ligands for CR3.
3. p150,95 appears to have a similar binding specificity to that of CR3 but its precise function is not known.

The members of this group of receptors differ somewhat in their tissue distributions: LFA-1 is present on B and T lymphocytes, polymorphs and monocytes; CR3 is found on monocytes, neutrophils, and NK cells, but not macrophages; p150,95 is found on monocytes and neutrophils, but also at high levels on tissue macrophages.

Deficiency of one or more members of the LFA-1 family of membrane glycoproteins has now been found in a number of patients with leucocyte adhesion deficiency (LAD) who suffer from recurrent bacterial and fungal infections (151, 152). In most cases, the patients appear to be deficient in all three complexes which suggests that the deficiency lies in the common β_2 chain. This suggestion is consistent with the observation that absence of the expression of the β chain on the cell-surface interferes with maturation and expression of the α chains. There is a heterogeneous collection of defects which are responsible for problems with the β_2 in LAD, and these include deletions and point mutations (152).

The cDNA for the common β_2 chain of the LFA-1 family has been cloned by two groups (153, 154). The derived protein sequence has a 22 residue signal peptide followed by an external domain of 669 residues that contains a 240 residue cysteine-rich (17 per cent) domain (most of the cysteine being found in three tandem repeating units), a single membrane-spanning domain, and an intracellular C-terminal domain of 46 amino acids. The entire sequence of the β_2 chain shows 47 per cent identity to the small chain of the fibronectin-binding protein, integrin, from chicken fibroblasts (155). Thus the β_2 chain appears to belong to an extended family of cell surface proteins which includes the fibronectin receptor, the vitronectin receptor, and the gpIIb/IIIa glycoprotein of platelets, all of which display a specific affinity for peptides containing the amino acid sequence –Arg–Gly–Asp– present in the binding region of the fibronectin molecule. The observation that a synthetic peptide derived from the known sequence of component C3, which contains the sequence –Arg–Gly–Asp–, can bind to CR3 (156) reinforces the view that the β chain of CR3 should be assigned on functional as well as structural terms to the integrin superfamily of cell adhesion molecules. However, there does not appear to be an absolute requirement for the presence of the –Arg–Gly–Asp sequence in potential ligands for CR3 (157) indicating that CR3 may contain at least two binding sites.

The α chains of the leucocyte integrins show a high degree of sequence identity when compared with each other, but little similarity to their common β_2 chain. Each of the α chains contains a region approximately 180 amino acids long in the N-terminal part of the protein, which has similarities to regions in von Willebrand factor and factor B. It is thought that this region may be important in ligand recognition since it is often recognized by monoclonal antibodies which block integrin function. This region of 180 amino acids has been found only in one other type of α-chain (in the β_1-integrin subfamily) and therefore is defined as an 'inserted' or 'I' domain which lies within repeats II and III of a stretch of seven repeating units of 60 amino acids (three of which, numbers V–VIII, show similarities to 'EF-hand' calcium binding proteins).

9. The terminal attack complex C5b–9 and the relationship of C8 and C9 to perforin from killer lymphocytes

A remarkable feature of the terminal components of the complement system is the manner by which the water soluble hydrophilic components C5b and C9 undergo a hydrophilic–amphiphilic transition via a non-enzymatic self-assembly mechanism to form the membrane attack complex (MAC). The MAC behaves as an integral membrane component, forming transmembrane channels and displacing lipid molecules and other constituents. This disrupts the phospholipid bilayer of the plasma membrane target cells and leads to cell lysis and death.

Freshly activated C5b, loosely bound to C3b, binds to C6 to form a C5b–6 complex and then to C7 to form a C5b–7 complex which has a metastable binding site for membrane surfaces (103). C6 and C7 are both single-chain glycoproteins showing similarity in molecular weight (about 115 000) and other physical parameters. Comparison of their derived amino acid sequences and studies of the genetic polymorphisms of C6 and C7 show they are closely linked. It is, therefore, probable that they evolved from a common ancestral gene. The C5b–7 complex appears to undergo a hydrophilic–amphiphilic transition upon binding of C7 which is considered to generate a metastable binding site for membranes and to cause the complex to become dissociated from C3b. If at this stage the C5b–7 complex fails to bind to a membrane surface, its potential cytolytic activity, via the generation of C5b–9, is lost and self-aggregation of the complex takes place in the fluid phase.

C8 is a protein of 151 000 mol. wt. It is composed of three chains (α, β, and γ) of 64 000, 64 000, and 22 000 mol. wt, the α and γ chains being disulfide bonded. Binding of the the C8 to the C5b–7 complex, probably via its β chain, may bring about a conformational change in C8 which allows the disulfide-linked α and γ

chains to penetrate into the hydrophobic core of the lipid bilayer of the membrane to which the C5b–7 complex is attached.

C5b–8 binds C9 and acts as a catalyst in the polymerization of C9 as seen in typical membrane lesions of target cells (Fig. 11a–c). The MAC has a composition C5b, C6, C7, C9, $C9_n$ (where n lies between 1 and 18). The minimum and maximum molecular weights of the complex are 660 000 and 1 850 000, respectively, C9 being a single chain protein of 71 000 mol. wt. The binding of the first molecule of C9 to the C5b–8 complex is mediated through the C8 component. This is known from the observation that this step is inhibited by antibody to C8. Furthermore, C8 and C9 interact in free solution. The mechanism by which C5b–8 binding of one molecule of C9 allows high-affinity C9–C9 interaction is not absolutely clear. However, it is the relatively hydrophobic C-terminal half of each C9 molecule which is considered to become inserted into the phospholipid membrane during

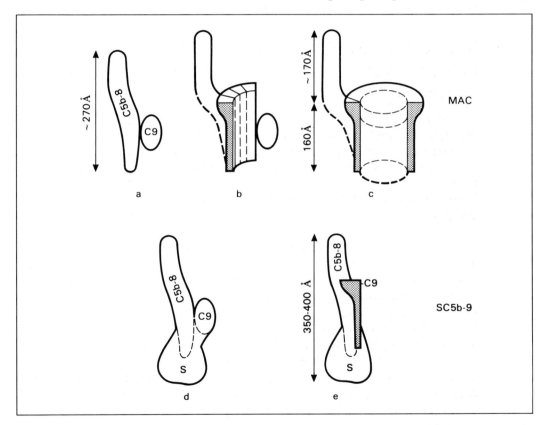

Fig. 11 Models for C5b–9 complex (MAC) and the fluid phase complex of S-protein plus C5b–9. Panels (a) to (c): assembly of MAC represented as C5b–8 complex in association with poly C9. After binding of the first molecule of C9 in its soluble globular form (a) a conformational change takes place causing it to behave like an integral membrane protein capable of disrupting the membrane and allowing further C9 molecules to polymerize (b) to eventually produce a cylindrical lesion (c). Panels (d) and (e): models for S–C5b–9 containing C9 in globular (d) or partially unfolded form (e) as described in the text. (This figure is from ref. 177.)

the formation of membrane lesions. The type of lesion seen is dependent upon the availability of monomeric C9. If enough C9 is present (for example at a ratio of six C9 molecules to one C5b–8 complex) then discrete typical cylinder-like membrane lesions are seen (Fig. 11c). At low levels of C9 relative to C5b–8 (for example a ratio of 1:1) no typical membrane lesions are seen, only a network of protein aggregates. However, membrane damage takes place even in the apparent absence of cylindrical lesions. This has been confirmed by showing that C9 cleaved by thrombin in only one position was as effective as intact C9 in cell lysis yet produced no 'ring-like' structures, only 'string-like' structures (158). Thus it is possible that the classical complement lesion is not always a prerequisite for cell lysis and that complexes carrying low numbers of C9 produce smaller hydrophobic membrane channels.

C9 also has the capacity to undergo a self-polymerization process, but what induces this procedure *in vivo* is not clear since rather non-physiological conditions (high concentrations of heavy metals, such as Zn^{2+}) and high temperatures (greater than 37°C) are required *in vitro*. Self-polymerizing C9 can attack model lipid vesicles in the absence of C5b–8 but it cannot lyse cells unless C5b–8 is present on the surface. This observation is consistent with the view that C5b–8 is not only involved with the binding and polymerization of C9 but also plays an important role in the formation of transmembrane channels.

The self-polymerization of C9 is similar in several respects to the Ca^{2+}-induced polymerization of perforin, a 70 000 mol. wt glycoprotein present in granules of cytotoxic T cells and NK-like lymphocytes (159, 160). Purified perforin produces structural and functional lesions in target membranes similar to those caused by the intact killer cells or isolated granules. The essential functional difference between C9 and perforin lies in the observation that purified perforin will lyse erythrocytes and tumour cells in the presence of Ca^{2+} while C9 requires the prior assembly of C5b–8 on the target membrane, as outlined above, before efficient cell lysis will take place.

As judged by their protein sequences derived from cDNA studies, C7, C8α C8β, and C9 show approximately 21–26 per cent identity within any pair of sequences C6 (161, 162) C7 (163), C8α (164), C8β (165), and C9 (166, 167). Each of these terminal components shows a mosaic structure with all of them containing a large internal domain (of approximately 240 amino acids) almost free of Cys residues. This domain is also found in perforin, the pore-forming protein of killer T cells (160, 168). It is possible that this domain is important for the capacity of C9 and perforin to polymerize, form tubules, and become inserted in membranes. Perforin does not show sequence similarity to the terminal complement proteins in regions outside the central C9/perforin domain. Other types of domain structures found in the terminal components include: one type of thrombospondin repeat (of the same type as found in properdin); the low-density lipoprotein receptor motif (LDLr motif); the epidermal growth factor motif (EGF motif); the complement control protein type repeat (CCP or SCR repeat). The C6 protein is a good example of the mosaic nature of the terminal components since it is composed of two, N-terminal,

TSRs followed by an LDLr motif, the perforin/C9 type domain, an EGF-motif, another TSR, two SCRs, and finally, at the C-terminal end, two motifs of a type as yet found only in C6, C7, and factor I. The TSR found in the C-terminal half of C6 is also present in C5, C7, C8α, and C8γ but not in C9 or perforin. It has been suggested that the lack of this motif in C9 or perforin may account for their ability to form circular polymers (160).

The human C6 and C7 genes are each approximately 80 kb long and contain 18 and 17 exons, respectively (169). They are closely linked, being only 170 kb apart, in a 3′ to 3′ orientation on chromosome 5q. The γ chain of C8 does not show similarity to any of the other chains. Its derived amino acid sequence indicates that it belongs to the lipcalin protein family which binds small hydrophobic ligands such as retinol and retinoic acid. Many of the domains/motifs seen in C6 and C7 are, unexpectedly, not precisely encoded by discrete exons since many of the exons code for sections of more than one type of domain/motif. This is also a feature of the C9 gene structure (170). There is a strong conservation of intron/exon boundaries between the genes for C6, C7, and C9, and preliminary data suggests that this is also the case for the C8α and C8β genes. The C8α and C8β genes lie within 2.5 kb of each other on chromosome 1 at 1p34 (171). The C9 gene, like the C6 and C7 genes, has been assigned to chromosome 5, but its position (at 5p13) is quite distant from the C6/C7 locus. The C8γ gene has been assigned to chromosome 9q (171). The human perforin locus maps to chromosome 10q22 and, therefore, is not linked to any of the genes encoding the terminal components of complement (172).

Control of the formation of the C5b–9 complex is partly carried out by members of the RCA cluster and factor I, which prevent the efficient activation of C5b. However, direct control at the level of the terminal components is carried out by CD59, 60 kDa homologous restriction factor, S-protein and SP40,40. CD59 is a 20 000 mol. wt, cell-surface protein which is found on a wide variety of cells and which is linked to the membrane by a glycosylphospholipidyl-inositol anchor. It is of prime importance in the prevention of cell lysis by homologous complement and is considered to act at the C5b–8 to C5b–9 stages by limiting C9 input into the complex (173). Several laboratories have been involved in the characterization of CD59 and, therefore, it has acquired a number of trivial names which include HRF20, MACIF, protectin, MIRL, and MEM-43-Ag. It is 103 amino acids long and shows no similarity to any of the other complement regulatory proteins. However, the human CD59 does show 25 per cent identity to the glycolipid-anchored mouse Ly6 antigen (T cell activating protein) with conservation of 10 out of the 11 cysteines (174) The molecular basis for a genetic defect in CD59 has been established in a patient suffering from paroxysmal nocturnal haemoglobinuria (PNH) (175). Since the patient showed no deficiency in any other GPI-anchored proteins, this finding emphasizes the important protective role which CD59, rather than, for example, DAF, plays in preventing the destruction of red blood cells by homologous human serum. The 65 000 mol. wt, homologous restriction factor (HRF) is also considered to be an important self-protection molecule which prevents the MAC from attacking red blood cells or lymphocytes (176). It is found in both soluble and membrane-bound forms and it is considered to inhibit the action of perforin, as well as C9, and thus the presence of the soluble form in the granules of killer lymphocytes may be of importance in the self-protection of these lymphocytes. To date, no protein or cDNA sequence has been reported for the 65 000 mol. wt HRF.

S-protein is a single-chain plasma glycoprotein of 80 000 mol wt. Up to three molecules of the S-protein can bind to the metastable C5b–7 complex which prevents the complex from binding to the cell surface (177). This protects bystander cells against lysis by the MAC. The resulting fluid phase S–C5b–7 can bind C8 (to form S–C5b–8) and C9 (to form S–C5b–9), but polymerization of C9 does not take place (Fig. 11d and e). The inhibitory effect of S-protein may be mediated by one of the following two mechanisms:

- by preventing the conformational change in C9 that appears to be a prerequisite for polymerization (Fig. 11d);
- if the required conformational change has taken place, by steric hindrance of the polymerization steps (Fig. 11e).

Clones for S-protein were isolated from a human liver cDNA expression library (178) which allowed derivation of the sequence of the leader peptide (19 residues) and the mature chain (459 residues). This showed that S-protein is identical to plasma vitronectin which had previously been characterized as a member of the family of substrate adhesion molecules such as collagen, fibronectin, and laminin. Thus it is clear that S-protein, as well as interacting with C5b–7, can bind to a wide variety of cell surfaces and subcellular matrices. However, it is possible that S-protein/vitronectin may play a relatively minor role in regulation of C5b–9 formation, and that the SP40,40 protein, also known as CL1, may be of more physiological importance in this area. SP40,40/CL1 is initially synthesized as a 427-amino-acid-long polypeptide which is converted to the two chain form found in blood plasma (50 μg/ml) and, in high levels (7 mg/ml) in seminal plasma (179, 180). The SP40,40/CL1 protein shows extensive (77 per cent) sequence identity to the rat Sertoli cell glycoprotein SGP-2 but no similarity to S-protein. SP40,40/CL1 is considered to act after the fluid-phase assembly of C5b–6 and prevents the hydrophilic–amphiphatic transition which C5b–7 undergoes prior to its becoming inserted in target membranes (181).

References

References marked * are recommended for further reading.

1. Schumaker, V. N., Závodszky, P., and Poon, P. H. (1987). Activation of the first component of complement. *Annu. Rev. Immunol.*, **5**, 21.
2. Reid, K. B. M. and Porter, R. R. (1976). Subunit composition and structure of subcomponent C1q of the first component of human complement. *Biochem. J.*, **155**, 19.
*3. Reid, K. B. M. (1983). Proteins involved in the activation and control of the two pathways of human complement. *Biochem. Soc. Trans.*, **11**, 1.

4. Perkins, S. J. (1985). Molecular modelling of human complement subcomponent C1q and its complex with C1r2 C1s2 derived from neutron-scattering curves and hydrodynamic properties. *Biochem. J.*, **228**, 13.

5. Colomb, M. G., Arlaud, G. J., and Villiers, C. L. (1984). Activation of C1. *Philos. Trans. R. Soc. London Biol.*, **306**, 283.

6. Cooper, N. R. (1985). The classical complement pathways: activation and regulation of the first complement component. *Adv. Immunol.*, **37**, 151.

7. Bock, S. C., Skriver, K., Nielsen, E., Thogersen, H. C., Wiman, B., Donaldson, V. H., Eddy, R. L., Marrinan, J., Radziejewska, E., Huber, R., Shows, T. B., and Magnusson, S. (1986). Human C1 inhibitor. Primary structure, cDNA cloning and chromosomal localization. *Biochemistry*, **25**, 4292.

8. Thiel, S. and Reid, K. B. M. (1989). Structures and functions associated with the group of mammalian lectins containing collagen-like sequences. *FEBS Lett.*, **250**, 78.

9. Lu, J., Thiel, S., Wiedemann, H., Timpl, R., and Reid, K. B. M. (1990). Binding of the pentamer/hexamer forms of mannan-binding protein to zymosan activates the pro-enzyme $C1r_2 C1s_2$ complex, of the classical pathway of complement, without involvement of C1q. *J. Immunol.*, **144**, 2287.

10. Ikeda, K., Sannoh, T., Kawasaki, N., Kawasaki, T., and Yamashina, I. (1987). Serum lectin with known structure activates complement through the classical pathway. *J. Biol. Chem.*, **262**, 7451.

11. Ohta, M., Okada, M., Yamashina, I., and Kawasaki, T. (1990). The mechanism of carbohydrate mediated-complement activation by the serum mannan-binding protein. *J. Biol. Chem.*, **265**, 1980.

12. Matsushita, M. and Fujita, T. (1992). Activation of the classical complement pathway by mannose-binding protein in association with a novel C1s-like serine protease. *J. Exp. Med.*, **176**, 1497.

*13. Ji, Y. H., Fujita, T., Hatsuse, H., Takahashi, A., Matsushita, M., and Kawakami, M. (1993). Activation of the C4 and C2 components of complement by a proteinase in serum bactericidal factor, Ra reactive factor. *J. Immunol.*, **150**, 571.

14. Takayama, Y., Takada, F., Takahashi, A., and Kawakami, M. (1994). A 100-kDa protein in the C4-activating component of Ra-reactive factor is a new serine protease having module organization similar to C1r and C1s. *J. Immunol.*, **152**, 2308.

15. Sato, T., Endo, Y., Matsushita, M., and Fujita, T. (1994). Molecular characterisation of a novel serine protease involved in activation of the complement system by mannose-binding protein. *Int. Immunol.*, **6**, 665.

16. Reid, K. B. M. and Turner, M. W. (1994). Mammalian lectins in activation and clearance mechanisms involving the complement system. *Springer Seminars in Immunopathology*, **15**, 307.

17. Reid, K. B. M. (1985). Molecular cloning and characterization of the complementary DNA and gene coding for the B-chain of subcomponent C1q of the human complement system. *Biochem. J.*, **231**, 729.

18. Tenner, A. J. and Volkin, D. B. (1986). Complement subcomponent C1q secreted by cultured human monocytes has a subunit structure identical with that of serum C1q. *Biochem. J.*, **233**, 451.

*19. Sellar, G. C., Blake, D. J., and Reid, K. B. M. (1991). Characterization and organization of the genes encoding the A-, B- and C-chains of human complement subcomponent C1q: The complete derived amino acid sequence of human C1q. *Biochem. J.*, **274**, 481.

20. Sellar, G. C., Cockburn, D., and Reid, K. B. M. (1992). Localization of the gene cluster

encoding the A, B and C chains of human C1q to 1p34. 1–1p36.3. *Immunogenetics*, **35**, 214.

21. White, R. T., Damm, D., Miller, J., Spratt, K., Schilling, J., Hawgood, S., Benson, B., and Cordell, B. (1985). Isolation and characterization of the human pulmonary surfactant apoprotein gene. *Nature*, **317**, 361.

22. Drickamer, K. and Taylor, M. E. (1993). Biology of animal lectins. *Annu. Rev. Cell Biol.*, **9**, 237.

23. McAdam, R. A., Goundis, D., and Reid, K. B. M. (1988). A homozygous point mutation results in a stop codon in the C1q B-chain of a C1q-deficient individual. *Immunogenetics*, **27**, 259.

24. Hoekzema, R., Hannema, A. J., Swaak, T. J. G., Paardekooper, J., and Hack, C. E. (1985). Low molecular weight C1q in systemic lupus erythematosus. *J. Immunol.*, **135**, 265.

25. Leytus, S. P., Kurachi, K., Sakariassen, K. S., and Davie, E. W. (1986). Nucleotide sequence of the cDNA coding for human complement C1r. *Biochemistry*, **25**, 4855.

26. Journet, A. and Tosi, M. (1986). Cloning and sequencing of full-length cDNA encoding the precursor of human complement component C1r. *Biochem. J.*, **240**, 783.

27. Kusumoto, H., Hirosawa, S., Salier, J. P., Hagen, F. S., and Kurachi, K. (1988). Human genes for complement components C1r and C1s in a close tail-to-tail arrangement. *Proc. Natl Acad. Sci. USA*, **85**, 7307.

28. Van Cong, N., Tosi, M., Gross, M. S., Cohen-Haguenauer, O., Jegon-Foubert, C., de Tand, M. F., Meo, T., and Frezal, J. (1988). Assignment of the complement serine protease genes C1r and C1s to chromosome 12 region 12p 13. *Hum. Genet.*, **78**, 363.

29. Drickamer, K., Dordal, M. S., and Reynolds, L. (1986). Mannose-binding proteins isolated from rat liver contain carbohydrate-recognition domains linked to collagenous tails. Complete primary structures and homology with pulmonary surfactant apoprotein. *J. Biol. Chem.*, **261**, 6878.

30. Kurata, H., Sannoh, T., Kozutsumi, Y., Yokota, Y., and Kawasaki, T. (1994). Structure and function of mannan-binding proteins isolated from human liver and serum. *J. Biochem.*, **115**, 1148.

31. Sastry, K., Zahedi, K., Lelias, J. M., Whitehead, A. S., and Ezekowitz, R. A. B. (1991). Molecular characterization of the mouse mannan-binding proteins. The mannose-binding protein A but not C is an acute phase reactant. *J. Immunol.*, **147**, 692.

32. Wada, M., Itoh, N., Ohta, M., and Kawasaki, T. (1992). Characterization of rat liver mannan-binding protein gene. *J. Biochem.*, **111**, 61.

33. Taylor, M. E., Brickell, P. M., Craig, R. K., and Summerfield, J. A. (1989). Structure and evolutionary origin of the gene encoding a human serum mannose-binding protein. *Biochem. J.*, **262**, 763.

34. Sastry, K., Herman, G. A., Day, L., Deignan, E., Bruns, G., Morton, C. C., and Ezekowitz, R. A. (1989). The human mannose-binding protein gene. Exon structure reveals its evolutionary relationship to a human pulmonary surfactant gene and localization to chromosome 10. *J. Exp. Med.*, **170**, 1175.

35. Kölble, K., Lu, J., Mole, S. E., Kaluz, S., and Reid, K. B. M. (1993). Assignment of the human pulmonary surfactant protein D gene (SFTP4) to 10q22–q23 close to the surfactant protein A gene cluster. *Genomics*, **17**, 294.

36. Ogata, R. T., Low, P. J., and Kawakami, M. (1995). Substrate specificities of the protease of mouse serum Ra-reactive factor. *J. Immunol.*, **154**, 2351.

37. Takada, F., Takayama, Y., Hatsuse, H., and Kawakami, M. (1993). A new member of the C1s family of complement proteins found in a bactericidal factor, Ra-reactive factor, in human serum. *Biochem. Biophys. Res. Commun.*, **196**, 1003.

38. Carter, P. E., Dunbar, B., and Fothergill, J. E. (1988). Genomic and cDNA cloning of the human C1 inhibitor. *Eur. J. Biochem.*, **173**, 163.

39. Carter, P. E., Duponchel, C., Tosi, M., and Fothergill, J. E. (1991). Complete nucleotide sequence of the gene for human C1 inhibitor with an unusually high density of Alu elements. *Eur. J. Biochem.*, **197**, 301.

40. Stoppa-Lyonnet, D., Carter, P. E., Meo, T., and Tosi, M. (1990). Clusters of intragenic Alu repeats predispose the human C1 inhibitor locus to deleterious rearrangements. *Proc. Natl Acad. Sci. USA*, **87**, 1551.

41. Whaley, K. 1987, Complement and immune complex diseases. In *Complement in health and disease* (ed. K. Whaley), pp. 163–83. MTP Press, Lancaster.

42. Verpy, E., Couture-Tosi, E., and Tosi, M. (1993). C1 inhibitor mutations which affect intracellular transport and secretion in type I hereditary angioedema. *Behring. Inst. Mitt.*, **93**, 120.

*43. Pangburn, M. K., Schreiber, R. D., and Müller-Eberhard, H. J. (1981). Formation of the initial C3 convertase of the alternative complement pathway. Acquisition of C3b-like activities by spontaneous hydrolysis of the putative thiolester in native C3. *J. Exp. Med.*, **154**, 856.

*44. White, R. T., Damm, D., Hancock, N., Rosen, B. S., Bradford, B. L., Usher, P., Flier, J. S., and Spiegelman, B. M. (1992). Human adipsin is identical to complement factor D and is expressed at high levels in adipose tissue. *J. Biol. Chem.*, **267**, 9210.

45. Barnum, S. R. and Volanakis, J. E. (1985). Biosynthesis of complement protein D by HepG2 cells: A comparison of D produced by HepG2 cells, U937 cells and blood monocytes. *Eur. J. Immunol.*, **15**, 1148.

46. Flier, J. S., Cook, K. S., Usher, P., and Spiegelman, B. M. (1987). Severely impaired adipsin expression in genetic and acquired obesity. *Science*, **237**, 405.

47. Rosen, B. S., Cook, K. S., Yaglon, J., Groves, D. L., Volanakis, J. E., Damm, D., White, T., and Spiegelman, B. M. (1989). Adipsin and complement factor D activity: an immune related defect in obesity. *Science*, **244**, 1483.

48. Platt, K. A., Min, H. Y., Ross, S. R., and Spiegelman, B. M. (1989). Obesity-linked regulation of the adipsin gene promotor in transgenic mice. *Proc. Natl Acad. Sci. USA*, **86**, 7490.

49. Hiemstra, P. S., Langeler, E., Compier, B., Keepers, Y., Leijh, P. C., van den Barselaar, M. T., Overbosch, D., and Daha, M. R. (1989). Complete and partial deficiencies of complement factor D in a Dutch family. *J. Clin. Invest.*, **84**, 1957.

50. Farries, T. C., Lachmann, P. J., and Harrison, R. A. (1988). Analysis of the interaction between properdin and factor B, components of the alternative-pathway C3 convertase of complement. *Biochem. J.*, **253**, 667.

51. Nolan, K. F., Schwaeble, W., Kaluz, S., Dierich, M. P., and Reid, K. B. M. (1991). Molecular cloning of the cDNA coding for properdin, a positive regulator of the alternative pathway of human complement. *Eur. J. Immunol.*, **21**, 771.

*52. Nolan, K. F., Kaluz, S., Higgins, J. M. G., Goundis, D., and Reid, K. B. M. (1992). Characterization of the human properdin gene. *Biochem. J.*, **287**, 291.

53. Frazier, W. A. (1991). Thrombospondins. *Curr. Opin. Cell Biol.*, **3**, 792.

54. Goonewardena, P., Sjöholm, A. G., Nilsson, L.-Å., and Petterson, U. (1988). Linkage analysis of the properdin deficiency gene: Suggestion of a locus in the proximal part

of the short arm of the X chromosome. *Genomics*, **2**, 115.

55. Coleman, M. P., Murray, J. C., Willard, H. F., Nolan, K. F., Reid, K. B. M., Blake, D. J., Lindsay, S., Bhattacharaya, S. S., Wright, A., and Davies, K. E. (1991), Genetic and physical mapping around the properdin P gene. *Genomics*, **11**, 991.

56. Williams, A. F. and Barclay, A. N. (1988). The immunoglobulin super family — domains for cell surface recognition. *Annu. Rev. Immunol.*, **6**, 381.

57. Chan, A. C. and Atkinson, J. P. (1985). Oligosaccharide structure of human C4. *J. Immunol.*, **134**, 1790.

58. Belt, K. T., Carroll, M. C., and Porter, R. R. (1984). The structural basis of the multiple forms of human complement component C4. *Cell*, **36**, 907.

59. Belt, K. T., Yu, C. Y., Carroll, M. C., and Porter, R. R. (1985). Polymorphism of human complement component C4. *Immunogenetics*, **21**, 173.

60. Nonaka, M., Nakayama, K., Yuel, Y. D., and Takahashi, M. (1985). Complete nucleotide and derived amino acid sequences of the fourth component of mouse complement (C4): evolutionary aspects. *J. Biol. Chem.*, **260**, 10936.

61. Sepich, D. S., Noonan, D. J., and Ogata, R. T. (1985). Complete cDNA sequence of the fourth component of murine complement. *Proc. Natl Acad. Sci. USA*, **82**, 5895.

*62. Law, S. K. A., Lichtenberg, N. A., and Levine, R. P. (1979). Evidence for an ester linkage between the labile binding site of C3b and receptive surfaces. *J. Immunol.*, **123**, 1388.

63. Dodds, A. W., Law, S. K. A., and Porter, R. R. (1985). The origin of the very variable haemolytic activities of the common human complement component C4 allotypes including C4-A6. *EMBO J.*, **4**, 2239.

64. Dodds, A. W., Law, S. K. A., and Porter, R. R. (1986). The purification and properties of some less common allotypes of the fourth component of human complement. *Immunogenetics*, **24**, 279.

65. Sottrup-Jensen, L., Stepanik, T. M., Kristensen, T., Lonblad, P. B., Jones, C. M., Wierzbick, D. M., Magnusson, S., Domdey, H., Wetsel, R. A., Lundwall, A., Tack, B. F., and Fey, G. H. (1985). Common evolutionary origin of α2-macroglobulin and complement components C3 and C4. *Proc. Natl Acad. Sci. USA*, **82**, 9.

66. Ueda, A., Kearney, J. F., Roux, K. H., and Volanakis, J. E. (1987). Probing functional sites on complement protein B with monoclonal antibodies: evidence for C3b-binding sites on Ba. *J. Immunol.*, **138**, 1143.

67. Bentley, D. R. and Campbell, R. D. (1986). C2 and factor B: Structure and genetics. *Biochem. Soc. Symp.*, **51**, 7.

68. Bentley, D. R. (1986). Primary structure of human complement component C2. Homology to two unrelated protein families. *Biochem. J.*, **239**, 339.

69. Morley, B. J. and Campbell, R. D. (1984). Internal homologies of the Ba fragment from human complement component factor B, a class III MHC antigen. *EMBO J.*, **3**, 153.

70. Campbell, R. D. and Porter, R. R. (1983). Molecular cloning and characterisation of the gene coding for human complement protein factor B. *Proc. Natl Acad. Sci. USA*, **80**, 4464.

71. Pangburn, M. K. and Müller-Eberhard, H. J. (1984). The alternative pathway of complement. *Springer Semin. Immunopathol.*, **7**, 163.

72. Carroll, M. C. and Alper, C. A. (1987). Polymorphism and molecular genetics of human C4. *B. Med. Bull.*, **43**, 50.

73. Hauptmann, G., Goetz, J., Uring-Lambert, B., and Grosshans, E. (1986). Component deficiencies. 2. The fourth component. *Progr. Allergy*, **39**, 232.

74. Law, S. K. A., Dodds, A. W., and Porter, R. R. (1984). A comparison of the properties of the two classes, C4A and C4B, of the human complement component C4. *EMBO J.*, **3**, 1819.

75. Isenman, D. E. and Young, J. R. (1986). Covalent binding properties of the C4A and C4B isotypes of the fourth component of human complement on several C1-bearing cell-surfaces. *J. Immunol.*, **136**, 2542.

76. Yu, C. Y. (1991). The complete exon–intron structure of a human complement component C4A gene. DNA sequences, polymorphism, and linkage to the 21-hydroxylase gene. *J. Immunol.*, **146**, 1057.

77. Yu, C. Y., Belt, K. T., Giles, C. M., Campbell, R. D., and Porter, R. R. (1986). Structural basis of the polymorphism of human complement components C4A and C4B: gene size, reactivity and antigenicity. *EMBO J.*, **5**, 2873.

78. Carroll, M. C., Fathallah, D. M., Bergamaschini, L., Alicot, E. M., and Isenman, D. E. (1990). Substitution of a single amino acid (aspartic acid for histidine) converts the functional activity of human complement C4B to C4A. *Proc. Natl Acad. Sci. USA*, **87**, 6868.

*79. Sepp, A., Dodds, A. W., Anderson, M. J., Campbell, R. D., Willis, A. C., and Law, S. K. A. (1993). Covalent binding properties of the human complement protein C4 and hydrolysis rate of the internal thioester upon activation. *Protein Sci.*, **2**, 706.

80. Porter, R. R. (1983). Complement polymorphism, the major histocompatibility complex and associated diseases: a speculation. *Mol. Biol. Med.*, **1**, 161.

81. Schifferli, J. A., Steiger, G., Paccaud, J. P., Sjöholm, A. G., and Hauptmann, G. (1986). Comparison of inhibition of immune precipitation by C4A and C4B alleles. *Clin. Exp. Immunol.*, **63**, 473.

82. Sim, E. and Law, S. K. A. (1985). Hydralazine binds covalently to complement component C4: different reactivity of C4A and C4B gene products. *FEBS Lett.*, **184**, 323.

83. Nakayama, K., Nonaka, M., Yokoyama, S., Yeul, Y. D., Pattanakitsakul, S. N., and Takahashi, M. (1987). Recombination of two homologous MHC class III genes of the mouse (C4 and Slp) that accounts for the loss of testosterone dependence of sex-limited protein expression. *J. Immunol.*, **138**, 620.

84. Ogata, R. T. and Sepich, D. S. (1985). Murine sex-limited protein complete cDNA sequence and comparison with murine fourth complement component. *J. Immunol.*, **135**, 4239.

85. Wu, L. C., Morley, B. J., and Campbell, R. D. (1987). Cell-specific expression of the human complement protein factor B gene: evidence for the role of two distinct 5' flanking elements. *Cell*, **48**, 331.

86. Colten, H. R. and Dowton, S. B. (1986). Regulation of complement gene expression. *Biochem. Soc. Symp.*, **51**, 37.

87. Perlmutter, D. H., Goldberger, G., Dinarello, C. A., Mizel, S. B., and Colten, H. R. (1986). Regulation of class III major histocompatibility complex gene product by interleukin-1. *Science*, **232**, 850.

88. Strunk, R., Cole, S., Perlmutter, D., and Colten, H. (1985). c-Interferon increases expression of class III complement genes C2 and Factor B in human monocytes and murine fibroblasts transfected with human C2 and Factor B genes. *J. Biol. Chem.*, **260**, 15280.

89. Johnston, C. A., Densen, P., Hurford, R. K., Colten, H. R., and Wetsel, R. A. (1992).

Type 1 human complement C2 deficiency. A 28-base pair gene deletion causes skipping of exon 6 during RNA splicing. *J. Biol. Chem.*, **267**, 9347.

90. Dunham, I., Sargent, C. A., Trowsdale, J., and Campbell, R. D. (1987). Molecular mapping of the human major histocompatibility complex by pulsed-field gel electrophoresis. *Proc. Natl Acad. Sci. USA*, **84**, 7237.

91. Milner, C. M. and Campbell, R. D. (1992). Genes, genes and more genes in the human major histocompatibility complex. *Bioessays*, **14**, 565.

92. Sim, R. B., Malhotra, V., Ripoche, J., Day, A. J., Micklem, K. J., and Sim, E. (1986). Complement receptors and related complement control proteins. *Biochem. Soc. Symp.*, **51**, 83.

93. de Bruijn, M. H. L. and Fey, G. H. (1985). Human complement component C3: cDNA coding sequence and derived primary structure. *Proc. Natl Acad. Sci. USA*, **82**, 708.

94. Wetsel, R. A., Lundwall, A., Davidson, F., Gibson, T., Tack, B. F., and Fey, G. H. (1984). Structure of murine complement component C3: II nucleotide sequence of cloned complementary DNA coding for the α chain. *J. Biol. Chem.*, **259**, 13857.

95. Müller-Eberhard, H. J., Dalmasso, A. P., and Calcott, M. A. (1966). The reaction mechanism of B1C-globulin (C′3) in immune hemolysis. *J. Exp. Med.*, **124**, 33.

96. Reid, K. B. M. and Porter, R. R. (1981). The proteolytic activation systems of complement. *Annu. Rev. Biochem.*, **50**, 433.

97. Law, S. K. A. (1983). The covalent binding reaction of C3 and C4. *Ann. NY Acad. Sci.*, **421**, 246.

98. Fong, K. Y., Botto, M., Walport, M. J., and So, A. K. (1990). Genomic organization of human complement component C3. *Genomics*, **7**, 579.

99. Vik, D. P., Amiguet, P., Moffat, G. J., Fey, M., Amiguet-Barras, F., Wetsel, R. A., and Tack, B. F. (1991). Structural features of the human C3 gene: intron/exon organization, transcriptional start site, and promoter region sequence. *Biochemistry*, **30**, 1080.

100. Vik, D. P., Tack, B. F., and Wong, W. W. (1991). Structure of the complement receptor type 1 (CR1) gene and sequence of the s allele. *Complement Inflammation*, **8**, 238 (Abstract).

101. Botto, M., Fong, K. Y., So, A. K., Barlow, R., Routier, R., Morley, B. J., and Walport, M. J. (1992). Homozygous hereditary C3 deficiency due to a partial gene deletion. *Proc. Natl Acad. Sci. USA*, **89**, 4957.

102. Haviland, D. L., Haviland, J. C., Fleischer, D. T., Hunt, A., and Wetsel, R. A. (1991). Complete cDNA sequence of human complement pro-C5. Evidence of truncated transcripts derived from a single copy gene. *J. Immunol.*, **146**, 362.

*103. Müller-Eberhard, H. J. (1986). The membrane attack complex of complement. *Annu. Rev. Immunol.*, **4**, 503.

104. Kim, Y. U., Carroll, M. C., Isenman, D. E., Nonaka, M., Pramoonjago, P., Takeda, J., Inoue, K., and Kinoshita, T. (1992). Covalent binding of C3b to C4b within the classical complement pathway C5 convertase. Determination of amino acid residues involved in ester linkage formation. *J. Biol. Chem.*, **267**, 4171.

105. Kinoshita, T., Takata, Y., Kozono, H., Takeda, J., Hong, K., and Inoue, K. (1988). C5 convertase of the alternative complement pathway: covalent linkage between two C3b molecules within the trimolecular complex enzyme. *J. Immunol.*, **141**, 3895.

106. Wetsel, R. A., Lemons, R. S., Le Beau, M. M., Barnum, S. R., Noack, D., and Tack, B. F. (1988). Molecular analysis of human complement component C5: localisation of the structural gene to chromosome 9. *Biochemistry*, **27**, 1474.

107. Rodriguez de Cordoba, S., Lublin, D. M., Rubinstein, P., and Atkinson, J. P. (1985).

Human genes for three complement components that regulate the activation of C3 are tightly linked. *J. Exp. Med.*, **161**, 1189.

108. Reid, K. B. M., Bentley, D. R., Campbell, R. D., Chung, L. P., Sim, R. B., Kristensen, T., and Tack, B. F. (1986). Complement system proteins which interact with C3b or C4b. A superfamily of structurally related proteins. *Immunol. Today*, **7**, 230.

109. Reid, K. B. M. and Day, A. J. (1989). Structure–function relationships of the complement components. *Immunol. Today*, **10**, 177.

110. Norman, D. G., Barlow, P. N., Baron, M., Day, A. J., Sim, R. B., and Campbell, I. D. (1991). Three-dimensional structure of a complement control protein module in solution. *J. Mol. Biol.*, **219**, 717.

111. Chung, L. P., Bentley, D. R., and Reid, K. B. M. (1985). Molecular cloning and characterisation of the cDNA coding for C4b-binding protein, a regulatory protein of the classical pathway of the human complement system. *Biochem. J.*, **230**, 133.

112. Pardo-Manuel, F., Rey-Campos, J., Hillarp, A., Dahlbäck, B., and Rodriguez de Cordoba, S. (1990). Human genes for the α and β chains of complement C4b-binding protein are closely linked in a head-to-tail arrangement. *Proc. Natl Acad. Sci. USA*, **87**, 4529.

113. Catterall, C. F., Lyons, A., Sim, R. B., Day, A. J., and Harris, T. J. R. (1987). Characterization of the primary amino acid sequence of human complement control protein factor I from an analysis of cDNA clones. *Biochem. J.*, **242**, 849.

114. Shiang, R., Murray, J. C., Morton, C. C., Betow, K. H., Wasmuth, J. J., Olney, A. H., Sanger, W. G., and Goldberger, G. (1989). Mapping of the human factor I gene to 4q25. *Genomics*, **4**, 82.

115. Hillarp, A. and Dahlbäck, B. (1990). Cloning of cDNA coding for the β chain of human complement component C4b-binding protein: Sequence homology with the α chain. *Proc. Natl Acad. Sci. USA*, **87**, 1183.

116. Dahlback, B., Smith, C. A., and Müller-Eberhard, H. J. (1983). Visualization of human C4b-binding protein and its complexes with vitamin K-dependent protein S and complement protein C4b. *Proc. Natl Acad. Sci. USA*, **80**, 3461.

117. Dahlback, B. and Müller-Eberhard, H. J. (1984). Ultrastructure of C4b-binding protein fragments formed by limited protein fragments formed by limited proteolysis using chymotrypsin. *J. Biol. Chem.*, **259**, 11631.

118. Ziccardi, R. J., Dahlback, B., and Müller-Eberhard, H. J. (1984). Characterization of the interaction of human C4b-binding protein with physiological ligands. *J. Biol. Chem.*, **259**, 13674.

119. Kristensen, T., Ogata, R. T., Chung, L. P., Reid, K. B. M., and Tack, B. F. (1987). cDNA structure of murine C4b-binding protein, a regulatory component of the serum complement system. *Biochemistry*, **26**, 4668.

120. Rodriguez de Cordoba, S., Sanchez-Corral, P., and Rey-Campos, J. (1991). Structure of the gene coding for the α polypeptide chain of the human complement component C4b-binding protein. *J. Exp. Med.*, **173**, 1073.

121. Ripoche, J., Day, A. J., Harris, T. J. R., and Sim, R. B. (1988). The complete amino acid sequence of human complement factor H. *Biochem. J.*, **249**, 593.

122. Kristensen, T. and Tack, B. F. (1986). Murine protein H is comprised of 20 repeating units, 61 amino acids in length. *Proc. Natl Acad. Sci. USA*, **83**, 3963.

123. Alsenz, J., Schulz, T. F., Lambris, J. D., Sim, R. B., and Dierich, M. P. (1985). Structural and functional analysis of the complement component factor H with the use of different enzymes and monoclonal antibodies to Factor H. *Biochem. J.*, **232**, 841.

124. Ripoche, J., Day, A. J., Moffatt, B., and Sim, R. B. (1987). mRNA coding for a truncated form of human complement factor H. *Biochem. Soc. Trans.*, **15**, 651.

125. Estaller, C., Schwaeble, W., Dierich, M., and Weiss, E. H. (1991). Human complement factor H: two factor H proteins are derived from alternatively spliced transcripts. *Eur. J. Immunol.*, **21**, 799.

126. Skerka, C., Timmann, C., Horstmann, R. D., and Zipfel, P. F. (1992). Two additional human serum proteins structurally related to complement factor H. *J. Immunol.*, **148**, 3313.

127. Skerka, C., Kühn, S., Günther, K., Lingelbach, K., and Zipfel, P. F. (1993). A novel short consensus repeat-containing molecule is related to human complement factor H. *J. Biol. Chem.*, **268**, 2904.

128. Klickstein, L. B., Wong, W. W., Smith, J. A., Weis, J. H., Wilson, J. G., and Fearon, D. T. (1987). Human C3b/C4b receptor (CR1). Demonstration of long homologous repeating domains that are composed of the short consensus repeats characteristic of C3/C4 binding proteins. *J. Exp. Med.*, **165**, 1095.

129. Klickstein, L. B., Bartow, T. J., Miletic, V., Rabson, L. D., Smith, J. A., and Fearon, D. T. (1988). Identification of distinct C3b and C4b recognition sites in the human C3b/C4b receptor (CR1, CD35) by deletion mutagenesis. *J. Exp. Med.*, **168**, 1699.

130. Krych, M., Hourcade, D., and Atkinson, J. P. (1991). Sites within the complement C3b/C4b receptor important for the specificity of ligand binding. *Proc. Natl Acad. Sci. USA*, **88**, 4353.

131. Holers, V. M., Cole, J. L., Lublin, D. M., Seya, T., and Atkinson, J. P. (1985). Human C3b- and C4b-regulatory proteins: a new multi-gene family. *Immunol. Today*, **6**, 188.

132. Wong, W. W., Cahill, J. M., Rosen, M. D., Kennedy, C. A., Bonaccio, E. T., Morris, M. J., Wilson, J. G., Klickstein, L. B., and Fearon, D. T. (1989). Structure of the human CR1 gene. Molecular basis of the structural and quantitative polymorphisms and identification of a new CR1-like allele. *J. Exp. Med.*, **169**, 847.

133. Wilson, J. G., Murphy, E. E., Wong, W. W., Klickstein, L. B., Weis, J. H., and Fearon, D. T. (1986). Identification of a restriction fragment length polymorphism by a CR1 cDNA that correlates with the number of CR1 on erthyrocytes. *J. Exp. Med.*, **164**, 50.

134. Cooper, N. R., Moore, M. D., and Nemerow, G. R. (1988). Immunobiology of CR2, the B lymphocyte receptor for Epstein–Barr virus and the C3d complement fragment. *Annu. Rev. Immunol.*, **6**, 85.

135. Changelian, P. S. and Fearon, D. T. (1986). Tissue-specific phosphorylation of complement receptors CR1 and CR2. *J. Exp. Med.*, **163**, 101.

136. Fujisaku, A., Harley, J. B., Frank, M. B., Gruner, B. A., Frazier, B., and Holers, V. M. (1989). Genomic organisation and polymorphisms of the human C3d/Epstein–Barr virus receptor. *J. Biol. Chem.*, **264**, 2118.

137. Martin, D. R., Yuryev, A., Kalli, K. R., Fearon, D. T., and Ahearn, J. M. (1991). Determination of the structural basis for selective binding of Epstein–Barr virus to human complement receptor type 2. *J. Exp. Med.*, **174**, 1299.

138. Weis, J. H., Morton, C. C., Bruns, G. A. P., Weis, J. J., Klickstein, L. B., Wong, W. W., and Fearon, D. T. (1987). A complement receptor locus: genes encoding C3b/C4b receptor and C3d/Epstein–Barr virus receptor map to 1q32[1]. *J. Immunol.*, **138**, 312.

139. Seya, T., Okada, M., Matsumoto, M., Hong, K. S., Kinoshita, T., and Atkinson, J. P. (1991). Preferential inactivation of the C5 convertase of the alternative complement pathway by factor I and membrane co-factor protein (MCP). *Mol. Immunol.*, **28**, 1137.

140. Lublin, D. M., Liszewski, M. K., Post, T. W., Arce, M. A., Le Beau, M. M., Rebentisch,

M. B., Lemons, L. S., Seya, T., and Atkinson, J. P. (1988). Molecular cloning and chromosomal localization of the human membrane co-factor protein (MCP). Evidence for inclusion in the multi-gene family of complement-regulatory proteins. *J. Exp. Med.*, **168**, 181.

141. Post, T. W., Liszewski, M. K., Adams, E. M., Tedja, I., Miller, E. A., and Atkinson, J. P. (1991). Membrane co-factor protein of the complement system: alternative splicing of serine/threonine/proline-rich exons and cytoplasmic tails produces multiple isoforms that correlate with protein phenotype. *J. Exp. Med.*, **174**, 93.

142. Purcell, D. F. J., Russell, S. M., Deacon, N. J., Brown, M. A., Hooker, D. J., and McKenzie, I. F. (1991). Alternatively spliced RNAs encode several isoforms of CD46 (MCP), a regulator of complement activation. *Immunogenetics*, **33**, 335.

*143. Liszewski, M. K., Post, T. W., and Atkinson, J. P. (1991). Membrane co-factor protein (MCP or CD46): newest member of the regulators of complement activation gene cluster. *Annu. Rev. Immunol.*, **9**, 431.

144. Caras, I. W., Davitz, M. A., Rhee, L., Weddell, G., Martin, D. W., and Nussenzweig, V. (1987). Cloning of decay-accelerating factor suggests novel use of splicing to generate two proteins. *Nature*, **325**, 545.

145. Post, T. W., Arce, M. A., Liszewski, M. K., Thompson, E. S., Atkinson, J. P., and Lublin, D. M. (1990). Structure of the gene for human complement protein decay accelerating factor. *J. Immunol.*, **144**, 740.

146. Gerard, N. P. and Gerard, C. (1991). The chemotactic receptor for human C5a anaphylatoxin. *Nature*, **349**, 614.

147. Boulay, F., Mery, L., Tardif, M., Brouchon, L., and Vignais, P. (1991). Expression cloning of a receptor for C5a anaphylatoxin on differentiated HL-60 cells. *Biochemistry*, **30**, 2993.

148. Rollins, T. E., Siciliano, S., Kobayashi, S., Cianciarulo, D. N., Bonilla-Argudo, V., Collier, K., and Springer, M. S. (1991). Purification of the active C5a receptor from human polymorphonuclear leukocytes as a receptor-G_i complex. *Proc. Natl Acad. Sci. USA*, **88**, 971.

149. Hynes, R. O. (1992). Integrins: Versatility, modulation and signaling in cell adhesion. *Cell*, **69**, 11.

150. Barclay, A. N., Beyers, A. D., Birkeland, M. L., Brown, M. H., Davis, S. J., Somoza, C., and Williams, A. F. (1993). *The leucocyte antigen facts book*. Academic Press. Harcourt Brace & Co., London.

151. Springer, T. A., Dustin, M. L., Kishimoto, T. K., and Marlin, S. D. (1987). The lymphocyte function-associated LFA-1, CD2 and LFA-3 molecules: cell adhesion receptors of the immune system. *Annu. Rev. Immunol.*, **5**, 223.

152. Arnaout, M. A. (1990). Leukocyte adhesion molecules deficiency: its structural basis, pathophysiology and implications for modulating the inflammatory response. *Immunol. Rev.*, **114**, 145.

153. Law, S. K. A., Gagnon, J., Hildreth, J. E. K., Wells, C. E., Willis, A. C., and Wong, A. J. (1987). The primary structure of the b-subunit of the cell-surface adhesion glycoproteins LFA-1, CR3 and p150,95 and its relationship to the fibronectin receptor. *EMBO J.*, **6**, 915.

154. Kishimoto, T. K., O'Connor, K., Lee, A., Roberts, T. M., and Springer, T. A. (1987). Cloning of the β subunit of the leukocyte adhesion proteins: homology to an extracellular matrix receptor defines a novel supergene family. *Cell*, **48**, 681.

155. Tamkun, J. W., DeSimone, D. W., Fonda, D., Patel, R. S., Buck, C., Horwitz, A. F.,

and Hynes, R. O. (1986). Structure of integrin, a glycoprotein involved in the transmembrane linkage between fibronectin and actin. *Cell*, **46**, 271.

156. Wright, S. D., Reddy, P. A., Jong, M. T., and Erickson, B. W. (1987). C3bi receptor (complement receptor type 3) recognizes a region of complement protein C3 containing the sequence Arg–Gly–Asp. *Proc. Natl Acad. Sci. USA*, **84**, 1965.

157. Altieri, D. C., Agbanyo, F. R., Plescia, J., Ginsberg, M. H., Edgington, T. S., and Plow, E. F. (1990). A unique recognition site mediates the interaction of fibrinogen with the leukocyte integrin Mac-1 (CD11b/CD18). *J. Biol. Chem.*, **265**, 12119.

158. Dankert, J. R. and Esser, A. F. (1985). Proteolytic modification of human complement protein C9. Loss of poly (C9) and circular lesion formation without impairment of function. *Proc. Natl Acad. Sci. USA*, **82**, 2128.

159. Young, J. D., Cohen, Z. A., and Podack, E. R. (1986). The ninth component of complement and the pore-forming protein (Perforin 1) from cytotoxic T cells: structural, immunological and functional studies. *Science*, **233**, 184.

*160. Tschopp, J. and Nabholz, M. (1990). Perforin-mediated target lysis by cytolytic T-lymphocytes. *Annu. Rev. Immunol.*, **8**, 279.

161. DiScipio, R. G. and Hugli, T. E. (1989). The molecular architecture of human complement component C6. *J. Biol. Chem.*, **264**, 16197.

162. Haefliger, J. A., Tschopp, J., Vial, N., and Jenne, D. E. (1989). Complete primary structure and functional characterization of the sixth component of the human complement system. Identification of the C5b-binding domain in complement C6. *J. Biol. Chem.*, **264**, 18041.

163. DiScipio, R. G., Chakravarti, D. N., Müller-Eberhard, H. J., and Fey, G. H. (1988). The structure of human complement component C7 and C5b–7 complex. *J. Biol. Chem.*, **263**, 549.

164. Rao, A. G., Howard, O. M. Z., Ng, S. C., Whitehead, A. S., Colten, H. R., and Sodetz, J. M. (1987). Complementary DNA and derived amino acid sequence for the α subunit of human complement protein C8: evidence for the existence of a separate α subunit mRNA. *Biochemistry*, **26**, 3556.

165. Howard, O. M. Z., Rao, A. G., and Sodetz, J. M. (1987). Complementary DNA and derived amino acid sequence of the β-subunit of C8: identification of a close structural and ancestral relationship to the α-subunit and C9. *Biochemistry*, **26**, 3565.

166. DiScipio, R. G., Gehring, M. R., Podack, E. R., Kan, C. C., Hugli, T. E., and Fey, G. F. (1984). Nucleotide sequence of cDNA and derived amino acid sequence of human complement component C9. *Proc. Natl Acad. Sci. USA*, **81**, 7298.

167. Stanley, K. K., Kocher, H. P., Luzio, J. P., Jackson, P., and Tschopp, J. (1985). The sequence and topology of human complement component C9. *EMBO J.*, **4**, 375.

168. Lichtenheld, M. G., Olsen, K. J., Lu, P., Lowry, D. M., Hamed, A., Hengartner, H., and Podack, E. R. (1988). Structure and function of human perforin. *Nature*, **325**, 448.

169. Hobart, M. J., Fernie, B. A., and DiScipio, R. D. (1993). The structure of the human C6 gene. *Biochemistry*, **32**, 6198.

170. Marazziti, D., Eggertsen, G., Fey, G. H., and Stanley, K. K. (1988). Relationships between the gene and protein structure in human complement component C9. *Biochemistry*, **27**, 6529.

171. Kaufman, K. M., Snider, J. V., Spurr, N. K., Schwartz, C. E., and Sodetz, J. M. (1989). Chromosomal assignment of genes encoding the α, β and γ subunits of human complement protein C8: identification of a close physical linkage between the α and the β loci. *Genomics*, **5**, 475.

172. Fink, T. M., Zimmer, M., Weitz, S., Tschopp, J., Jenne, D. E., and Lichter, P. (1992). Human perforin (PRF1) maps to 10q22, a region that is syntenic with mouse chromosome 10. *Genomics*, **13**, 1300.

*173. Meri, S., Morgan, B. P., Davies, A., Daniels, R. H., Olavesen, M. G., Waldmann, H., and Lachmann, P. J. (1990). Human protectin (CD59), an 18,000–20,000 MW complement lysis restricting factor, inhibits C5b–8 catalysed insertion of C9 into lipid bilayers. *Immunology*, **71**, 1.

174. Davies, A., Simmons, D. L., Hale, G., Harrison, R. A., Tighe, H., Lachmann, P. J., and Waldmann, H. (1989). CD59, an LY-6-like protein expressed in human lymphoid cells, regulates the action of the complement membrane attack complex on homologous cells. *J. Exp. Med.*, **170**, 637.

175. Motoyama, N., Okada, N., Yamashina, M., and Okada, H. (1992). Paroxysmal nocturnal hemoglobinuria due to a hereditary nucleotide deletion in the HRF20 (CD59) gene. *Eur. J. Immunol.*, **22**, 2669.

176. Zalman, L. S., Brothers, M. A., and Müller-Eberhard, H. J. (1988). Self-protection of cytotoxic lymphocytes: A soluble form of homologous restriction factor in cytoplasmic granules. *Proc. Natl Acad. Sci. USA*, **85**, 4827.

177. Podack, E. R., Preissner, K. T., and Müller-Eberhard, H. J. (1984). Inhibition of C9 polymerization within the SC5b–9 complex of complement by S-protein. *Acta Pathol. Microbiol. Scand.*, **92**, 89.

178. Jenne, D. and Stanley, K. K. (1985). Molecular cloning of S-protein, a link between complement, coagulation and cell–substrate adhesion. *EMBO J.*, **4**, 3153.

179. Jenne, D. E. and Tschopp, J. (1989). Molecular structure and functional characterization of a human complement cytolysis inhibitor found in blood and seminal plasma: Identity to sulphated glycoprotein 2, a constituent of rat testis fluid. *Proc. Natl Acad. Sci. USA*, **86**, 7123.

180. Kirszbaum, L., Sharpe, J. A., Murphy, B., d'Apice, A. J., Classon, B., Hudson, P., and Walker, I. D. (1989). Molecular cloning and characterization of the novel, human complement-associated protein, SP-40,40: a link between the complement and reproductive systems. *EMBO J.*, **8**, 711.

181. Murphy, B. F., Saunders, J. R., O'Bryan, M. K., Kirszbaum, L., Walker, I. D., and d'Apice, A. J. (1989). SP-40,40 is an inhibitor of C5b–6 initiated haemolysis. *Int. Immunol.*, **1**, 551.

Index

allelic exclusion
  of Ig genes
    HC allelic exclusion 41–2
    LC allelic exclusion 42–3
    models of 40–1
  of TCR genes 108–9
antibody engineering
  applications
    'antigenized' antibodies 308
    bifunctional antibodies 307
    bispecific antibodies 306–7
    bivalent antibodies 306–7
    catalytic antibodies 309–10
    'genuine' human antibodies 304–5
    'humanized' antibodies 302–4
    insertion of functional sites 308
    minibodies 308
    molecular recognition units (MRUs) 309
    structure–function studies 301–2
  combinatorial libraries 294–6
  PCR assembly 295–6
  phage display of antibody fragments
    affinity maturation 300
    to bypass immunization 299–300
    to create new specificities 300
    limitations to repertoire cloning 300–1
    overview 296–9
  recombinant Ig gene expression in animal cells 285–7
    B. subtilis 291–2
    E. coli 287–91
    filamentous fungi 293
    plants 294–5
    Staphylococcus 292
    Streptomyces 292
    yeast 292–3
    see also recombinant Ig gene expression

B cell activation
  B cell receptor complex (BCR), structure 251–3
  CD antigens 249–50
  CD19/CD21 complex 256–7, 258
  CD20 263–4
  CD22 264
  CD23 259–60
  CD38 264
  CD40 262–3
  CD45 261–2
  CD72 262
  Class II MHC role 260, 265–9
  FcγRII (CD32) 257–9
  leucocyte common antigen 261–2
  murine Lyb2 263
  murine Lyb8 264
  murine Lys 261–2
  responses to polyclonal antigens 248–9
  responses to specific antigens 248
  role of PI-derived second messengers 255–6
  signal transduction
    coupling of BCR complex to PLC 253–5
    role of PI-derived second messengers 251–3
    second messengers and sIg receptors 253
    structure of BCR complex 251–3
    via sIg receptors 251–6
  surface immunoglobulin receptors (sIg)
    relationship to BCR 251–3
    role of 250–6
    second messenger invoked by 253
    signal transduction via 251–6
  T cell dependent B cell activation
    B cells as antigen-presenting cells 265–6
    class II unrestricted, contact-dependent B cell activation 268–9
    cytokine action on resting B cells 270

cytokine-driven B cell profileration and differentiation 269–71
cytokine production by T_H subsets 270–1
events in T cell–B cell interaction 266–9
high affinity MHC-restricted antigen recognition 267–8
low affinity reversible interactions 267
MHC restricted 265–6, 267–8
unrestricted 265–6, 268–9
B cells
  activation, see B cell activation
  clonal selection 2–3
  differentiation 1–3, 250, 269–71

class switch recombination (CSR)
  accessibility model for control 60–1
  B cell specificity of 59
  control by additional sequence elements 63–4
  control of 59–64
  CSR substrate studies 58
  directed heavy chain class switch recombination 54–6
  induction of 59–60
  mechanism of 55–8
  non-deletional CSR 58–9
  overview 52–3
  potential control by enhancers 63–4
  potential role of I region sequences 62
  potential role of locus control regions 63–4
  potential role of S region transcription 62
  role of DNA replication 57
  role of germline transcription 61–2
  sequential switching 55–6
  S regions, role in 55–7
  S regions, structure 53–5
  topology of CSR 55

complement
  activation of the alternative
    pathway
    factor D 338–9
    overview 326–30, 338
    properdin 339
  activation of the classical pathway
    cloning of C1-Inh 337–8
    cloning of C1q 334–5
    cloning of C1r 335–6
    cloning of C1s 335–6
    cloning of MASP 336–7
    cloning of MBP 336
    model for C1 complex 332
    model for C1r$_2$–C1s$_2$ 332
    overview 326–30
    structure, activation, and control
      of C1 330–3
    structure of human C1q 330–2
    via mannan-binding protein
      (MBP) 333–4
  C3a, C4a, C5a anaphylotoxins 364
  complement C3 349–53
  complement C5 354
  complement class III products of
    MHC
    C2 molecular genetics 346–8
    C2 structure, activation 340–3
    C4 molecular genetics 343–6
    C4 structure, activation 340,
      341
    factor B molecular genetics
      346–9
    factor B, structure, activation
      340–3
    formation of C3 convertases
      340–3
    human factor B gene
      organization 348
    molecular map of 349, 350
    overview 339–40
  complement receptor type 3
    364–6
  leucocyte-specific β integrins
    364–6
  LFA-1 364–6
  membrane associated molecules as
    regulators/receptors 329
  p150, 95 364–6
  plasma proteins involved in
    complement activation/
    control 328
  regulation of complement
    activation via
    C4b-binding protein 358–60

complement receptor type 1
    360–2
  complement receptor type 2 362
  control of C3 and C5
    convertases 354–64
  DAF 363–4
  factor H 358–60
  factor I 354–8
  members of the RCA 354–6,
    357
  membrane associated regulatory
    proteins 360–4
  membrane co-factor protein
    362–3
  relationship of C8 and C9 to
    perforin 366–70
  terminal attack complex C6b-9
    366–70

differential splicing of heavy chain
    transcripts 6–7

engineering of, see antibody
    engineering

gene expression
  of D J$_H$ rearrangements 34
  of H chain genes 6–7, 30–1
  Ig H class switching, see class
    switch recombination
  of Ig κ LC in the germline 35
  of incompletely-assembled H
    chain genes 33–5
  of J$_H$ locus 30–1, 33–4
  regulation of HC gene expression
    32–3
  regulation of LC gene expression
    33
  of V$_H$ segments in the germline
    34–5
generation of antibody repertoire
  primary diversification
    mechanisms 31–2
  somatic mechanisms 32
gene rearrangement
  deletion vs inversion mechanisms
    12–14
  double-strand breaks 15
  hairpins 15, 16–17
  homology-mediated joining 16
  Ku 20–1
  N-regions 15–16
  overview of V(D)J recombination
    8–10

palindromic sequences 16–17
P-nucleotides 16–17
pseudonormal V(D)J joining
    13–14
recombination activating genes
    (Rag) 17–18, 45–6
recombination recognition
    sequences 10–14
regulation of, see regulation of
    gene assembly
scid defect 19–20
terminal deoxynucleotidyl
    transferase (TdT) 18–19
XR-1 20–1
Xrs-6 20–1

Immunoglobulins
  gene organization 4–6, 21–30
  gene sequence diversity 113
  heavy chain genes
    human 24–5
    murine 21–3
    overview 4–6
  light chain genes
    human κ LC locus 26–7
    human λ LC locus 27–8
    murine κ LC locus 26
    murine λ LC locus 27
    overview 4–6
    surrogate LC genes 28–30
  protein structure 3–4, 283–5
  regulation of gene assembly 6–21,
    35–52
    correlation with chromosome
      structure methylation 48–9
    correlation with gene
      transcription 47–8, 49–50
    HC allelic exclusion 41–2
    H chain gene assembly 37
    Ig receptor editing 43
    κ vs λ L chain gene assembly
      38–9
    LC allelic exclusion 42–3
    L chain gene assembly 37–8
    LC isotype exclusion 38–9
    models of allelic exclusion 40–1
    model systems for study 46–7
    ordered assembly and
      expression 35–6
    potential restrictions 43–5
    potential role of SLC receptors
      40
    regulated models for control
      35–6

regulation via accessibility
45–6, 47–52
salvaging undesired
rearrangements 43
surface markers to monitor
39–40
isotypic exclusion 38–9

MHC
class I (endogenous) antigen
processing pathway
determinant selection 224–6
implications of *Tap* mutants
220–1
MHC-linked proteasome
subunit genes 221–3
MHC-linked transport protein
(*Tap*) genes 214–21
polymorphism 224–6
role of molecular chaperones
223–4
selectivity in antigen processing
224–6
specificity of 224–6
Tap gene defects 214–16
Tap gene organization 215–16
TAP-independent transport
218–20
TAP protein function in peptide
transport 216–18
TAP protein structure 216
class II (exogenous) antigen
processing pathway
different structural forms of I_i
229–31
generation of different I_i forms
229–31
I_i as inhibitor of peptide
binding 231–2
I_i as a molecular chaperone 228
I_i as a targeting signal 228–9
overview 226
presentation of endogenous
antigen 233–4
processing mutants 232–3
proteases involved 226–7
role in B cell activation 260,
265–9
role in invariant chain (I_i)
227–32
class III products 339–49; see also
complement
comparison of class I versus class
II structures 204–8

gene organization 191
genetic map for human MHC
segments 191
overview 189–91, 212–14
pathway for class I presentation of
peptide fragments 191–4
pathway for class II presentation
of peptide fragments 191–4
peptide–MHC class I interactions
198–204
peptide sequence motifs for MHC
class I antigens 194
single peptide–MHC class I
structures 199
structure of MHC class I antigens
194–8

proteasomes 221–3

Rag genes 17–18
recombinant Ig gene expression
in animal cells
CHO cells 287
COS cells 287
insect cells 287
myeloma cells 285–6
in *B. subtilis* 291–2
in *E. coli*
domain antibodies (dAbs)
290–1
Fab 288–9
factors influencing expression
291
Fv and scFv 289–90
overview 287–8
in filamentous fungi 293
in plants 293–4
in *Staphylococcus* 292
in *Streptomyces* 292
in yeast 292–3
recombination recognition sequences
(RSSs)
involvement in V(D)J
recombination 12–14
structure 10–12

scid mice 19–20

T cell activation
antigen recognition 132–7
initial events 132–6

major T cell co-stimulatory
structures 135
positive selection
role of CD4 and CD8 165–6
signal transduction
alternative co-simulatory
structures 170–2
CD28–B7 interactions 168–70
CD40–gp39 (CD40L) interaction
171
GTP binding by CD3 152–4
induction by triggering of CD28
and CTLA-4 166–70
involvement of CD27 171–2
involvement of CD95 (Fas) 172
PI-3 kinase 151–2
primary co-stimulatory events
160–74
role of CD4 and CD8 161–6
role of phosphotyrosine
phosphatase CD45 172–4
specificity of 154
src-family tyrosine kinases
147–50
syk tyrosine kinase 150–1
TCR/FCε R1γ$^+$ receptors 156–9
use of CD3-ξ$^{-/-}$ mice 154–9
use of transgenic mice 159–60
ZAP70 tyrosine kinase 150–1
structure of TCR 137, 102–3
TCR/CD3 complex
assembly in ER 139–45
components of 113, 138–9
model for assembly 144
role in signal transduction
145–60; see also signal
transduction
sequences of CD3 polypeptides
142–3
T cell receptor (TCR) genes
affinity of αβ TCRs for peptide/
MHC ligands 116–18
αβ TCR recognition of peptide/
MHC complexes 113–16
chromosomal locations of TCR
loci 105
γδ T cell recognition
function of γδ T cells 122
monomorphic γδ T cells 122–3
specificity of γδ T cells
119–22
gene organization 103–5
gene rearrangement
allelic exclusion 108–9
αβ TCR rearrangement 107–8

T cell receptor (TCR) genes (*cont.*)
  gene rearrangement (*cont.*)
    choice of γδ versus αβ lineage 109
    correlation with chromatin
      structure/methylation 48–9
    correlation with gene
      transcription 47–8, 49–50
    γδ TCR rearrangement 108
    model systems 46–7
    negative selection 110–11
    non-productive rearrangements
      110
    overview 106–8
    positive selection 111–13
    regulation via accessibility
      45–6, 47–52
    TCR repertoire selection
      109–13
  mapping peptide–TCR contacts
    115–16
  sequence diversity in TCR genes
    113
  superantigens 118–19
  TCR polypeptides 102–3, 137
T cells
  differentiation 106–8, 154–5, 158,
    164–5
  role of CD4 and CD8 in T cell
    development 164–5
  tissue distribution 157